Global
Environment
Outlook 3
Data
Compendium

United Nations Environment Programme
PO Box 30552, Nairobi, Kenya
Tel: +254 2 621234
Fax: +254 2 623943/44
E-mail: geo@unep.org
Web: www.unep.org/geo
            www.unep.net

DISCLAIMER

The designation employed and the presentations do not imply the expression of any opinion whatsoever on the part of UNEP or cooperating agencies concerning the legal status of any country, territory, city, or area of its authorities, or the declination of its frontiers or boundaries. Mention of a commercial company or product in this report does not imply endorsement by the United Nations Environment Programme. The use of information from this publication concerning proprietary products for publicity or advertising is not permitted.

Contents

UNEP's third Global Environment Outlook report (GEO-3) was published in May 2002. This Data Compendium presents major statistical data sets underlying the integrated analysis of the environment at global and regional levels in GEO-3. As such, the Data Compendium supports the scientific and empirical nature of the GEO process and provides background information to other assessment programmes and data users. The Data Compendium is a product of the wider GEO data process, being a collaborative effort of UNEP, various other UN organizations and key data partners around the world.

UNEP recognizes the need to have a harmonized collection of core data sets, in order to perform in-depth analysis, derive sound conclusions and present tangible recommendations for improving the environmental decision-making process. Despite the huge volume of data that is now available worldwide, it is evident that serious shortcomings and critical data gaps still persist - in particular in areas relating to the impacts of environmental change on humans and ecosystems, as well as measurements of the impacts of policy responses on the environment, cultural and institutional characteristics of societies. These broader data issues were highlighted in the second edition of GEO (GEO-2000) and further explored in a technical report entitled "Data Issues of Global Environmental Reporting: Experiences from GEO-2000" (UNEP/RIVM1999 – UNEP/DEIA&EW/TR.99-3). A second technical report, drawing on the experience of GEO-3, is under preparation and will help point the way forward towards better observation and monitoring of the global environment and easily-accessible harmonized core databases.

The data in this Compendium have been extracted from the wider GEO Data Portal during May 2002. The GEO Data Portal is the on-line, up-to-date and comprehensive reference data system developed for GEO assessment and reporting purposes. It also allows the user to explore data by means of maps, graphs and tables. Whilst the GEO Data Portal is constantly being updated, this Compendium represents a 'frozen' copy of selected statistics from the Data Portal, specifically produced for the third GEO report.

The Compendium does not only give data for major environmental issues, such as climate change, deforestation or biodiversity, but also for relevant characteristics of the society and economy, including economic growth, energy use, and life expectancy. The data are presented for the GEO regions and sub-regions, as aggregated from national figures by UNEP/DEWA/GRID-Geneva. National statistics are given in cases where there was enough space for the country list of a given variable. The CD-ROM included in this volume gives access to the full compendium tables, that is for all the years available and including all national statistics. The GEO-3 Data Compendium is also available on the Internet at http://geocompendium.grid.unep.ch. Still, for additional and up-to-date statistical and geospatial data, the reader is directed to the GEO Data Portal at http://geodata.grid.unep.ch.

We hope that you find this GEO-3 Data Compendium useful, and would very much welcome any feedback you may have.

Daniel van R. Claasen
Acting Director
Division of Early Warning and Assessment
United Nations Environment Programme

About GEO-3

The Global Environment Outlook (GEO) series of reports are the flagship publications of UNEP. They present comprehensive and authoritative reviews and analyses of worldwide environmental conditions and trends, and the policies and policy instruments available to address them. GEO-3 comes 10 years after the Earth Summit held in Rio de Janeiro in 1992 and 30 years after the Stockholm Conference on the Human Environment in 1972. By tracking and analysing important environmental issues over the period 1972-2002, GEO-3 provides an integrated explanation of major trends that have shaped our environmental inheritance.

Continuing the global and regional focus of previous GEO reports, GEO-3 presents the 30-year retrospective analysis of environmental state and trends, and associated policy responses under themes of land, forests, biodiversity, freshwater, coastal and marine areas, atmosphere, urban areas and disasters.

A special focus on human vulnerability to environmental change highlights the increasing risks and impacts on people. The outlook section of the report, spanning the next 30 years, is presented through four scenarios. This innovative approach reveals salutary lessons for all who strive towards a more desirable future.

The GEO-3 Data Compendium - Printed Edition

The GEO-3 Data Compendium aims to provide an overview of national, regional, sub-regional and global statistical data sets used to prepare UNEP's third **Global Environment Outlook (GEO-3)** report. Such data have been made available during the assessment and reporting process for analysis purposes, and have also served as a basis for many graphics in the final GEO-3 report.

What is to be found in this Data Compendium?

The compendium holds data tables for environmental issues as addressed by UNEP's GEO, such as climate change, water stress and deforestation, as well as a wide variety of socio-economic variables. Examples of data sets are CO_2 emissions, water consumption, forest cover change, population growth, land use change and GDP development. Documentation on data sources, definitions and other parameters is also provided in the form of meta-data for each data set. Most of the tables are available for the GEO regions and sub-regions, and for the world as a whole. Where possible, the data cover the time period 1972-2002, which is the time-frame of the retrospective chapter of GEO-3. Because of space constraints, this printed compendium does not feature national statistics in most cases, nor are data for each and every year included. Maps showing the GEO regions and sub-regions are available on the following pages.

The data for the GEO-3 Data Compendium have been extracted from the wider GEO Data Portal available at http://geodata.grid.unep.ch, through the GEO-3 site at http://www.unep.org/geo[1] or through the UNEP.Net site at http://www.unep.net.

Contact information

The GEO-3 Data Compendium was developed by UNEP/DEWA/GRID-Geneva (http://www.grid.unep.ch). If you have any questions or remarks about it, please contact geo@grid.unep.ch.

■ Regional data aggregation:	Andrea de Bono, Ola Nordbeck
■ Meta-data supervision and support:	Gregory Giuliani, Stefan Schwarzer
■ Design and development:	Stéphane Kluser
■ Overall coordination:	Jaap van Woerden

Acknowledgments

UNEP acknowledges the contributions made by the many individuals and institutions that have contributed to the Data Compendium of the Global Environment Outlook, most notably the GEO Team at UNEP headquarters in Nairobi, the key data providers as mentioned in the section on the Data Sources and all GEO Collaborating Centres.

---

[1] This site is mirrored in Japan at http://www-cger.nies.go.jp/geo, in Mexico at http://www.rolac.unep.mx/geo, in Norway at http://www.grida.no/geo, in Switzerland at http://www.grid.unep.ch/geo, in the United Kingdom at http://geo.unep-wcmc.org and in the United States at http://grid2.cr.usgs.gov/geo.

Data Sources

The data sets and documentation on the GEO-3 Data Compendium have been extracted from the GEO Data Portal, focusing on those data sets that have actually been used for the 1972-2002 state of the environment analysis and associated graphics presented in GEO-3.

Many organizations have contributed to the data sets and documentation available in this report. It contains readily available, globally harmonized primary data sources such as FAO's FAOSTAT, the World Bank's World Development Indicators, or the UN Common Database. Also included are various regional or other specific data sources, reflecting the broad scope and regional focus of the GEO process.

For the GEO-3 Data Compendium, key data providers are in alphabetical order:

Full Name	Abbreviation	URL
Carbon Dioxide Information Analysis Center	CDIAC	http://cdiac.esd.ornl.gov/trends/emis/em_cont.htm
Center for International Earth Science Information Network	CIESIN	http://www.ciesin.org
Environmental Systems Research Institute	ESRI	http://www.esri.com
Food and Agriculture Organization of the United Nations	FAO	http://apps.fao.org/page/collections
Forest Stewardship Council	FSC	http://www.fscoax.org
Global Resource Information Database, Arendal	GRID-Arendal	http://www.grida.no
Global Resource Information Database, Geneva	GRID-Geneva	http://www.grid.unep.ch
Global Resource Information Database, Sioux Falls	GRID-Sioux Falls	http://grid2.cr.usgs.gov
International Soil Reference and Information Centre	ISRIC	http://www.isric.org
International Union for the Conservation of Nature, The World Conservation Union	IUCN	http://www.iucn.org
International Energy Agency	IEA	http://www.iea.org
National Institute for Public Health and the Environment	RIVM	http://www.rivm.nl
NOAA National Geophysical Data Center	NOAA/NGDC	http://www.nesdis.noaa.gov
Office of U.S. Foreign Disaster Assistance - The Centre for Research on the Epidemiology of Disasters	OFDA/CRED	http://www.cred.be/emdat
Organization for Economic Co-operation and Development	OECD	http://www.oecd.org
The Ramsar Convention on Wetlands Secretariat		http://www.ramsar.org
The World Bank		http://www.worldbank.org/data/wdi2000
The World Fish Centre, formerly known as International Centre for Living Aquatic Resources Management	ICLARM	http://www.iclarm.org
UNEP - The World Conservation Monitoring Centre	WCMC	http://www.unep-wcmc.org
UNEP/WMO Intergovernmental Panel on Climate Change Data Distribution Centre	DDC	http://www.ipcc.ch
United Nations Children's Fund	UNICEF	http://www.unicef.org
United Nations Development Programme	UNDP	http://www.undp.org
United Nations Educational, Scientific and Cultural Organization	UNESCO	http://www.unesco.org
United Nations Environment Programme Ozone Secretariat		http://www.unep.ch/ozone
United Nations Framework Convention on Climate Change Secretariat	UNFCCC	http://unfccc.int.index.html
United Nations Human Settlements Programme, formerly known as UNCHS – Habitat	UN-HABITAT	http://www.unchs.org/habrdd/statprog.htm
United Nations International Labour Organization	ILO	http://www.ilo.org
United Nations International Telecommunications Union	ITU	http://www.itu.org
United Nations Population Division		http://www.un.org/esa/population/unpop.htm
United Nations Statistics Division	UN-STAT	http://www.un.org/depts/unsd
United Nations World Health Organization	WHO	http://www.who.int/whr/2000/en/statistics.htm
United States Geological Survey - EROS Data Centre	EDC	http://edc.usgs.gov
World Resources Institute	WRI	http://www.wri.org/wr2000
World Energy Council - Conseil Mondial de l'Energie	WEC	http://www.worldenergy.org/wec-geis
World Wildlife Fund	WWF	http://www.wwf.org

Regional breakdown

AFRICA	ASIA AND THE PACIFIC	EUROPE
Central Africa	Australia and New Zealand	Central Europe
Eastern Africa	Central Asia	Eastern Europe
Northern Africa	Northwest Pacific and East Asia	Western Europe
Southern Africa	South Asia	
Western Africa	Southeast Asia	
Western Indian Ocean	South Pacific	

LATIN AMERICA AND THE CARIBBEAN	NORTH AMERICA	WEST ASIA	POLAR
Caribbean	North America	Arabian Peninsula	The Arctic
Meso-America		Mashriq	The Antarctic
South America			

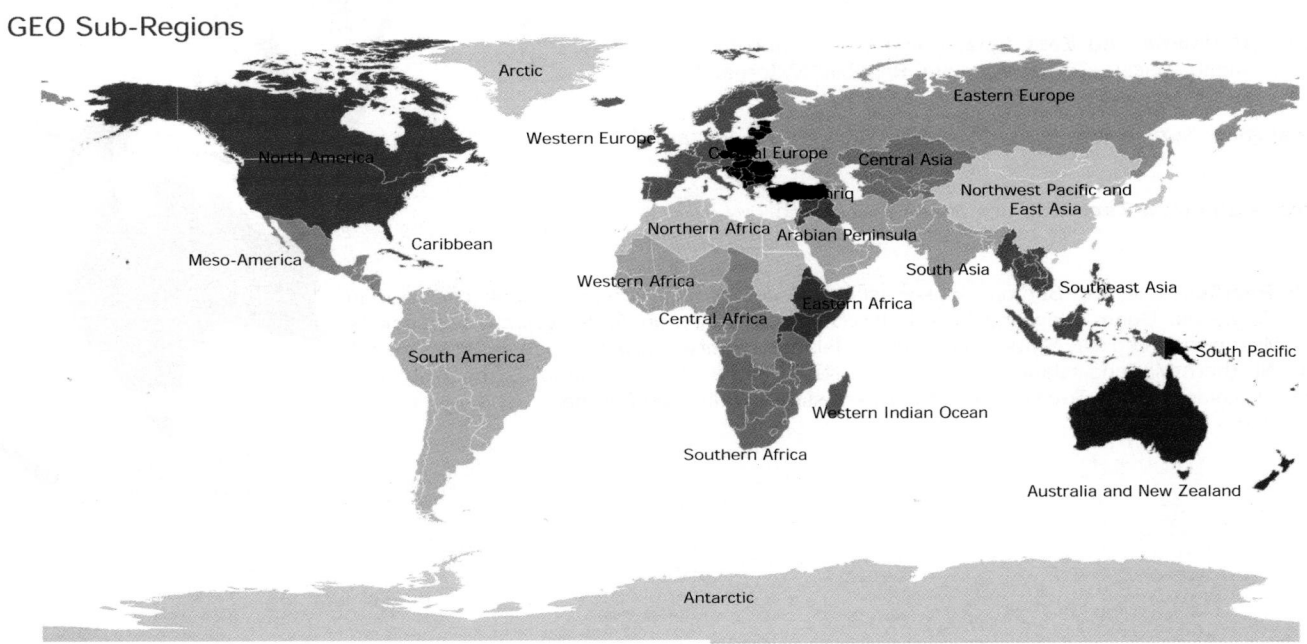

GEO Regions

Copyright © 2002 United Nations Environment Programme/DEWA/GRID-Geneva

GEO Sub-Regions

Copyright © 2002 United Nations Environment Programme/DEWA/GRID-Geneva

Sub-regional breakdown

AFRICA

Northern Africa: Algeria, Egypt, Libyan Arab Jamahiriya, Morocco, The Sudan, Tunisia, Western Sahara

Central Africa: Cameroon, Central African Republic, Chad, Congo, Democratic Republic of Congo, Equatorial Guinea, Gabon, Sao Tome and Principe

Western Africa: Benin, Burkina Faso, Cape Verde, Cote d'Ivoire, Gambia, Ghana, Guinea, Guinea-Bissau, Liberia, Mali, Mauritania, Niger, Nigeria, Senegal, Sierra Leone, Togo

Eastern Africa: Burundi, Djibouti, Eritrea, Ethiopia, Kenya, Rwanda, Somalia, Uganda

Western Indian Ocean: Comoros, Glorioso Islands, Juan De Nova Island, Madagascar, Mauritius, Mayotte, Reunion (France), Seychelles

Southern Africa: Angola, Botswana, Lesotho, Malawi, Mozambique, Namibia, South Africa, Swaziland, United Republic of Tanzania, Zambia, Zimbabwe

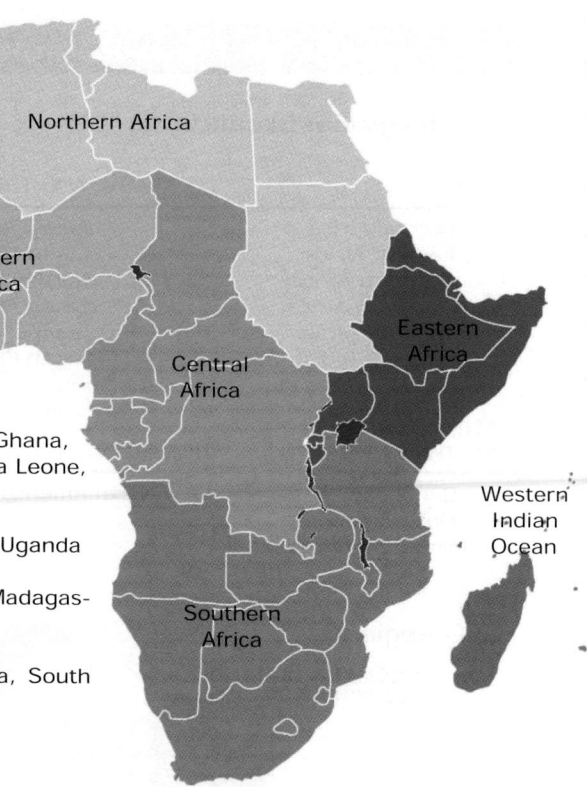

ASIA AND THE PACIFIC

South Asia: Afghanistan, Bangladesh, Bhutan, British Indian Ocean Territory, India, Iran (Islamic Republic of), Maldives, Nepal, Pakistan, Sri Lanka

Southeast Asia: Brunei Darussalam, Cambodia, Christmas Island, Indonesia, Lao People's Democratic Republic, Malaysia, Myanmar, Paracel Islands, Philippines, Singapore, Spratly Islands, Thailand, Viet Nam

Northwest Pacific and East Asia: China, Democratic People's Republic of Korea, Japan, Macau, China, Mongolia, Republic of Korea, Taiwan

Central Asia: Kazakhstan, Kyrgyzstan, Tajikistan, Turkmenistan, Uzbekistan

Australia and New Zealand: Australia, New Zealand

South Pacific: American Samoa, Baker Island, Cocos (Keeling) Island, Cook Islands, Fiji, French Polynesia, Guam, Howland Island, Jarvis Island, Johnston Atoll, Kiribati, Marshall Islands, Micronesia (Federal States of), Midway Islands, Nauru, New Caledonia, Niue, Norfolk Island, Northern Mariana Islands, Palau, Papua New Guinea, Pitcairn Island, Samoa, Solomon Islands, Tokelau, Tonga, Tuvalu, Vanuatu, Wake Island, Wallis and Futuna

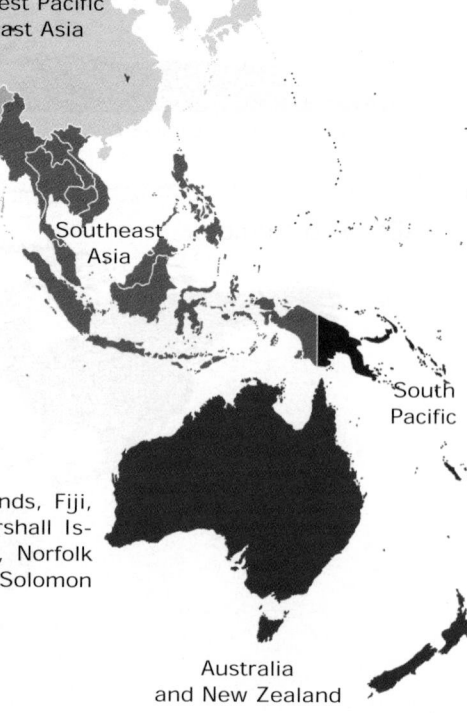

GEO REGIONS AND SUB-REGIONS

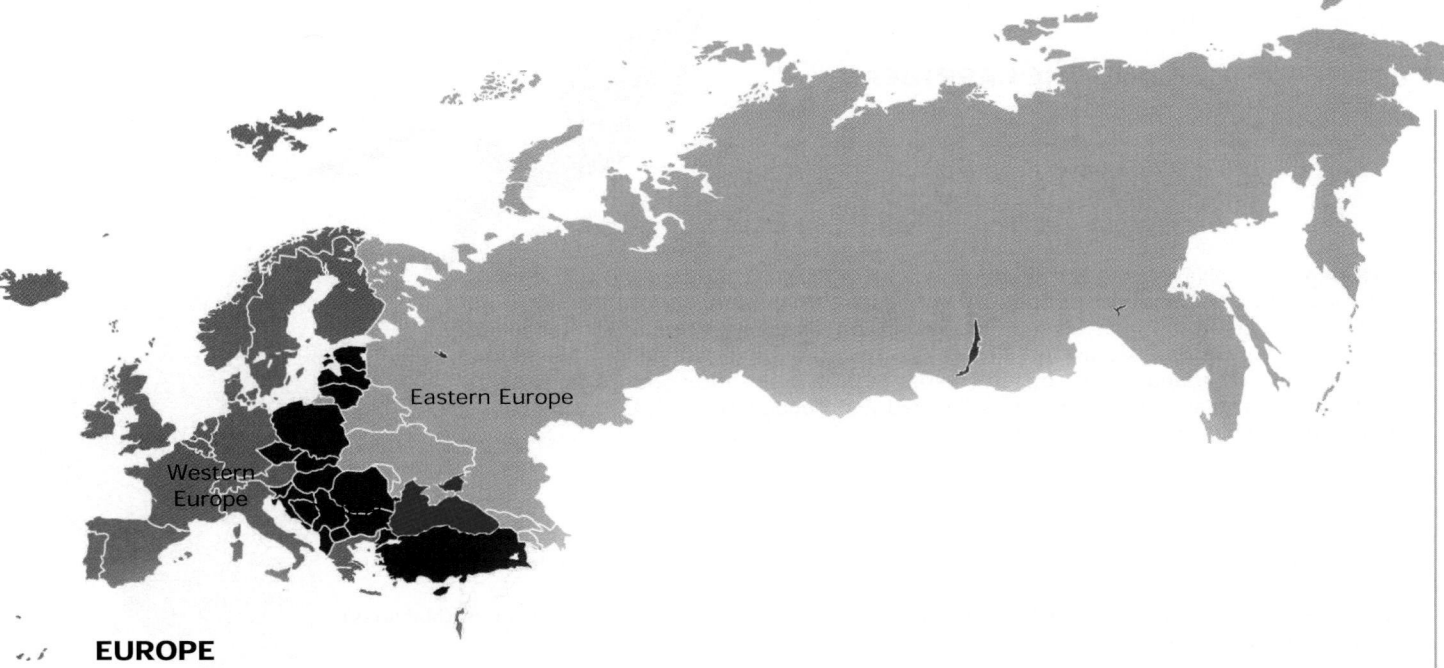

EUROPE

Western Europe: Andorra, Austria, Belgium, Denmark, Feroe Islands, Finland, France, Germany, Gibraltar, Greece, Guernsey, Holy See, Iceland, Ireland, Isle of Man, Israel, Italy, Jersey, Liechtenstein, Luxembourg, Malta, Monaco, Netherlands, Norway, Portugal, San Marino, Spain, Svalbard and Jan Mayen Islands, Sweden, Switzerland, United Kingdom of Great Britain and Northern Ireland

Central Europe: Albania, Bosnia and Herzegovina, Bulgaria, Croatia, Cyprus, Czech Republic, Estonia, Hungary, Latvia, Lithuania, Poland, Romania, Slovakia, Slovenia, The Former Yugoslav Republic of Macedonia, Turkey, Yugoslavia

Eastern Europe: Armenia, Azerbaijan, Belarus, Georgia, Republic of Moldova, Russian Federation, Ukraine

WEST ASIA

Arabian Peninsula: Bahrain, Kuwait, Oman, Qatar, Saudi Arabia, United Arab Emirates, Yemen

Mashriq: Iraq, Jordan, Lebanon, Occupied Palestinian Territories, Syrian Arab Republic

NORTH AMERICA

North America: Canada, United States of America

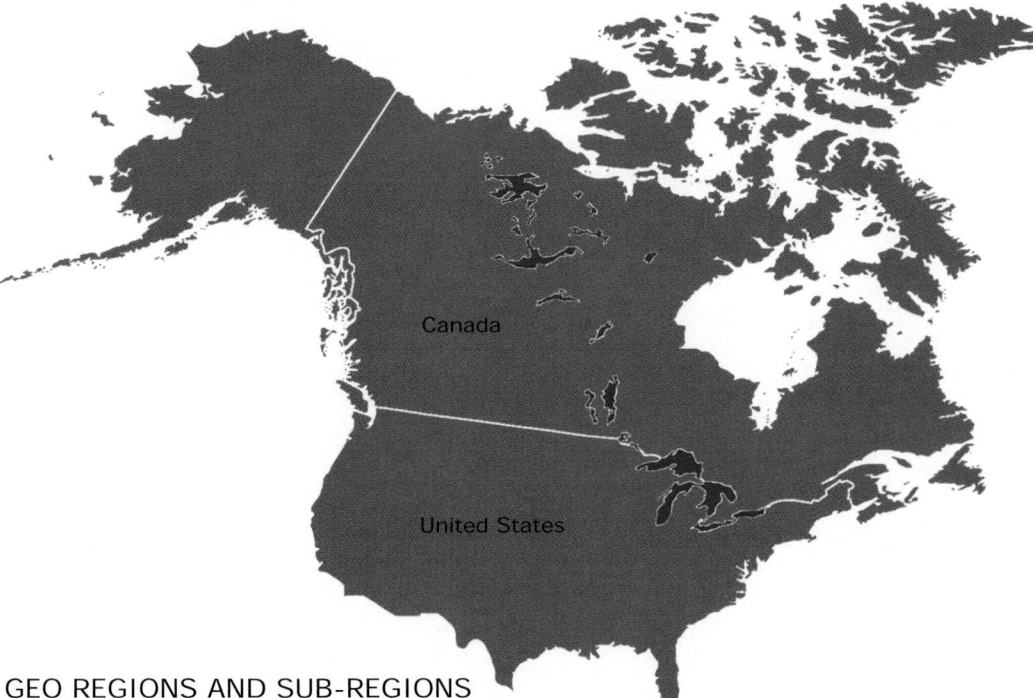

LATIN AMERICA AND THE CARRIBEAN

Meso-America   Caribbean

South America

Caribbean: Anguilla (United Kingdom), Antigua and Barbuda, Aruba (The Netherlands), Bahamas, Barbados, Bermuda, British Virgin Islands , United Kingdom), Cayman Islands (United Kingdom), Cuba, Dominica, Dominican Republic, Grenada, Guadeloupe (France), Haiti, Jamaica, Martinique (France), Montserrat (United Kingdom), Netherlands Antilles (The Netherlands), Puerto Rico (United States), Saint Kitts and Nevis, Saint Lucia, Saint Vincent and the Grenadines, Trinidad and Tobago, Turks and Caicos (United Kingdom), Virgin Islands (United States)

Meso-America: Belize, Costa Rica, El Salvador, Guatemala, Honduras, Mexico, Nicaragua, Panama

South America: Argentina, Bolivia, Brazil, Chile, Colombia, Ecuador, Falkland Islands (Malvinas), French Guyana (France), Guyana, Paraguay, Peru , Suriname, Uruguay, Venezuela

POLAR

The Arctic: for GEO-3, the eight Arctic countries are Canada, Greenland (Denmark), Finland, Iceland, Norway, Russia, Sweden and Alaska (United States). However, when Arctic data are shown in this publication, they only include Greenland since this is the only territory from which Arctic data are specifically available. Arctic data could not be separated out for any other individual territoty/country in the Arctic region.

The Antarctic: Antarctic data shown in this publication have been aggregated from the following territories: Bouvet Island, French Southern and Antarctic Territories, Heard Island and McDonald Islands, South Georgia and the South Sandwich Islands.

1 Aggregated Data Sets

Table Legend

---- : "No Data" or less than 2/3 of national data available / **** : data "Not Available"

For all aggregations, copyright © 2002, United Nations Environment Programme/DEWA/GRID-Geneva.

The vertical grey lines in the table indicates that a time span has been deleted due to space constraints. To get full time coverage for each variable, see the data files on the enclosed CD ROM (in Excel format).

1.1 Socio-economic

1.1.1 Economy

Foreign Direct Investment - Net Inflows

Units: million constant 1995 US$

Data Source: World Development Indicators 2001

Data Provider: The World Bank

Years: 1970-1999

	1970	1980	1990	1991	1992	1993	1994	1995	1996	1997	1998	1999
AFRICA	722.30	1786.82	2407.70	3409.54	3734.30	4459.02	7083.89	6899.32	7622.35	11956.41	10027.98	11691.04
Central Africa	----	----	----	----	----	----	----	----	----	----	----	----
Eastern Africa	33.12	120.39	106.01	39.88	25.62	92.76	145.26	244.70	194.22	342.00	521.21	419.87
Northern Africa	104.75	1540.90	1230.45	893.50	1790.78	1969.98	2848.33	1210.26	1204.45	1689.82	2667.87	2266.17
Southern Africa	136.62	252.53	10.75	1341.84	602.62	384.26	1009.48	3182.39	2325.61	6517.95	4095.19	6220.40
Western Africa	411.41	-400.24	961.22	1082.62	1059.38	2015.08	3052.49	2094.06	2847.20	2949.17	2243.96	2076.24
Western Indian Ocean	14.64	12.50	103.75	68.65	55.01	62.22	69.91	87.84	99.32	157.75	108.53	212.39
ASIA AND PACIFIC	1516.41	6422.60	35164.41	33367.91	44388.19	65720.69	81802.76	101745.66	105425.85	118568.41	106315.46	109234.20
Australia and New Zealand	1133.58	2584.66	11610.12	7136.22	9186.82	8053.34	8987.57	20305.92	10639.35	13048.41	8734.41	7136.09
Central Asia	----	----	----	----	176.68	258.71	357.40	1502.28	1720.36	2274.38	2044.44	2321.45
North West Pacific and East Asia	0.00	360.93	7638.08	8621.38	18483.61	35626.51	44821.05	47544.50	53915.69	63486.53	66192.40	76256.60
South Asia	122.11	233.85	223.57	491.55	947.38	1415.33	2005.82	3731.10	4462.31	6249.17	4508.62	3981.86
South East Asia	252.64	3098.91	15343.07	16909.78	15273.66	20102.28	25426.98	27951.97	34487.99	33364.29	24519.08	19154.42
South Pacific	8.08	144.25	349.57	208.99	320.04	264.52	203.94	709.88	200.15	145.63	316.51	383.76
EUROPE	----	27179.78	131954.04	107543.12	110820.93	112041.60	108965.88	175112.65	171618.50	207421.13	355552.87	501668.86
Central Europe	73.20	142.57	1486.15	4366.29	5850.68	7981.43	7671.51	16770.25	14119.34	16707.15	23264.56	25446.29
Eastern Europe	0.00	0.00	0.00	0.00	8.83	22.21	1079.64	3390.74	4770.11	11290.61	6620.20	6029.58
Western Europe	----	27037.21	130467.89	103176.83	104961.41	104037.96	100214.72	154951.66	152729.05	179423.37	325668.11	470192.99
LATIN AMERICA AND CARIBBEAN	1448.35	7782.34	10509.80	16440.90	18945.28	17383.39	35811.54	37700.73	54825.45	82682.48	91138.71	114651.14
Caribbean	----	----	----	----	----	----	----	----	----	----	----	----
Meso-America	554.40	2895.12	3836.48	6575.65	6248.29	6358.97	14913.18	13111.04	12998.40	18860.34	19174.07	16906.26
South America	479.52	4391.51	5850.63	8871.17	11893.34	10028.23	19650.60	23142.43	40645.11	61015.05	69107.12	93550.54
NORTH AMERICA	3894.61	28702.06	70769.93	32869.28	31044.95	70813.17	68583.09	84674.07	121326.29	148654.39	262485.90	379438.48
North America	3894.61	28702.06	70769.93	32869.28	31044.95	70813.17	68583.09	84674.07	121326.29	148654.39	262485.90	379438.48
POLAR	----	----	----	----	----	----	----	----	----	----	----	----
Antarctic	----	----	----	----	----	----	----	----	----	----	----	----
Arctic	----	----	----	----	----	----	----	----	----	----	----	----
WEST ASIA	----	----	----	----	----	----	----	----	----	----	----	----
Arabian Peninsula	----	----	----	----	----	----	----	----	----	----	----	----
Mashriq	----	----	----	----	----	----	----	----	----	----	----	----
GLOBAL TOTALS	----	----	----	----	----	----	----	----	----	----	----	----
REPORTED GLOBAL TOTALS FROM ORIGINAL DATA SOURCES	****	****	****	****	****	****	****	****	****	****	****	****

Comments:

Foreign direct investment (FDI) is net inflows of investment to acquire a lasting management interest (10 percent or more of voting stock) in an enterprise operating in an economy other than that of the investor. It is the sum of equity capital, reinvestment of earnings, other long-term capital, and short-term capital as shown in the balance of payments. This series shows total net, that is, net FDI in the reporting economy less net FDI by the reporting economy.

General Government Final Consumption Expenditure

Units: million constant 1995 US$

Data Source: World Development Indicators 2001

Data Provider: The World Bank

Years: 1960-1999

	1960	1970	1980	1990	1995	1996	1997	1998	1999
AFRICA	----	25291.32	47865.22	65854.33	68206.91	70190.56	73275.68	75563.15	77624.49
Central Africa	958.75	2002.40	2204.52	2708.04	2146.96	2335.30	2426.24	2461.79	2415.64
Eastern Africa	----	----	1866.04	2591.50	2872.10	3041.69	3288.08	3707.95	4207.73
Northern Africa	----	6704.80	13495.66	18626.21	21616.44	22297.25	23221.21	24248.22	25287.36
Southern Africa	----	----	----	32668.77	33041.74	34111.96	35209.58	35251.14	34745.78
Western Africa	1558.70	2937.89	7170.42	8440.74	7651.77	7509.57	8188.33	8889.66	9979.47
Western Indian Ocean	216.72	302.26	565.28	819.08	877.91	894.80	942.24	1004.39	988.51
ASIA AND PACIFIC	167963.24	308639.80	513660.28	715782.47	853169.45	883090.18	907215.39	929538.80	958371.95
Australia and New Zealand	20222.25	34652.98	52218.94	70830.69	79089.55	79848.83	83271.82	85105.93	88357.46
Central Asia	----	----	----	----	7181.20	7700.68	7440.33	6908.04	7120.62
North West Pacific and East Asia	134158.20	236573.81	397681.86	542975.85	644246.64	665570.13	681205.14	696980.77	713434.64
South Asia	6215.80	23475.43	31917.09	49587.05	60612.47	63800.88	68036.37	75528.61	81261.92
South East Asia	----	12928.99	30782.65	46959.57	60860.46	64803.91	65942.24	63653.11	66900.90
South Pacific	----	----	----	----	----	----	----	----	----
EUROPE	----	----	----	1829493.29	2015624.18	2050914.70	2063158.87	2086000.49	2119319.19
Central Europe	----	----	----	70487.57	74964.01	76427.69	78154.95	81285.36	82790.06
Eastern Europe	----	----	----	----	58323.13	58147.53	57356.43	57348.43	56504.21
Western Europe	502744.02	1000472.33	1390933.70	1740256.39	1882337.04	1916339.48	1927647.49	1947366.70	1980024.92
LATIN AMERICA AND CARIBBEAN	68263.78	105001.09	184281.53	274293.99	247539.49	265491.50	283860.14	287285.01	274596.46
Caribbean	----	----	----	----	----	----	----	----	----
Meso-America	4733.07	10764.53	24632.67	31738.63	35078.40	34952.08	36003.28	36956.59	37398.15
South America	62924.80	93059.79	157307.33	240131.06	209652.52	227451.21	244318.50	246605.62	233315.58
NORTH AMERICA	600744.02	860882.39	953406.13	1238248.28	1257644.06	1261991.08	1280692.57	1300101.23	1307341.33
North America	600744.02	860882.39	953406.13	1238248.28	1257644.06	1261991.08	1280692.57	1300101.23	1307341.33
POLAR	----	----	----	----	----	----	----	----	----
Antarctic	----	----	----	----	----	----	----	----	----
Arctic	----	----	----	----	----	----	----	----	----
WEST ASIA	----	----	----	----	----	----	----	----	----
Arabian Peninsula	----	----	----	----	----	----	----	----	----
Mashriq	----	----	----	----	----	----	----	----	----
GLOBAL TOTALS	----	----	3099320.15	4128769.82	4448720.48	4538908.04	4616105.88	4686613.51	4745885.04
REPORTED GLOBAL TOTALS FROM ORIGINAL DATA SOURCES	1628628.06	2427904.33	3258372.59	4281526.12	4566000.07	4660389.74	4741305.20	4815177.38	----

Comments:

General government final consumption expenditure (general government consumption) includes all government current expenditures for purchases of goods and services (including compensation of employees). It also includes most expenditures on national defence and security, but excludes government military expenditures that are part of government capital formation.

Gross Domestic Product

Units: million constant 1995 US$

Data Source: World Development Indicators 2001

Data Provider: The World Bank

Years: 1960-1999

	1960	1970	1980	1990	1995	1996	1997	1998	1999
AFRICA	122959.62	209261.87	334162.24	429162.40	460113.32	484555.24	499129.86	515259.84	528627.48
Central Africa	11124.14	14949.29	20701.64	26242.24	23846.11	24379.60	25047.60	26088.70	26009.17
Eastern Africa	----	----	15296.70	21303.00	23933.16	25577.25	26649.68	27312.65	28472.99
Northern Africa	----	51589.47	93522.53	138226.43	152380.34	162260.71	166424.34	175783.29	182592.38
Southern Africa	55952.88	101402.30	151390.34	178045.25	186289.97	195099.15	200731.67	203006.65	206261.92
Western Africa	20385.18	33264.55	47980.80	58346.33	65807.69	69065.40	71729.21	74130.69	76026.69
Western Indian Ocean	2901.29	3897.26	5270.23	6999.15	7856.05	8173.13	8547.36	8937.85	9264.33
ASIA AND PACIFIC	1222009.64	2825158.69	4407951.08	6973562.94	8176398.63	8640694.64	8900519.89	8782228.33	8996962.92
Australia and New Zealand	126156.62	205703.50	277827.63	369796.66	436668.42	452517.27	472152.15	490737.80	512333.16
Central Asia	----	----	----	69992.16	45876.91	46106.79	46769.18	47567.82	49541.36
North West Pacific and East Asia	889203.03	2281163.98	3599718.40	5633957.34	6473930.84	6839714.88	7026984.16	6909759.71	7042736.10
South Asia	135277.35	206489.88	264164.26	437995.18	557529.91	591776.82	616062.53	649560.99	685063.00
South East Asia	69735.79	125365.91	257162.30	448969.99	646869.50	694584.00	722648.57	668763.21	691055.14
South Pacific	----	6435.43	9078.48	12851.62	15523.05	15994.88	15903.31	15838.80	16234.15
EUROPE	----	5640893.99	7637865.01	9741384.27	10082341.78	10241071.20	10502242.30	10760704.26	10997706.99
Central Europe	----	----	----	503399.20	507204.41	533787.70	559984.03	576341.67	575206.71
Eastern Europe	----	----	559492.84	698829.16	417807.01	402179.05	406490.89	391842.16	402854.92
Western Europe	3555289.55	5283229.69	6879834.64	8539155.90	9157330.35	9305104.46	9535767.37	9792520.43	10019645.36
LATIN AMERICA AND CARIBBEAN	428963.14	721061.14	1272991.90	1431928.64	1716455.57	1778877.05	1869228.07	1905624.14	1907195.98
Caribbean	----	----	----	----	----	----	----	----	----
Meso-America	73594.90	139506.89	257257.67	304760.71	336254.42	352104.56	374919.88	393538.77	407735.54
South America	336970.36	547643.95	963736.33	1059167.10	1304001.62	1347771.75	1412298.52	1427015.34	1410954.02
NORTH AMERICA	2560250.26	3762208.00	5170728.80	7056939.90	7917632.10	8188490.80	8551323.20	8918022.90	9245369.60
North America	2560250.26	3762208.00	5170728.80	7056939.90	7917632.10	8188490.80	8551323.20	8918022.90	9245369.60
POLAR	----	----	----	----	----	----	----	----	----
Antarctic	----	----	----	----	----	----	----	----	----
Arctic	----	----	----	----	----	----	----	----	----
WEST ASIA	----	----	194540.69	207728.16	256753.23	263621.04	271329.88	272542.86	----
Arabian Peninsula	----	73293.73	181257.58	185077.21	218791.21	224656.94	230893.04	230119.11	140579.09
Mashriq	----	----	----	----	----	----	----	----	----
GLOBAL TOTALS	7922493.43	13235495.54	19018239.72	25840706.31	28609694.62	29597309.96	30593773.18	31154382.32	31847703.83
REPORTED GLOBAL TOTALS FROM ORIGINAL DATA SOURCES	7811724.00	13202570.00	19204160.00	26229070.00	29077750.00	30088790.00	31111250.00	31690230.00	32512230.00

Comments:

Gross domestic product (GDP) is the sum of gross value added by all resident producers in the economy plus any product taxes and minus any subsidies not included in the value of the products. It is calculated without making deductions for depreciation of fabricated assets or for depletion and degradation of natural resources. Data are in constant 1995 U.S. dollars. Dollar figures for GDP are converted from domestic currencies using 1995 official exchange rates. For a few countries where the official exchange rate does not reflect the rate effectively applied to actual foreign exchange transactions, an alternative conversion factor is used.

Data sources: GDP is estimated by World Bank staff based on national accounts data collected by Bank staff during economic missions or reported by national statistical offices to other international organizations such as the Organisation for Economic Co-operation and Development.

Gross Domestic Product - Annual Percentage Growth Rate

Units: percent

Data Source: World Development Indicators 2001

Data Provider: The World Bank

Years: 1961-1999

	1961	1970	1980	1990	1991	1992	1993	1994	1995	1996	1997	1998	1999
AFRICA	----	----	----	2.11	1.22	0.14	0.95	2.97	3.12	5.46	3.19	3.31	2.72
Central Africa	----	----	----	----	----	----	----	----	----	----	----	----	----
Eastern Africa	----	----	----	3.53	0.68	-0.11	3.89	1.19	7.45	7.10	4.26	2.58	4.40
Northern Africa	-4.30	6.40	5.17	3.49	2.12	2.11	0.52	4.03	1.73	6.58	2.66	5.63	3.94
Southern Africa	----	----	----	0.50	0.02	-2.15	0.89	3.02	3.30	4.75	2.90	1.16	1.62
Western Africa	3.13	15.06	0.26	4.17	3.22	2.10	1.48	1.88	3.91	4.96	3.94	3.43	2.60
Western Indian Ocean	7.91	3.98	-3.21	5.37	-0.72	4.22	4.04	1.79	2.91	4.07	4.59	4.59	3.67
ASIA AND PACIFIC	8.21	9.76	2.95	5.20	4.33	2.89	2.57	3.16	3.72	5.72	3.03	-1.04	2.54
Australia and New Zealand	2.48	3.60	2.72	-0.24	0.16	3.35	4.41	4.64	4.43	3.63	4.35	3.96	4.40
Central Asia	----	----	----	-1.20	-4.72	-11.64	-7.73	-11.04	-5.19	0.60	1.64	1.82	4.29
North West Pacific and East Asia	9.57	10.78	2.74	5.28	4.55	2.44	1.91	2.46	3.05	5.70	2.75	-1.45	2.02
South Asia	4.27	4.24	2.54	6.50	3.11	5.76	4.14	5.65	6.36	6.16	4.12	5.47	5.49
South East Asia	6.55	7.95	6.74	8.42	7.25	6.61	7.92	8.24	8.18	7.40	4.17	-7.11	3.38
South Pacific	----	14.66	-1.55	1.20	5.03	4.89	6.15	3.26	1.20	3.16	-0.50	-0.23	3.25
EUROPE	----	----	----	1.89	0.06	0.30	-0.63	1.90	2.20	1.62	2.59	2.49	2.23
Central Europe	----	----	----	----	-6.75	-0.39	3.05	0.56	6.17	5.56	5.16	3.00	-0.01
Eastern Europe	----	----	----	-3.56	-5.49	-14.05	-9.36	-14.01	-5.34	-3.65	1.14	-3.48	2.83
Western Europe	5.72	5.00	1.83	2.43	0.85	1.27	-0.32	2.76	2.32	1.63	2.50	2.70	2.33
LATIN AMERICA AND CARIBBEAN	6.62	6.29	6.58	-0.45	4.32	3.65	4.24	5.22	1.73	3.67	5.12	1.97	0.11
Caribbean	----	----	----	----	----	----	----	----	----	----	----	----	----
Meso-America	4.86	6.24	7.89	4.93	4.20	4.04	2.42	4.35	-4.59	4.73	6.48	4.97	3.62
South America	7.03	6.19	6.42	-2.15	4.53	3.62	4.86	5.63	3.26	3.39	4.83	1.07	-1.06
NORTH AMERICA	2.47	0.38	-0.11	1.65	-0.58	2.90	2.63	4.09	2.67	3.42	4.43	4.29	3.67
North America	2.47	0.38	-0.11	1.65	-0.58	2.90	2.63	4.09	2.67	3.42	4.43	4.29	3.67
POLAR	----	----	----	----	----	----	----	----	----	----	----	----	----
Antarctic	----	----	----	----	----	----	----	----	----	----	----	----	----
Arctic	----	----	----	----	----	----	----	----	----	----	----	----	----
WEST ASIA	----	----	----	10.59	8.89	----	----	----	----	----	----	----	----
Arabian Peninsula	----	----	----	----	7.80	3.16	3.94	1.92	1.96	2.82	2.68	-0.04	0.52
Mashriq	----	-3.81	14.08	11.44	16.49	----	----	----	----	----	----	----	----
GLOBAL TOTALS	4.97	4.88	2.09	2.72	1.39	1.95	1.52	3.09	2.75	3.50	3.40	1.99	2.62
REPORTED GLOBAL TOTALS FROM ORIGINAL DATA SOURCES	4.74	4.74	2.03	2.65	1.38	1.86	1.47	3.00	2.72	3.48	3.40	1.86	2.59

Comments:

Annual percentage growth rate of GDP at market prices based on constant local currency. Aggregates are based on constant 1995 U.S. dollars. GDP is the sum of gross value added by all resident producers in the economy plus any product taxes and minus any subsidies not included in the value of the products. It is calculated without making deductions for depreciation of fabricated assets or for depletion and degradation of natural resources.

The weighting factor, GDP Constant US$ 1995, is used in the World Development Indicators 2001.

Gross Domestic Product - from Agriculture

Units: percent

Data Source: World Development Indicators 2001

Data Provider: The World Bank

Years: 1960-1999

	1960	1970	1980	1990	1991	1992	1993	1994	1995	1996	1997	1998	1999
AFRICA	----	24.59	16.38	17.99	17.50	16.49	16.46	16.16	15.98	17.33	17.09	18.16	18.02
Central Africa	----	37.17	21.16	20.63	21.65	21.22	21.74	20.87	19.87	19.28	19.58	21.95	20.66
Eastern Africa	----	42.09	37.77	38.92	38.48	39.68	41.83	40.16	40.00	39.06	38.34	38.28	38.27
Northern Africa	----	18.10	10.96	14.87	14.55	15.68	15.07	15.35	14.27	16.08	15.04	15.77	15.22
Southern Africa	14.17	9.16	9.07	9.01	9.40	6.76	7.74	7.31	6.84	8.18	8.16	8.49	8.28
Western Africa	51.99	40.19	25.31	31.83	31.17	28.74	29.20	30.96	32.62	32.47	33.08	35.34	36.98
Western Indian Ocean	----	22.88	25.84	22.65	21.46	21.29	21.61	21.73	19.41	19.81	18.90	18.66	17.01
ASIA AND PACIFIC	26.73	22.55	13.91	8.87	7.78	7.29	6.53	6.75	6.76	7.35	7.46	7.85	7.53
Australia and New Zealand	----	----	----	3.52	3.42	3.65	3.97	3.43	3.88	3.75	3.58	3.44	----
Central Asia	----	----	----	29.42	31.64	31.00	25.38	28.55	23.51	20.18	21.99	20.30	23.04
North West Pacific and East Asia	19.04	15.46	8.21	5.39	4.79	4.46	3.90	4.20	4.32	4.78	4.81	5.08	4.62
South Asia	45.03	44.20	32.14	29.14	29.41	29.11	28.41	28.22	26.92	26.99	26.20	27.31	26.33
South East Asia	29.72	32.44	21.91	15.75	15.22	15.16	14.08	13.96	13.84	13.39	13.00	13.07	13.33
South Pacific	----	----	----	13.73	13.48	12.90	15.19	16.42	14.65	15.16	16.08	14.45	14.26
EUROPE	----	----	----	4.97	4.28	3.51	3.48	3.17	3.23	3.26	3.12	2.91	2.77
Central Europe	----	----	----	15.24	12.26	11.29	11.88	10.76	11.08	11.09	10.07	10.67	9.18
Eastern Europe	----	----	----	18.44	16.07	10.57	11.09	8.72	9.98	9.14	8.98	7.70	7.89
Western Europe	----	----	----	3.00	2.80	2.63	2.48	2.48	2.49	2.47	2.32	2.19	2.08
LATIN AMERICA AND CARIBBEAN	22.57	12.78	10.17	8.63	8.29	7.76	7.40	8.53	8.57	8.28	7.78	7.92	7.54
Caribbean	----	9.30	8.72	5.61	6.05	5.45	5.08	----	----	----	----	----	----
Meso-America	----	15.01	10.46	9.00	8.56	7.69	7.29	6.87	7.23	7.58	7.00	6.67	6.12
South America	17.57	12.28	10.13	8.72	8.36	7.97	7.62	9.07	8.75	8.29	7.87	8.17	7.91
NORTH AMERICA	4.34	3.08	2.62	2.03	1.87	1.92	1.75	1.84	1.65	1.85	1.78	1.78	1.80
North America	4.34	3.08	2.62	2.03	1.87	1.92	1.75	1.84	1.65	1.85	1.78	1.78	1.80
POLAR	----	----	----	----	----	----	----	----	----	----	----	----	----
Antarctic	----	----	----	----	----	----	----	----	----	----	----	----	----
Arctic	----	----	----	----	----	----	----	----	----	----	----	----	----
WEST ASIA	----	9.74	4.53	9.43	9.28	----	----	6.38	5.57	5.27	5.36	6.04	----
Arabian Peninsula	----	3.33	0.95	4.96	4.88	5.10	5.21	5.19	4.95	4.62	4.71	5.37	----
Mashriq	----	----	----	----	----	----	----	----	----	----	----	----	----
GLOBAL TOTALS	13.17	9.61	7.19	5.52	4.95	4.46	4.26	4.35	4.34	4.53	4.43	4.35	4.34
REPORTED GLOBAL TOTALS FROM ORIGINAL DATA SOURCES	----	----	----	6.13	5.74	5.49	5.60	5.45	5.37	5.22	5.20	----	----

Comments:

GDP - from agriculture measures the output of the agricultural sector from forestry, hunting, and fishing as well as cultivation of crops and livestock production.

GDP measures the total output of goods and services for final use occurring within the domestic territory of a given country. GDP at purchaser values (market prices) is the sum of gross value added by all resident and non-resident producers in the economy plus any taxes and minus any subsidies not included in the value of the products. It is calculated without making deductions for depreciation of fabricated assets or for depletion and degradation of natural resources. The weighting factor, Current US$, is used in the World Development Indicators 2001. Current US$ is based on the World Bank's currency index for the period 1970 to 2010 with the base year 1995.

AGGREGATIONS

Gross Domestic Product - from Industry

Units: percent

Data Source: World Development Indicators 2001

Data Provider: The World Bank

Years: 1960-1999

	1960	1970	1980	1990	1991	1992	1993	1994	1995	1996	1997	1998	1999
AFRICA	----	30.38	43.06	34.90	34.98	34.72	33.55	33.48	33.52	33.46	33.08	31.17	31.53
Central Africa	----	35.19	34.92	31.36	27.82	26.33	24.85	27.94	28.94	30.70	30.73	27.05	26.38
Eastern Africa	----	17.35	17.92	16.49	16.23	15.45	14.54	15.47	14.86	15.66	15.77	15.91	16.04
Northern Africa	----	35.74	49.94	37.11	38.74	36.12	35.47	34.68	36.19	35.64	36.13	34.74	35.28
Southern Africa	37.05	38.53	45.89	39.17	37.34	36.86	35.51	35.32	34.88	33.90	32.98	31.83	33.59
Western Africa	9.17	15.73	37.03	29.44	30.81	37.35	35.06	34.26	32.79	35.26	33.93	28.41	----
Western Indian Ocean	----	17.22	18.02	21.61	22.78	22.92	22.66	23.60	24.04	23.48	23.61	22.99	23.52
ASIA AND PACIFIC	35.52	37.88	39.55	38.41	38.89	38.63	38.28	37.75	37.96	37.89	37.68	36.65	37.51
Australia and New Zealand	----	----	----	26.39	25.72	25.53	25.36	25.53	25.23	24.93	25.13	24.56	----
Central Asia	----	----	----	----	----	40.57	36.79	33.64	32.70	31.41	27.67	29.34	29.35
North West Pacific and East Asia	43.76	44.31	42.83	41.34	41.52	40.92	40.21	39.52	39.70	39.83	39.81	38.93	38.70
South Asia	18.62	19.69	26.46	27.16	26.12	26.47	27.13	27.87	28.41	28.12	27.60	26.18	26.70
South East Asia	23.52	25.17	39.74	37.66	38.43	37.73	37.70	37.83	38.40	39.41	39.57	38.68	39.25
South Pacific	----	----	----	----	----	----	----	----	----	----	----	----	----
EUROPE	----	----	----	32.13	31.24	30.21	29.20	28.84	28.54	28.02	27.70	27.17	26.88
Central Europe	----	----	----	40.28	39.05	36.14	35.45	34.71	34.28	32.91	33.07	31.30	30.31
Eastern Europe	----	----	----	47.64	47.80	43.14	41.76	42.59	39.25	36.84	34.43	36.50	37.81
Western Europe	----	----	----	30.08	29.33	29.01	27.99	27.82	27.74	27.25	26.95	26.52	26.08
LATIN AMERICA AND CARIBBEAN	----	36.19	38.90	35.91	33.60	33.63	33.78	33.39	33.29	30.56	30.52	29.58	30.26
Caribbean	----	----	----	38.63	37.66	37.87	37.80	38.34	37.95	37.75	37.58	37.26	37.59
Meso-America	----	30.87	32.70	27.87	27.60	27.76	26.60	26.56	27.36	27.81	28.02	27.93	28.02
South America	----	38.23	41.93	38.69	35.92	36.09	36.98	35.98	34.53	30.89	30.87	29.63	30.63
NORTH AMERICA	38.00	33.84	33.15	28.06	26.96	26.00	26.01	27.03	27.08	26.19	26.23	----	----
North America	38.00	33.84	33.15	28.06	26.96	26.00	26.01	27.03	27.08	26.19	26.23	----	----
POLAR	----	----	----	----	----	----	----	----	----	----	----	----	----
Antarctic	----	----	----	----	----	----	----	----	----	----	----	----	----
Arctic	----	----	----	----	----	----	----	----	----	----	----	----	----
WEST ASIA	----	----	66.92	45.59	38.97	45.67	44.27	43.92	49.09	50.73	50.46	46.05	----
Arabian Peninsula	----	----	78.77	52.85	44.37	53.42	51.89	51.95	53.14	54.96	54.60	50.25	----
Mashriq	----	----	----	----	----	----	----	----	----	----	----	----	----
GLOBAL TOTALS	----	----	----	----	----	31.48	31.19	31.31	31.35	30.62	30.35	31.05	----
REPORTED GLOBAL TOTALS FROM ORIGINAL DATA SOURCES	----	----	----	----	----	34.40	33.80	33.30	33.10	32.30	32.20	31.40	----

Comments:

GDP from industry comprises value added in mining, manufacturing (also reported as a separate subgroup), construction, electricity, water, and gas. Value added is the net output of a sector after adding up all outputs and subtracting intermediate inputs. It is calculated without making deductions for depreciation of fabricated assets or depletion and degradation of natural resources.

GDP measures the total output of goods and services for final use occurring within the domestic territory of a given country. GDP at purchaser values (market prices) is the sum of gross value added by all resident and non-resident producers in the economy plus any taxes and minus any subsidies not included in the value of the products. It is calculated without making deductions for depreciation of fabricated assets or for depletion and degradation of natural resources. The weighting factor, Current US$, is used in the World Development Indicators 2001. Current US$ is based on the World Bank's currency index for the period 1970 to 2010 with the base year 1995.

AGGREGATIONS

Gross Domestic Product - from Manufacturing

Units: percent

Data Source: World Development Indicators 2001

Data Provider: The World Bank

Years: 1960-1999

	1960	1970	1980	1990	1991	1992	1993	1994	1995	1996	1997	1998	1999
AFRICA	----	12.30	12.29	15.16	15.06	15.24	15.07	15.23	15.68	14.70	14.70	14.71	14.76
Central Africa	----	----	11.06	11.13	10.42	9.52	9.99	7.88	8.10	7.83	8.02	8.41	8.25
Eastern Africa	----	----	----	10.00	9.74	8.98	8.43	9.24	8.59	9.17	9.32	9.40	9.47
Northern Africa	----	----	9.03	13.86	13.44	14.87	14.41	14.79	15.94	15.14	15.27	15.87	16.16
Southern Africa	20.03	20.67	20.44	20.86	20.44	20.35	19.46	19.33	19.58	18.26	17.99	17.22	16.95
Western Africa	----	6.11	8.85	9.68	9.56	8.58	9.07	8.58	9.15	8.78	8.78	9.47	9.62
Western Indian Ocean	----	----	11.51	16.39	16.77	16.77	16.34	17.20	17.63	17.48	17.59	17.24	17.62
ASIA AND PACIFIC	26.42	28.29	26.66	25.73	26.12	25.69	24.97	24.46	24.89	24.87	24.97	24.46	24.44
Australia and New Zeeland	----	----	----	13.38	13.23	13.34	13.62	13.73	13.48	13.25	13.35	13.07	13.00
Central Asia	----	----	----	----	----	14.51	----	----	----	----	----	----	----
North West Pacific and East Asia	----	33.78	30.90	28.71	28.66	27.93	26.71	25.99	26.46	26.71	27.02	26.59	26.26
South Asia	13.44	13.44	14.18	15.70	15.52	15.73	15.66	16.22	16.84	16.46	16.26	15.88	15.92
South East Asia	16.08	15.88	18.88	23.81	24.56	24.38	24.57	24.80	25.09	25.60	25.77	25.24	26.75
South Pacific	----	----	----	----	----	----	----	----	----	----	----	----	----
EUROPE	----	----	----	----	----	----	----	----	----	----	----	----	----
Central Europe	----	----	----	----	24.24	22.47	21.93	22.25	22.30	20.53	20.55	19.08	18.11
Eastern Europe	----	----	----	----	----	----	----	----	----	----	----	----	----
Western Europe	----	----	----	----	----	----	----	----	----	----	----	----	----
LATIN AMERICA AND CARIBBEAN	----	25.27	25.65	24.69	23.21	22.23	21.46	20.80	21.38	21.22	21.15	20.88	20.79
Caribbean	----	21.37	23.77	28.97	28.70	29.39	29.62	29.36	28.98	28.65	28.23	28.04	27.80
Meso-America	----	22.44	21.75	20.42	20.27	20.00	18.86	18.57	20.26	20.89	20.83	20.71	20.80
South America	----	26.65	27.76	25.96	24.07	22.75	22.12	21.18	21.20	20.86	20.84	20.49	20.30
NORTH AMERICA	----	24.48	21.36	18.42	17.81	17.48	17.51	18.06	18.34	18.08	----	----	----
North America	----	24.48	21.36	18.42	17.81	17.48	17.51	18.06	18.34	18.08	----	----	----
POLAR	----	----	----	----	----	----	----	----	----	----	----	----	----
Antarctic	----	----	----	----	----	----	----	----	----	----	----	----	----
Arctic	----	----	----	----	----	----	----	----	----	----	----	----	----
WEST ASIA	----	----	6.60	9.94	9.31	9.90	9.85	11.44	10.66	10.64	11.02	11.69	----
Arabian Peninsula	----	6.75	4.90	8.17	7.73	8.38	8.58	9.42	9.68	9.81	10.16	10.68	----
Mashriq	----	----	----	----	----	----	----	----	----	----	----	----	----
GLOBAL TOTALS	----	----	----	----	22.32	21.69	20.84	20.88	21.10	20.72	20.19	19.83	----
REPORTED GLOBAL TOTALS FROM ORIGINAL DATA SOURCES	****	****	****	****	****	****	****	****	****	****	****	****	****

Comments:

GDP from manufacturing refers to industries belonging to the International Standard Industrial Classification (ISIC) divisions 15-37. Value added is the net output of a sector after adding up all outputs and subtracting intermediate inputs. It is calculated without making deductions for depreciation of fabricated assets or depletion and degradation of natural resources. The origin of value added is determined by the ISIC, revision 3.

GDP measures the total output of goods and services for final use occurring within the domestic territory of a given country. Gross domestic product at purchaser values (market prices) is the sum of gross value added by all resident and non-resident producers in the economy plus any taxes and minus any subsidies not included in the value of the products. It is calculated without making deductions for depreciation of fabricated assets or for depletion and degradation of natural resources. The weighting factor, Current US$, is used in the World Development Indicators 2001. Current US$ is based on the World Bank's currency index for the period 1970 to 2010 with the base year 1995.

Gross Domestic Product - from Services

Units: percent

Data Source: World Development Indicators 2001

Data Provider: The World Bank

Years: 1960-1999

	1960	1970	1980	1990	1991	1992	1993	1994	1995	1996	1997	1998	1999
AFRICA	----	46.55	40.41	46.81	47.06	48.17	49.28	49.78	49.96	48.62	49.27	50.15	49.57
Central Africa	----	45.00	40.80	45.16	45.10	44.13	43.06	36.07	36.92	35.06	34.32	36.33	36.84
Eastern Africa	----	----	----	41.78	42.46	42.93	40.74	42.56	43.99	44.33	46.19	45.86	46.43
Northern Africa	----	44.18	38.12	46.90	45.73	47.55	48.63	49.00	48.75	47.67	47.69	48.53	48.54
Southern Africa	50.79	52.31	45.04	51.82	53.26	56.38	56.75	57.37	58.29	57.93	58.86	59.68	58.14
Western Africa	33.48	43.63	37.94	37.38	36.69	33.52	35.32	35.95	35.46	33.19	33.94	36.99	36.74
Western Indian Ocean	----	59.71	56.14	55.74	55.76	55.78	55.73	54.67	56.55	56.71	57.49	58.35	59.47
ASIA AND PACIFIC	37.90	39.60	46.55	52.77	53.30	54.07	55.18	55.49	55.27	54.76	54.85	55.49	55.61
Australia and New Zealand	----	----	----	70.08	70.86	70.83	70.67	71.04	70.89	71.29	71.18	71.84	----
Central Asia	----	----	----	36.61	30.40	28.43	37.67	37.80	43.80	48.41	50.35	50.37	47.60
North West Pacific and East Asia	37.20	40.24	48.96	53.27	53.69	54.62	55.88	56.28	55.98	55.40	55.38	56.00	57.49
South Asia	36.55	36.10	41.40	43.70	44.46	44.42	44.45	43.90	44.67	44.89	46.20	46.50	46.96
South East Asia	47.96	42.90	38.46	46.59	46.35	47.11	48.22	48.22	47.76	47.20	47.43	48.25	47.43
South Pacific	----	----	----	----	----	----	----	----	----	----	----	----	----
EUROPE	----	----	60.63	62.53	64.19	66.21	67.26	67.92	68.21	68.73	69.18	69.91	70.34
Central Europe	----	----	----	44.29	48.69	52.57	52.64	54.53	54.85	56.19	57.05	58.21	60.68
Eastern Europe	----	----	----	33.95	36.30	46.37	47.20	48.67	50.77	54.01	56.59	55.80	54.30
Western Europe	----	----	61.28	66.50	67.52	68.29	69.46	69.61	69.75	70.28	70.72	71.27	71.82
LATIN AMERICA AND CARIBBEAN	----	51.19	50.96	55.54	58.17	58.67	58.88	58.42	58.48	61.48	61.94	62.77	62.47
Caribbean	----	58.17	51.81	57.05	57.41	57.86	58.46	58.01	58.07	58.19	57.61	58.07	57.63
Meso-America	----	54.01	56.84	63.13	63.84	64.55	66.11	66.57	65.41	64.61	64.98	65.40	65.87
South America	----	49.54	47.92	52.59	55.72	55.94	55.39	54.95	56.72	60.82	61.26	62.20	61.47
NORTH AMERICA	57.97	63.08	63.79	69.91	71.01	71.98	71.96	71.88	71.81	71.77	71.74	----	----
North America	57.97	63.08	63.79	69.91	71.01	71.98	71.96	71.88	71.81	71.77	71.74	----	----
POLAR	----	----	----	----	----	----	----	----	----	----	----	----	----
Antarctic	----	----	----	----	----	----	----	----	----	----	----	----	----
Arctic	----	----	----	----	----	----	----	----	----	----	----	----	----
WEST ASIA	----	----	28.56	45.01	51.79	45.46	47.16	47.45	44.56	43.51	44.31	48.30	----
Arabian Peninsula	----	27.83	20.28	42.19	50.74	41.47	42.91	43.17	42.51	41.37	41.91	46.16	----
Mashriq	----	----	----	----	----	----	----	----	----	----	----	----	----
GLOBAL TOTALS	----	----	----	----	62.78	63.95	64.40	64.51	64.45	64.80	65.15	63.40	----
REPORTED GLOBAL TOTALS FROM ORIGINAL DATA SOURCES	----	----	----	----	59.00	59.87	60.69	61.05	61.56	62.31	62.60	63.36	----

Comments:

GDP from services include value added in wholesale and retail trade (including hotels and restaurants), transport, and government, financial, professional, and personal services such as education, health care, and real estate services. Also included are imputed bank service charges, import duties, and any statistical discrepancies noted by national compilers as well as discrepancies arising from rescaling. Value added is the net output of a sector after adding up all outputs and subtracting intermediate inputs. It is calculated without making deductions for depreciation of fabricated assets or depletion and degradation of natural resources. The industrial origin of value added is determined by the ISIC, revision 3.

GDP measures the total output of goods and services for final use occurring within the domestic territory of a given country. Gross domestic product at purchaser values (market prices) is the sum of gross value added by all resident and non-resident producers in the economy plus any taxes and minus any subsidies not included in the value of the products. It is calculated without making deductions for depreciation of fabricated assets or for depletion and degradation of natural resources. The weighting factor, Current US$, is used in the World Development Indicators 2001. Current US$ is based on the World Bank's currency index for the period 1970 to 2010 with the base year 1995.

Gross Domestic Product - per Capita

Units: constant 1995 US$ per person

Data Source: World Development Indicators 2001

Data Provider: The World Bank

Years: 1960-1999

	1960	1970	1980	1990	1991	1992	1993	1994	1995	1996	1997	1998	1999
AFRICA	----	----	855.03	727.35	715.19	694.65	680.91	681.62	683.80	702.30	705.54	711.09	748.70
Central Africa	416.62	437.24	462.70	429.23	405.87	376.22	344.33	330.20	331.58	329.40	329.29	334.09	----
Eastern Africa	----	----	----	----	----	195.17	198.46	194.80	202.60	210.58	212.50	212.51	214.05
Northern Africa	606.28	757.98	1085.26	1250.55	1248.03	1245.87	1226.90	1250.23	1245.85	1303.26	1313.91	1364.31	1393.75
Southern Africa	----	----	2185.41	1550.94	1508.13	1434.20	1403.76	1405.57	1411.94	1441.12	1446.99	1429.80	1420.14
Western Africa	299.94	350.65	382.50	335.24	335.53	332.42	327.35	324.16	327.42	334.17	337.96	340.27	340.19
Western Indian Ocean	477.65	502.47	511.34	524.16	506.65	513.81	519.26	512.61	512.97	518.08	525.54	533.03	535.60
ASIA AND PACIFIC	755.93	1394.00	1801.49	2311.94	2372.81	2402.00	2425.36	2461.74	2511.86	2617.54	2658.30	2588.08	2618.39
Australia and New Zealand	9975.21	13421.41	15601.52	18043.66	17846.76	18227.75	18841.96	19503.00	20114.85	20556.22	21194.03	21785.15	22505.90
Central Asia	----	----	----	1519.14	1428.89	1240.44	1125.11	991.13	865.29	863.72	872.05	881.72	910.03
North West Pacific and East Asia	1130.21	2333.97	3068.48	4178.64	4318.27	4374.84	4411.58	4470.17	4555.72	4770.80	4856.30	4738.89	4793.72
South Asia	186.98	222.08	285.12	381.62	385.96	400.89	409.72	424.67	443.30	462.11	472.41	488.86	505.94
South East Asia	----	----	----	1127.14	1184.23	1237.75	1311.19	1393.98	1483.37	1565.49	1602.16	1458.61	1488.85
South Pacific	----	1839.74	1925.04	2192.07	2249.01	2301.50	2377.86	2398.71	2366.71	2391.13	2327.72	2274.00	2294.19
EUROPE	----	----	----	12252.69	12218.74	12180.26	12056.28	12227.53	12475.29	12649.55	12949.68	13252.65	13521.40
Central Europe	----	----	----	2906.71	2685.23	2648.49	2713.18	2666.08	2822.07	2961.07	3088.39	3163.32	3139.43
Eastern Europe	----	----	----	3099.54	2857.63	2448.63	2219.61	1907.74	1807.92	1745.78	1770.57	1714.13	1770.51
Western Europe	8844.60	13213.36	16850.07	22370.38	22549.51	22712.43	22526.46	23056.64	23520.40	23819.94	24337.64	24934.57	25441.10
LATIN AMERICA AND CARIBBEAN	2020.52	2603.60	3598.84	3328.58	3404.13	3459.46	3541.93	3660.27	3655.51	3727.97	3855.13	3871.15	3819.14
Caribbean	----	----	----	----	----	----	----	----	----	----	----	----	----
Meso-America	1493.06	2066.23	2863.39	2736.75	2794.57	2849.52	2860.48	2926.48	2734.42	2813.15	2945.57	3042.16	3101.74
South America	2275.54	2846.84	3978.45	3591.50	3683.00	3742.97	3859.74	4009.52	4070.84	4142.90	4275.27	4255.06	4145.31
NORTH AMERICA	12926.69	16621.68	20540.43	25470.42	25039.71	25468.03	25828.55	26556.95	26927.16	27502.38	28369.77	29234.47	29953.61
North America	12926.69	16621.68	20540.43	25470.42	25039.71	25468.03	25828.55	26556.95	26927.16	27502.38	28369.77	29234.47	29953.61
POLAR	----	----	----	----	----	----	----	----	----	----	----	----	----
Antarctic	----	----	----	----	----	----	----	----	----	----	----	----	----
Arctic	----	----	----	----	----	----	----	----	----	----	----	----	----
WEST ASIA	----	----	----	----	----	----	----	3413.36	3370.40	3323.52	3292.48	3216.98	3151.56
Arabian Peninsula	----	----	----	5418.00	5516.69	5855.60	5771.07	5632.66	5494.02	5387.26	5287.73	5072.57	4884.95
Mashriq	----	----	----	----	----	----	----	----	----	----	----	----	----
GLOBAL TOTALS	----	3609.60	4322.98	4982.54	4970.60	4985.62	4981.68	5053.03	5113.52	5214.30	5312.86	5336.27	5402.85
REPORTED GLOBAL TOTALS FROM ORIGINAL DATA SOURCES	2587.29	3591.58	4334.97	4993.68	4981.54	4997.64	4999.38	5077.38	5141.75	5245.68	5347.68	5373.44	5438.67

Comments:

GDP per capita is gross domestic product divided by midyear population. GDP is the sum of gross value added by all resident producers in the economy plus any product taxes and minus any subsidies not included in the value of the products. It is calculated without making deductions for depreciation of fabricated assets or for depletion and degradation of natural resources. Data are in constant 1995 U.S.$.

AGGREGATIONS

Gross Domestic Product - Purchasing Power Parity Revisited

Units: Million Constant 1995 US$

Data Source: World Development Indicators 2001

Data Provider: The World Bank

Years: 1975-1999

	1975	1980	1990	1995	1996	1997	1998	1999
AFRICA	413981.84	695897.31	1337563.47	1609066.51	1707292.01	1751166.99	1793521.83	1845394.66
Central Africa	42398.00	60679.22	104919.76	100758.67	103164.56	103167.92	103470.13	----
Eastern Africa	----	----	89301.85	113733.14	123409.57	128630.97	129265.77	137895.62
Northern Africa	101916.60	206607.96	438377.00	542357.29	581585.18	593825.13	616668.50	654757.22
Southern Africa	180488.81	278079.94	476369.94	566081.23	598630.78	614964.07	622698.05	658727.15
Western Africa	71823.10	119576.02	208120.04	260963.72	274230.88	283425.88	293196.71	307540.21
Western Indian Ocean	6993.01	10935.39	20474.88	25172.46	26271.04	27153.02	28222.68	29645.38
ASIA AND PACIFIC	2192833.72	3790398.83	9715799.38	14480373.72	15652964.44	16362406.95	16476381.09	17612101.80
Australia and New Zealand	133433.63	218121.37	433518.88	572934.35	601970.81	623408.24	635296.02	680093.02
Central Asia	----	----	251366.48	178233.28	181777.82	184448.81	184653.91	198593.66
North West Pacific and East Asia	1234027.31	2180904.17	5604860.63	8628350.36	9363814.18	9837393.16	9958146.78	10643514.25
South Asia	572936.56	882786.76	2177169.68	3135976.06	3373494.63	3505639.62	3657827.92	3936477.54
South East Asia	245222.26	497240.14	1226751.72	1933533.25	2098515.96	2178568.98	2008393.11	2119793.75
South Pacific	7213.96	11346.40	22132.01	31346.42	33391.04	32948.15	32063.36	33629.58
EUROPE	----	----	12133004.79	13196046.27	13521948.10	13824348.43	13939678.19	14570583.22
Central Europe	----	----	1380844.60	1534676.06	1625123.64	1681956.16	1696805.93	1731407.87
Eastern Europe	----	----	2535884.95	1718176.01	1671503.02	1691003.42	1637802.52	1751411.43
Western Europe	2782916.51	4538622.99	8216275.24	9943194.19	10225321.43	10451388.84	10605069.74	11087763.93
LATIN AMERICA AND CARIBBEAN	921324.93	1704467.50	2820804.23	3766116.65	3953558.30	4141652.11	4180863.43	4283857.40
Caribbean	----	----	----	----	----	----	----	----
Meso-America	234788.73	452613.90	778010.30	980203.08	1035135.54	1090217.06	1129509.69	1191732.35
South America	662379.97	1208233.28	1966886.01	2693243.34	2821301.89	2950065.94	2946123.66	2978956.74
NORTH AMERICA	2449556.78	4101627.50	8127162.92	10285013.43	10781468.80	11214711.56	11508247.85	12201138.85
North America	2449556.78	4101627.50	8127162.92	10285013.43	10781468.80	11214711.56	11508247.85	12201138.85
POLAR	----	----	----	----	----	----	----	----
Antarctic	----	----	----	----	----	----	----	----
Arctic	----	----	----	----	----	----	----	----
WEST ASIA	----	----	----	----	----	----	----	----
Arabian Peninsula	94925.70	199789.64	279889.67	374938.99	391162.44	399824.80	395057.35	402324.80
Mashriq	----	----	----	----	----	----	----	----
GLOBAL TOTALS	----	----	34474094.76	43809201.34	46110572.91	47803498.95	48413256.15	51053859.58
REPORTED GLOBAL TOTALS FROM ORIGINAL DATA SOURCES	****	****	****	****	****	****	****	****

Comments:

GDP - purchasing power parity is gross domestic product converted to international dollars using purchasing power parity rates. An international dollar has the same purchasing power over GDP as the U.S. dollar has in the United States. GDP is the sum of gross value added by all resident producers in the economy plus any product taxes and minus any subsidies not included in the value of the products. It is calculated without making deductions for depreciation of fabricated assets or for depletion and degradation of natural resources.

Data sources: purchasing power parity conversion factors are estimates by World Bank staff based on data collected by the International Comparison Programme.

AGGREGATIONS

Units: gigawatt-hour

Data Source: Energy Statistics of OECD Countries (1960-1999), and Non-OECD Countries (1971-1999)

Data Provider: International Energy Agency (IEA)

Years: 1960-1999

	1960	1970	1980	1990	1995	1996	1997	1998	1999
AFRICA	----	----	178398	313249	359522	378594	397039	404335	413171
Central Africa	----	----	6583	9755	9457	9745	10011	9968	10149
Eastern Africa	----	----	----	----	----	----	----	----	----
Northern Africa	----	----	39883	92115	112052	117133	123351	130735	139081
Southern Africa	----	----	114884	185840	207775	219752	230068	231200	229824
Western Africa	----	----	----	21303	24792	26128	27560	26205	27730
Western Indian Ocean	----	----	----	----	----	----	----	----	----
ASIA AND PACIFIC	----	----	1492127	2695180	3594781	3772762	3950862	4056323	4248168
Australia and New Zealand	28427	63601	118704	187343	208771	214221	220174	233752	241482
Central Asia	----	----	144856	193328	151031	143315	134194	130498	130615
North West Pacific and East Asia	----	----	1004441	1759070	2400373	2524115	2635218	2683317	2807715
South Asia	----	----	160850	397982	573348	601737	640843	681801	727299
South East Asia	----	----	63276	157457	261258	289374	320433	326955	341057
South Pacific	----	----	----	----	----	----	----	----	----
EUROPE	----	----	3388813	4398936	4214403	4278219	4297371	4368697	4433215
Central Europe	----	----	436188	540342	548125	574724	579420	588774	579916
Eastern Europe	----	----	1121496	1496773	1114536	1090547	1073494	1060314	1080616
Western Europe	566668	1204664	1831129	2361821	2551742	2612948	2644457	2719609	2772683
LATIN AMERICA AND CARIBBEAN	----	----	361794	606824	766526	811973	859636	892597	922118
Caribbean	----	----	18029	25703	29576	32091	34284	35266	35879
Meso-America	----	----	76422	137136	172313	183265	197817	205403	217609
South America	----	----	267343	443985	564637	596617	627535	651928	668630
NORTH AMERICA	915795	1833542	2800699	3679411	4142095	4249932	4271385	4392108	4517341
North America	915795	1833542	2800699	3679411	4142095	4249932	4271385	4392108	4517341
POLAR	----	----	----	----	----	----	----	----	----
Antarctic	----	----	----	----	----	----	----	----	----
Arctic	----	----	----	----	----	----	----	----	----
WEST ASIA	----	----	60343	155669	224262	238288	252399	269410	281089
Arabian Peninsula	----	----	41178	114920	168871	179380	190096	204498	214977
Mashriq	----	----	19165	40749	55391	58908	62303	64912	66112
GLOBAL TOTALS	----	----	8282174	11849269	13301589	13729768	14028692	14383470	14815102
REPORTED GLOBAL TOTALS FROM ORIGINAL DATA SOURCES	----	----	8296607	11870159	13324938	13753121	14051976	14407120	14838752

Comments:

Gross electricity production is measured at the terminals of all alternator sets in a station; it therefore includes the energy taken by station auxiliaries and losses in transformers that are considered integral parts of the station.

The difference between gross and net production is generally calculated as 7 percent for conventional thermal stations, 1 percent for hydro stations, and 6 percent for nuclear, geothermal and solar stations. Hydro stations' production includes production from pumped storage plants.

A gigawatt-hour is defined as 85.98 toe (tonnes of oil equivalents) or 3.6 TJ (terajoules).

Military Expenditures - Percent of Central Government Expenditures

Units: percent

Data Source: World Development Indicators 2001

Data Provider: The World Bank

Years: 1985-1997

	1985	1986	1987	1988	1989	1990	1991	1992	1993	1994	1995	1996	1997
AFRICA	12.21	----	13.52	15.20	13.09	13.34	11.09	9.83	9.45	9.93	8.97	9.18	8.64
Central Africa	----	----	----	----	----	----	----	----	----	----	----	----	----
Eastern Africa	17.16	17.60	23.19	21.85	21.15	22.59	20.29	13.27	11.07	9.55	11.62	12.05	12.45
Northern Africa	----	12.41	13.04	12.43	12.58	10.16	11.42	8.61	9.40	9.65	9.47	10.73	11.09
Southern Africa	12.73	----	14.41	18.61	13.72	16.10	10.98	9.69	9.69	10.01	8.05	7.64	6.28
Western Africa	9.69	7.12	7.16	9.93	10.48	----	9.66	11.82	8.52	11.17	9.70	9.95	9.10
Western Indian Ocean	5.78	5.63	----	4.09	3.63	3.89	3.83	3.08	3.37	2.65	2.64	3.17	3.81
ASIA AND PACIFIC	13.62	13.36	11.38	10.83	10.40	10.54	9.69	9.62	9.57	9.45	9.57	9.87	9.89
Australia and New Zeeland	9.40	8.61	9.06	8.63	8.58	8.91	9.12	8.69	9.16	8.89	8.40	8.34	8.07
Central Asia	----	----	----	----	----	----	----	----	----	7.69	5.97	----	----
North West Pacific and East Asia	10.85	9.00	8.52	8.62	8.75	9.20	8.96	9.08	8.74	8.65	8.86	9.18	9.34
South Asia	25.06	32.95	25.84	23.47	20.56	18.77	15.18	14.87	15.05	15.20	13.84	14.63	14.70
South East Asia	12.95	12.28	13.48	13.53	12.55	13.06	13.40	13.00	14.16	13.42	14.73	13.63	12.30
South Pacific	----	----	----	----	----	----	----	----	----	----	----	----	----
EUROPE	----	----	8.46	8.19	8.26	7.92	7.66	7.68	7.93	6.97	6.41	6.37	6.67
Central Europe	20.97	22.03	21.18	19.40	----	----	----	12.30	10.50	10.28	11.12	10.08	9.49
Eastern Europe	----	----	----	----	----	----	----	26.06	33.35	24.11	23.39	21.28	24.67
Western Europe	8.13	7.84	8.10	7.90	7.96	7.54	7.37	6.60	6.28	6.04	5.70	5.69	5.74
LATIN AMERICA AND CARIBBEAN	6.31	6.06	6.91	6.64	5.84	6.10	6.48	5.93	6.44	6.48	6.64	----	----
Caribbean	----	----	----	----	----	----	----	----	----	----	----	----	----
Meso-America	5.27	4.53	4.75	4.17	4.38	4.66	3.93	4.21	4.20	4.57	4.25	3.83	6.03
South America	7.05	6.59	7.57	7.58	6.25	6.36	7.18	6.42	7.28	7.13	7.06	----	----
NORTH AMERICA	24.04	25.41	25.44	24.37	23.56	21.71	18.14	19.50	18.52	17.60	17.40	16.50	16.30
North America	24.04	25.41	25.44	24.37	23.56	21.71	18.14	19.50	18.52	17.60	17.40	16.50	16.30
POLAR	----	----	----	----	----	----	----	----	----	----	----	----	----
Antarctic	----	----	----	----	----	----	----	----	----	----	----	----	----
Arctic	----	----	----	----	----	----	----	----	----	----	----	----	----
WEST ASIA	----	----	41.19	33.63	36.41	65.78	----	68.66	37.94	36.32	35.41	33.25	31.79
Arabian Peninsula	28.13	31.21	41.36	33.85	35.90	68.78	----	70.99	38.89	37.49	36.92	34.79	33.73
Mashriq	----	----	----	----	----	----	----	----	----	----	----	----	----
GLOBAL TOTALS	16.94	16.72	15.95	15.00	14.64	14.03	12.42	12.54	11.84	11.01	10.45	10.53	10.74
REPORTED GLOBAL TOTALS FROM ORIGINAL DATA SOURCES	17.06	16.85	16.45	15.52	15.29	14.58	11.96	12.69	11.98	11.14	10.25	10.47	10.75

Comments:

Military expenditures for North Atlantic Treaty Organization (NATO) countries are based on the NATO definition, which covers military-related expenditures of the defence ministry (including recruiting, training, construction, and the purchase of military supplies and equipment) and other ministries. Civilian-type expenditures of the defence ministry are excluded. Military assistance is included in the expenditures of the donor country, and purchases of military equipment on credit are included at the time the debt is incurred, not at the time of payment. Data for other countries generally cover expenditures of the ministry of defence (excluded are expenditures on public order and safety, which are classified separately)

Data sources: the data on military expenditures are from the Bureau of Verification and Compliance's World Military Expenditures and Arms Transfers 1998 (U.S. Department of State 1999).

AGGREGATIONS

Units: million constant 1995 US$

Data Source: World Development Indicators 2001

Data Provider: The World Bank

Years: 1960-1999

	1960	1970	1980	1990	1991	1992	1993	1994	1995	1996	1997	1998	1999
AFRICA	1202.69	1957.17	12780.54	30818.66	30243.21	29796.00	25876.60	27789.06	26316.83	23295.35	20635.45	20400.02	18000.46
Central Africa	90.66	284.90	1261.90	2994.36	2299.87	2145.99	1819.98	2495.11	1802.05	2113.07	1690.47	1261.02	1396.55
Eastern Africa	116.27	264.89	1894.45	5189.36	4534.43	5320.04	5388.74	5483.44	4891.74	3792.59	3033.75	2963.11	2923.45
Northern Africa	889.89	639.81	4207.42	10074.03	9905.32	7434.58	5224.02	5376.52	3959.31	4421.77	3825.77	4084.31	3584.95
Southern Africa	----	----	----	----	----	----	6357.87	6372.04	7764.82	5636.16	5597.26	5467.07	4602.11
Western Africa	64.58	516.73	2731.56	6427.59	6687.41	6632.07	6404.13	7481.84	7284.69	6619.81	5197.00	5743.34	4804.62
Western Indian Ocean	5.22	83.49	442.94	793.58	860.76	694.44	681.86	580.10	614.22	711.96	1291.20	881.17	688.77
ASIA AND PACIFIC	2244.83	3907.83	11582.14	18165.01	20516.41	21659.68	19481.48	22106.16	20869.64	18229.18	15603.86	18430.17	19170.54
Australia and New Zealand	----	----	----	----	----	----	----	----	----	----	----	----	----
Central Asia	----	----	----	----	----	39.72	208.50	428.64	646.67	717.42	747.18	870.38	889.02
North West Pacific and East Asia	325.68	348.92	278.22	2771.06	2717.74	3971.76	4284.21	4227.68	4815.66	3441.11	2812.68	3330.10	3411.85
South Asia	1312.72	1727.99	6691.86	7686.26	10263.93	8500.28	6708.91	9070.50	6753.49	6731.21	5673.51	6548.58	5572.60
South East Asia	596.23	1500.23	3377.18	6093.45	5785.93	7380.37	6425.56	6254.99	6429.81	5292.79	4562.84	5803.72	7710.30
South Pacific	----	330.69	1234.89	1614.24	1608.09	1767.57	1854.31	2124.35	2224.00	2046.64	1807.66	1877.39	1586.76
EUROPE	----	----	----	----	----	----	----	----	----	----	----	----	----
Central Europe	----	----	----	3704.93	7728.92	4067.60	3173.80	4741.22	7876.39	4972.22	4142.44	5212.72	6801.61
Eastern Europe	----	----	----	685.27	1415.71	3544.60	4031.22	3565.02	3475.41	3145.70	2016.73	2349.02	3825.69
Western Europe	----	----	----	----	----	----	----	----	----	----	----	----	----
LATIN AMERICA AND CARIBBEAN	187.46	1177.41	2547.47	5950.12	6639.24	6160.10	6110.80	6332.21	7303.47	7086.39	5773.72	5736.21	6065.16
Caribbean	----	125.29	704.46	1042.83	804.56	657.08	554.11	1224.13	1644.35	1105.07	966.78	1119.84	928.18
Meso-America	52.44	257.11	850.80	2327.37	2772.16	2828.61	2374.34	2485.46	2549.48	2674.55	1756.83	1748.75	2591.57
South America	123.08	795.01	992.21	2579.92	3062.52	2674.41	3182.35	2622.63	3109.64	3306.78	3050.10	2867.62	2545.42
NORTH AMERICA	----	----	----	----	----	----	----	----	----	----	----	----	----
North America	----	----	----	----	----	----	----	----	----	----	----	----	----
POLAR	----	----	----	----	----	----	----	----	----	----	----	----	----
Antarctic	----	----	----	----	----	----	----	----	----	----	----	----	----
Arctic	----	----	----	----	----	----	----	----	----	----	----	----	----
WEST ASIA	147.34	199.02	5230.09	3215.00	3000.89	1622.04	1958.33	3017.95	2777.28	2874.67	2843.30	2466.86	2505.21
Arabian Peninsula	29.11	34.40	1169.21	825.42	494.55	504.47	605.77	408.70	383.33	535.25	665.02	549.18	688.38
Mashriq	118.22	164.62	4060.89	2389.57	2506.34	1117.58	1352.56	2609.25	2393.95	2339.42	2178.28	1917.67	1816.83
GLOBAL TOTALS	4132.50	7575.53	34608.31	64321.92	71778.81	69463.01	62265.10	69166.46	69054.37	62492.15	52553.52	55967.90	57543.29
REPORTED GLOBAL TOTALS FROM ORIGINAL DATA SOURCES	4132.50	4132.50	4132.50	4132.50	4132.50	4132.50	4132.50	4132.50	4132.50	4132.50	4132.50	4132.50	4132.50

Comments:

Official development assistance and net official aid (ODA) is the net amount of disbursed grants and concessional loans received by a country less repayments of concessional loans. Grants include gifts of money, goods, or services for which no repayment is required. A concessional loan has a grant element of 25 percent or more. The grant element is the amount by which the face value of the loan exceeds its present market value because of below-market interest rates, favourable maturity schedules, or repayment grace periods. Nonconcessional loans are not a component of ODA.

ODA receipts are shown as positive numbers; for some developing countries (e.g., Republic of Korea) the data are shown as negative numbers because of net repayments of concessional loans. The ratios include aid allocated by country only (e.g. for administrative costs, research into development issues, aid to nongovernmental organizations). Regional allocated aid is not included.

Total Debt Service - Percent of Exports of Goods and Services

Units: percent

Data Source: World Development Indicators 2001

Data Provider: The World Bank

Years: 1970-1999

	1970	1980	1990	1991	1992	1993	1994	1995	1996	1997	1998	1999
AFRICA	----	16.29	25.23	26.87	26.78	24.37	20.16	18.67	17.13	16.91	17.60	17.26
Central Africa	4.00	17.43	16.30	14.21	13.33	10.68	17.62	13.15	13.57	11.36	13.06	----
Eastern Africa	----	16.50	35.63	33.38	29.81	26.80	30.47	24.75	27.24	17.60	19.03	24.45
Northern Africa	----	23.07	38.03	39.53	40.65	39.08	29.36	23.85	21.15	20.00	21.87	21.71
Southern Africa	----	----	----	----	----	----	10.55	15.28	13.20	16.18	14.63	15.04
Western Africa	----	9.13	23.91	22.89	25.42	16.74	21.40	17.01	17.34	14.23	16.79	13.06
Western Indian Ocean	----	15.41	15.80	13.38	10.66	8.45	8.32	8.61	7.40	13.52	11.93	11.10
ASIA AND PACIFIC	----	----	14.57	12.11	12.02	12.81	11.38	11.69	12.22	11.09	12.85	16.05
Australia and New Zealand	----	----	----	----	----	----	----	----	----	----	----	----
Central Asia	----	----	----	----	----	----	----	----	6.51	9.33	14.64	19.42
North West Pacific and East Asia	----	----	11.44	9.25	8.97	9.95	8.04	9.00	9.00	8.00	10.72	16.03
South Asia	32.00	9.53	19.58	17.95	17.90	18.33	22.29	26.04	22.69	23.11	18.75	17.00
South East Asia	----	----	21.10	17.60	18.20	18.21	16.18	14.41	16.34	14.72	15.97	15.87
South Pacific	----	10.60	25.45	20.86	21.58	22.86	23.52	15.71	11.46	13.25	6.75	7.55
EUROPE	----	----	----	----	----	----	----	----	----	----	----	----
Central Europe	----	----	18.95	19.93	22.00	15.99	18.54	16.65	15.33	14.97	17.86	20.32
Eastern Europe	----	----	----	----	2.00	2.82	4.37	6.82	6.56	6.70	11.20	12.82
Western Europe	----	----	----	----	----	----	----	----	----	----	----	----
LATIN AMERICA AND CARIBBEAN	----	34.79	23.59	23.83	25.64	27.14	24.69	26.18	31.11	35.10	31.78	40.85
Caribbean	----	----	----	----	----	----	----	----	----	----	----	----
Meso-America	----	34.90	19.74	22.53	30.87	31.23	24.82	24.96	32.32	29.95	18.36	22.64
South America	----	40.66	26.63	25.37	22.99	26.26	26.27	29.10	32.61	42.26	47.51	65.02
NORTH AMERICA	----	----	----	----	----	----	----	----	----	----	----	----
North America	----	----	----	----	----	----	----	----	----	----	----	----
POLAR	----	----	----	----	----	----	----	----	----	----	----	----
Antarctic	----	----	----	----	----	----	----	----	----	----	----	----
Arctic	----	----	----	----	----	----	----	----	----	----	----	----
WEST ASIA	----	----	----	----	----	----	----	----	----	----	----	----
Arabian Peninsula	----	----	----	----	----	----	----	----	----	----	----	----
Mashriq	----	----	20.80	14.13	10.70	8.27	9.17	7.64	9.23	12.38	11.02	----
GLOBAL TOTALS	----	----	----	----	----	----	----	----	----	----	----	----
REPORTED GLOBAL TOTALS FROM ORIGINAL DATA SOURCES	----	----	----	----	----	----	----	----	----	----	----	----

Comments:

Total debt service is the sum of principal repayments and interest actually paid in foreign currency, goods, or services on long-term debt, interest paid on short-term debt, and repayments (repurchases and charges) to the International Monetary Fund.

Total External Debt

Units: million constant 1995 US$

Data Source: World Development Indicators 2001

Data Provider: The World Bank

Years: 1971-1999

	1971	1980	1990	1991	1992	1993	1994	1995	1996	1997	1998	1999
AFRICA	16960.40	135185.31	327806.65	332275.38	328186.25	332929.86	382571.64	406099.74	399333.65	380064.67	392033.61	370726.88
Central Africa	----	----	----	----	----	----	----	----	----	----	----	----
Eastern Africa	1233.35	7504.48	28351.59	30032.32	29944.99	31018.32	32337.49	33415.36	32607.18	32410.18	33730.74	27451.53
Northern Africa	6948.70	71249.62	135962.96	133912.33	131497.75	129377.46	140053.10	147825.51	145032.19	136872.35	139847.77	133119.55
Southern Africa	----	----	----	----	----	----	78534.89	85084.92	84873.92	82304.74	85004.41	81438.50
Western Africa	4245.37	32882.67	94231.14	96690.53	92253.97	96473.21	98852.21	103636.20	101546.96	93043.60	96457.18	93028.45
Western Indian Ocean	659.90	2326.62	6355.56	6693.52	6706.90	6504.98	7372.60	8128.92	7972.05	8753.23	9168.94	9146.09
ASIA AND PACIFIC	----	----	508521.39	560905.62	620190.97	687873.79	807705.36	909472.92	978297.22	1047649.04	1091640.46	1073976.78
Australia and New Zealand	----	----	----	----	----	----	----	----	----	----	----	----
Central Asia	----	----	----	----	136.42	4683.41	6927.88	9061.92	9934.46	13676.93	17940.09	18853.90
North West Pacific and East Asia	----	----	113920.49	126190.03	147570.96	168493.91	218745.61	257985.23	309381.20	358772.71	371541.89	359540.28
South Asia	17331.80	52282.14	168897.88	179355.19	193888.75	209624.00	226133.36	219330.42	210351.80	204078.90	216904.86	213778.51
South East Asia	11802.48	69552.35	221522.02	250957.29	273012.25	299987.25	351489.84	419059.97	444654.97	467072.51	481006.99	477591.16
South Pacific	537.99	1367.88	4181.01	4403.12	5582.58	5085.23	4408.67	4035.37	3974.80	4047.99	4246.63	4212.93
EUROPE	----	----	----	----	----	----	----	----	----	----	----	----
Central Europe	----	----	177609.71	191578.03	193217.88	232186.42	234669.16	264563.92	278321.61	292903.64	338171.95	343214.15
Eastern Europe	----	----	----	----	100102.34	149257.50	165964.99	169658.99	177190.23	181256.64	247449.81	244545.81
Western Europe	----	----	----	----	----	----	----	----	----	----	----	----
LATIN AMERICA AND CARIBBEAN	46251.29	306336.62	561462.29	579306.84	596431.04	646849.59	696453.25	777367.25	807465.59	853278.19	961000.25	982031.85
Caribbean	----	----	----	----	----	----	----	----	----	----	----	----
Meso-America	11176.52	86964.93	169971.34	182590.70	180563.56	206092.18	218851.50	250388.62	232834.21	223310.52	239928.16	251545.24
South America	33079.29	212703.03	374389.34	379967.38	399199.12	424191.14	461223.01	510021.52	558658.25	614091.59	704598.70	713205.28
NORTH AMERICA	----	----	----	----	----	----	----	----	----	----	----	----
North America	----	----	----	----	----	----	----	----	----	----	----	----
POLAR	----	----	----	----	----	----	----	----	----	----	----	----
Antarctic	----	----	----	----	----	----	----	----	----	----	----	----
Arctic	----	----	----	----	----	----	----	----	----	----	----	----
WEST ASIA	----	----	----	----	----	----	----	----	----	----	----	----
Arabian Peninsula	----	----	----	----	----	----	----	----	----	----	----	----
Mashriq	----	----	----	----	----	----	----	----	----	----	----	----
GLOBAL TOTALS	----	----	----	----	----	----	----	----	----	----	----	----
REPORTED GLOBAL TOTALS FROM ORIGINAL DATA SOURCES	****	****	****	****	****	****	****	****	****	****	****	****

Comments:

Total external debt is debt owed to non-residents repayable in foreign currency, goods, or services. Total external debt is the sum of public, publicly guaranteed, and private non-guaranteed long-term debt, use of International Monetary Fund (IMF) credit, and short-term debt. Short-term debt includes all debt having an original maturity of one year or less and interest in arrears on long-term debt.

For this variable, regional aggregations are not meaningful. For national-level data, see the enclosed CD ROM.

1.1.2 Education

Total enrolment in a specific level of education, regardless of age, expressed as a percentage of the official school-age population corresponding to the same level of education in a given school-year. Gross Enrolment Ratio is widely used to show the general level of participation in a given level of education. It indicates the capacity of the education system to enrol students of a particular age-group. It is used as a substitute indicator to net enrolment ratio (NER) when data on enrolment by single years of age are not available. Furthermore, it can also be a complementary indicator to NER by indicating the extent of over-aged and under-aged enrolment.

Calculation methods: divide the number of pupils enrolled in a given level of education regardless of age by the population of the age-group which officially corresponds to the given level of education, and multiply the result by 100.

Enrolment Ratios in Primary Education

Units: percent

Data Source: World Education Report 2000

Data Provider: The Institute for Statistics (UIS) of the United Nations Organization for Education, Science and Culture (UNESCO)

Years: 1990, 1996

Gross Enrolment Ratios in Primary Education

	1990
AFRICA	78.78
Central Africa	76.43
Eastern Africa	63.51
Northern Africa	84.81
Southern Africa	94.31
Western Africa	73.16
Western Indian Ocean	101.88
ASIA AND PACIFIC	104.57
Australia and New Zealand	107.66
Central Asia	86.27
North West Pacific and East Asia	122.02
South Asia	90.36
South East Asia	109.12
South Pacific	83.14
EUROPE	100.72
Central Europe	96.48
Eastern Europe	100.99
Western Europe	103.94
LATIN AMERICA AND CARIBBEAN	104.76
Caribbean	----
Meso-America	107.31
South America	105.37
NORTH AMERICA	102.10
North America	102.10
POLAR	----
Antarctic	----
Arctic	----
WEST ASIA	97.30
Arabian Peninsula	----
Mashriq	110.14
GLOBAL TOTALS	99.59
REPORTED GLOBAL TOTALS FROM ORIGINAL DATA SOURCES	99.20

Net Enrolment Ratios in Primary Education

	1990
AFRICA	----
Central Africa	----
Eastern Africa	----
Northern Africa	----
Southern Africa	70.20
Western Africa	----
Western Indian Ocean	----
ASIA AND PACIFIC	----
Australia and New Zealand	99.17
Central Asia	----
North West Pacific and East Asia	97.37
South Asia	----
South East Asia	----
South Pacific	----
EUROPE	----
Central Europe	----
Eastern Europe	----
Western Europe	----
LATIN AMERICA AND CARIBBEAN	87.24
Caribbean	----
Meso-America	97.50
South America	84.94
NORTH AMERICA	96.10
North America	96.10
POLAR	----
Antarctic	----
Arctic	----
WEST ASIA	----
Arabian Peninsula	----
Mashriq	----
GLOBAL TOTALS	----
REPORTED GLOBAL TOTALS FROM ORIGINAL DATA SOURCES	----

Comments:
Primary education provides children with basic reading, writing, and mathematics skills along with an elementary understanding of such subjects as history, geography, natural science, social science, art, and music.

Enrolment Ratios in Secondary Education

Units: percent

Data Source: World Education Report 2000

Data Provider: The Institute for Statistics (UIS) of the United Nations Organization for Education, Science and Culture (UNESCO)

Years: 1990, 1996

Gross Enrolment Ratios in Secondary Education

	1990	1996
AFRICA	30.06	34.30
Central Africa	22.06	25.47
Eastern Africa	15.01	14.28
Northern Africa	55.42	60.11
Southern Africa	30.93	38.70
Western Africa	21.30	26.39
Western Indian Ocean	20.47	19.33
ASIA AND PACIFIC	46.37	60.01
Australia and New Zealand	83.38	141.31
Central Asia	99.13	88.48
North West Pacific and East Asia	55.15	73.90
South Asia	39.63	51.00
South East Asia	41.09	51.40
South Pacific	19.48	----
EUROPE	87.47	96.71
Central Europe	68.15	74.61
Eastern Europe	92.58	----
Western Europe	94.56	111.93
LATIN AMERICA AND CARIBBEAN	50.47	60.62
Caribbean	----	----
Meso-America	48.87	58.95
South America	50.42	61.00
NORTH AMERICA	93.79	97.78
North America	93.79	97.78
POLAR	----	----
Antarctic	----	----
Arctic	----	----
WEST ASIA	49.83	49.77
Arabian Peninsula	----	55.49
Mashriq	51.12	45.07
GLOBAL TOTALS	51.61	61.47
REPORTED GLOBAL TOTALS FROM ORIGINAL DATA SOURCES	51.80	----

Net Enrolment Ratios in Secondary Education

	1990	1996
AFRICA	----	----
Central Africa	----	----
Eastern Africa	----	----
Northern Africa	----	----
Southern Africa	----	----
Western Africa	----	----
Western Indian Ocean	----	----
ASIA AND PACIFIC	----	----
Australia and New Zealand	80.18	89.59
Central Asia	----	----
North West Pacific and East Asia	----	----
South Asia	----	----
South East Asia	----	----
South Pacific	----	----
EUROPE	----	----
Central Europe	----	66.45
Eastern Europe	----	----
Western Europe	----	----
LATIN AMERICA AND CARIBBEAN	34.04	----
Caribbean	----	----
Meso-America	43.89	49.75
South America	23.95	----
NORTH AMERICA	86.30	90.10
North America	86.30	90.10
POLAR	----	----
Antarctic	----	----
Arctic	----	----
WEST ASIA	----	----
Arabian Peninsula	----	----
Mashriq	----	----
GLOBAL TOTALS	----	
REPORTED GLOBAL TOTALS FROM ORIGINAL DATA SOURCES	----	----

Comments:

Secondary education completes the provision of basic education that began at the primary level, and aims at laying the foundations for lifelong learning and human development, by offering more subject- or skill-oriented instruction using more specialized teachers.

Enrolment in Tertiary Education

Units: thousand people

Data Source: World Education Report 2000

Data Provider: The Institute for Statistics (UIS) of the United Nations Organization for Education, Science and Culture (UNESCO)

Years: 1990, 1996

	1990	1996
AFRICA	2524.84	----
Central Africa	132.52	----
Eastern Africa	89.43	----
Northern Africa	1561.80	1990.89
Southern Africa	569.52	911.15
Western Africa	----	----
Western Indian Ocean	40.53	34.16
ASIA AND PACIFIC	20662.46	28660.15
Australia and New Zealand	596.52	1203.97
Central Asia	1383.39	----
North West Pacific and East Asia	8354.24	12611.07
South Asia	6408.53	7367.56
South East Asia	3914.30	6830.54
South Pacific	----	----
EUROPE	19953.69	23396.07
Central Europe	2355.15	3903.43
Eastern Europe	7571.02	6719.89
Western Europe	10027.53	12772.75
LATIN AMERICA AND CARIBBEAN	6887.15	7752.16
Caribbean	----	----
Meso-America	1592.93	2102.97
South America	5021.86	5327.70
NORTH AMERICA	15658.25	16275.01
North America	15658.25	16275.01
POLAR	----	----
Antarctic	----	----
Arctic	----	----
WEST ASIA	----	----
Arabian Peninsula	249.87	393.63
Mashriq	----	----
GLOBAL TOTALS	66246.75	80096.56
REPORTED GLOBAL TOTALS FROM ORIGINAL DATA SOURCES	68600.00	----

Comments:

Enrolment in tertiary education indicates the number of students enrolled in tertiary or higher education.

Tertiary education whether or not leading to an advanced research qualification, normally requires, as a minimum condition of admission, the successful completion of education at the secondary level.

AGGREGATIONS

Illiteracy Rate - Ages 15 and over

Units: percent

Data Source: World Development Indicators 2001

Data Provider: The World Bank

Years: 1970-1999

	1970	1980	1990	1991	1992	1993	1994	1995	1996	1997	1998	1999
AFRICA	71.52	61.58	50.26	49.09	47.95	46.72	45.63	44.52	43.38	42.22	41.10	40.02
Central Africa	77.04	65.20	51.25	49.75	48.27	46.79	45.32	43.90	42.50	41.13	39.78	38.45
Eastern Africa	76.37	66.78	56.07	54.80	53.66	52.14	51.17	50.06	48.92	47.68	46.47	45.26
Northern Africa	72.87	62.82	51.89	50.80	49.72	48.65	47.62	46.56	45.48	44.41	43.36	42.37
Southern Africa	50.50	41.67	32.37	31.47	30.56	29.64	28.82	28.07	27.22	26.46	25.73	25.05
Western Africa	81.83	71.16	57.91	56.54	55.18	53.84	52.51	51.13	49.73	48.28	46.90	45.58
Western Indian Ocean	58.19	49.45	39.99	39.12	38.27	37.44	36.66	35.86	35.09	34.32	33.55	32.79
ASIA AND PACIFIC	52.80	41.74	32.27	31.51	30.80	30.06	29.37	28.70	28.07	27.44	26.87	26.26
Australia and New Zealand	----	----	----	----	----	----	----	----	----	----	----	----
Central Asia	----	----	----	----	----	----	----	----	----	----	----	----
North West Pacific and East Asia	47.45	33.49	22.33	21.55	20.77	19.99	19.31	18.63	17.95	17.27	16.68	16.00
South Asia	68.34	60.56	52.38	51.50	50.75	49.92	49.10	48.27	47.46	46.65	45.85	45.03
South East Asia	33.42	23.62	16.44	15.81	15.23	14.66	14.09	13.59	13.11	12.62	12.17	11.79
South Pacific	52.43	43.94	36.80	36.16	35.47	34.83	34.22	33.59	32.96	32.38	31.78	31.24
EUROPE	----	----	----	----	----	----	----	----	----	----	----	----
Central Europe	14.24	11.05	8.66	8.41	8.17	7.93	7.67	7.45	7.24	7.00	6.81	6.59
Eastern Europe	1.92	1.23	0.80	0.72	0.71	0.70	0.62	0.61	0.60	0.53	0.52	0.51
Western Europe	----	----	----	----	----	----	----	----	----	----	----	----
LATIN AMERICA AND CARIBBEAN	26.16	20.05	15.20	14.78	14.36	13.96	13.62	13.26	12.89	12.56	12.23	11.92
Caribbean	29.59	23.84	18.93	18.58	18.29	18.00	17.71	17.40	17.18	16.92	16.68	16.45
Meso-America	29.24	21.91	16.03	15.60	15.09	14.66	14.30	13.88	13.56	13.24	12.87	12.56
South America	24.74	18.98	14.47	14.05	13.65	13.25	12.91	12.58	12.17	11.83	11.51	11.19
NORTH AMERICA	----	----	----	----	----	----	----	----	----	----	----	----
North America	----	----	----	----	----	----	----	----	----	----	----	----
POLAR	----	----	----	----	----	----	----	----	----	----	----	----
Antarctic	----	----	----	----	----	----	----	----	----	----	----	----
Arctic	----	----	----	----	----	----	----	----	----	----	----	----
WEST ASIA	66.84	54.58	42.19	41.56	40.42	39.32	38.32	37.29	36.27	35.33	34.36	33.47
Arabian Peninsula	73.16	57.45	42.78	42.77	41.52	40.35	39.38	38.36	37.26	36.29	35.30	34.49
Mashriq	62.07	52.18	41.63	40.42	39.39	38.36	37.35	36.33	35.36	34.46	33.51	32.56
GLOBAL TOTALS	----	----	----	----	----	----	----	----	----	----	----	----
REPORTED GLOBAL TOTALS FROM ORIGINAL DATA SOURCES	----	----	----	----	----	----	----	----	----	----	----	----

Comments:

Adult illiteracy rate is the percentage of people ages 15 and above who cannot, with understanding, read and write a short, simple statement on their everyday life.

The data are based on UNESCO's 1999 literacy estimates and projections.

AGGREGATIONS

Illiterate People - Ages 15 and over

Units: thousand people

Data Source: World Development Indicators 2001

Data Provider: The World Bank

Years: 1970-1999

	1970	1980	1990	1991	1992	1993	1994	1995	1996	1997	1998	1999
AFRICA	136097.64	153500.47	166923.51	167970.38	168981.20	168676.34	169156.45	169756.20	170309.77	170783.27	170951.55	171048.40
Central Africa	14840.94	15899.72	16532.73	16519.27	16495.85	16459.29	16413.25	16367.29	16309.85	16244.76	16171.25	16081.47
Eastern Africa	25105.04	28951.04	31942.00	32299.71	32691.31	31848.36	31783.08	32005.37	32191.57	32445.93	32440.18	32469.68
Northern Africa	33556.54	38545.92	42498.35	42905.09	43281.61	43610.02	43915.22	44129.68	44313.88	44475.38	44630.43	44766.17
Southern Africa	17487.68	18840.18	19526.97	19519.21	19488.14	19430.97	19439.22	19476.34	19473.96	19491.15	19493.48	19544.94
Western Africa	42653.38	48431.50	53334.73	53676.68	53998.37	54318.38	54607.00	54795.33	55010.28	55087.85	55152.78	55096.72
Western Indian Ocean	2454.06	2832.12	3088.74	3050.42	3025.91	3009.33	2998.68	2982.20	3010.22	3038.20	3063.43	3089.42
ASIA AND PACIFIC	595201.32	615348.40	618978.78	617672.55	616229.93	613726.05	611638.20	609331.93	607850.04	605981.09	604833.23	602521.34
Australia and New Zealand	----	----	----	----	----	----	----	----	----	----	----	----
Central Asia	----	----	----	----	----	----	----	----	----	----	----	----
North West Pacific and East Asia	243103.14	220678.83	190682.59	187435.52	183712.42	179704.72	176402.56	172853.78	168611.61	164166.35	160437.95	155636.66
South Asia	294913.56	341357.79	379730.69	382285.91	385137.74	387266.28	389178.68	391025.94	394251.27	397453.12	400565.87	403387.79
South East Asia	53617.71	49872.64	45456.34	44867.70	44318.74	43720.95	43051.29	42478.26	42035.31	41434.21	40924.02	40622.86
South Pacific	927.43	981.27	1044.82	1051.90	1057.48	1063.29	1070.00	1075.98	1082.45	1089.01	1095.15	1098.98
EUROPE	----	----	----	----	----	----	----	----	----	----	----	----
Central Europe	12077.52	10762.16	9617.69	9467.93	9286.35	9123.96	8930.01	8773.00	8609.84	8438.93	8300.95	8108.76
Eastern Europe	2741.06	1983.84	1353.26	1229.09	1210.38	1200.80	1070.23	1047.52	1039.26	913.03	903.67	894.36
Western Europe	----	----	----	----	----	----	----	----	----	----	----	----
LATIN AMERICA AND CARIBBEAN	42595.32	43477.20	42532.48	42406.41	42238.13	42058.66	42009.60	41881.21	41624.89	41447.30	41221.48	41040.24
Caribbean	4095.51	4159.45	4070.70	4068.43	4074.85	4073.63	4074.32	4068.74	4089.69	4092.26	4103.25	4118.37
Meso-America	10563.62	10809.88	10767.07	10784.01	10743.02	10739.28	10777.20	10757.12	10762.82	10759.85	10694.51	10674.71
South America	27936.19	28507.87	27694.72	27553.97	27420.26	27245.76	27158.09	27055.35	26772.37	26595.19	26423.73	26247.17
NORTH AMERICA	----	----	----	----	----	----	----	----	----	----	----	----
North America	----	----	----	----	----	----	----	----	----	----	----	----
POLAR	----	----	----	----	----	----	----	----	----	----	----	----
Antarctic	----	----	----	----	----	----	----	----	----	----	----	----
Arctic	----	----	----	----	----	----	----	----	----	----	----	----
WEST ASIA	12291.21	14415.89	16841.39	17284.54	17413.93	17503.87	17513.59	17549.62	17738.28	17929.17	18062.51	18215.46
Arabian Peninsula	5787.78	6901.11	8322.65	8635.21	8655.01	8652.81	8586.85	8567.89	8659.78	8748.73	8814.38	8898.76
Mashriq	6503.43	7514.79	8518.74	8649.33	8758.91	8851.06	8926.74	8981.73	9078.49	9180.44	9248.13	9316.70
GLOBAL TOTALS	----	----	----	----	----	----	----	----	----	----	----	----
REPORTED GLOBAL TOTALS FROM ORIGINAL DATA SOURCES	----	----	----	----	----	----	----	----	----	----	----	----

Comments:

Adult illiteracy rate is the percentage of people ages 15 and above who cannot, with understanding, read and write a short, simple statement on their everyday life.

The data are based on UNESCO's 1999 literacy estimates and projections.

1.1.3 Energy Consumption and Production

Energy Capacity – Nuclear (Projection)

Units: megawatt-electricity (MWe)

Data Source: 19th WEC Survey of Energy Resources, London (2001)

Data Provider: World Energy Council

Years: 1996, 2010

	1996	2010
AFRICA	1840	1840
Central Africa	0	----
Eastern Africa	0	----
Northern Africa	0	----
Southern Africa	1840	1840
Western Africa	0	----
Western Indian Ocean	0	----
ASIA AND PACIFIC	60185	132077
Australia and New Zealand	0	----
Central Asia	70	500
North West Pacific and East Asia	58295	123939
South Asia	1820	7638
South East Asia	0	----
South Pacific	0	----
EUROPE	170122	182021
Central Europe	12187	16833
Eastern Europe	32339	41547
Western Europe	125596	123641
LATIN AMERICA AND CARIBBEAN	2870	6036
Caribbean	0	----
Meso-America	1309	1309
South America	1561	4727
NORTH AMERICA	114986	103768
North America	114986	103768
POLAR	0	----
Antarctic	0	----
Arctic	0	----
WEST ASIA	0	----
Arabian Peninsula	0	----
Mashriq	0	----
GLOBAL TOTALS	350003	----
REPORTED GLOBAL TOTALS FROM ORIGINAL DATA SOURCES	350003	----

Comments:

Energy capacity - nuclear is the actual capacity of the nuclear electric power industry to describe the size of generating plants. "MWe" is the symbol for the actual output of a generating station in megawatts of electricity. Figures for 1996 are estimates and those for year 2010 are projections.

Data Source: The majority of the data were provided by World Energy Council (WEC) Member Committees in 1997. If information was not available from this source, data have been derived from the following published sources:

- *Nuclear Power Generation and Fuel Cycle Report 1997* (September 1997) Energy Information Administration, US Department of Energy.
- *Nuclear Power Reactors in the World* (April 1997) International Atomic Energy Agency, Vienna.
- *Energy, Electricity and Nuclear Power Estimates for the Period up to 2015* (1997) International Atomic Energy Agency, Vienna.
- *Les Centrales Nucléaires dans le Monde* (1997) Commissariat à l'Energie Atomique, France.

For more information see http://www.worldenergy.org.

Units: thousand tonnes of oil equivalent (ktoe)

Data Source: Energy Balances of OECD Countries (1960-1999) and Non-OECD Countries (1971-1999)

Data Provider: International Energy Agency (IEA)

Years: 1992-1999

	1992	1993	1994	1995	1996	1997	1998	1999
AFRICA	206307.87	210137.79	215113.98	220203.52	228240.47	234429.68	240263.54	244846.73
Central Africa	16263.52	16755.06	16954.21	17455.79	17992.52	18527.25	19099.55	19591.50
Eastern Africa	----	----	----	----	----	----	----	----
Northern Africa	11695.07	11891.68	12098.81	12215.18	15341.43	15790.01	16257.86	16595.35
Southern Africa	43738.25	44606.55	44927.70	45670.46	46431.00	47195.04	48131.03	49145.82
Western Africa	----	----	----	----	----	----	----	----
Western Indian Ocean	----	----	----	----	----	----	----	----
ASIA AND PACIFIC	536316.81	540255.88	544952.09	550370.55	554869.33	561542.18	563926.39	572593.80
Australia and New Zealand	4083.00	4775.27	4688.86	4879.98	5502.60	5682.65	5854.83	5958.32
Central Asia	119.33	82.33	80.63	82.78	77.86	77.86	77.12	77.12
North West Pacific and East Asia	210832.78	211764.26	212740.92	213623.67	214668.08	216031.70	214793.66	218566.71
South Asia	219675.79	222985.96	225557.08	228149.61	230647.58	233918.15	237081.16	240907.18
South East Asia	99026.58	97994.55	99145.79	100804.98	101098.43	102951.30	103342.77	104307.61
South Pacific	----	----	----	----	----	----	----	----
EUROPE	57054.48	60800.59	56791.27	57119.26	61708.87	61659.48	61417.55	64662.42
Central Europe	12890.63	15440.48	15534.82	15491.54	19544.54	18918.86	18629.45	17981.59
Eastern Europe	12212.84	11955.91	7764.81	7034.68	6143.55	5860.73	4500.84	6083.81
Western Europe	31951.01	33404.20	33491.64	34593.03	36020.78	36879.88	38287.26	40597.02
LATIN AMERICA AND CARIBBEAN	77756.02	76433.08	78865.03	78619.18	79155.23	78808.64	79472.53	81118.78
Caribbean	7849.57	6576.74	6719.45	6035.76	6783.92	6668.79	6235.19	6401.29
Meso-America	15504.23	15551.61	15282.48	15687.06	15851.99	16409.96	16379.38	16197.79
South America	53826.77	53728.91	56274.58	56272.48	55888.58	55095.37	56208.23	57869.97
NORTH AMERICA	72098.50	63696.64	65864.62	68123.48	70116.34	67883.59	69714.20	75999.81
North America	72098.50	63696.64	65864.62	68123.48	70116.34	67883.59	69714.20	75999.81
POLAR	----	----	----	----	----	----	----	----
Antarctic	----	----	----	----	----	----	----	----
Arctic	----	----	----	----	----	----	----	----
WEST ASIA	214.56	218.19	223.75	228.77	231.16	233.66	236.24	236.24
Arabian Peninsula	77.39	77.39	77.39	77.39	77.39	77.39	77.39	77.39
Mashriq	137.17	140.80	146.36	151.38	153.77	156.28	158.86	158.86
GLOBAL TOTALS	949748.23	951542.16	961810.74	974664.76	994321.39	1004557.23	1015030.45	1039457.77
REPORTED GLOBAL TOTALS FROM ORIGINAL DATA SOURCES	949748.23	949748.23	949748.23	949748.23	949748.23	949748.23	949748.23	949748.23

Comments:

Production refers to the quantities of fuels extracted or produced, calculated after any operation for removal of inert matter or impurities (for example sulphur from natural gas).

Combustible renewables refers to the quantities of fuels extracted or produced from primary solid biomass and biogas. Primary solid biomass is defined as any plant matter used directly as fuel or converted into other forms before combustion. Included are wood, vegetal waste (including wood waste and crops used for energy production), animal materials/wastes, sulphite lyes, also known as "black liquor" (an alkaline spent liquor from the digesters in the production of sulphate or soda pulp during the manufacture of paper where the energy content derives from the lignin removed from the wood pulp) and other solid biomass.

This category contains only primary solid biomass. This includes inputs to charcoal production but not the actual production of charcoal (this would be double counting since charcoal is a secondary product). Biogas is derived principally from the anaerobic fermentation of biomass and solid wastes and combusted to produce heat and/or power. Included in this category are landfill gas and sludge gas (sewage gas and gas from animal slurries) and other biogas.

A ktoe (thousand tonne of oil equivalent) is defined as 41.868 TJ (terajoules) or 11.630 GWh (gigawatt-hour).

Units: million tonnes of oil equivalent (Mtoe)

Data Source: Energy Balances of Organization for Economic Co-operation and Development (OECD) Countries (1960-1999) and Non-OECD Countries (1971-1999)

Data Provider: International Energy Agency (IEA)

Years: 1960-1999

	1960	1970	1980	1990	1991	1992	1993	1994	1995	1996	1997	1998	1999
AFRICA	----	----	528.13	655.75	681.82	689.45	700.14	703.76	730.48	760.41	794.43	803.02	806.13
Central Africa	----	----	28.87	47.59	48.59	49.18	51.61	49.98	53.77	56.25	58.53	59.50	58.89
Eastern Africa	----	----	----	----	----	----	----	----	----	----	----	----	----
Northern Africa	----	----	212.39	246.70	259.60	258.84	258.81	254.13	262.82	272.02	283.46	293.09	299.53
Southern Africa	----	----	111.38	174.88	179.73	181.04	186.12	193.28	203.43	207.35	217.38	221.52	223.35
Western Africa	----	----	----	----	----	----	----	----	----	----	----	----	----
Western Indian Ocean	----	----	----	----	----	----	----	----	----	----	----	----	----
ASIA AND PACIFIC	----	----	1516.60	2312.05	2376.32	2414.18	2482.57	2537.94	2670.78	2753.41	2781.32	2787.09	2779.20
Australia and New Zealand	23.79	51.09	91.58	169.94	175.96	183.56	187.35	186.81	200.15	204.26	216.48	228.10	227.35
Central Asia	----	----	170.50	204.00	194.07	181.66	181.79	151.71	145.68	145.96	136.62	138.83	148.79
North West Pacific and East Asia	----	----	696.52	1066.28	1080.19	1098.69	1135.00	1189.94	1273.98	1317.12	1311.12	1287.73	1258.86
South Asia	----	----	339.14	568.66	598.84	611.78	631.18	646.59	671.71	686.17	700.92	709.99	709.34
South East Asia	----	----	218.86	303.17	327.26	338.49	347.26	362.90	379.25	399.90	416.17	422.45	434.86
South Pacific	----	----	----	----	----	----	----	----	----	----	----	----	----
EUROPE	----	----	2116.61	2523.65	2455.84	2373.65	2301.57	2251.39	2247.81	2297.42	2263.92	2241.96	2272.76
Central Europe	----	----	289.17	270.30	256.99	252.30	250.71	242.68	247.22	255.25	249.87	234.75	220.40
Eastern Europe	----	----	1179.60	1411.37	1342.68	1252.76	1167.46	1086.49	1055.87	1047.13	1021.54	1029.72	1056.47
Western Europe	360.27	417.00	647.84	841.98	856.18	868.59	883.40	922.22	944.73	995.04	992.51	977.49	995.89
LATIN AMERICA AND CARIBBEAN	----	----	475.65	612.43	636.53	645.68	663.44	686.94	716.08	772.01	811.66	830.39	816.04
Caribbean	----	----	20.80	21.63	20.37	21.22	19.47	20.42	19.87	21.32	21.14	21.98	25.03
Meso-America	----	----	157.29	204.43	211.92	211.62	214.54	214.96	212.70	224.37	234.90	240.26	233.80
South America	----	----	297.56	386.37	404.24	412.84	429.43	451.56	483.51	526.32	555.63	568.15	557.21
NORTH AMERICA	1021.97	1595.63	1760.68	1923.75	1923.64	1938.68	1914.21	2000.20	2010.82	2045.13	2048.62	2068.68	2054.44
North America	1021.97	1595.63	1760.68	1923.75	1923.64	1938.68	1914.21	2000.20	2010.82	2045.13	2048.62	2068.68	2054.44
POLAR	----	----	----	----	----	----	----	----	----	----	----	----	----
Antarctic	----	----	----	----	----	----	----	----	----	----	----	----	----
Arctic	----	----	----	----	----	----	----	----	----	----	----	----	----
WEST ASIA	----	----	908.90	735.57	725.57	808.66	855.11	874.08	887.98	900.94	949.26	1020.46	988.92
Arabian Peninsula	----	----	762.57	605.98	681.99	756.10	798.80	812.20	822.10	834.18	848.25	873.69	822.51
Mashriq	----	----	146.32	129.59	43.57	52.57	56.31	61.88	65.88	66.76	101.01	146.77	166.41
GLOBAL TOTALS	----	----	7306.56	8763.21	8799.72	8870.29	8917.02	9054.31	9263.95	9529.32	9649.21	9751.60	9717.50
REPORTED GLOBAL TOTALS FROM ORIGINAL DATA SOURCES	----	----	7341.42	8809.31	8847.08	8919.01	8966.64	9105.29	9315.66	9581.46	9704.92	9809.70	9776.75

Comments:

Total indigenous production is the production of primary energy, from, the total of all energy sources: hard coal, lignite/brown coal, peat, crude oil, NGLs, natural gas, combustible renewables and wastes, nuclear, hydro, geothermal, solar and the heat from heat pumps that is extracted from the ambient environment. Production is calculated after removal of impurities (for example sulphur from natural gas).

A toe (tonne of oil equivalents) is defined as 41.868 gigajoules or 10^7 kilocalories. One terawatt-hour = 0.086 Mtoe.

Energy Production - Waste Combustion

Units: thousand tonnes of oil equivalent (ktoe)

Data Source: Energy Balances of OECD Countries (1960-1999) and Non-OECD Countries (1971-1999)

Data Provider: International Energy Agency (IEA)

Years: 1992-1999

	1992	1993	1994	1995	1996	1997	1998	1999
AFRICA	0.00	0.00	0.00	0.00	0.00	0.00	0.00	0.00
Central Africa	0.00	0.00	0.00	0.00	0.00	0.00	0.00	0.00
Eastern Africa	----	----	----	----	----	----	----	----
Northern Africa	0.00	0.00	0.00	0.00	0.00	0.00	0.00	0.00
Southern Africa	0.00	0.00	0.00	0.00	0.00	0.00	0.00	0.00
Western Africa	----	----	----	----	----	----	----	----
Western Indian Ocean	----	----	----	----	----	----	----	----
ASIA AND PACIFIC	683.74	712.62	1452.47	1685.30	2026.37	2508.17	3086.10	3561.65
Australia and New Zealand	270.37	275.96	257.36	283.34	267.46	373.41	462.02	483.47
Central Asia	0.00	0.00	0.00	0.00	0.00	0.00	0.00	0.00
North West Pacific and East Asia	413.37	436.66	1195.11	1401.95	1758.91	2134.76	2624.08	3078.17
South Asia	0.00	0.00	0.00	0.00	0.00	0.00	0.00	0.00
South East Asia	0.00	0.00	0.00	0.00	0.00	0.00	0.00	0.00
South Pacific	----	----	----	----	----	----	----	----
EUROPE	8094.94	8103.42	9512.28	11012.97	10342.19	11372.86	11762.13	12695.50
Central Europe	1042.01	780.60	947.43	1235.43	585.87	576.38	484.38	707.01
Eastern Europe	1696.79	1527.90	1719.40	2343.39	1646.60	1985.76	2048.03	2547.58
Western Europe	5356.14	5794.93	6845.44	7434.15	8109.73	8810.71	9229.72	9440.91
LATIN AMERICA AND CARIBBEAN	0.00	0.00	0.00	0.00	0.00	0.00	0.00	0.00
Caribbean	0.00	0.00	0.00	0.00	0.00	0.00	0.00	0.00
Meso-America	0.00	0.00	0.00	0.00	0.00	0.00	0.00	0.00
South America	0.00	0.00	0.00	0.00	0.00	0.00	0.00	0.00
NORTH AMERICA	9605.12	9062.22	9689.69	9522.05	9789.48	9805.41	13656.52	13781.41
North America	9605.12	9062.22	9689.69	9522.05	9789.48	9805.41	13656.52	13781.41
POLAR	----	----	----	----	----	----	----	----
Antarctic	----	----	----	----	----	----	----	----
Arctic	----	----	----	----	----	----	----	----
WEST ASIA	0.00	0.00	0.00	0.00	0.00	0.00	0.00	0.00
Arabian Peninsula	0.00	0.00	0.00	0.00	0.00	0.00	0.00	0.00
Mashriq	0.00	0.00	0.00	0.00	0.00	0.00	0.00	0.00
GLOBAL TOTALS	18383.80	17878.26	20654.44	22220.31	22158.04	23686.44	28504.75	30038.55
REPORTED GLOBAL TOTALS FROM ORIGINAL DATA SOURCES	18383.80	18383.80	18383.80	18383.80	18383.80	18383.80	18383.80	18383.80

Comments:

Production refers to the quantities of fuels extracted or produced, calculated after any operation for removal of inert matter or impurities (for example sulphur from natural gas).

Waste refers to the quantities of fuels extracted or produced from industrial waste and municipal waste. Industrial waste consists of solid and liquid products (e.g. tyres) combusted directly, usually in specialised plants, to produce heat and/or power and that are not reported in the category solid biomass.

Municipal waste consists of products that are combusted directly to produce heat and/or power and comprises wastes produced by the residential, commercial and public services sectors that are collected by local authorities for disposal in a central location. Hospital waste is included in this category.

A ktoe (thousand tonne of oil equivalent) is defined as 41.868 TJ (terajoules) or 11.630 GWh (gigawatt-hour).

Total Final Energy Consumption - per Capita

Units: tonnes of oil equivalent (ktoe) per person

Data Source: Energy Balances of OECD Countries (1960-1999) and Non-OECD Countries (1971-1999)

Data Provider: International Energy Agency (IEA)

Years: 1960-1999

	1960	1970	1980	1990	1991	1992	1993	1994	1995	1996	1997	1998	1999
AFRICA	----	----	0.25	0.24	0.24	0.24	0.23	0.49	0.49	0.50	0.50	0.50	0.50
Central Africa	----	----	0.07	0.06	0.05	0.05	0.05	0.31	0.31	0.31	0.31	0.31	0.31
Eastern Africa	----	----	----	----	----	----	----	----	----	----	----	----	----
Northern Africa	----	----	0.28	0.36	0.36	0.36	0.36	0.39	0.40	0.41	0.42	0.43	0.44
Southern Africa	----	----	0.62	0.56	0.53	0.51	0.49	0.74	0.75	0.75	0.75	0.75	0.73
Western Africa	----	----	----	----	----	----	----	----	----	----	----	----	----
Western Indian Ocean	----	----	----	----	----	----	----	----	----	----	----	----	----
ASIA AND PACIFIC	----	----	0.38	0.48	0.49	0.50	0.50	0.65	0.67	0.68	0.67	0.67	0.66
Australia and New Zealand	2.10	2.64	3.09	3.36	3.30	3.29	3.37	3.42	3.50	3.58	3.61	3.63	3.65
Central Asia	----	----	2.17	2.05	1.94	2.01	1.71	1.46	1.41	1.30	1.23	2.41	2.16
North West Pacific and East Asia	----	----	0.52	0.67	0.69	0.70	0.72	0.89	0.91	0.94	0.91	0.89	0.88
South Asia	----	----	0.12	0.17	0.18	0.19	0.19	0.37	0.38	0.39	0.38	0.38	0.38
South East Asia	----	----	0.29	0.38	0.38	0.40	0.42	0.48	0.50	0.53	0.54	0.52	0.53
South Pacific	----	----	----	----	----	----	----	----	----	----	----	----	----
EUROPE	----	----	2.53	2.58	2.53	2.52	2.45	2.30	2.30	2.30	2.26	2.26	2.27
Central Europe	----	----	1.74	1.51	1.38	1.30	1.28	1.22	1.27	1.35	1.32	1.27	1.22
Eastern Europe	----	----	3.20	3.47	3.33	3.39	3.17	2.70	2.61	2.37	2.28	2.25	2.32
Western Europe	1.27	2.24	2.50	2.56	2.62	2.60	2.59	2.59	2.63	2.72	2.71	2.75	2.77
LATIN AMERICA AND CARIBBEAN	----	----	0.82	0.80	0.80	0.81	0.81	0.83	0.85	0.86	0.87	0.88	0.87
Caribbean	----	----	0.80	0.82	0.72	0.72	0.63	0.66	0.67	0.72	0.73	0.75	0.79
Meso-America	----	----	0.89	0.91	0.93	0.93	0.91	0.93	0.91	0.88	0.87	0.86	0.84
South America	----	----	0.79	0.76	0.77	0.77	0.79	0.82	0.85	0.87	0.89	0.90	0.89
NORTH AMERICA	4.30	5.72	5.79	5.20	5.13	5.10	5.15	5.22	5.25	5.36	5.33	5.24	5.34
North America	4.30	5.72	5.79	5.20	5.13	5.10	5.15	5.22	5.25	5.36	5.33	5.24	5.34
POLAR	----	----	----	----	----	----	----	----	----	----	----	----	----
Antarctic	----	----	----	----	----	----	----	----	----	----	----	----	----
Arctic	----	----	----	----	----	----	----	----	----	----	----	----	----
WEST ASIA	----	----	1.19	1.48	1.42	1.51	1.59	1.63	1.60	1.64	1.65	1.68	1.64
Arabian Peninsula	----	----	1.88	2.15	2.16	2.29	2.39	2.43	2.36	2.48	2.46	2.51	2.42
Mashriq	----	----	0.66	0.88	0.76	0.82	0.89	0.92	0.92	0.89	0.92	0.94	0.92
GLOBAL TOTALS	----	----	1.09	1.06	1.05	1.05	1.04	1.14	1.15	1.16	1.15	1.14	1.14
REPORTED GLOBAL TOTALS FROM ORIGINAL DATA SOURCES	----	----	----	----	----	----	----	----	----	----	----	----	----

Comments:

Total final energy consumption (TFC) is the sum of consumption by the different end-use sectors. In final consumption, petrochemical feedstocks are shown under industry, while non-energy use of such oil products as white spirit, lubricants, bitumen, paraffin waxes and other products are shown under non-energy use, and are included in Total final consumption only. Backflows from the petrochemical industry are not included in final consumption.

A ktoe (thousand tonne of oil equivalent) is defined as 41.868 TJ (terajoules) or 11.630 GWh (gigawatt-hour).

Total Final Energy Consumption – Total

Units: thousand tonnes of oil equivalent (ktoe)

Data Source: Energy Balances of OECD Countries (1960-1999) and Non-OECD Countries (1971-1999)

Data Provider: International Energy Agency (IEA)

Years: 1960-1999

	1960	1970	1980	1990	1995	1996	1997	1998	1999
AFRICA	----	----	99550.13	130247.52	285532.87	294961.00	304395.18	311752.75	316759.22
Central Africa	----	----	2832.40	2967.95	19267.01	19779.01	20383.63	20971.02	21286.26
Eastern Africa	----	----	-11246.71	-11805.87	-15485.47	-17308.66	-17760.00	-17869.94	-18179.82
Northern Africa	----	----	31129.93	51710.05	62924.98	65913.73	68755.68	72153.63	75147.81
Southern Africa	----	----	49537.75	57832.27	91064.20	93079.79	95468.39	96962.55	96483.59
Western Africa	----	----	----	----	----	----	----	----	----
Western Indian Ocean	----	----	----	----	----	----	----	----	----
ASIA AND PACIFIC	----	----	982287.40	1464319.99	2201086.60	2283708.77	2266996.44	2255621.56	2269145.56
Australia and New Zealand	26610.34	40520.82	54619.53	67979.70	75924.71	78449.76	80062.34	81400.84	82877.58
Central Asia	----	----	89534.55	103657.84	75983.92	70618.58	67553.11	72204.79	70273.11
North West Pacific and East Asia	----	----	626948.11	930302.42	1331156.53	1381508.90	1351708.93	1334606.37	1328198.10
South Asia	----	----	109710.45	201376.70	483044.79	502553.96	510048.70	514944.21	522965.66
South East Asia	----	----	101474.76	161003.33	234976.65	250577.57	257623.36	252465.35	264831.11
South Pacific	----	----	----	----	----	----	----	----	----
EUROPE	----	----	1912478.90	2061882.68	1870531.40	1871612.63	1844158.66	1845676.35	1857410.03
Central Europe	----	----	295040.55	282543.50	242083.06	259249.43	255407.01	247019.02	237139.63
Eastern Europe	----	----	692768.21	802046.30	602587.19	546346.12	524886.72	515045.79	528806.54
Western Europe	412869.19	787071.24	924670.14	977292.88	1025861.15	1066017.08	1063864.93	1083611.54	1091463.86
LATIN AMERICA AND CARIBBEAN	----	----	290758.49	347916.46	402086.13	414546.33	427637.18	436317.20	437489.13
Caribbean	----	----	19419.16	23346.18	20146.84	21979.07	22571.61	23282.05	24779.81
Meso-America	----	----	80180.25	101654.45	111409.17	109757.70	111306.41	111768.06	111606.30
South America	----	----	191159.08	222915.83	270530.12	282809.56	293759.16	301267.09	301103.02
NORTH AMERICA	877157.28	1325498.01	1475699.02	1467592.08	1565439.08	1616361.40	1624297.90	1612229.70	1661638.94
North America	877157.28	1325498.01	1475699.02	1467592.08	1565439.08	1616361.40	1624297.90	1612229.70	1661638.94
POLAR	----	----	----	----	----	----	----	----	----
Antarctic	----	----	----	----	----	----	----	----	----
Arctic	----	----	----	----	----	----	----	----	----
WEST ASIA	----	----	58704.63	104274.38	130964.84	139238.64	144161.30	151701.89	152456.77
Arabian Peninsula	----	----	41261.01	72867.34	92695.88	100881.59	103691.82	109423.56	109770.23
Mashriq	----	----	17443.62	31407.04	38268.96	38357.05	40469.48	42278.33	42686.54
GLOBAL TOTALS	----	----	4819478.57	5576233.11	6455640.92	6620428.77	6611646.66	6613299.45	6694899.65
REPORTED GLOBAL TOTALS FROM ORIGINAL DATA SOURCES	----	----	4830756.80	5590170.78	6509176.03	6675065.04	6667782.59	6670565.08	6753275.03

Comments:

Total final energy consumption (TFC) is the sum of consumption by the different end-use sectors. In final consumption, petrochemical feedstocks are shown under industry, while non-energy use of such oil products as white spirit, lubricants, bitumen, paraffin waxes and other products are shown under non-energy use, and are included in Total final consumption only. Backflows from the petrochemical industry are not included in final consumption.

A ktoe (thousand tonne of oil equivalent) is defined as 41.868 TJ (terajoules) or 11.630 GWh (gigawatt-hour).

Units: thousand tonnes of oil equivalent (ktoe)

Data Source: Energy Balances of OECD Countries (1960-1999) and Non-OECD Countries (1971-1999)

Data Provider: International Energy Agency (IEA)

Years: 1960-1999

	1960	1970	1980	1990	1995	1996	1997	1998	1999
AFRICA	----	----	51292.67	72957.79	83903.52	85525.64	86851.14	84771.53	88130.63
Central Africa	----	----	208.13	220.13	240.90	214.06	216.01	216.61	216.61
Eastern Africa	----	----	----	----	----	----	----	----	----
Northern Africa	----	----	1124.18	2585.15	2969.48	3469.30	3338.62	3710.79	3726.22
Southern Africa	----	----	49853.17	70018.76	80545.50	81699.56	83151.94	80761.71	84106.00
Western Africa	----	----	----	----	----	----	----	----	----
Western Indian Ocean	----	----	----	----	----	----	----	----	----
ASIA AND PACIFIC	----	----	555964.04	909759.76	1100350.46	1150481.65	1130348.39	1107089.03	1088289.80
Australia and New Zealand	17462.51	22597.82	28338.42	36098.21	38730.52	41960.90	46568.54	45839.08	48445.31
Central Asia	----	----	55073.09	48668.85	29819.88	27115.84	23287.53	22811.02	21695.69
North West Pacific and East Asia	----	----	410117.84	702898.70	857559.47	893968.05	871303.23	851216.26	830638.11
South Asia	----	----	58953.82	109623.63	155448.90	164205.99	164610.50	163769.13	160844.35
South East Asia	----	----	3480.87	12470.37	18791.69	23230.87	24578.59	23453.54	26666.34
South Pacific	----	----	----	----	----	----	----	----	----
EUROPE	----	----	765042.47	721614.13	570559.85	567436.90	540290.82	520364.65	506152.55
Central Europe	----	----	193158.81	182777.58	154135.27	161090.65	158657.37	149063.01	136059.59
Eastern Europe	----	----	265816.51	234905.70	171120.36	165889.39	151523.13	145612.42	156557.88
Western Europe	333998.94	328177.54	306067.14	303930.85	245304.22	240456.86	230110.32	225689.22	213535.08
LATIN AMERICA AND CARIBBEAN	----	----	13735.13	20285.56	24987.84	26732.28	28118.62	28643.53	27522.61
Caribbean	----	----	89.80	191.67	144.04	114.34	119.36	232.63	212.33
Meso-America	----	----	2483.37	3204.22	5463.19	6461.97	6618.76	6724.54	6560.95
South America	----	----	11161.96	16889.67	19380.61	20155.97	21380.50	21686.36	20749.33
NORTH AMERICA	235426.65	308523.11	397454.48	480929.48	500659.33	523402.06	557018.38	566691.50	567285.88
North America	235426.65	308523.11	397454.48	480929.48	500659.33	523402.06	557018.38	566691.50	567285.88
POLAR	----	----	----	----	----	----	----	----	----
Antarctic	----	----	----	----	----	----	----	----	----
Arctic	----	----	----	----	----	----	----	----	----
WEST ASIA	----	----	6.44	0.00	120.28	133.50	133.50	133.50	133.50
Arabian Peninsula	----	----	0.00	0.00	0.00	0.00	0.00	0.00	0.00
Mashriq	----	----	6.44	0.00	120.28	133.50	133.50	133.50	133.50
GLOBAL TOTALS	----	----	1783495.23	2205546.72	2280581.28	2353712.03	2342760.85	2307693.74	2277514.97
REPORTED GLOBAL TOTALS FROM ORIGINAL DATA SOURCES	----	----	1784084.00	2206518.00	2281505.00	2354579.00	2343684.00	2308703.00	2278524.00

Comments:

Total primary energy supply (TPES) is made up of production + imports - exports - international marine bunkers ± stock changes.

Coal and coal products includes all coal, both primary (including hard coal and lignite/brown coal) and derived fuels (including patent fuel, coke oven coke, gas coke, BKB, coke oven gas, and blast furnace gas). Peat is also included in this category.

A ktoe (thousand tonne of oil equivalent) is defined as 41.868 TJ (terajoules) or 11.630 GWh (gigawatt-hour).

Total Primary Energy Supply (TPES) - Combustible Renewables and Waste

Units: thousand tonnes of oil equivalent (ktoe)

Data Source: Energy Balances of OECD Countries (1960-1999) and Non-OECD Countries (1971-1999)

Data Provider: International Energy Agency (IEA)

Years: 1960-1999

	1960	1970	1980	1990	1995	1996	1997	1998	1999
AFRICA	----	----	151687.82	196130.75	219929.17	227960.57	234144.59	239973.38	245710.91
Central Africa	----	----	11362.24	15294.00	17452.44	17989.07	18523.70	19095.89	19587.74
Eastern Africa	----	----	----	----	----	----	----	----	----
Northern Africa	----	----	9046.90	11243.29	12195.57	15320.88	15769.05	16236.51	16573.61
Southern Africa	----	----	33326.87	41924.96	45446.91	46203.63	46963.85	47895.98	48906.90
Western Africa	----	----	----	67350.46	77330.60	79381.38	81681.41	83812.96	85902.23
Western Indian Ocean	----	----	----	----	----	----	----	----	----
ASIA AND PACIFIC	----	----	437961.51	519964.04	549017.10	553812.98	560954.60	564013.99	573139.70
Australia and New Zealand	4093.10	3539.61	4145.47	4640.63	5162.34	5768.96	6054.89	6315.64	6440.56
Central Asia	----	----	134.00	139.11	82.77	77.85	77.85	77.11	77.11
North West Pacific and East Asia	----	----	180866.83	208145.60	214993.87	216394.22	218131.88	217382.59	221608.91
South Asia	----	----	176071.30	211759.75	228105.77	230603.27	233873.20	237035.61	240860.89
South East Asia	----	----	76743.91	95278.95	100672.35	100968.68	102816.78	103203.04	104152.23
South Pacific	----	----	----	----	----	----	----	----	----
EUROPE	----	----	50706.01	67712.03	68472.98	72464.42	73228.47	73986.36	77877.78
Central Europe	----	----	13578.82	13239.70	16603.75	19951.69	19253.55	18825.78	18375.97
Eastern Europe	----	----	16901.94	17546.42	9409.87	7853.94	7604.65	6942.18	8708.43
Western Europe	9184.98	14063.09	20225.25	36925.91	42459.36	44658.79	46370.27	48218.40	50793.38
LATIN AMERICA AND CARIBBEAN	----	----	77838.87	84769.30	85933.95	86763.04	86132.70	86751.32	88420.47
Caribbean	----	----	7287.34	8289.08	6091.79	6839.81	6726.01	6282.10	6450.75
Meso-America	----	----	15348.79	15386.12	15684.05	15848.95	16406.81	16376.25	16194.69
South America	----	----	54556.98	60514.42	63532.87	63442.19	62364.01	63441.89	65123.95
NORTH AMERICA	36417.10	42588.52	62135.91	70391.08	80357.36	81806.67	80116.12	86048.94	92588.87
North America	36417.10	42588.52	62135.91	70391.08	80357.36	81806.67	80116.12	86048.94	92588.87
POLAR	----	----	----	----	----	----	----	----	----
Antarctic	----	----	----	----	----	----	----	----	----
Arctic	----	----	----	----	----	----	----	----	----
WEST ASIA	----	----	207.21	230.40	271.88	260.66	259.16	260.89	261.15
Arabian Peninsula	----	----	67.04	100.15	119.78	104.71	100.70	99.85	99.85
Mashriq	----	----	140.17	130.25	152.10	155.95	158.46	161.04	161.30
GLOBAL TOTALS	----	----	751139.02	902145.80	962589.26	980696.49	990910.68	1005797.28	1031602.03
REPORTED GLOBAL TOTALS FROM ORIGINAL DATA SOURCES	----	----	782909.35	941665.67	1006811.41	1025942.56	1037715.62	1053811.19	1080775.19

Comments:

TPES is made up of production + imports - exports - international marine bunkers ± stock changes.

Combustible renewables and waste comprises solid biomass and animal products, gas/liquids from biomass, industrial waste and municipal waste. Biomass is defined as any plant matter used directly as fuel or converted into fuels (e.g. charcoal) or electricity and/or heat. Included here are wood, vegetal waste (including wood waste and crops used for energy production), ethanol, animal materials/wastes and sulphite lyes (sulphite lyes are also known as "black liquor" and are an alkaline spent liquor from the digesters in the production of sulphate or soda pulp during the manufacture of paper. The energy is derived from the lignin removed from the wood pulp.). Municipal waste comprises wastes produced by the residential, commercial and public service sectors that are collected by local authorities for disposal in a central location for the production of heat and/or power. Hospital waste is included in this category.

Data under this heading are often based on small sample surveys or other incomplete information. Thus the data give only a broad impression of developments, and are not strictly comparable between countries. In some cases complete categories of vegetal fuel are omitted due to lack of information. It is therefore important to refer to the data of the individual country when consulting the regional aggregation.

A ktoe (thousand tonne of oil equivalent) is defined as 41.868 TJ (terajoules) or 11.630 GWh (gigawatt-hour).

Units: thousand tonnes of oil equivalent (ktoe)

Data Source: Energy Balances of OECD Countries (1960-1999) and Non-OECD Countries (1971-1999)

Data Provider: International Energy Agency (IEA)

Years: 1960-1999

	1960	1970	1980	1990	1995	1996	1997	1998	1999
AFRICA	----	----	70859.62	113835.14	124845.66	124482.03	136204.18	130097.82	125767.14
Central Africa	----	----	2059.81	2358.49	2535.28	2673.26	2693.71	2607.55	2360.32
Eastern Africa	----	----	-10953.88	-10816.45	-10723.82	-10738.07	-10988.50	-10596.57	-10596.57
Northern Africa	----	----	39697.21	75716.95	81949.76	80491.80	87500.55	88764.02	86116.36
Southern Africa	----	----	14240.51	15311.03	19974.19	19821.21	25621.21	23078.73	20781.43
Western Africa	----	----	----	----	----	----	----	----	----
Western Indian Ocean	----	----	----	----	----	----	----	----	----
ASIA AND PACIFIC	----	----	581175.77	685557.95	856675.71	908527.74	961567.21	944092.07	969641.27
Australia and New Zealand	11678.93	27987.38	34849.13	37287.45	40245.13	42945.68	42819.39	43182.50	42041.05
Central Asia	----	----	42507.51	40718.43	22554.79	22965.05	22757.22	24388.67	22266.72
North West Pacific and East Asia	----	----	357358.92	382388.44	501670.08	524661.73	568827.36	553526.70	575530.99
South Asia	----	----	70335.33	108329.45	131781.02	138004.72	137459.46	143771.99	159632.90
South East Asia	----	----	76124.88	116834.18	160424.69	179950.56	189703.78	179222.21	170169.61
South Pacific	----	----	----	----	----	----	----	----	----
EUROPE	----	----	1211660.89	1113703.29	960338.87	966061.98	976916.40	986233.60	953864.99
Central Europe	----	----	123927.85	114592.58	97813.58	97242.29	98096.92	100309.12	93541.21
Eastern Europe	----	----	433206.64	414973.54	229421.55	217557.72	216327.69	205576.10	210336.92
Western Europe	186762.45	636139.90	654526.40	584137.17	633103.74	651261.97	662491.79	680348.38	649986.86
LATIN AMERICA AND CARIBBEAN	----	----	287386.89	282665.97	299534.00	308981.13	316657.76	329588.36	330432.64
Caribbean	----	----	49770.38	24691.60	24664.31	25053.23	23090.89	25527.19	23017.05
Meso-America	----	----	73652.41	87422.48	89003.57	86438.31	86011.36	88586.03	90203.23
South America	----	----	163964.10	170551.89	185866.12	197489.59	207555.51	215475.14	217212.36
NORTH AMERICA	486893.87	686921.65	885493.76	866353.99	904599.12	927751.94	958435.04	966108.23	962163.78
North America	486893.87	686921.65	885493.76	866353.99	904599.12	927751.94	958435.04	966108.23	962163.78
POLAR	----	----	----	----	----	----	----	----	----
Antarctic	----	----	----	----	----	----	----	----	----
Arctic	----	----	----	----	----	----	----	----	----
WEST ASIA	----	----	98557.36	176737.54	220031.76	231837.73	220651.86	219630.90	218626.65
Arabian Peninsula	----	----	77848.27	141172.04	179857.00	190986.01	179649.42	177888.77	177720.02
Mashriq	----	----	20709.09	35565.50	40174.76	40851.72	41002.44	41742.13	40906.63
GLOBAL TOTALS	----	----	3135134.29	3238853.88	3366025.12	3467642.55	3570432.45	3575750.98	3560496.47
REPORTED GLOBAL TOTALS FROM ORIGINAL DATA SOURCES	----	----	3147420.00	3241267.00	3368742.00	3470307.00	3573274.00	3578339.00	3563084.00

Comments:

TPES is made up of production + imports - exports - international marine bunkers ± stock changes.

Crude oil, natural gas liquids, refinery feedstocks comprises crude oil, natural gas liquids, refinery feedstocks, and additives as well as other hydrocarbons (which are shown separately in the row "from other sources" in the Energy Statistics of OECD Countries publication).

A ktoe (thousand tonne of oil equivalent) is defined as 41.868 TJ (terajoules) or 11.630 GWh (gigawatt-hour).

Total Primary Energy Supply (TPES) – Geothermal

Units: thousand tonnes of oil equivalent (ktoe)

Data Source: Energy Balances of OECD Countries (1960-1999) and Non-OECD Countries (1971-1999)

Data Provider: International Energy Agency (IEA)

Years: 1960-1999

	1960	1970	1980	1990	1991	1992	1993	1994	1995	1996	1997	1998	1999
AFRICA	----	----	0.00	280.36	255.42	233.92	233.92	224.46	249.40	335.40	337.98	314.76	357.76
Central Africa	----	----	0.00	0.00	0.00	0.00	0.00	0.00	0.00	0.00	0.00	0.00	0.00
Eastern Africa	----	----	----	----	----	----	----	----	----	----	----	----	----
Northern Africa	----	----	0.00	0.00	0.00	0.00	0.00	0.00	0.00	0.00	0.00	0.00	0.00
Southern Africa	----	----	0.00	0.00	0.00	0.00	0.00	0.00	0.00	0.00	0.00	0.00	0.00
Western Africa	----	----	----	----	----	----	----	----	----	----	----	----	----
Western Indian Ocean	----	----	----	----	----	----	----	----	----	----	----	----	----
ASIA AND PACIFIC	----	----	3579.32	9481.32	9692.69	9783.17	9669.35	11062.09	12308.63	13253.37	14351.44	15775.25	17279.74
Australia and New Zealand	350.02	1094.78	1019.10	2315.80	2314.75	2412.11	2329.25	2355.11	2228.21	2298.73	2401.50	2582.29	2621.94
Central Asia	----	----	0.00	0.00	0.00	0.00	0.00	0.00	0.00	0.00	0.00	0.00	0.00
North West Pacific and East Asia	----	----	774.00	1497.26	1524.78	1536.82	1528.22	1894.92	2915.76	3397.82	3484.96	3277.16	3200.88
South Asia	----	----	0.00	0.00	0.00	0.00	0.00	0.00	0.00	0.00	0.00	0.00	0.00
South East Asia	----	----	1786.22	5668.26	5853.16	5834.24	5811.88	6812.06	7164.66	7556.82	8464.98	9915.80	11456.92
South Pacific	----	----	----	----	----	----	----	----	----	----	----	----	----
EUROPE	----	----	2489.05	3309.82	3284.83	3469.90	3592.89	3501.98	3599.01	3839.81	3972.66	4461.94	5087.42
Central Europe	----	----	0.00	84.80	85.66	90.19	97.07	122.46	150.55	174.81	191.95	244.73	251.75
Eastern Europe	----	----	0.00	0.00	0.00	24.94	24.08	26.66	25.80	24.08	24.94	25.80	24.08
Western Europe	1613.49	2215.96	2489.05	3225.02	3199.17	3354.77	3471.74	3352.86	3422.66	3640.92	3755.77	4191.41	4811.59
LATIN AMERICA AND CARIBBEAN	----	----	1210.02	5098.94	5433.48	5730.18	5730.18	5473.90	5925.40	5974.42	5766.30	5866.06	6129.22
Caribbean	----	----	0.00	0.00	0.00	0.00	0.00	0.00	0.00	0.00	0.00	0.00	0.00
Meso-America	----	----	1210.02	5098.94	5433.48	5730.18	5730.18	5473.90	5925.40	5974.42	5766.30	5866.06	6129.22
South America	----	----	0.00	0.00	0.00	0.00	0.00	0.00	0.00	0.00	0.00	0.00	0.00
NORTH AMERICA	30.10	483.32	4602.72	13770.32	13989.62	14764.48	15285.64	15031.94	12849.26	13541.56	12820.02	13217.34	14947.66
North America	30.10	483.32	4602.72	13770.32	13989.62	14764.48	15285.64	15031.94	12849.26	13541.56	12820.02	13217.34	14947.66
POLAR	----	----	----	----	----	----	----	----	----	----	----	----	----
Antarctic	----	----	----	----	----	----	----	----	----	----	----	----	----
Arctic	----	----	----	----	----	----	----	----	----	----	----	----	----
WEST ASIA	----	----	0.00	0.00	0.00	0.00	0.00	0.00	0.00	0.00	0.00	0.00	0.00
Arabian Peninsula	----	----	0.00	0.00	0.00	0.00	0.00	0.00	0.00	0.00	0.00	0.00	0.00
Mashriq	----	----	0.00	0.00	0.00	0.00	0.00	0.00	0.00	0.00	0.00	0.00	0.00
GLOBAL TOTALS	----	----	11881.11	31940.76	32656.04	33981.65	34511.98	35294.37	34931.70	36944.56	37248.40	39635.35	43801.80
REPORTED GLOBAL TOTALS FROM ORIGINAL DATA SOURCES	----	----	11881.00	31941.00	32656.00	33982.00	34512.00	35294.00	34932.00	36945.00	37248.00	39635.00	43802.00

Comments:

TPES is made up of production + imports - exports - international marine bunkers ± stock changes.

Geothermal includes production of geothermal for electricity generation and primary energy used directly by end-users (e.g. geothermal heating of buildings and greenhouses). For electricity generation, unless the actual efficiency of the geothermal process is known, the quantity of geothermal energy entering electricity generation is inferred from the electricity production at geothermal plants assuming an average thermal efficiency of 10 per cent.

A ktoe (thousand tonne of oil equivalent) is defined as 41.868 TJ (terajoules) or 11.630 GWh (gigawatt-hour).

AGGREGATIONS

Total Primary Energy Supply (TPES) – Hydro

Units: thousand tonnes of oil equivalent (ktoe)

Data Source: Energy Balances of OECD Countries (1960-1999) and Non-OECD Countries (1971-1999)

Data Provider: International Energy Agency (IEA)

Years: 1960-1999

	1960	1970	1980	1990	1995	1996	1997	1998	1999
AFRICA	----	----	3872.40	4373.70	4717.69	5020.06	5198.09	5451.37	5701.26
Central Africa	----	----	513.25	814.94	790.34	813.13	833.08	825.60	843.23
Eastern Africa	----	----	-10048.19	-10081.39	-10082.59	-10091.36	-10089.64	-10088.70	-10094.03
Northern Africa	----	----	1046.53	1085.24	1178.98	1299.03	1325.25	1364.82	1549.03
Southern Africa	----	----	1370.32	1173.81	1209.23	1308.22	1383.73	1915.82	1958.55
Western Africa	----	----	----	----	----	----	----	----	----
Western Indian Ocean	----	----	----	----	----	----	----	----	----
ASIA AND PACIFIC	----	----	26744.18	40259.01	46294.31	45681.82	46530.37	48829.23	48290.27
Australia and New Zealand	822.42	1743.31	2740.82	3223.97	3710.38	3590.24	3502.95	3457.28	3456.86
Central Asia	----	----	2771.08	3495.55	3460.39	3520.93	3175.80	3095.75	3387.45
North West Pacific and East Asia	----	----	14956.43	22559.87	26430.72	26010.95	27530.06	28921.80	27901.23
South Asia	----	----	5431.41	8563.80	9349.41	9003.17	9258.59	10189.36	9890.00
South East Asia	----	----	844.44	2415.82	3343.41	3556.53	3062.97	3165.04	3654.73
South Pacific	----	----	----	----	----	----	----	----	----
EUROPE	----	----	52911.60	57745.26	63446.08	59989.92	61934.80	64510.24	64618.82
Central Europe	----	----	5881.76	5772.04	7908.05	8509.20	8543.16	9156.16	8546.75
Eastern Europe	----	----	12901.95	16275.05	16726.40	14745.91	15135.31	15857.02	15606.07
Western Europe	19110.47	27354.62	34127.90	35698.17	38811.63	36734.81	38256.33	39497.06	40466.00
LATIN AMERICA AND CARIBBEAN	----	----	18688.73	33223.87	42289.91	44561.59	46049.40	46801.03	47582.18
Caribbean	----	----	85.57	86.60	88.66	115.24	95.46	105.27	135.97
Meso-America	----	----	1933.28	3035.20	3356.76	3869.23	3512.41	3237.91	4100.14
South America	----	----	16669.88	30102.07	38844.49	40577.12	42441.53	43457.85	43346.07
NORTH AMERICA	21865.67	35141.06	45573.90	49010.45	55903.36	60765.10	58545.19	53758.43	54523.66
North America	21865.67	35141.06	45573.90	49010.45	55903.36	60765.10	58545.19	53758.43	54523.66
POLAR	----	----	----	----	----	----	----	----	----
Antarctic	----	----	----	----	----	----	----	----	----
Arctic	----	----	----	----	----	----	----	----	----
WEST ASIA	----	----	352.69	753.28	709.41	716.72	764.20	807.55	827.66
Arabian Peninsula	----	----	0.00	0.00	0.00	0.00	0.00	0.00	0.00
Mashriq	----	----	352.69	753.28	709.41	716.72	764.20	807.55	827.66
GLOBAL TOTALS	----	----	148143.50	185365.57	213360.76	216735.21	219022.05	220157.85	221543.85
REPORTED GLOBAL TOTALS FROM ORIGINAL DATA SOURCES	----	----	148552.00	186105.00	214125.00	217512.00	219702.00	220837.00	222223.00

Comments:

TPES is made up of production + imports - exports - international marine bunkers ± stock changes.

A ktoe (thousand tonne of oil equivalent) is defined as 41.868 TJ (terajoules) or 11.630 GWh (gigawatt-hour).

Total Primary Energy Supply (TPES) - Natural Gas

Units: thousand tonnes of oil equivalent (ktoe)

Data Source: Energy Balances of OECD Countries (1960-1999) and Non-OECD Countries (1971-1999)

Data Provider: International Energy Agency (IEA)

Years: 1960-1999

	1960	1970	1980	1990	1995	1996	1997	1998	1999
AFRICA	----	----	12264.91	32039.31	39968.05	40968.93	43020.22	44475.17	49102.05
Central Africa	----	----	12.53	68.77	66.81	71.28	67.95	67.95	67.95
Eastern Africa	----	----	----	----	----	----	----	----	----
Northern Africa	----	----	10950.04	26754.07	33250.64	34030.11	35815.23	37001.20	40525.66
Southern Africa	----	----	64.49	1945.28	2172.34	2000.87	1843.73	1634.19	1975.56
Western Africa	----	----	----	----	----	----	----	----	----
Western Indian Ocean	----	----	----	----	----	----	----	----	----
ASIA AND PACIFIC	----	----	98544.55	212840.95	275581.01	292784.01	312592.41	329473.59	348536.45
Australia and New Zealand	58.13	1289.43	8255.88	18658.69	20564.35	21197.15	21633.22	21868.39	23015.36
Central Asia	----	----	35059.60	62086.93	55516.82	52651.39	51677.84	58238.35	59014.59
North West Pacific and East Asia	----	----	34941.77	63548.90	82564.01	92944.24	98592.52	103585.56	110982.88
South Asia	----	----	10965.75	41418.35	67915.50	73247.34	81284.83	85608.12	91448.58
South East Asia	----	----	9321.55	27128.08	49020.33	52743.89	59404.00	60173.17	64075.04
South Pacific	----	----	----	----	----	----	----	----	----
EUROPE	----	----	527476.70	796083.46	754129.79	797938.52	779684.80	783480.34	804172.32
Central Europe	----	----	64350.68	75948.53	66829.64	72907.75	69543.82	68170.67	67536.16
Eastern Europe	----	----	278502.62	493199.39	406748.44	414386.04	401459.97	393225.35	401061.94
Western Europe	10168.05	62795.66	184623.40	226935.53	280551.71	310644.73	308681.01	322084.32	335574.22
LATIN AMERICA AND CARIBBEAN	----	----	50787.79	75946.91	90807.17	99951.02	104579.66	112825.86	113891.77
Caribbean	----	----	2450.57	4723.85	5044.50	5780.81	6063.41	7188.92	8355.90
Meso-America	----	----	19130.63	23119.89	25353.52	27777.80	28823.23	31451.28	30952.41
South America	----	----	29206.59	48103.17	60409.15	66392.41	69693.02	74185.66	74583.46
NORTH AMERICA	292527.49	528206.28	522335.49	494079.28	575862.23	574986.36	579216.36	568102.53	592290.14
North America	292527.49	528206.28	522335.49	494079.28	575862.23	574986.36	579216.36	568102.53	592290.14
POLAR	----	----	----	----	----	----	----	----	----
Antarctic	----	----	----	----	----	----	----	----	----
Arctic	----	----	----	----	----	----	----	----	----
WEST ASIA	----	----	27312.89	59921.92	84523.84	91855.26	97954.18	102830.22	104350.34
Arabian Peninsula	----	----	25082.70	55053.45	78830.79	86096.65	90647.23	94311.05	95415.89
Mashriq	----	----	2230.19	4868.47	5693.05	5758.61	7306.95	8519.17	8934.45
GLOBAL TOTALS	----	----	1238722.33	1670911.83	1820872.09	1898484.10	1917047.63	1941187.71	2012343.07
REPORTED GLOBAL TOTALS FROM ORIGINAL DATA SOURCES	----	----	1238816.00	1671176.00	1821117.00	1898721.00	1917261.00	1941404.00	2012559.00

Comments:

TPES is made up of production + imports - exports - international marine bunkers ± stock changes.

Natural gas excludes natural gas liquids and includes gas works gas. The latter appears as a positive figure in the "gas works" row but is not part of production.

A ktoe (thousand tonne of oil equivalent) is defined as 41.868 TJ (terajoules) or 11.630 GWh (gigawatt-hour).

Units: thousand tonnes of oil equivalent (ktoe)

Data Source: Energy Balances of OECD Countries (1960-1999) and Non-OECD Countries (1971-1999)

Data Provider: International Energy Agency (IEA)

Years: 1960-1999

	1960	1970	1980	1995	1996	1997	1998	1999
AFRICA	----	----	0.00	2945.11	3068.64	3295.89	3544.50	3345.40
Central Africa	----	----	0.00	0.00	0.00	0.00	0.00	0.00
Eastern Africa	----	----	----	----	----	----	----	----
Northern Africa	----	----	0.00	0.00	0.00	0.00	0.00	0.00
Southern Africa	----	----	0.00	2945.11	3068.64	3295.89	3544.50	3345.40
Western Africa	----	----	----	----	----	----	----	----
Western Indian Ocean	----	----	----	----	----	----	----	----
ASIA AND PACIFIC	----	----	25349.16	108131.97	114094.63	119195.74	126445.55	126761.39
Australia and New Zealand	0.00	0.00	0.00	0.00	0.00	0.00	0.00	0.00
Central Asia	----	----	0.00	0.00	0.00	0.00	0.00	0.00
North West Pacific and East Asia	----	----	24566.56	105918.64	111604.80	116477.88	123255.21	123278.39
South Asia	----	----	782.60	2213.33	2489.83	2717.86	3190.34	3483.00
South East Asia	----	----	0.00	0.00	0.00	0.00	0.00	0.00
South Pacific	----	----	----	----	----	----	----	----
EUROPE	----	----	80681.67	280851.09	298078.35	300734.87	297901.87	304426.06
Central Europe	----	----	4039.01	18715.63	19967.90	20253.95	20739.29	19913.21
Eastern Europe	----	----	17770.35	44706.93	50111.88	49730.35	47806.06	51451.44
Western Europe	617.64	11613.40	58872.31	217428.53	227998.57	230750.57	229356.52	233061.41
LATIN AMERICA AND CARIBBEAN	----	----	609.82	4698.21	4629.93	5625.45	5207.70	5494.88
Caribbean	----	----	0.00	0.00	0.00	0.00	0.00	0.00
Meso-America	----	----	0.00	2200.30	2053.06	2724.90	2414.52	2606.58
South America	----	----	609.82	2497.91	2576.87	2900.55	2793.18	2888.30
NORTH AMERICA	144.38	6348.63	79769.48	211583.28	210624.45	195194.15	204741.24	221873.75
North America	144.38	6348.63	79769.48	211583.28	210624.45	195194.15	204741.24	221873.75
POLAR	----	----	----	----	----	----	----	----
Antarctic	----	----	----	----	----	----	----	----
Arctic	----	----	----	----	----	----	----	----
WEST ASIA	----	----	0.00	0.00	0.00	0.00	0.00	0.00
Arabian Peninsula	----	----	0.00	0.00	0.00	0.00	0.00	0.00
Mashriq	----	----	0.00	0.00	0.00	0.00	0.00	0.00
GLOBAL TOTALS	----	----	186410.13	608209.66	630496.00	624046.10	637840.86	661901.48
REPORTED GLOBAL TOTALS FROM ORIGINAL DATA SOURCES	----	----	186410.00	608210.00	630496.00	624046.00	637841.00	661901.00

Comments:

TPES is made up of production + imports - exports - international marine bunkers ± stock changes.

Nuclear power shows the primary heat equivalent of the electricity produced by a nuclear power plant with an average thermal efficiency of 33 percent.

A ktoe (thousand tonne of oil equivalent) is defined as 41.868 TJ (terajoules) or 11.630 GWh (gigawatt-hour).

Total Primary Energy Supply (TPES) - per Capita

Units: thousand tonnes of oil equivalent (ktoe)

Data Source: Energy Balances of OECD Countries (1960-1999) and Non-OECD Countries (1971-1999)

Data Provider: International Energy Agency (IEA)

Years: 1960-1999

	1960	1970	1980	1990	1991	1992	1993	1994	1995	1996	1997	1998	1999
AFRICA	----	----	0.66	0.69	0.70	0.68	0.67	0.67	0.68	0.68	0.68	0.69	0.68
Central Africa	----	----	0.39	0.37	0.37	0.36	0.35	0.34	0.34	0.34	0.34	0.34	0.34
Eastern Africa	----	----	----	----	----	----	----	----	----	----	----	----	----
Northern Africa	----	----	0.47	0.63	0.65	0.64	0.65	0.63	0.64	0.66	0.68	0.71	0.69
Southern Africa	----	----	1.25	1.27	1.27	1.19	1.21	1.22	1.23	1.22	1.21	1.20	1.18
Western Africa	----	----	----	----	----	----	----	----	----	----	----	----	----
Western Indian Ocean	----	----	----	----	----	----	----	----	----	----	----	----	----
ASIA AND PACIFIC	----	----	0.68	0.82	0.83	0.84	0.86	0.88	0.91	0.93	0.94	0.94	0.94
Australia and New Zealand	2.82	3.81	4.50	5.01	4.91	4.95	5.10	5.08	5.10	5.38	5.50	5.42	5.56
Central Asia	----	----	3.19	3.10	3.04	2.88	2.53	2.29	2.09	1.94	1.81	3.72	3.43
North West Pacific and East Asia	----	----	0.87	1.10	1.10	1.13	1.17	1.21	1.27	1.31	1.31	1.27	1.28
South Asia	----	----	0.35	0.43	0.44	0.44	0.45	0.46	0.48	0.49	0.49	0.49	0.49
South East Asia	----	----	0.44	0.56	0.58	0.61	0.64	0.66	0.69	0.71	0.75	0.72	0.73
South Pacific	----	----	----	----	----	----	----	----	----	----	----	----	----
EUROPE	----	----	3.50	3.71	3.69	3.52	3.43	3.25	3.26	3.32	3.26	3.25	3.27
Central Europe	----	----	2.39	2.22	2.04	1.94	1.92	1.83	1.91	2.02	1.98	1.91	1.81
Eastern Europe	----	----	4.52	5.05	5.01	4.60	4.32	3.75	3.63	3.57	3.44	3.36	3.49
Western Europe	1.72	2.98	3.41	3.63	3.70	3.65	3.64	3.65	3.70	3.81	3.79	3.85	3.86
LATIN AMERICA AND CARIBBEAN	----	----	1.08	1.05	1.06	1.07	1.06	1.10	1.10	1.13	1.15	1.18	1.17
Caribbean	----	----	1.27	1.17	1.06	1.03	0.99	1.01	1.02	1.09	1.06	1.10	1.11
Meso-America	----	----	1.25	1.25	1.28	1.29	1.26	1.28	1.23	1.24	1.26	1.30	1.29
South America	----	----	1.00	0.97	0.98	0.99	1.00	1.05	1.06	1.10	1.12	1.15	1.13
NORTH AMERICA	5.38	7.32	7.86	7.56	7.52	7.59	7.69	7.74	7.78	7.88	7.93	7.94	8.08
North America	5.38	7.32	7.86	7.56	7.52	7.59	7.69	7.74	7.78	7.88	7.93	7.94	8.08
POLAR	----	----	----	----	----	----	----	----	----	----	----	----	----
Antarctic	----	----	----	----	----	----	----	----	----	----	----	----	----
Arctic	----	----	----	----	----	----	----	----	----	----	----	----	----
WEST ASIA	----	----	1.74	2.07	2.13	2.32	2.38	2.40	2.41	2.51	2.40	2.39	2.37
Arabian Peninsula	----	----	2.93	3.17	3.43	3.68	3.78	3.77	3.75	3.97	3.71	3.64	3.62
Mashriq	----	----	0.81	1.08	0.97	1.10	1.13	1.20	1.23	1.22	1.23	1.26	1.23
GLOBAL TOTALS	----	----	1.62	1.64	1.64	1.62	1.62	1.60	1.62	1.65	1.64	1.64	1.64
REPORTED GLOBAL TOTALS FROM ORIGINAL DATA SOURCES	****	****	****	****	****	****	****	****	****	****	****	****	****

Comments:

TPES is made up of indigenous production + imports - exports - international marine bunkers ± stock changes. For the global total, TPES excludes international marine bunkers.

Indigenous production is the production of primary energy, in other words hard coal, lignite, peat, crude oil, NGLs, natural gas, combustible renewables and waste, nuclear, hydro, geothermal, solar and the heat from heat pumps that is extracted from the ambient environment. Production is calculated after removal of impurities.

A toe (tonne of oil equivalent) is defined as 41.868 gigajoules or 10^7 kilocalories. One terawatt-hour = 0.086 Mtoe.

Units: thousand tonnes of oil equivalent (ktoe)

Data Source: Energy Balances of OECD Countries (1960-1999) and Non-OECD Countries (1971-1999)

Data Provider: International Energy Agency (IEA)

Years: 1960-1999

	1960	1970	1980	1990	1995	1996	1997	1998	1999
AFRICA	----	----	-10157.28	-31079.32	-37250.95	-37134.59	-42440.81	-29914.34	-34897.10
Central Africa	----	----	587.91	509.09	-202.10	-338.37	-230.32	-92.76	-47.45
Eastern Africa	----	----	-10360.58	-11033.70	-10838.66	-10798.36	-10904.32	-11085.56	-11252.41
Northern Africa	----	----	-9524.00	-27248.18	-29768.33	-28411.31	-31072.73	-28933.26	-30646.30
Southern Africa	----	----	-332.74	-835.32	-4060.47	-3433.83	-8639.70	-3210.14	-4827.07
Western Africa	----	----	----	----	----	----	----	----	----
Western Indian Ocean	----	----	----	----	----	----	----	----	----
ASIA AND PACIFIC	----	----	8116.24	54112.82	68032.41	62267.75	45631.97	26002.83	38397.61
Australia and New Zealand	1205.92	307.73	212.97	-788.52	-201.02	178.43	-917.28	-1613.95	-15.38
Central Asia	----	----	-1396.22	-1670.41	770.91	-653.16	-1135.95	-2700.26	-2418.76
North West Pacific and East Asia	----	----	21893.36	66606.93	66250.84	68848.15	50774.87	38429.66	47861.34
South Asia	----	----	2293.91	12816.93	19956.22	24482.41	22293.96	19795.63	11243.02
South East Asia	----	----	-14887.78	-22852.11	-18744.54	-30588.08	-25383.63	-27908.25	-18272.61
South Pacific	----	----	----	----	----	----	----	----	----
EUROPE	----	----	-45830.82	-56226.57	-52949.16	-63254.68	-79165.44	-78307.10	-50813.96
Central Europe	----	----	-3955.76	-1215.24	3491.72	8597.14	6969.59	5533.74	9096.08
Eastern Europe	----	----	-44301.84	-53001.72	-37812.39	-45987.56	-50291.25	-45027.37	-47990.92
Western Europe	-644.21	-36420.75	2426.79	-2009.61	-18628.49	-25864.26	-35843.78	-38813.47	-11919.12
LATIN AMERICA AND CARIBBEAN	----	----	-65481.33	-48238.44	-32387.79	-31951.79	-29832.33	-29163.28	-29915.98
Caribbean	----	----	-28957.98	-4843.14	-5246.39	-4699.17	-3497.07	-5134.98	-3101.26
Meso-America	----	----	-1823.22	1257.67	4395.58	7127.68	11322.43	14079.50	13558.59
South America	----	----	-34700.13	-44652.97	-31536.98	-34380.30	-37657.69	-38107.80	-40373.31
NORTH AMERICA	25113.15	88643.51	7328.99	-19041.02	-24385.57	-19782.50	-25378.15	-15963.16	5595.96
North America	25113.15	88643.51	7328.99	-19041.02	-24385.57	-19782.50	-25378.15	-15963.16	5595.96
POLAR	----	----	----	----	----	----	----	----	----
Antarctic	----	----	----	----	----	----	----	----	----
Arctic	----	----	----	----	----	----	----	----	----
WEST ASIA	----	----	-40563.76	-91848.78	-107380.37	-111518.70	-109701.20	-107964.00	-103293.63
Arabian Peninsula	----	----	-38740.00	-89050.81	-111605.08	-116025.19	-114423.26	-113447.00	-109378.50
Mashriq	----	----	-1823.76	-2797.97	4224.71	4506.49	4722.06	5483.00	6084.87
GLOBAL TOTALS	----	----	-146587.96	-192321.31	-186321.43	-201374.51	-240885.96	-235309.05	-174927.10
REPORTED GLOBAL TOTALS FROM ORIGINAL DATA SOURCES	----	----	-147015.00	-182567.00	-176619.00	-191492.00	-231344.00	-225503.00	-165121.00

Comments:

TPES is made up of production + imports - exports - international marine bunkers ± stock changes.

Petroleum products comprise refinery gas, ethane, LPG, aviation gasoline, motor gasoline, jet fuels, kerosene, gas/diesel oil, heavy fuel oil, naphtha, white spirit, lubricants, bitumen, paraffin waxes, petroleum coke and other petroleum products.

A ktoe (thousand tonne of oil equivalent) is defined as 41.868 TJ (terajoules) or 11.630 GWh (gigawatt-hour).

Total Primary Energy Supply (TPES) - Solar, Wind, Tide and Wave

Units: thousand tonnes of oil equivalent (ktoe)

Data Source: Energy Balances of OECD Countries (1960-1999) and Non-OECD Countries (1971-1999)

Data Provider: International Energy Agency (IEA)

Years: 1960-1999

	1960	1970	1980	1990	1991	1992	1993	1994	1995	1996	1997	1998	1999
AFRICA	----	----	0.00	0.00	0.00	0.00	0.00	0.00	0.00	0.00	0.00	0.00	0.00
Central Africa	----	----	0.00	0.00	0.00	0.00	0.00	0.00	0.00	0.00	0.00	0.00	0.00
Eastern Africa	----	----	----	----	----	----	----	----	----	----	----	----	----
Northern Africa	----	----	0.00	0.00	0.00	0.00	0.00	0.00	0.00	0.00	0.00	0.00	0.00
Southern Africa	----	----	0.00	0.00	0.00	0.00	0.00	0.00	0.00	0.00	0.00	0.00	0.00
Western Africa	----	----	----	----	----	----	----	----	----	----	----	----	----
Western Indian Ocean	----	----	----	----	----	----	----	----	----	----	----	----	----
ASIA AND PACIFIC	----	----	19.46	87.01	88.23	91.10	92.10	106.02	139.38	254.73	258.40	1135.83	1122.76
Australia and New Zealand	0.00	0.00	19.46	81.31	81.38	81.48	81.38	81.82	84.05	87.78	90.78	93.16	100.48
Central Asia	----	----	0.00	0.00	0.00	0.00	0.00	0.00	0.00	0.00	0.00	0.00	0.00
North West Pacific and East Asia	----	----	0.00	2.86	3.41	5.06	5.73	8.46	11.04	14.22	17.21	889.77	869.38
South Asia	----	----	0.00	2.75	3.35	4.47	4.90	15.65	44.20	97.35	89.18	89.18	89.18
South East Asia	----	----	0.00	0.09	0.09	0.09	0.09	0.09	0.09	55.38	61.23	63.72	63.72
South Pacific	----	----	----	----	----	----	----	----	----	----	----	----	----
EUROPE	----	----	44.71	615.07	683.67	786.83	899.23	1038.75	1213.79	1364.72	1675.46	2086.62	2425.01
Central Europe	----	----	0.00	20.99	26.98	32.00	37.99	44.99	83.08	96.14	113.30	134.95	149.38
Eastern Europe	----	----	0.00	0.00	0.00	0.00	0.00	0.00	0.00	0.00	0.00	0.26	0.43
Western Europe	0.00	42.48	44.71	594.08	656.69	754.83	861.24	993.76	1130.71	1268.58	1562.16	1951.41	2275.20
LATIN AMERICA AND CARIBBEAN	----	----	0.00	0.00	0.00	0.00	0.00	0.34	0.52	3.44	8.60	42.46	49.27
Caribbean	----	----	0.00	0.00	0.00	0.00	0.00	0.00	0.00	0.00	0.00	0.00	0.00
Meso-America	----	----	0.00	0.00	0.00	0.00	0.00	0.34	0.52	2.41	6.88	39.10	46.00
South America	----	----	0.00	0.00	0.00	0.00	0.00	0.00	0.00	1.03	1.72	3.36	3.27
NORTH AMERICA	0.00	0.00	0.00	256.97	297.99	323.45	348.21	378.92	354.23	379.52	372.55	344.43	485.39
North America	0.00	0.00	0.00	256.97	297.99	323.45	348.21	378.92	354.23	379.52	372.55	344.43	485.39
POLAR	----	----	----	----	----	----	----	----	----	----	----	----	----
Antarctic	----	----	----	----	----	----	----	----	----	----	----	----	----
Arctic	----	----	----	----	----	----	----	----	----	----	----	----	----
WEST ASIA	----	----	0.00	58.11	58.11	61.36	61.96	62.58	62.63	64.08	67.60	70.44	71.31
Arabian Peninsula	----	----	0.00	0.00	0.00	0.00	0.00	0.00	0.00	0.00	0.00	0.00	0.00
Mashriq	----	----	0.00	58.11	58.11	61.36	61.96	62.58	62.63	64.08	67.60	70.44	71.31
GLOBAL TOTALS	----	----	64.17	1017.16	1128.00	1262.74	1401.50	1586.61	1770.55	2066.49	2382.61	3679.78	4153.74
REPORTED GLOBAL TOTALS FROM ORIGINAL DATA SOURCES	----	----	64.00	1017.00	1128.00	1263.00	1402.00	1587.00	1771.00	2066.00	2383.00	3680.00	4154.00

Comments:

TPES is made up of production + imports - exports - international marine bunkers ± stock changes.

Solar, wind, tide and wave - indigenous production of solar, wind, wide and wave energy and the use of these energy forms for electricity generation. Other uses shown in this column relate only to solar heat. For solar, wind, tide and wave energy, the quantities entering electricity generation are equal to the electrical energy generated.

A ktoe (thousand tonne of oil equivalent) is defined as 41.868 TJ (terajoules) or 11.630 GWh (gigawatt-hour).

Units: million tonnes of oil equivalent (ktoe)

Data Source: Energy Balances of OECD Countries (1960-1999) and Non-OECD Countries (1971-1999)

Data Provider: International Energy Agency (IEA)

Years: 1960-1999

	1960	1970	1980	1990	1991	1992	1993	1994	1995	1996	1997	1998	1999
AFRICA	----	----	251.88	354.10	368.86	369.97	373.53	384.00	398.49	408.23	423.05	433.91	437.69
Central Africa	----	----	14.74	19.27	19.62	19.94	20.47	20.20	20.81	21.34	22.02	22.64	22.95
Eastern Africa	----	----	----	----	----	----	----	----	----	----	----	----	----
Northern Africa	----	----	52.35	90.14	94.35	95.85	99.37	97.45	101.78	106.19	112.69	118.20	117.99
Southern Africa	----	----	99.33	131.54	137.55	132.31	138.12	143.20	148.17	150.47	153.40	155.39	156.40
Western Africa	----	----	----	----	----	----	----	----	----	----	----	----	----
Western Indian Ocean	----	----	----	----	----	----	----	----	----	----	----	----	----
ASIA AND PACIFIC	----	----	1735.04	2512.38	2576.39	2661.62	2763.65	2871.34	3017.05	3141.93	3191.67	3163.33	3212.00
Australia and New Zealand	35.67	58.56	79.58	101.52	100.89	103.07	107.74	108.68	110.52	118.02	122.15	121.73	126.11
Central Asia	----	----	131.71	156.84	156.07	150.12	133.62	122.20	112.50	105.95	99.89	106.15	104.25
North West Pacific and East Asia	----	----	1045.45	1522.71	1549.73	1606.96	1681.44	1758.58	1858.37	1938.16	1955.16	1920.47	1941.89
South Asia	----	----	324.83	494.31	515.97	533.46	553.77	581.51	614.95	642.27	651.70	663.56	677.60
South East Asia	----	----	153.47	237.00	253.73	268.01	287.08	300.37	320.71	337.53	362.77	351.42	362.15
South Pacific	----	----	----	----	----	----	----	----	----	----	----	----	----
EUROPE	----	----	2647.57	2965.32	2960.03	2838.51	2773.08	2634.56	2649.26	2703.82	2659.10	2654.77	2667.81
Central Europe	----	----	404.49	415.14	384.07	366.99	364.88	348.63	365.87	388.60	381.74	371.58	352.97
Eastern Europe	----	----	979.23	1166.01	1160.35	1067.47	1001.30	868.82	839.28	823.64	790.68	769.48	795.03
Western Europe	560.79	1046.00	1263.84	1384.17	1415.61	1404.05	1406.90	1417.11	1444.11	1491.58	1486.68	1513.71	1519.81
LATIN AMERICA AND CARIBBEAN	----	----	384.07	456.35	467.49	479.48	486.53	513.74	521.11	544.79	562.63	586.02	588.98
Caribbean	----	----	30.72	33.15	30.49	29.88	29.29	30.17	30.79	33.20	32.60	34.20	35.07
Meso-America	----	----	111.99	139.19	145.00	148.94	149.04	154.24	151.32	155.55	161.34	168.90	170.40
South America	----	----	241.36	284.01	292.00	300.66	308.20	329.33	339.00	356.04	368.69	382.92	383.51
NORTH AMERICA	1098.36	1696.82	2004.65	2134.67	2147.28	2190.34	2242.30	2283.60	2317.92	2373.51	2416.20	2443.00	2511.76
North America	1098.36	1696.82	2004.65	2134.67	2147.28	2190.34	2242.30	2283.60	2317.92	2373.51	2416.20	2443.00	2511.76
POLAR	----	----	----	----	----	----	----	----	----	----	----	----	----
Antarctic	----	----	----	----	----	----	----	----	----	----	----	----	----
Arctic	----	----	----	----	----	----	----	----	----	----	----	----	----
WEST ASIA	----	----	85.87	145.85	155.51	174.42	184.22	191.90	198.36	213.41	210.18	215.83	221.06
Arabian Peninsula	----	----	64.26	107.27	119.97	132.59	139.88	143.43	147.19	161.16	155.97	158.85	163.87
Mashriq	----	----	21.61	38.58	35.54	41.83	44.34	48.47	51.17	52.25	54.21	56.98	57.19
GLOBAL TOTALS	----	----	7109.08	8568.67	8675.56	8714.34	8823.31	8879.14	9102.19	9385.69	9462.83	9496.86	9639.30
REPORTED GLOBAL TOTALS FROM ORIGINAL DATA SOURCES	----	----	7153.83	8622.19	8730.07	8769.58	8879.35	8936.28	9160.67	9445.35	9523.79	9559.20	9702.79

Comments:

TPES is made up of indigenous production + imports - exports - international marine bunkers ± stock changes.

Indigenous production is the production of primary energy, i.e. hard coal, lignite, peat, crude oil, NGLs, natural gas, combustible renewables and waste, nuclear, hydro, geothermal, solar and the heat from heat pumps that is extracted from the ambient environment. Production is calculated after removal of impurities.

A toe (tonne of oil equivalent) is defined as 41.868 gigajoules or 10^7 kilocalories. One terawatt-hour = 0.086 Mtoe.

1.1.4 Health

Crude Birth Rate (Projection)

Units: births per thousand people

Data Source: World Population Prospects: The 2000 Revision

Data Provider: United Nations Population Division/Department of Economic and Social Affairs

Years: 1950-1955, 1955-1960, ..., 2045-2050

	1950-55	1970-75	1990-95	1995-00	2000-05	2020-25	2025-30	2030-35	2035-40	2040-45	2045-50
AFRICA	48.98	46.45	40.26	38.71	37.35	29.50	27.34	25.14	23.29	21.61	19.81
Central Africa	45.70	46.71	46.18	45.22	44.65	36.42	33.65	30.57	27.38	24.18	20.75
Eastern Africa	51.21	49.16	44.95	43.70	43.11	34.57	32.51	30.42	28.20	25.58	22.85
Northern Africa	48.72	42.91	30.23	27.57	25.31	18.05	16.79	15.77	15.14	14.63	14.20
Southern Africa	47.12	45.13	39.27	38.08	36.15	29.18	27.14	24.95	23.02	21.46	19.90
Western Africa	50.19	48.94	43.93	42.24	40.83	30.62	27.64	24.70	22.78	21.60	20.16
Western Indian Ocean	50.09	43.51	41.61	41.08	38.88	30.80	28.26	25.48	22.61	19.66	18.13
ASIA AND PACIFIC	42.71	33.53	24.28	21.95	20.12	15.74	14.87	14.21	13.74	13.39	13.05
Australia and New Zealand	23.52	19.78	15.12	13.69	12.87	11.90	11.73	11.58	11.58	11.71	11.92
Central Asia	38.46	31.66	27.45	22.87	19.92	16.31	15.13	14.57	14.26	13.85	13.36
North West Pacific and East Asia	40.76	27.72	17.49	15.61	13.88	11.67	11.03	10.90	10.95	10.81	10.50
South Asia	45.55	39.69	30.92	28.09	26.03	18.90	17.73	16.59	15.63	15.08	14.62
South East Asia	43.90	38.18	26.83	23.81	21.43	16.59	15.60	14.79	14.26	13.90	13.62
South Pacific	43.43	39.82	35.28	32.30	30.49	23.88	22.00	20.07	18.05	16.19	16.00
EUROPE	22.77	17.00	12.75	11.27	10.44	9.61	9.40	9.30	9.40	9.60	9.71
Central Europe	30.27	22.83	16.80	14.82	13.44	12.03	11.58	11.34	11.41	11.63	11.68
Eastern Europe	26.36	15.94	11.47	9.23	8.77	8.23	7.83	7.93	8.38	8.75	8.74
Western Europe	18.01	15.10	11.53	10.73	9.90	9.03	9.00	8.86	8.77	8.86	9.03
LATIN AMERICA AND CARIBBEAN	41.95	35.37	25.13	23.10	21.44	16.50	15.70	15.02	14.49	14.06	13.70
Caribbean	37.45	31.34	22.68	20.80	19.71	15.96	15.02	14.37	13.86	13.42	13.16
Meso-America	46.40	43.08	28.81	26.47	24.07	17.46	16.45	15.55	14.83	14.29	13.82
South America	41.16	33.14	24.00	22.05	20.60	16.16	15.47	14.87	14.41	14.03	13.71
NORTH AMERICA	24.57	15.71	15.46	14.22	12.88	12.51	12.34	12.25	12.26	12.36	12.47
North America	24.57	15.71	15.46	14.22	12.88	12.51	12.34	12.25	12.26	12.36	12.47
POLAR	----	----	----	----	----	----	----	----	----	----	----
Antarctic	----	----	----	----	----	----	----	----	----	----	----
Arctic	----	----	----	----	----	----	----	----	----	----	----
WEST ASIA	48.60	47.07	38.18	36.09	34.71	26.91	25.02	23.62	22.43	20.92	19.66
Arabian Peninsula	50.60	49.32	40.55	38.98	37.85	32.15	30.35	28.39	26.28	24.05	22.52
Mashriq	47.17	45.54	36.06	33.49	31.75	21.03	18.66	17.57	17.26	16.50	15.41
GLOBAL TOTALS	37.51	30.88	24.29	22.49	21.14	17.56	16.71	15.99	15.43	14.96	14.44
REPORTED GLOBAL TOTALS FROM ORIGINAL DATA SOURCES	37.54	30.91	24.32	22.53	21.18	17.58	16.74	16.01	15.45	14.97	14.44

Crude Death Rate (Projection)

Units: deaths per thousand people

Data Source: World Population Prospects: The 2000 Revision

Data Provider: United Nations Population Division/Department of Economic and Social Affairs

Years: 1950-1955, 1955-1960, ..., 2045-2050

	1950-55	1970-75	1990-95	1995-00	2000-05	2020-25	2025-30	2030-35	2035-40	2040-45	2045-50
AFRICA	26.83	19.11	14.31	14.05	13.88	9.72	8.84	8.18	7.68	7.36	7.15
Central Africa	26.95	19.97	15.38	15.62	14.51	8.92	7.71	6.85	6.15	5.81	5.65
Eastern Africa	28.62	20.31	19.03	17.80	17.68	11.48	10.18	9.05	8.13	7.43	6.89
Northern Africa	24.50	16.52	8.36	7.46	6.83	6.12	6.29	6.66	7.14	7.63	8.18
Southern Africa	25.40	17.50	14.12	15.93	17.91	13.24	11.78	10.57	9.60	8.85	8.29
Western Africa	28.84	21.58	15.90	15.06	14.08	9.29	8.47	7.88	7.45	7.18	6.99
Western Indian Ocean	25.43	17.65	14.62	13.59	12.35	8.47	7.66	7.02	6.57	6.31	6.31
ASIA AND PACIFIC	23.78	11.23	8.52	7.98	7.69	7.72	8.08	8.53	9.06	9.66	10.23
Australia and New Zealand	9.37	8.51	7.09	7.17	7.43	8.74	9.28	9.90	10.49	10.92	11.14
Central Asia	14.60	9.39	7.74	7.59	7.19	6.84	7.00	7.35	7.86	8.45	9.00
North West Pacific and East Asia	23.11	6.42	7.11	7.02	7.09	8.62	9.39	10.26	11.14	12.16	13.03
South Asia	25.54	16.43	10.33	9.27	8.56	7.28	7.39	7.57	7.88	8.23	8.64
South East Asia	23.27	14.39	8.02	7.37	7.04	6.82	7.07	7.49	8.05	8.68	9.30
South Pacific	25.31	14.63	10.05	8.95	8.18	6.42	6.17	6.12	6.27	6.50	6.84
EUROPE	11.28	10.11	10.78	10.97	11.21	11.89	12.32	12.88	13.54	14.19	14.75
Central Europe	13.94	10.14	9.74	9.46	9.41	10.03	10.38	10.89	11.49	12.08	12.60
Eastern Europe	10.11	9.02	12.90	13.80	14.64	14.08	14.36	14.85	15.54	16.23	16.90
Western Europe	10.88	10.73	10.01	10.06	10.17	11.77	12.34	13.00	13.71	14.41	15.00
LATIN AMERICA AND CARIBBEAN	15.62	9.87	6.74	6.48	6.46	6.83	7.21	7.71	8.24	8.80	9.30
Caribbean	15.46	9.47	8.24	8.17	8.23	8.96	9.27	9.65	10.06	10.68	11.16
Meso-America	17.81	10.03	5.58	5.36	5.30	5.83	6.20	6.68	7.25	7.86	8.57
South America	14.93	9.87	7.01	6.73	6.73	7.02	7.42	7.94	8.47	9.01	9.42
NORTH AMERICA	9.41	9.04	8.80	8.39	8.31	8.59	9.13	9.74	10.33	10.75	10.92
North America	9.41	9.04	8.80	8.39	8.31	8.59	9.13	9.74	10.33	10.75	10.92
POLAR	----	----	----	----	----	----	----	----	----	----	----
Antarctic	----	----	----	----	----	----	----	----	----	----	----
Arctic	----	----	----	----	----	----	----	----	----	----	----
WEST ASIA	24.74	15.28	7.63	6.69	5.58	4.27	4.26	4.35	4.47	4.64	4.86
Arabian Peninsula	28.74	18.31	7.37	6.36	5.69	4.28	4.15	4.08	4.02	4.02	4.06
Mashriq	21.88	13.21	7.86	6.99	5.47	4.26	4.39	4.69	5.08	5.52	6.04
GLOBAL TOTALS	19.70	11.59	9.40	9.05	8.88	8.37	8.51	8.75	9.05	9.41	9.73
REPORTED GLOBAL TOTALS FROM ORIGINAL DATA SOURCES	19.70	11.59	9.41	9.05	8.88	8.36	8.51	8.75	9.05	9.40	9.72

Comments:

Assumptions about the future paths of fertility and mortality:

Medium-fertility assumptions.

General-mortality assumption.

AGGREGATIONS

Fertility (Projection)

Units: number of children per woman

Data Source: World Population Prospects: The 2000 Revision

Data Provider: United Nations Population Division/Department of Economic and Social Affairs

Years: 1950-1955, 1955-1960, ..., 2045-2050

	1950-55	1970-75	1990-95	1995-00	2000-05	2020-25	2025-30	2030-35	2035-40	2040-45	2045-50
AFRICA	6.72	6.69	5.67	5.36	5.08	3.71	3.38	3.06	2.81	2.59	2.38
Central Africa	5.82	6.27	6.41	6.30	6.22	4.56	4.09	3.62	3.17	2.74	2.32
Eastern Africa	7.15	7.20	6.55	6.32	6.20	4.62	4.23	3.86	3.48	3.09	2.72
Northern Africa	6.80	6.36	4.09	3.59	3.15	2.23	2.14	2.10	2.10	2.10	2.10
Southern Africa	6.51	6.42	5.32	5.09	4.78	3.50	3.21	2.92	2.69	2.51	2.34
Western Africa	6.85	7.04	6.36	5.96	5.59	3.80	3.34	2.92	2.65	2.50	2.37
Western Indian Ocean	6.76	6.15	5.74	5.65	5.29	3.81	3.43	3.04	2.66	2.28	2.09
ASIA AND PACIFIC	5.87	5.07	2.96	2.69	2.52	2.12	2.09	2.06	2.04	2.04	2.04
Australia and New Zealand	3.28	2.59	1.90	1.80	1.79	1.85	1.88	1.92	1.96	1.99	2.03
Central Asia	5.23	5.10	3.34	2.78	2.34	2.06	2.06	2.06	2.06	2.06	2.06
North West Pacific and East Asia	5.73	4.53	1.88	1.76	1.76	1.87	1.88	1.89	1.89	1.90	1.90
South Asia	6.12	5.64	4.04	3.63	3.31	2.35	2.26	2.18	2.14	2.13	2.12
South East Asia	5.97	5.55	3.27	2.85	2.53	2.10	2.09	2.08	2.08	2.08	2.08
South Pacific	6.30	5.74	4.70	4.35	4.10	3.07	2.82	2.58	2.33	2.10	2.10
EUROPE	2.83	2.35	1.71	1.51	1.42	1.57	1.64	1.70	1.76	1.82	1.85
Central Europe	3.85	3.03	2.17	1.85	1.66	1.78	1.84	1.91	1.97	2.02	2.04
Eastern Europe	2.90	2.17	1.60	1.28	1.16	1.38	1.47	1.56	1.65	1.73	1.76
Western Europe	2.41	2.16	1.55	1.48	1.45	1.56	1.61	1.66	1.71	1.76	1.78
LATIN AMERICA AND CARIBBEAN	5.94	5.07	2.99	2.72	2.52	2.16	2.13	2.11	2.10	2.09	2.10
Caribbean	5.24	4.40	2.79	2.59	2.46	2.17	2.12	2.08	2.05	2.02	2.02
Meso-America	6.87	6.43	3.45	3.08	2.79	2.20	2.15	2.11	2.10	2.10	2.10
South America	5.74	4.67	2.84	2.59	2.42	2.14	2.12	2.11	2.10	2.10	2.10
NORTH AMERICA	3.47	2.01	2.02	2.00	1.90	1.93	1.98	2.03	2.06	2.07	2.08
North America	3.47	2.01	2.02	2.00	1.90	1.93	1.98	2.03	2.06	2.07	2.08
POLAR	----	----	----	----	----	----	----	----	----	----	----
Antarctic	----	----	----	----	----	----	----	----	----	----	----
Arctic	----	----	----	----	----	----	----	----	----	----	----
WEST ASIA	7.16	7.21	5.85	5.42	5.11	3.66	3.34	3.09	2.89	2.67	2.51
Arabian Peninsula	7.39	7.36	6.58	6.30	6.02	4.52	4.16	3.80	3.43	3.06	2.80
Mashriq	7.00	7.12	5.21	4.63	4.25	2.70	2.36	2.19	2.16	2.13	2.09
GLOBAL TOTALS	5.11	4.55	3.09	2.87	2.73	2.36	2.30	2.24	2.19	2.16	2.12
REPORTED GLOBAL TOTALS FROM ORIGINAL DATA SOURCES	5.01	4.48	3.01	2.82	2.68	2.39	2.34	2.28	2.24	2.20	2.15

Comments:

Fertility assumptions are described in terms of the following groups of countries:

High-fertility countries: countries that until 2000 have had no fertility reduction or only an incipient decline; fertility in high-fertility countries is generally assumed to decline at an average pace of nearly one child per decade starting in 2005 or later. Consequently, some of these countries do not reach replacement level by 2050.

Medium-fertility countries: countries where fertility has been declining but whose level is still above replacement level (2.1 children per woman); Fertility in medium-fertility countries is assumed to reach replacement level before 2050.

Low-fertility countries: countries with fertility at or below replacement level (2.1 children per woman) plus a few with levels very close to replacement levels that are expected to fall below replacement level in the near future. Fertility in low-fertility countries is generally assumed to remain below replacement level during most of the projection period. It will be reaching by 2045-2050 the fertility of the cohort of women born in the early 1960s or, if that information is lacking, reaching 1.7 children per woman if current fertility is below 1.5 children per woman or 1.9 children per woman if current fertility is equal to or higher than 1.5 children per woman.

Infant Mortality Rate

Units: deaths per thousand people

Data Source: World Population Prospects: The 2000 Revision

Data Provider: United Nations Population Division/Department of Economic and Social Affairs

Years: 1950-1955, 1955-1960, ..., 2045-2050

	1950-55	1970-75	1990-95	1995-00	2000-05	2020-25	2025-30	2030-35	2035-40	2040-45	2045-50
AFRICA	181.13	134.73	97.81	91.20	83.02	52.33	45.60	39.60	34.22	29.93	26.16
Central Africa	177.73	130.96	96.55	93.16	81.88	48.91	41.21	35.20	29.57	26.22	23.23
Eastern Africa	186.64	135.38	114.86	105.00	96.78	60.66	52.51	44.77	38.36	33.24	28.78
Northern Africa	188.06	137.03	67.19	57.65	49.00	26.19	22.12	19.11	16.78	14.89	13.50
Southern Africa	154.96	120.45	96.36	94.90	88.55	58.60	51.77	45.53	39.73	34.44	29.87
Western Africa	190.60	143.91	104.93	96.02	86.62	53.78	47.39	41.75	36.52	32.10	28.07
Western Indian Ocean	171.79	124.42	102.27	95.37	87.34	57.07	49.23	42.11	35.68	30.25	26.34
ASIA AND PACIFIC	181.45	98.40	65.29	59.79	53.87	33.28	30.34	26.97	24.47	22.19	20.41
Australia and New Zealand	24.16	16.50	6.77	5.61	5.35	4.53	4.36	4.18	4.00	3.83	3.67
Central Asia	100.68	65.68	45.88	45.50	41.27	27.99	24.98	22.25	19.94	17.97	16.10
North West Pacific and East Asia	181.54	56.44	43.26	38.56	33.90	20.80	18.61	16.56	14.69	13.56	12.46
South Asia	190.61	134.96	83.67	77.12	69.39	43.74	40.15	35.84	32.94	29.70	27.20
South East Asia	167.91	108.33	55.20	47.33	40.45	21.65	18.74	16.45	14.65	13.21	12.10
South Pacific	141.56	95.10	62.77	55.28	49.77	30.55	25.91	22.39	19.79	17.51	15.71
EUROPE	85.54	38.20	19.04	16.00	14.40	9.41	8.43	7.70	7.04	6.51	6.11
Central Europe	137.18	75.23	34.93	30.46	26.18	15.14	13.22	11.72	10.30	9.27	8.67
Eastern Europe	91.31	27.15	20.68	17.35	17.05	11.80	10.73	9.73	8.83	8.08	7.49
Western Europe	47.69	20.64	6.70	5.46	5.08	4.28	4.15	4.03	3.90	3.77	3.64
LATIN AMERICA AND CARIBBEAN	126.24	80.80	40.07	35.62	32.01	19.49	17.06	14.84	12.89	11.21	9.85
Caribbean	124.49	76.46	40.77	37.91	35.07	22.98	20.14	17.81	15.33	13.10	11.15
Meso-America	127.17	75.17	36.73	32.87	29.74	19.44	17.47	15.73	14.23	12.89	11.73
South America	126.13	83.93	41.52	36.67	32.74	19.17	16.59	14.18	12.10	10.35	8.96
NORTH AMERICA	28.55	17.94	8.35	7.44	6.72	5.17	5.02	4.86	4.71	4.56	4.41
North America	28.55	17.94	8.35	7.44	6.72	5.17	5.02	4.86	4.71	4.56	4.41
POLAR	----	----	----	----	----	----	----	----	----	----	----
Antarctic	----	----	----	----	----	----	----	----	----	----	----
Arctic	----	----	----	----	----	----	----	----	----	----	----
WEST ASIA	183.52	109.48	68.88	53.62	41.83	20.13	17.46	15.17	13.32	11.88	10.58
Arabian Peninsula	221.56	138.77	58.78	49.23	41.69	21.79	18.96	16.48	14.53	13.01	11.64
Mashriq	154.51	87.65	79.04	58.26	41.98	17.26	14.53	12.46	10.82	9.53	8.28
GLOBAL TOTALS	157.30	93.69	64.19	59.65	54.55	34.71	31.06	27.34	24.24	21.60	19.36
REPORTED GLOBAL TOTALS FROM ORIGINAL DATA SOURCES	157.24	93.65	64.17	59.63	54.53	34.69	31.04	27.32	24.22	21.59	19.35

AGGREGATIONS

Life Expectancy

Units: years

Data Source: World Population Prospects: The 2000 Revision

Data Provider: United Nations Population Division/Department of Economic and Social Affairs

Years: 1950-1955, 1955-1960, ..., 2045-2050

	1950-55	1970-75	1990-95	1995-00	2000-05	2020-25	2025-30	2030-35	2035-40	2040-45	2045-50
AFRICA	38.17	46.75	52.64	52.47	52.61	60.46	62.63	64.61	66.48	68.10	69.61
Central Africa	37.36	45.02	50.75	49.67	50.83	60.09	62.77	64.96	67.10	68.61	69.92
Eastern Africa	36.10	44.50	45.80	45.65	45.36	55.03	57.73	60.38	62.84	64.99	66.99
Northern Africa	41.92	51.40	62.98	65.08	66.86	72.32	73.50	74.53	75.45	76.30	77.01
Southern Africa	39.21	48.21	51.98	48.58	45.33	53.80	56.69	59.34	61.78	64.03	66.16
Western Africa	35.67	43.59	49.66	50.35	51.56	60.20	62.33	64.21	66.02	67.63	69.18
Western Indian Ocean	39.00	47.79	52.52	54.07	55.83	62.93	65.02	67.00	68.85	70.48	71.75
ASIA AND PACIFIC	42.18	57.33	64.73	66.32	67.82	72.91	73.85	74.77	75.59	76.30	76.94
Australia and New Zealand	69.57	71.69	77.35	78.43	78.96	80.80	81.22	81.63	82.04	82.44	82.85
Central Asia	56.05	63.85	66.60	66.60	67.88	72.84	73.94	74.98	75.91	76.74	77.56
North West Pacific and East Asia	43.96	64.20	69.49	70.79	72.09	76.28	77.07	77.85	78.62	79.12	79.60
South Asia	38.97	49.75	59.46	61.60	63.53	69.95	71.09	72.22	73.16	74.09	74.91
South East Asia	41.25	52.21	63.80	65.67	67.29	72.94	74.04	75.01	75.86	76.62	77.28
South Pacific	39.37	49.78	58.08	60.01	61.78	68.43	70.17	71.58	72.71	73.74	74.66
EUROPE	64.79	70.27	72.48	73.14	73.82	77.53	78.23	78.91	79.54	80.12	80.64
Central Europe	58.66	66.92	69.86	71.03	72.06	75.80	76.58	77.29	77.94	78.50	78.99
Eastern Europe	64.81	69.78	67.37	67.11	67.11	72.71	73.76	74.78	75.74	76.59	77.29
Western Europe	67.13	71.98	76.84	77.71	78.49	80.88	81.32	81.77	82.21	82.65	83.10
LATIN AMERICA AND CARIBBEAN	52.03	61.22	68.31	69.49	70.53	74.39	75.13	75.84	76.53	77.18	77.82
Caribbean	53.60	64.25	68.01	68.51	68.94	71.54	72.29	73.14	74.00	74.74	75.52
Meso-America	49.35	61.15	70.00	71.04	71.88	75.22	75.94	76.57	77.17	77.77	78.27
South America	52.67	60.86	67.70	69.00	70.18	74.35	75.09	75.80	76.51	77.16	77.85
NORTH AMERICA	68.89	71.67	75.86	76.72	77.66	80.69	81.08	81.47	81.87	82.26	82.66
North America	68.89	71.67	75.86	76.72	77.66	80.69	81.08	81.47	81.87	82.26	82.66
POLAR	----	----	----	----	----	----	----	----	----	----	----
Antarctic	----	----	----	----	----	----	----	----	----	----	----
Arctic	----	----	----	----	----	----	----	----	----	----	----
WEST ASIA	41.92	54.56	64.30	66.03	68.62	73.94	74.84	75.60	76.36	77.05	77.69
Arabian Peninsula	36.26	49.70	64.70	66.87	68.43	73.37	74.28	75.07	75.85	76.56	77.21
Mashriq	45.96	57.87	63.95	65.27	68.79	74.58	75.50	76.27	77.03	77.74	78.40
GLOBAL TOTALS	49.41	59.85	65.30	66.32	67.34	71.92	72.89	73.84	74.73	75.52	76.26
REPORTED GLOBAL TOTALS FROM ORIGINAL DATA SOURCES	46.52	58.03	63.94	64.98	65.99	71.26	72.35	73.42	74.40	75.25	76.04

Mortality due to Respiratory Infections

Units: number of people

Data Source: WHO Mortality Data Bank (data as of February 2001)

Data Provider: World Health Organization (WHO)

Years: 1990-1999

	1990	1991	1992	1993	1994	1995	1996	1997	1998	1999
AFRICA	----	----	----	----	----	----	----	----	----	----
Central Africa	----	----	----	----	----	----	----	----	----	----
Eastern Africa	----	----	----	----	----	----	----	----	----	----
Northern Africa	----	----	----	----	----	----	----	----	----	----
Southern Africa	----	----	----	----	----	----	----	----	----	----
Western Africa	----	----	----	----	----	----	----	----	----	----
Western Indian Ocean	----	----	----	----	----	----	----	----	----	----
ASIA AND PACIFIC	----	----	----	----	----	----	----	----	----	----
Australia and New Zealand	2887	2953	3195	2738	3003	2965	3194	6067	----	----
Central Asia	33398	----	----	----	----	----	----	----	----	----
North West Pacific and East Asia	----	----	----	----	----	----	----	----	----	----
South Asia	----	----	----	----	----	----	----	----	----	----
South East Asia	----	----	----	----	----	----	----	----	----	----
South Pacific	----	----	----	----	----	----	----	----	----	----
EUROPE	172667	162581	167604	215495	207913	218226	212858	193838	----	----
Central Europe	----	----	----	----	----	----	----	----	----	----
Eastern Europe	27935	28125	30150	39260	44478	45474	39725	35939	32429	----
Western Europe	121792	112986	110314	148600	135980	144577	142573	135540	----	----
LATIN AMERICA AND CARIBBEAN	91725	80514	81296	87216	87639	----	----	----	----	----
Caribbean	----	----	----	----	----	----	----	----	----	----
Meso-America	27278	23890	22894	22853	22649	22276	----	----	----	----
South America	59606	51804	53111	59562	59935	----	----	----	----	----
NORTH AMERICA	87249	85701	83363	90950	89733	91152	92034	95276	----	----
North America	87249	85701	83363	90950	89733	91152	92034	95276	----	----
POLAR	----	----	----	----	----	----	----	----	----	----
Antarctic	----	----	----	----	----	----	----	----	----	----
Arctic	----	----	----	----	----	----	----	----	----	----
WEST ASIA	----	----	----	----	----	----	----	----	----	----
Arabian Peninsula	----	----	----	----	----	----	----	----	----	----
Mashriq	----	----	----	----	----	----	----	----	----	----
GLOBAL TOTALS	----	----	----	----	----	----	----	----	----	----
REPORTED GLOBAL TOTALS FROM ORIGINAL DATA SOURCES	****	****	****	****	****	****	****	****	****	****

Comments:

Mortality due to respiratory infections: effects are generally irritation and reduced lung function with increased incidence of respiratory disease, especially in more susceptible members of the population such as young children, the elderly and asthmatics. Pollutants of concern include sulphur and nitrogen oxides, photochemical oxidants (for example ozone), particulates and mixtures of pollutants.

For more information see: http://www.who.int/whr/2001/archives/2000/en/statistics.htm.

AGGREGATIONS

Rural Population Water Comfort

The Assessment 2000 did not provide standardized definitions of urban and rural, as none could be found that would be consistent with the range of definitions adopted locally.

Units: percent

Data Source: Global Water Supply and Sanitation Assessment 2000 Report

Data Provider: World Health Organization
and United Nations Children's Fund Joint Monitoring Programme for Water Supply and Sanitation (JMP)

Years: 1990, 2000

Rural Population with Access to Improved Sanitation

	1990	2000
AFRICA	51.33	45.80
Central Africa	----	19.30
Eastern Africa	43.04	33.91
Northern Africa	60.41	66.59
Southern Africa	----	63.24
Western Africa	42.64	42.07
Western Indian Ocean	32.37	36.38
ASIA AND PACIFIC	22.17	30.30
Australia and New Zealand	100.00	100.00
Central Asia	----	99.46
North West Pacific and East Asia	18.00	24.54
South Asia	13.69	22.35
South East Asia	56.44	62.20
South Pacific	----	71.65
EUROPE	----	----
Central Europe	----	----
Eastern Europe	----	----
Western Europe	----	----
LATIN AMERICA AND CARIBBEAN	38.93	48.70
Caribbean	----	50.71
Meso-America	36.33	50.36
South America	41.85	47.24
NORTH AMERICA	99.91	99.90
North America	99.91	99.90
POLAR	----	----
Antarctic	----	----
Arctic	----	----
WEST ASIA	----	52.74
Arabian Peninsula	----	43.43
Mashriq	----	64.06
GLOBAL TOTALS	34.28	38.53
REPORTED GLOBAL TOTALS FROM ORIGINAL DATA SOURCES	35.00	38.00

Rural Population with Access to Water Supply

	1990	2000
AFRICA	43.84	48.68
Central Africa	----	29.13
Eastern Africa	24.20	28.68
Northern Africa	74.39	81.23
Southern Africa	----	55.83
Western Africa	40.00	48.80
Western Indian Ocean	37.25	37.23
ASIA AND PACIFIC	66.09	74.63
Australia and New Zealand	100.00	100.00
Central Asia	----	77.57
North West Pacific and East Asia	60.00	66.35
South Asia	74.85	85.51
South East Asia	59.40	64.76
South Pacific	----	41.87
EUROPE	----	----
Central Europe	----	----
Eastern Europe	----	96.56
Western Europe	----	----
LATIN AMERICA AND CARIBBEAN	55.62	61.76
Caribbean	----	61.72
Meso-America	62.12	70.15
South America	52.69	56.44
NORTH AMERICA	99.91	99.90
North America	99.91	99.90
POLAR	----	----
Antarctic	----	----
Arctic	----	----
WEST ASIA	----	62.08
Arabian Peninsula	----	63.19
Mashriq	----	60.73
GLOBAL TOTALS	65.26	71.52
REPORTED GLOBAL TOTALS FROM ORIGINAL DATA SOURCES	66.00	71.00

Comments:

The access to sanitation is defined in terms of the types of technology and levels of service afforded. This includes connection to a public sewer, connection to septic system, pour-flush latrine, simple pit latrine, ventilated improved pit latrine allowance was also made for other locally-defined technologies.

Access to sanitation, does not imply that the level of service is "adequate" or "safe"; these terms were replaced with "improved".

Comments:

The access to water supply is defined in terms of the types of technology and levels of service afforded. This included house connections, public standpipes, boreholes with handpumps, protected dug wells, protected springs and rainwater collection; allowance was also made for other locally-defined technologies. "Reasonable access" was broadly defined as the availability of at least 20 litters per person per day from a source within one kilometre of the user's dwelling.

Access to water, does not imply that the level of service or quality of water is "adequate" or "safe"; these terms were replaced with "improved".

Total Population Water Comfort

The Assessment 2000 did not provide standardized definitions of urban and rural, as none could be found that would be consistent with the range of definitions adopted locally.

Units: percent

Data Source: Global Water Supply and Sanitation Assessment 2000 Report

Data Provider: World Health Organization
and United Nations Children's Fund Joint Monitoring Programme for Water Supply and Sanitation (JMP)

Years: 1990, 2000

Total Population with Access to Improved Sanitation

	1990	2000
AFRICA	62.09	60.66
Central Africa	----	35.16
Eastern Africa	48.40	42.67
Northern Africa	75.30	80.72
Southern Africa	----	74.67
Western Africa	53.24	57.64
Western Indian Ocean	43.41	47.93
ASIA AND PACIFIC	34.19	45.45
Australia and New Zealand	100.00	100.00
Central Asia	----	99.65
North West Pacific and East Asia	29.00	39.85
South Asia	27.68	38.96
South East Asia	63.81	70.34
South Pacific	----	75.37
EUROPE	----	----
Central Europe	----	----
Eastern Europe	----	----
Western Europe	----	----
LATIN AMERICA AND CARIBBEAN	70.68	77.66
Caribbean	----	70.81
Meso-America	70.01	76.16
South America	74.03	78.99
NORTH AMERICA	100.00	100.00
North America	100.00	100.00
POLAR	----	----
Antarctic	----	----
Arctic	----	----
WEST ASIA	----	80.88
Arabian Peninsula	----	75.03
Mashriq	----	86.28
GLOBAL TOTALS	54.07	60.31
REPORTED GLOBAL TOTALS FROM ORIGINAL DATA SOURCES	55.00	60.00

Total Population with Access to Water Supply

	1990	2000
AFRICA	56.77	61.01
Central Africa	----	47.61
Eastern Africa	33.13	36.66
Northern Africa	82.56	87.93
Southern Africa	----	67.05
Western Africa	52.50	57.30
Western Indian Ocean	50.08	52.39
ASIA AND PACIFIC	74.21	80.67
Australia and New Zealand	100.00	100.00
Central Asia	----	86.25
North West Pacific and East Asia	71.00	75.97
South Asia	79.84	87.85
South East Asia	68.29	73.41
South Pacific	----	50.91
EUROPE	----	----
Central Europe	----	----
Eastern Europe	----	99.09
Western Europe	----	----
LATIN AMERICA AND CARIBBEAN	80.35	84.86
Caribbean	----	76.82
Meso-America	82.12	86.24
South America	81.70	85.21
NORTH AMERICA	100.00	100.00
North America	100.00	100.00
POLAR	----	----
Antarctic	----	----
Arctic	----	----
WEST ASIA	----	82.87
Arabian Peninsula	----	79.98
Mashriq	----	85.54
GLOBAL TOTALS	77.25	81.48
REPORTED GLOBAL TALS FROM ORIGINAL DATA SOURCES	79.00	81.00

Comments:

The access to sanitation is defined in terms of the types of technology and levels of service afforded. This includes connection to a public sewer, connection to septic system, pour-flush latrine, simple pit latrine, ventilated improved pit latrine allowance was also made for other locally-defined technologies.

Access to sanitation, does not imply that the level of service is "adequate" or "safe"; these terms were replaced with "improved".

Under-five Mortality Rate (Projection)

Units: deaths per thousand births

Data Source: World Population Prospects: The 2000 Revision

Data Provider: United Nations Population Division/Department of Economic and Social Affairs

Years: 1950-1955, 1955-1960, ..., 2045-2050

	1990-95	1995-00	2000-05	2005-10	2010-15	2015-20	2020-25	2025-30	2030-35	2035-40	2040-45	2045-50
AFRICA	161.53	151.60	138.31	124.25	109.57	95.77	83.29	70.98	60.28	50.77	43.34	37.10
Central Africa	161.51	157.01	136.96	121.65	105.26	89.80	76.40	62.34	51.55	41.86	36.57	32.12
Eastern Africa	197.94	181.08	165.89	148.26	130.32	114.34	99.28	84.12	70.06	58.30	49.08	41.41
Northern Africa	93.79	78.46	65.45	55.31	47.56	39.78	33.25	27.18	22.84	19.53	16.96	15.19
Southern Africa	156.98	158.25	152.39	139.49	124.42	110.43	97.62	84.82	73.53	63.11	53.55	45.33
Western Africa	178.26	163.47	146.22	129.56	113.10	98.40	85.49	73.58	63.42	54.16	46.47	39.94
Western Indian Ocean	166.82	154.15	139.53	125.44	111.83	98.74	86.14	72.53	60.18	49.05	39.65	33.22
ASIA AND PACIFIC	90.43	81.02	71.83	62.51	53.79	47.84	42.49	38.04	33.53	29.88	26.93	24.39
Australia and New Zealand	8.26	6.95	6.71	6.42	6.17	5.94	5.74	5.55	5.35	5.13	4.93	4.74
Central Asia	62.41	62.20	56.14	50.01	44.86	40.46	36.72	32.45	28.64	25.47	22.81	20.31
North West Pacific and East Asia	51.08	44.86	38.73	33.93	29.90	26.43	23.15	20.60	18.24	16.12	14.85	13.60
South Asia	121.40	108.70	95.90	84.01	73.03	65.33	57.83	51.86	45.89	41.25	36.94	33.17
South East Asia	76.87	64.57	54.09	45.03	37.54	31.75	26.83	22.90	19.87	17.51	15.66	14.28
South Pacific	87.75	76.18	67.50	59.53	52.18	45.27	38.67	32.16	27.35	23.85	20.89	18.55
EUROPE	24.58	20.40	18.14	16.26	14.48	13.05	11.70	10.46	9.54	8.73	8.06	7.56
Central Europe	45.91	39.28	33.08	27.92	23.55	20.70	18.10	15.62	13.79	12.07	10.81	10.08
Eastern Europe	26.10	21.88	21.40	20.10	18.40	16.77	15.47	14.15	12.87	11.71	10.72	9.93
Western Europe	8.42	6.78	6.37	6.14	5.90	5.66	5.48	5.32	5.16	5.01	4.84	4.68
LATIN AMERICA AND CARIBBEAN	50.33	44.84	40.30	35.84	31.85	28.11	24.81	22.13	19.43	17.05	14.99	13.28
Caribbean	62.67	59.07	55.06	50.53	45.49	40.49	35.95	32.41	28.69	25.01	21.74	19.00
Meso-America	46.23	41.41	37.54	33.89	30.54	27.49	24.76	22.28	20.10	18.20	16.49	15.01
South America	50.91	44.97	40.07	35.22	31.05	27.14	23.74	21.08	18.28	15.84	13.76	12.05
NORTH AMERICA	10.11	8.96	8.12	7.33	6.96	6.58	6.29	6.11	5.93	5.75	5.57	5.38
North America	10.11	8.96	8.12	7.33	6.96	6.58	6.29	6.11	5.93	5.75	5.57	5.38
POLAR	----	----	----	----	----	----	----	----	----	----	----	----
Antarctic	----	----	----	----	----	----	----	----	----	----	----	----
Arctic	----	----	----	----	----	----	----	----	----	----	----	----
WEST ASIA	87.58	67.43	52.99	40.01	33.42	28.10	23.89	20.54	17.74	15.51	13.76	12.23
Arabian Peninsula	80.84	66.21	54.91	45.41	37.41	30.98	25.94	22.26	19.09	16.65	14.78	13.14
Mashriq	94.36	68.73	50.83	33.48	28.09	23.74	20.34	17.18	14.94	13.15	11.66	10.27
GLOBAL TOTALS	93.33	86.43	78.94	70.50	61.99	55.13	48.83	42.91	37.22	32.30	28.35	24.96
REPORTED GLOBAL TOTALS FROM ORIGINAL DATA SOURCES	93.31	86.40	78.91	70.48	61.97	55.11	48.81	42.89	37.20	32.29	28.34	24.94

AGGREGATIONS

1.1.5 Infrastructure

Internet Hosts per 10000 People

Units: number of computers

Data Source: Internet Software Consortium (http://www.isc.org)

Data Provider: The World Bank

Years: 1994-1999

	1994	1995	1996	1997	1998	1999
AFRICA	0.41	0.72	1.47	1.71	2.00	1.99
Central Africa	0.00	0.00	0.00	0.01	0.01	0.03
Eastern Africa	0.00	0.01	0.02	0.04	0.06	0.06
Northern Africa	0.01	0.06	0.16	0.18	0.16	0.16
Southern Africa	2.12	3.68	7.40	8.58	10.11	10.06
Western Africa	0.00	0.00	0.03	0.05	0.05	0.06
Western Indian Ocean	0.00	0.13	0.09	0.17	0.24	0.43
ASIA AND PACIFIC	0.98	2.13	4.54	6.30	7.76	10.85
Australia and New Zealand	89.70	167.27	272.21	387.15	411.51	478.39
Central Asia	0.00	0.04	0.20	0.27	0.40	0.87
North West Pacific and East Asia	0.82	2.12	5.82	7.82	10.77	16.57
South Asia	0.00	0.01	0.03	0.05	0.10	0.16
South East Asia	0.19	0.73	1.55	2.57	2.83	4.10
South Pacific	0.01	0.18	0.67	1.21	3.27	6.11
EUROPE	12.73	27.23	45.15	54.86	79.82	104.63
Central Europe	2.02	4.76	10.35	11.18	19.06	27.56
Eastern Europe	0.31	1.07	2.84	4.05	6.49	9.56
Western Europe	25.37	53.72	87.13	106.06	152.40	197.65
LATIN AMERICA AND CARIBBEAN	0.43	1.18	3.38	3.48	7.65	14.85
Caribbean	0.05	0.25	0.92	0.40	2.19	3.97
Meso-America	0.63	1.28	2.85	3.34	6.98	17.67
South America	0.40	1.25	3.86	3.87	8.52	14.95
NORTH AMERICA	115.95	218.65	360.00	415.46	897.90	1376.26
North America	115.95	218.65	360.00	415.46	897.90	1376.26
POLAR	----	----	----	----	----	386.61
Antarctic	----	----	----	----	----	----
Arctic	----	----	----	----	----	386.61
WEST ASIA	0.03	0.23	0.71	1.37	2.23	2.31
Arabian Peninsula	0.07	0.47	1.35	----	4.32	4.13
Mashriq	0.00	0.02	0.14	0.24	0.33	0.64
GLOBAL TOTALS	8.63	16.85	28.42	33.62	63.34	93.75
REPORTED GLOBAL TOTALS FROM ORIGINAL DATA SOURCES	8.61	16.81	28.91	34.79	63.09	94.44

Comments:

Internet hosts are computers connected directly to the worldwide network, each allowing many computer users to access the Internet. Hosts are assigned to countries on the basis of the host's country code, though this does not necessarily indicate that the host is physically located in that country. All hosts lacking a country code identification are assigned to the United States. The Internet Software Consortium changed the methods used in its Internet domain survey beginning in July 1998. The new survey is believed to be more reliable and to avoid the undercounting that occurs when organizations restrict download access to their domain data. Nevertheless, some measurement problems remain, so the number of Internet hosts shown for each country should be considered an approximation. In particular, most hosts are now under generic top-level domains (for example, .com, .net, and .org), which, unlike country code top-level domains (.de, .uk), have never had a geographic designation. For detailed analysis of Internet trends by country, it is best to use the original source data.

Internet Users

Units: number of people

Data Source: International Telecommunication Union's (ITU) World Telecommunication Development Report 2000
and Challenges to the Network: Internet for Development (1999)

Data Provider: The World Bank

Years: 1990-1999

	1990	1991	1992	1993	1994	1995	1996	1997	1998	1999
AFRICA	----	----	----	----	----	----	692325	953375	1682905	2664550
Central Africa	----	----	----	----	----	----	----	2200	5705	27000
Eastern Africa	----	----	----	----	----	910	4800	16325	38000	76450
Northern Africa	----	----	----	----	4750	22500	44550	71700	154000	312000
Southern Africa	----	----	----	----	----	----	625116	817650	1310200	1939000
Western Africa	----	----	----	----	----	----	14550	37000	139800	241300
Western Indian Ocean	----	----	----	----	----	----	3100	8500	35200	68800
ASIA AND PACIFIC	----	----	----	----	1934044	3817250	8810959	18914946	30768676	64037500
Australia and New Zealand	----	190000	320000	372500	515000	680000	900000	2150000	3600000	6700000
Central Asia	----	----	----	----	----	----	----	----	28500	91500
North West Pacific and East Asia	----	77000	213000	692000	1322150	2627350	6694455	14317600	23080400	49296000
South Asia	----	----	----	----	10750	253960	475575	804600	1603400	3133500
South East Asia	----	----	----	----	----	----	730100	1618700	2432500	4781500
South Pacific	----	----	----	----	----	----	4829	11546	23876	35000
EUROPE	----	----	----	2028850	4135446	7562410	13807200	25348000	44665200	74505000
Central Europe	----	----	----	160250	435950	700150	1378000	2480300	4786200	7397000
Eastern Europe	----	----	----	20400	87496	244910	458700	814700	1380500	3033000
Western Europe	----	860300	1405600	1848200	3612000	6617350	11970500	22053000	38498500	64075000
LATIN AMERICA AND CARIBBEAN	----	----	----	----	215100	494676	1580482	2883130	5969470	9741600
Caribbean	----	----	----	----	----	24756	69272	138640	275230	432600
Meso-America	----	----	15036	47700	90300	159770	318710	716000	1563000	2174000
South America	----	----	27550	66800	121900	310150	1192500	2028490	4131240	7135000
NORTH AMERICA	----	3160000	4760000	5840000	9190000	21220000	32000000	44500000	67500000	85100000
North America	----	3160000	4760000	5840000	9190000	21220000	32000000	44500000	67500000	85100000
POLAR	----	----	----	----	36	30	1000	----	----	----
Antarctic	----	----	----	----	----	----	----	----	----	----
Arctic	----	----	----	----	36	30	1000	----	----	----
WEST ASIA	----	----	----	----	----	----	----	256900	514800	1254000
Arabian Peninsula	----	----	----	----	----	----	39770	179500	344000	914000
Mashriq	----	----	----	----	----	----	----	----	----	----
GLOBAL TOTALS	----	----	----	----	15812786	33595396	56938736	92856351	151101051	237302650
REPORTED GLOBAL TOTALS FROM ORIGINAL DATA SOURCES	45000	4324650	6894796	9327200	15992786	33845396	57542336	94352280	154119984	241863648

Comments:

Internet users are people with access to the worldwide network.

Data on Internet users are based on reported estimates, derived from reported counts of Internet service provider (ISP) subscribers, or calculated by multiplying the number of hosts by an estimated multiplier.

AGGREGATIONS

Mobile Phones Subscribers per 1000 People

Units: number of subscribers

Data Source: International Telecommunication Union's (ITU) World Telecommunication Development Report 2000
and Challenges to the Network: Internet for Development (1999)

Data Provider: The World Bank

Years: 1990-1999

	1980	1990	1991	1992	1993	1994	1995	1996	1997	1998	1999
AFRICA	0.00	0.02	0.03	0.05	0.12	0.57	0.89	1.54	2.70	4.65	----
Central Africa	0.00	0.00	0.00	0.00	0.02	0.06	0.09	0.17	0.21	0.19	----
Eastern Africa	0.00	0.00	0.00	0.01	0.01	0.02	0.04	0.06	0.10	0.35	0.77
Northern Africa	0.00	0.05	0.08	0.10	0.14	0.17	0.29	0.44	1.10	1.76	6.40
Southern Africa	0.00	0.05	0.06	0.10	0.31	2.53	3.94	6.83	11.55	19.68	38.20
Western Africa	0.00	0.00	0.00	0.00	0.06	0.09	0.11	0.25	0.49	1.11	----
Western Indian Ocean	0.00	0.16	0.18	0.20	0.27	0.39	0.84	1.50	2.96	4.51	7.05
ASIA AND PACIFIC	0.00	0.47	0.77	1.10	1.63	3.16	7.05	14.21	21.47	29.82	42.48
Australia and New Zealand	0.00	11.67	17.52	28.46	39.28	67.92	119.90	204.06	231.27	272.61	346.80
Central Asia	0.00	0.00	0.00	0.00	0.01	0.02	0.16	0.36	0.57	1.11	1.75
North West Pacific and East Asia	0.00	0.72	1.17	1.58	2.33	4.89	11.93	25.72	40.42	58.56	84.11
South Asia	0.00	0.00	0.01	0.01	0.02	0.05	0.15	0.40	1.03	1.50	2.19
South East Asia	0.00	0.51	0.88	1.46	2.37	3.86	7.02	11.01	15.01	16.60	24.50
South Pacific	0.00	0.00	0.14	0.32	0.56	1.16	1.97	3.18	4.69	7.82	14.59
EUROPE	0.03	4.36	5.94	7.80	11.56	18.86	30.18	47.92	76.89	131.36	223.15
Central Europe	0.00	0.20	0.35	0.62	1.11	2.58	5.41	10.96	25.63	51.69	108.26
Eastern Europe	0.00	0.00	0.00	0.03	0.04	0.15	0.50	1.22	2.74	4.43	8.39
Western Europe	0.06	9.04	12.27	15.99	23.57	37.97	59.88	93.53	145.48	244.42	403.98
LATIN AMERICA AND CARIBBEAN	0.00	0.29	0.66	1.41	2.46	4.83	8.21	13.63	25.26	42.99	82.21
Caribbean	0.00	0.82	1.41	2.49	4.14	7.18	12.18	14.43	18.00	28.52	43.01
Meso-America	0.00	0.58	1.38	2.65	3.23	4.77	6.00	8.98	15.08	29.24	67.46
South America	0.00	0.12	0.30	0.82	1.98	4.59	8.62	15.34	30.00	49.94	92.28
NORTH AMERICA	0.00	21.16	29.74	42.50	60.44	89.59	124.54	161.03	200.05	246.83	303.20
North America	0.00	21.16	29.74	42.50	60.44	89.59	124.54	161.03	200.05	246.83	303.20
POLAR	0.00	0.00	0.00	3.10	7.91	17.31	36.67	73.57	115.71	158.93	240.64
Antarctic	----	----	----	----	----	----	----	----	----	----	----
Arctic	0.00	0.00	0.00	3.10	7.91	17.31	36.67	73.57	115.71	158.93	240.64
WEST ASIA	0.00	1.33	1.91	2.09	2.54	3.05	5.62	9.90	16.39	24.67	31.23
Arabian Peninsula	0.00	2.78	4.01	4.40	5.36	6.46	8.27	15.21	23.76	37.28	47.65
Mashriq	0.00	0.03	0.03	0.03	0.03	0.03	3.26	5.17	9.78	13.23	16.18
GLOBAL TOTALS	0.01	2.12	3.01	4.21	6.11	9.83	15.87	24.99	36.45	53.01	79.85
REPORTED GLOBAL TOTALS FROM ORIGINAL DATA SOURCES	0.01	2.14	3.06	4.32	6.26	10.03	16.06	25.31	36.99	54.36	85.60

Comments:

Mobile phones refers to users of portable telephones subscribing to an automatic public mobile telephone service using cellular technology that provides access to the public switched telephone network.

Personal Computers per 1000 People

Units: number of personal computers

Data Source: International Telecommunication Union's (ITU) World Telecommunication Development Report 2000 and Challenges to the Network: Internet for Development (1999)

Data Provider: The World Bank

Years: 1980-1999

	1980	1990	1991	1992	1993	1994	1995	1996	1997	1998	1999
AFRICA	----	----	----	----	----	----	----	----	6.55	7.12	8.20
Central Africa	----	----	----	----	----	----	----	----	----	----	----
Eastern Africa	----	----	----	----	----	----	----	----	----	1.64	2.05
Northern Africa	----	----	----	----	----	2.63	3.28	4.39	5.26	7.10	9.20
Southern Africa	----	----	----	----	----	----	----	----	17.43	18.17	19.54
Western Africa	----	----	----	----	2.89	3.19	3.46	3.76	4.35	4.85	5.75
Western Indian Ocean	----	----	----	----	----	----	----	----	7.43	7.97	8.76
ASIA AND PACIFIC	----	4.69	5.42	6.25	7.25	8.86	11.33	14.63	18.24	21.85	25.86
Australia and New Zealand	----	137.44	149.63	173.03	196.91	230.01	266.03	304.91	345.85	391.39	445.87
Central Asia	----	----	----	----	----	----	----	----	----	----	----
North West Pacific and East Asia	----	7.14	8.09	9.00	10.29	12.55	16.22	21.84	28.05	34.22	41.74
South Asia	----	0.39	0.73	0.92	1.22	1.61	2.56	3.19	3.96	4.91	5.68
South East Asia	----	2.59	3.27	4.69	5.85	7.54	9.28	11.65	14.03	16.01	16.66
South Pacific	----	----	----	----	----	----	----	----	----	----	----
EUROPE	----	36.81	42.36	49.36	57.19	66.50	81.20	93.76	108.77	128.58	143.75
Central Europe	----	6.83	8.74	12.10	15.38	18.77	22.92	27.16	34.22	41.61	51.09
Eastern Europe	----	3.00	3.95	5.75	7.35	10.31	14.91	19.75	24.40	28.44	30.91
Western Europe	----	71.95	82.04	93.89	107.53	123.34	148.99	169.99	194.78	229.67	254.72
LATIN AMERICA AND CARIBBEAN	----	9.71	8.14	11.04	11.44	15.10	19.04	23.66	28.23	31.60	37.25
Caribbean	----	----	----	----	----	----	----	----	----	----	----
Meso-America	----	----	----	14.85	16.49	21.39	24.01	28.70	30.35	32.72	38.61
South America	----	4.74	6.31	8.30	10.26	13.66	18.36	23.19	28.90	32.41	38.10
NORTH AMERICA	----	206.03	222.88	242.41	261.99	287.05	317.26	352.84	393.60	443.63	495.82
North America	----	206.03	222.88	242.41	261.99	287.05	317.26	352.84	393.60	443.63	495.82
POLAR	----	----	----	----	----	----	107.33	----	----	----	----
Antarctic	----	----	----	----	----	----	----	----	----	----	----
Arctic	----	----	----	----	----	----	107.33	----	----	----	----
WEST ASIA	----	----	----	----	----	----	----	18.52	21.93	26.08	29.57
Arabian Peninsula	----	----	----	----	----	----	----	27.18	31.00	36.37	41.31
Mashriq	----	----	----	----	----	----	----	----	----	----	----
GLOBAL TOTALS	----	20.66	22.67	25.46	28.33	32.18	37.69	43.58	49.99	57.64	65.17
REPORTED GLOBAL TOTALS FROM ORIGINAL DATA SOURCES	----	25.01	27.91	30.56	32.95	36.43	41.92	48.20	53.64	60.79	68.35

Comments:

Personal computers are self-contained computers designed to be used by a single individual.

The estimates of personal computers are derived from an annual questionnaire, supplemented by other sources. In many countries mainframe computers are used extensively, and thousands of users can be connected to a single mainframe computer; thus the number of personal computers understates the total use of computers.

AGGREGATIONS

1.1.6 Population

Population

Units: thousand people
Data Source: World Population Prospects: The 2000 Revision
Data Provider: United Nations Population Division/Department of Economic and Social Affairs
Years: 1960-2000

	1960	1970	1980	1990	1995	1996	1997	1998	1999	2000
AFRICA	276686.05	356339.58	466871.48	619476.76	703487.39	720951.63	738674.77	756679.77	774991.33	793627.22
Central Africa	27017.36	34677.74	45122.22	61018.53	72395.65	74421.95	76320.05	78179.85	80131.37	82269.27
Eastern Africa	47818.94	62585.91	82491.88	111499.25	124947.42	128310.42	132000.87	135922.83	139928.31	143910.58
Northern Africa	67252.59	85964.97	110827.88	143166.22	158721.29	161783.10	164829.23	167885.02	170984.41	174149.60
Southern Africa	52524.14	68023.04	89757.96	118427.89	134963.99	138201.51	141380.27	144489.99	147523.08	150475.85
Western Africa	75426.69	96533.44	127700.74	171151.66	196207.73	201531.85	206975.28	212553.53	218285.00	224182.64
Western Indian Ocean	6646.33	8554.48	10970.80	14213.21	16251.31	16702.81	17169.07	17648.56	18139.16	18639.29
ASIA AND PACIFIC	1649602.24	2074513.27	2539402.35	3040544.38	3282974.45	3330003.83	3376675.91	3422933.10	3468677.08	3513848.34
Australia and New Zealand	12648.47	15354.93	17682.14	20248.33	21675.99	21940.68	22195.24	22441.06	22680.57	22915.65
Central Asia	24403.00	33177.50	41312.10	50622.90	53909.42	54508.35	55105.00	55692.65	56259.44	56798.10
North West Pacific and East Asia	792227.85	986776.77	1177958.20	1349961.21	1420185.31	1433023.39	1445643.99	1457954.02	1469797.46	1481075.13
South Asia	594952.76	750117.25	940304.26	1173908.03	1298627.36	1323651.14	1348666.25	1373708.29	1398831.20	1424070.36
South East Asia	222328.22	285242.14	357291.93	439722.59	481750.74	489901.27	497932.68	505849.48	513663.70	521384.20
South Pacific	3041.94	3844.68	4853.72	6081.31	6825.64	6979.01	7132.76	7287.60	7444.70	7604.90
EUROPE	645210.17	707764.20	756452.28	799311.25	812825.65	814444.46	815721.99	816679.50	817343.69	817741.84
Central Europe	133648.11	150674.99	169368.16	186803.18	191281.06	192182.50	193135.24	194108.74	195052.99	195931.65
Eastern Europe	183803.60	202740.20	216701.45	230986.70	231136.77	230539.04	229778.47	228866.95	227813.71	226632.11
Western Europe	327758.46	354349.01	370382.67	381521.37	390407.81	391722.92	392808.28	393703.81	394476.99	395178.08
LATIN AMERICA AND CARIBBEAN	218250.73	284802.42	361382.78	440412.48	479841.25	487690.25	495521.38	503331.17	511115.77	518871.75
Caribbean	20478.16	24906.04	29301.86	33966.41	36027.51	36427.33	36825.11	37220.80	37613.98	38004.39
Meso-America	49376.51	67503.48	89833.29	111409.34	123223.05	125615.85	128005.78	130389.50	132764.66	135129.03
South America	148396.06	192392.90	242247.64	295036.74	320590.68	325647.06	330690.50	335720.87	340737.12	345738.34
NORTH AMERICA	204072.25	231833.35	254928.59	282483.50	298104.84	301327.34	304563.48	307779.43	310932.71	313993.79
North America	204072.25	231833.35	254928.59	282483.50	298104.84	301327.34	304563.48	307779.43	310932.71	313993.79
POLAR	33.00	46.43	50.20	55.58	55.78	55.82	55.90	56.01	56.12	56.22
Antarctic	----	----	----	----	----	----	----	----	----	----
Arctic	33.00	46.43	50.20	55.58	55.78	55.82	55.90	56.01	56.12	56.22
WEST ASIA	25711.72	34898.74	49948.49	71653.55	83589.35	86226.88	88970.12	91807.74	94720.49	97694.24
Arabian Peninsula	10449.31	14097.30	21919.40	33874.89	39267.22	40614.85	42095.27	43677.10	45310.33	46958.69
Mashriq	15262.42	20801.44	28029.09	37778.66	44322.13	45612.02	46874.85	48130.64	49410.16	50735.55
GLOBAL TOTALS	3019566.17	3690197.98	4429036.17	5253937.49	5660878.69	5740700.21	5820183.56	5899266.72	5977837.19	6055833.42
REPORTED GLOBAL TOTALS FROM ORIGINAL DATA SOURCES	3020176.65	3690924.51	4429746.73	5254819.53	5661861.97	5741670.39	5821127.12	5900178.97	5978725.84	6056714.93

Population - Female

Units: thousand people

Data Source: World Population Prospects: The 2000 Revision

Data Provider: United Nations Population Division/Department of Economic and Social Affairs

Years: 1960-2000

	1960	1970	1980	1990	1995	1996	1997	1998	1999	2000
AFRICA	139827.42	179749.99	234634.95	310569.81	352402.00	361087.88	369897.58	378838.41	387917.34	397139.67
Central Africa	13962.08	17776.48	22970.79	30903.45	36562.97	37569.28	38511.19	39433.34	40400.24	41458.89
Eastern Africa	24137.99	31608.47	41664.71	56260.77	62975.98	64654.34	66496.29	68452.77	70448.25	72428.24
Northern Africa	33486.18	42892.85	54974.48	70772.55	78518.96	80044.84	81562.38	83083.43	84624.00	86194.63
Southern Africa	26642.73	34465.75	45484.26	59938.94	68232.93	69848.33	71430.80	72974.44	74474.56	75928.72
Western Africa	38258.38	48720.11	64044.18	85583.19	97975.38	100608.82	103301.04	106058.54	108888.93	111797.56
Western Indian Ocean	3340.05	4286.34	5496.53	7110.92	8135.77	8362.28	8595.88	8835.88	9081.36	9331.64
ASIA AND PACIFIC	805533.39	1012547.56	1238307.44	1482664.45	1603285.56	1626698.91	1649899.80	1672887.91	1695659.43	1718216.32
Australia and New Zealand	6261.95	7629.44	8858.45	10157.72	10907.54	11042.21	11169.30	11290.48	11408.44	11525.22
Central Asia	12623.32	17090.24	21162.74	25768.50	27367.72	27663.55	27960.47	28254.48	28538.73	28808.88
North West Pacific and East Asia	387234.62	482103.44	574228.85	657519.63	692545.28	699028.61	705412.26	711656.27	717694.49	723484.93
South Asia	286807.87	361449.07	453086.09	566656.04	628610.18	640961.91	653262.09	665551.04	677889.93	690321.04
South East Asia	111262.00	142564.52	178784.24	219885.75	240831.30	244907.08	248928.86	252897.74	256818.04	260692.97
South Pacific	1343.63	1710.85	2187.07	2676.81	3023.54	3095.55	3166.82	3237.90	3309.81	3383.28
EUROPE	339392.65	369096.76	392820.13	412854.98	419082.12	419821.41	420405.07	420841.79	421141.64	421315.44
Central Europe	68140.36	76529.65	85873.95	94492.21	96885.78	97358.59	97857.41	98367.07	98862.14	99324.08
Eastern Europe	101562.18	110185.65	116671.65	122835.95	122599.54	122285.13	121904.52	121459.88	120950.68	120379.43
Western Europe	169690.12	182381.47	190274.54	195526.83	199596.81	200177.69	200643.13	201014.83	201328.83	201611.93
LATIN AMERICA AND CARIBBEAN	108837.53	142116.06	180737.95	221157.59	241411.72	245439.80	249456.64	253461.16	257452.45	261429.56
Caribbean	9977.81	12168.79	14364.97	16718.77	17742.99	17943.63	18144.62	18345.52	18545.42	18743.68
Meso-America	24683.10	33674.99	44899.11	55951.50	61964.72	63182.80	64400.01	65614.70	66825.63	68031.65
South America	74176.62	96272.28	121473.87	148487.32	161704.02	164313.37	166912.01	169500.94	172081.40	174654.23
NORTH AMERICA	102859.40	117807.56	129628.49	143322.45	151139.06	152753.02	154372.68	155982.54	157563.51	159102.32
North America	102859.40	117807.56	129628.49	143322.45	151139.06	152753.02	154372.68	155982.54	157563.51	159102.32
POLAR	----	----	----	----	----	----	----	----	----	----
Antarctic	----	----	----	----	----	----	----	----	----	----
Arctic	----	----	----	----	----	----	----	----	----	----
WEST ASIA	12581.64	17108.64	24136.48	34225.02	40143.41	41437.38	42773.32	44149.43	45562.42	47009.82
Arabian Peninsula	5116.93	6912.60	10366.34	15628.97	18325.17	18983.15	19695.98	20452.22	21233.71	22027.31
Mashriq	7464.70	10196.04	13770.14	18596.05	21818.24	22454.24	23077.34	23697.21	24328.71	24982.51
GLOBAL TOTALS	1509032.02	1838426.57	2200265.44	2604794.30	2807463.87	2847238.39	2886805.09	2926161.23	2965296.80	3004213.12
REPORTED GLOBAL TOTALS FROM ORIGINAL DATA SOURCES	1509872.70	1839426.06	2201346.98	2606058.05	2808846.58	2848628.52	2888195.81	2927550.26	2966688.25	3005615.63

AGGREGATIONS

Population - Male

Units: thousand people

Data Source: World Population Prospects: The 2000 Revision

Data Provider: United Nations Population Division/Department of Economic and Social Affairs

Years: 1960-2000

	1960	1970	1980	1990	1995	1996	1997	1998	1999	2000
AFRICA	136747.97	176458.37	232074.14	308717.09	350878.28	359653.18	368563.17	377623.92	386853.09	396263.14
Central Africa	12991.28	16828.33	22057.62	30000.52	35706.90	36724.57	37678.39	38613.66	39595.87	40672.67
Eastern Africa	23680.95	30977.44	40827.18	55238.48	61971.44	63656.08	65504.58	67470.06	69480.07	71482.34
Northern Africa	33766.41	43072.12	55853.40	72393.67	80202.33	81738.26	83266.85	84801.59	86360.41	87954.97
Southern Africa	25876.40	33552.30	44268.38	58483.17	66725.02	68347.10	69943.33	71509.35	73042.28	74540.84
Western Africa	37168.30	47813.33	63656.56	85568.48	98232.35	100923.03	103674.24	106494.99	109396.07	112385.08
Western Indian Ocean	3264.62	4214.85	5411.00	7032.78	8040.24	8264.14	8495.78	8734.28	8978.40	9227.24
ASIA AND PACIFIC	843823.31	1061646.96	1300727.96	1557412.83	1679169.66	1702774.94	1726235.10	1749492.87	1772453.83	1795056.62
Australia and New Zealand	6386.53	7725.49	8823.69	10090.61	10768.45	10898.47	11025.95	11150.59	11272.13	11390.43
Central Asia	11779.68	16087.26	20149.36	24854.40	26541.70	26844.79	27144.53	27438.17	27720.71	27989.22
North West Pacific and East Asia	404993.23	504673.33	603729.36	692441.58	727640.04	733994.78	740231.72	746297.75	752102.97	757590.21
South Asia	308144.89	388668.18	487218.17	607251.99	670017.17	682689.23	695404.17	708157.24	720941.27	733749.32
South East Asia	111066.22	142677.62	178507.69	219836.85	240919.44	244994.19	249003.82	252951.74	256845.67	260691.24
South Pacific	1452.76	1815.08	2299.70	2937.41	3282.86	3353.48	3424.91	3497.38	3571.09	3646.22
EUROPE	305648.18	338466.14	363390.26	386179.62	393445.65	394319.85	395007.95	395522.69	395880.89	396099.12
Central Europe	65507.75	74145.34	83494.21	92310.98	94395.28	94823.91	95277.83	95741.66	96190.86	96607.58
Eastern Europe	82241.43	92554.56	100029.81	108150.76	108537.23	108253.91	107873.95	107407.07	106863.03	106252.68
Western Europe	157899.00	171766.24	179866.24	185717.89	190513.14	191242.02	191856.18	192373.96	192827.00	193238.87
LATIN AMERICA AND CARIBBEAN	108902.24	142108.58	180011.16	218576.96	237705.88	241518.73	245325.67	249123.93	252910.01	256681.04
Caribbean	9991.54	12161.53	14305.04	16571.74	17563.05	17754.19	17943.65	18131.46	18317.54	18501.87
Meso-America	24693.41	33828.49	44934.17	55457.84	61258.33	62433.05	63605.77	64774.81	65939.04	67097.37
South America	74217.29	96118.56	120771.95	146547.38	158884.49	161331.49	163776.25	166217.67	168653.43	171081.80
NORTH AMERICA	101207.85	114020.39	125294.10	139154.64	146959.18	148567.68	150184.10	151790.15	153362.41	154884.62
North America	101207.85	114020.39	125294.10	139154.64	146959.18	148567.68	150184.10	151790.15	153362.41	154884.62
POLAR	----	----	----	----	----	----	----	----	----	----
Antarctic	----	----	----	----	----	----	----	----	----	----
Arctic	----	----	----	----	----	----	----	----	----	----
WEST ASIA	13130.09	17790.09	25812.01	37428.53	43445.94	44789.49	46196.81	47658.31	49158.07	50684.42
Arabian Peninsula	5332.37	7184.70	11553.06	18245.93	20942.05	21631.71	22399.29	23224.88	24076.62	24931.38
Mashriq	7797.72	10605.40	14258.96	19182.61	22503.88	23157.79	23797.51	24433.43	25081.45	25753.04
GLOBAL TOTALS	1509459.63	1850490.52	2227309.63	2647469.68	2851604.58	2891623.87	2931512.80	2971211.87	3010618.29	3049668.98
REPORTED GLOBAL TOTALS FROM ORIGINAL DATA SOURCES	1510303.95	1851498.45	2228399.74	2648761.49	2853015.39	2893041.87	2932931.31	2972628.71	3012037.59	3051099.30

AGGREGATIONS

Population Density

Units: people per square kilometre

Data Source: World Population Prospects: The 2000 Revision

Data Provider: United Nations Population Division/Department of Economic and Social Affairs

Years: 1960-2000

	1960	1970	1980	1990	1991	1992	1993	1994	1995	1996	1997	1998	1999	2000
AFRICA	11.10	12.61	13.87	15.37	15.53	15.70	15.87	16.04	16.22	16.39	16.57	16.74	16.92	17.08
Central Africa	5.15	6.61	8.61	11.64	12.05	12.49	12.95	13.39	13.81	14.20	14.56	14.91	15.28	15.69
Eastern Africa	18.59	24.32	32.06	43.33	44.41	45.40	46.37	47.40	48.55	49.86	51.29	52.82	54.37	55.92
Northern Africa	8.02	10.26	13.22	17.08	17.46	17.83	18.20	18.57	18.94	19.30	19.67	20.03	20.40	20.78
Southern Africa	7.73	10.01	13.20	17.42	17.90	18.38	18.88	19.37	19.85	20.33	20.80	21.26	21.70	22.14
Western Africa	12.46	15.94	21.09	28.26	29.07	29.88	30.70	31.54	32.40	33.28	34.18	35.10	36.05	37.02
Western Indian Ocean	11.29	14.53	18.63	24.14	24.78	25.45	26.14	26.86	27.60	28.37	29.16	29.98	30.81	31.66
ASIA AND PACIFIC	10.81	14.11	17.91	21.82	22.21	22.61	23.00	23.39	23.78	24.17	24.55	24.94	25.33	25.71
Australia and New Zealand	1.59	1.93	2.22	2.55	2.58	2.62	2.66	2.69	2.73	2.76	2.79	2.82	2.85	2.88
Central Asia	6.26	8.51	10.59	12.97	13.17	13.35	13.51	13.66	13.81	13.96	14.12	14.27	14.41	14.55
North West Pacific and East Asia	68.80	85.72	102.34	117.28	118.66	119.93	121.12	122.26	123.39	124.50	125.60	126.67	127.71	128.69
South Asia	92.91	117.15	146.85	183.33	187.21	191.10	195.00	198.90	202.81	206.72	210.62	214.53	218.46	222.40
South East Asia	51.16	65.64	82.22	101.19	103.14	105.09	107.03	108.96	110.86	112.74	114.59	116.41	118.21	119.98
South Pacific	5.48	6.91	8.74	10.89	11.14	11.40	11.66	11.94	12.21	12.48	12.76	13.05	13.33	13.61
EUROPE	27.67	30.35	32.43	34.27	34.43	34.21	34.33	34.42	34.51	34.58	34.63	34.67	34.70	34.72
Central Europe	66.91	75.28	84.51	93.10	93.67	90.24	90.63	91.00	91.40	91.83	92.28	92.75	93.20	93.62
Eastern Europe	10.27	11.33	12.11	12.91	12.94	12.96	12.96	12.94	12.92	12.88	12.84	12.79	12.73	12.67
Western Europe	91.19	98.57	103.00	106.10	106.60	107.12	107.65	108.14	108.57	108.94	109.24	109.49	109.70	109.90
LATIN AMERICA AND CARIBBEAN	9.33	12.02	15.75	20.90	21.45	22.01	22.58	23.15	23.73	24.32	24.92	25.53	26.14	26.77
Caribbean	89.22	108.56	127.77	148.13	149.99	151.80	153.57	155.32	157.05	158.78	160.51	162.22	163.92	165.61
Meso-America	20.41	27.90	37.13	46.05	47.00	47.97	48.95	49.94	50.93	51.92	52.91	53.89	54.87	55.85
South America	8.47	10.98	13.82	16.83	17.13	17.42	17.71	18.00	18.29	18.58	18.87	19.15	19.44	19.72
NORTH AMERICA	47.56	59.82	73.23	87.68	89.14	90.56	91.94	93.32	94.68	96.04	97.38	98.72	100.04	101.34
North America	11.10	12.61	13.87	15.37	15.53	15.70	15.87	16.04	16.22	16.39	16.57	16.74	16.92	17.08
POLAR	6.63	9.09	13.04	18.70	19.30	19.90	20.51	21.13	21.78	22.47	23.17	23.91	24.66	25.43
Antarctic	0.00	0.00	0.00	0.00	0.00	0.00	0.00	0.00	0.00	0.00	0.00	0.00	0.00	0.00
Arctic	0.10	0.14	0.15	0.16	0.16	0.16	0.16	0.16	0.16	0.16	0.16	0.16	0.16	0.16
WEST ASIA	0.05	0.07	0.08	0.09	0.09	0.09	0.09	0.09	0.09	0.09	0.09	0.09	0.09	0.09
Arabian Peninsula	3.48	4.69	7.30	11.28	11.65	11.99	12.32	12.68	13.07	13.52	14.02	14.54	15.09	15.64
Mashriq	19.75	27.40	36.97	49.62	51.19	52.87	54.61	56.37	58.07	59.73	61.34	62.94	64.56	66.25
GLOBAL TOTALS	23.10	28.24	33.89	40.20	40.84	41.47	42.09	42.70	43.32	43.93	44.53	45.14	45.74	46.34
REPORTED GLOBAL TOTALS FROM ORIGINAL DATA SOURCES	23.10	23.10	23.10	23.10	23.10	23.10	23.10	23.10	23.10	23.10	23.10	23.10	23.10	23.10

Comments:

Population density is midyear population divided by land area in square kilometres. Population is based on the de facto defini-tion of population, which counts all residents regardless of legal status or citizenship--except for refugees not permanently set-tled in the country of asylum, who are generally considered part of the population of their country of origin.

Land area is a country's total area, excluding area under inland water bodies, national claims to continental shelf, and exclusive economic zones. In most cases the definition of inland water bodies includes major rivers and lakes (data source for land area: FAOSTAT, data as of July 2001).

AGGREGATIONS

Population Growth Rate

Units: percent

Data Source: World Population Prospects: The 2000 Revision

Data Provider: United Nations Population Division/Department of Economic and Social Affairs

Years: 1950-1955, 1955-1960, ..., 2045-2050

	1950-55	1970-75	1990-95	1995-00	2000-05	2020-25	2025-30	2030-35	2035-40	2040-45	2045-50
AFRICA	2.17	2.61	2.54	2.41	2.33	1.96	1.84	1.68	1.55	1.41	1.25
Central Africa	1.87	2.44	3.42	2.56	3.00	2.74	2.59	2.37	2.12	1.83	1.51
Eastern Africa	2.23	2.71	2.26	2.82	2.60	2.30	2.23	2.13	2.00	1.81	1.59
Northern Africa	2.27	2.43	2.06	1.85	1.81	1.15	1.01	0.88	0.77	0.67	0.57
Southern Africa	2.16	2.79	2.61	2.17	1.74	1.59	1.53	1.43	1.34	1.26	1.16
Western Africa	2.15	2.65	2.73	2.66	2.66	2.12	1.90	1.67	1.52	1.43	1.30
Western Indian Ocean	2.47	2.42	2.68	2.74	2.65	2.23	2.06	1.84	1.60	1.33	1.18
ASIA AND PACIFIC	1.90	2.22	1.53	1.36	1.21	0.78	0.65	0.54	0.44	0.35	0.26
Australia and New Zealand	2.26	2.02	1.36	1.11	0.95	0.63	0.55	0.46	0.40	0.36	0.35
Central Asia	2.95	2.38	1.25	1.04	0.85	0.82	0.69	0.60	0.52	0.43	0.33
North West Pacific and East Asia	1.75	2.11	1.01	0.83	0.66	0.29	0.15	0.05	-0.03	-0.15	-0.27
South Asia	2.00	2.32	2.02	1.84	1.71	1.14	1.01	0.88	0.75	0.66	0.58
South East Asia	2.08	2.34	1.83	1.58	1.38	0.92	0.80	0.68	0.57	0.47	0.38
South Pacific	1.88	2.34	2.31	2.16	2.10	1.66	1.50	1.32	1.11	0.91	0.85
EUROPE	1.08	0.72	0.33	0.12	-0.03	-0.18	-0.25	-0.31	-0.37	-0.41	-0.46
Central Europe	1.61	1.20	0.47	0.48	0.36	0.15	0.07	-0.01	-0.06	-0.10	-0.14
Eastern Europe	1.56	0.69	0.01	-0.39	-0.63	-0.63	-0.70	-0.74	-0.77	-0.80	-0.87
Western Europe	0.64	0.52	0.46	0.24	0.12	-0.13	-0.19	-0.27	-0.35	-0.41	-0.45
LATIN AMERICA AND CARIBBEAN	2.65	2.45	1.71	1.56	1.42	0.91	0.80	0.69	0.58	0.49	0.40
Caribbean	1.77	1.78	1.18	1.07	0.98	0.59	0.47	0.37	0.28	0.17	0.10
Meso-America	2.75	3.03	2.01	1.84	1.64	1.01	0.88	0.75	0.63	0.53	0.42
South America	2.75	2.32	1.66	1.51	1.38	0.91	0.80	0.69	0.59	0.50	0.43
NORTH AMERICA	1.70	0.97	1.08	1.04	0.88	0.73	0.65	0.57	0.50	0.46	0.45
North America	1.70	0.97	1.08	1.04	0.88	0.73	0.65	0.57	0.50	0.46	0.45
POLAR	3.21	1.31	0.07	0.16	0.16	0.18	0.18	0.19	0.19	0.20	0.20
Antarctic	----	----	----	----	----	----	----	----	----	----	----
Arctic	3.21	1.31	0.07	0.16	0.16	0.18	0.18	0.19	0.19	0.20	0.20
WEST ASIA	2.43	3.45	3.07	3.12	3.00	2.29	2.10	1.95	1.81	1.64	1.49
Arabian Peninsula	2.14	3.81	2.93	3.58	3.36	2.84	2.67	2.47	2.26	2.03	1.87
Mashriq	2.63	3.20	3.19	2.70	2.66	1.67	1.42	1.28	1.21	1.09	0.93
GLOBAL TOTALS	1.78	1.93	1.49	1.34	1.23	0.92	0.82	0.72	0.64	0.56	0.47
REPORTED GLOBAL TOTALS FROM ORIGINAL DATA SOURCES	1.79	1.93	1.49	1.35	1.23	0.92	0.82	0.73	0.64	0.56	0.47

AGGREGATIONS

1.1.7 Private Consumption

Household Final Consumption Expenditure - per Capita

Units: constant 1995 US$ per person

Data Source: World Development Indicators 2001

Data Provider: The World Bank

Years: 1960-1999

	1960	1970	1980	1990	1991	1992	1993	1994	1995	1996	1997	1998	1999
AFRICA	----	438.18	529.92	516.44	513.50	493.14	481.50	481.69	489.02	507.29	508.56	511.27	511.42
Central Africa	313.96	311.70	298.98	301.26	286.96	227.03	220.27	214.62	216.75	214.61	211.08	217.69	208.03
Eastern Africa	----	----	----	149.72	153.81	152.81	152.63	145.59	156.55	161.84	160.33	156.68	158.27
Northern Africa	----	481.09	714.60	786.35	797.11	797.10	786.76	815.63	808.97	843.26	837.74	864.22	867.98
Southern Africa	----	----	991.11	986.20	963.84	921.38	907.79	905.36	926.39	954.12	966.91	959.23	957.54
Western Africa	349.27	312.33	377.23	292.58	293.02	272.43	251.76	239.26	250.88	270.36	275.30	272.88	278.32
Western Indian Ocean	411.20	418.01	425.36	379.88	372.51	371.93	378.87	373.30	376.15	372.30	385.50	388.79	391.70
ASIA AND PACIFIC	----	846.77	1079.15	1347.56	1368.29	1392.98	1410.00	1435.23	1465.12	1514.65	1520.28	1486.97	572.55
Australia and New Zealand	6265.08	8046.94	9548.15	11042.65	11048.31	11101.49	11228.19	11668.03	12015.98	12197.92	12615.26	12985.36	----
Central Asia	----	----	----	----	----	----	----	----	----	----	----	----	----
North West Pacific and East Asia	744.08	1368.12	1831.73	2403.94	2454.49	2503.93	2530.44	2576.68	2636.26	2714.04	2719.73	2678.26	543.24
South Asia	----	191.14	205.87	255.97	256.90	264.13	269.34	276.01	283.35	300.83	304.54	299.53	316.23
South East Asia	----	----	483.66	685.92	709.25	736.57	778.97	821.65	857.90	936.58	961.19	879.95	891.26
South Pacific	----	----	----	----	----	----	----	----	----	----	----	----	----
EUROPE	----	----	----	6971.88	6958.17	7017.18	7016.46	7088.74	7195.91	7341.36	7497.02	7702.43	7879.43
Central Europe	----	----	----	1748.09	1655.12	1672.42	1756.76	1750.79	1805.65	1918.24	2004.98	2031.41	2026.88
Eastern Europe	----	----	----	----	----	1071.16	1130.12	1076.91	1065.78	1099.96	1163.52	1155.86	1119.72
Western Europe	5933.94	8301.68	10649.59	12954.05	13059.42	13217.39	13119.56	13283.58	13466.16	13675.22	13902.20	14304.07	14677.07
LATIN AMERICA AND CARIBBEAN	1270.73	1658.52	2233.00	2086.01	2124.32	2119.14	2195.94	2271.15	2290.20	2342.74	2410.99	2435.14	2368.49
Caribbean	----	----	----	----	----	----	----	----	----	----	----	----	----
Meso-America	1172.65	1526.41	1981.59	1896.85	1948.59	2002.42	2001.94	2052.46	1861.40	1869.70	1949.28	2016.75	2058.70
South America	1377.32	1819.81	2469.16	2280.61	2315.17	2281.69	2402.29	2497.50	2596.72	2666.83	2742.65	2748.11	2624.80
NORTH AMERICA	7872.72	10440.39	13147.83	16564.60	16347.46	16624.80	16982.33	17426.05	17744.92	18107.16	18562.33	19211.32	----
North America	7872.72	10440.39	13147.83	16564.60	16347.46	16624.80	16982.33	17426.05	17744.92	18107.16	18562.33	19211.32	----
POLAR	----	----	----	----	----	----	----	----	----	----	----	----	----
Antarctic	----	----	----	----	----	----	----	----	----	----	----	----	----
Arctic	----	----	----	----	----	----	----	----	----	----	----	----	----
WEST ASIA	----	----	----	----	----	----	----	----	----	----	----	----	----
Arabian Peninsula	----	----	----	----	----	----	----	----	----	----	----	----	----
Mashriq	----	----	----	----	----	----	----	----	----	----	----	----	----
GLOBAL TOTALS	----	2243.90	2645.43	2976.17	2963.44	2984.70	3004.75	3045.18	3083.40	3144.70	3185.54	3216.76	----
REPORTED GLOBAL TOTALS FROM ORIGINAL DATA SOURCES	1564.00	2144.00	2599.00	3015.00	3009.00	3035.00	3057.00	3102.00	3141.00	3202.00	3244.00	3276.00	----

Comments:

Household final consumption expenditure (private consumption) is the market value of all goods and services, including durable products (such as cars, washing machines, and home computers), purchased by households. It excludes purchases of dwellings but includes imputed rent for owner-occupied dwellings. It also includes payments and fees to governments to obtain permits and licenses. Here, household consumption expenditure includes the expenditures of non-profit institutions serving households, even when reported separately by the country.

Household final consumption expenditure per capita is calculated by UNEP/DEWA/GRID-Geneva using household final consumption expenditure from the World Bank and the total population from the World Population Prospects: The 2000 Revision.

Data sources: the national accounts indicators for most developing countries are collected from national statistical organizations and central banks by visiting and resident World Bank missions. Data for high-income economies come from Organisation for Economic Co-operation and Development (OECD) data files (see the OECD's National Accounts, 1988–1998, volumes 1 and 2). The United Nations Statistics Division publishes detailed national accounts for United Nations member countries in National Accounts Statistics: main aggregates and detailed tables and publishes updates in the Monthly Bulletin of Statistics.

Units: Million Constant 1995 US$

Data Source: World Development Indicators 2001

Data Provider: The World Bank

Years: 1960-1999

	1960	1970	1980	1990	1995	1996	1997	1998	1999
AFRICA	----	----	212255.25	280373.92	300399.15	319378.64	328213.94	337410.07	345583.26
Central Africa	6939.47	8966.66	11358.95	15739.36	13506.38	13745.46	13857.15	14628.90	14321.37
Eastern Africa	----	----	----	15081.23	17825.54	18931.73	19292.87	19403.12	20158.65
Northern Africa	----	33420.23	63113.11	89532.39	101769.97	108053.99	109296.53	114764.65	117300.23
Southern Africa	----	----	81400.32	106591.49	113746.36	119883.06	124205.94	125846.53	128165.83
Western Africa	20649.97	28888.46	46243.79	48259.53	47687.56	52797.47	55207.68	56176.93	58810.18
Western Indian Ocean	2218.31	2900.39	4424.28	5169.92	5863.34	5966.94	6353.77	6589.94	6827.01
ASIA AND PACIFIC	----	1647868.14	2635121.07	3989548.28	4677567.25	4899785.12	4984776.39	4929207.94	1597479.31
Australia and New Zealand	79243.67	123560.21	168831.63	223595.17	260458.18	267630.63	279998.73	291405.25	----
Central Asia	----	----	----	----	----	----	----	----	----
North West Pacific and East Asia	580131.64	1328266.04	2122692.82	3191919.53	3681277.76	3824053.89	3865814.00	3839336.71	716180.91
South Asia	----	115801.59	190177.73	296499.62	361970.32	391593.81	403858.14	404572.79	434923.38
South East Asia	----	----	151662.03	264207.77	361199.01	400836.43	417961.86	388597.59	399557.70
South Pacific	----	----	----	----	----	----	----	----	----
EUROPE	----	----	----	5515613.36	5813357.76	5941469.37	6076030.58	6249854.37	6399005.62
Central Europe	----	----	----	300074.22	318824.03	340341.39	357405.44	363774.58	364559.35
Eastern Europe	----	----	----	----	246340.04	253583.34	267351.90	264537.93	255087.79
Western Europe	1942041.36	2937508.45	3938394.61	4933996.22	5248193.69	5347544.63	5451273.24	5621541.86	5779358.48
LATIN AMERICA AND CARIBBEAN	249549.86	458862.94	791590.60	899683.59	1081795.23	1124475.83	1177143.06	1207177.07	1190912.42
Caribbean	----	----	----	----	----	----	----	----	----
Meso-America	56471.90	103038.20	178013.15	211327.27	229367.34	234864.35	249519.05	262963.06	273321.98
South America	188887.65	346745.68	597974.10	670010.35	832119.28	868058.44	906555.11	922167.85	893942.03
NORTH AMERICA	1606563.74	2420373.75	3351679.60	4679120.52	5289730.93	5456061.27	5653282.62	5912721.00	----
North America	1606563.74	2420373.75	3351679.60	4679120.52	5289730.93	5456061.27	5653282.62	5912721.00	----
POLAR	----	----	----	----	----	----	----	----	----
Antarctic	----	----	----	----	----	----	----	----	----
Arctic	----	----	----	----	----	----	----	----	----
WEST ASIA	----	----	----	----	----	----	----	----	----
Arabian Peninsula	----	----	----	----	----	----	----	----	----
Mashriq	----	----	----	----	----	----	----	----	----
GLOBAL TOTALS	----	----	11146939.37	15386709.01	17196012.59	17775679.00	18255175.65	18673787.30	----
REPORTED GLOBAL TOTALS FROM ORIGINAL DATA SOURCES	4722592.84	7879723.58	11513911.84	15836640.18	17761709.33	18366634.92	18872950.33	19317800.30	----

Comments:

Household final consumption expenditure (private consumption) is the market value of all goods and services, including durable products (such as cars, washing machines, and home computers), purchased by households. It excludes purchases of dwellings but includes imputed rent for owner-occupied dwellings. It also includes payments and fees to governments to obtain permits and licenses. Here, household consumption expenditure includes the expenditures of nonprofit institutions serving households, even when reported separately by the country.

Data sources: the national accounts indicators for most developing countries are collected from national statistical organizations and central banks by visiting and resident World Bank missions. Data for high-income economies come from Organisation for Economic Co-operation and Development (OECD) data files (see the OECD's National Accounts, 1988–1998, volumes 1 and 2). The United Nations Statistics Division publishes detailed national accounts for United Nations member countries in National Accounts Statistics: main aggregates and detailed tables and publishes updates in the Monthly Bulletin of Statistics.

1.1.8 Technological Hazards

According to the Centre for Research on the Epidemiology of Disasters (CRED) and the definition considered in "EM-DAT: the OFDA/CRED Disaster Database", a disaster is a situation or event, which overwhelms local capacity, necessitating a request to national or international level for external assistance; an unforeseen and often sudden event that causes great damage, destruction and human suffering. Though often caused by nature, disasters can have human origins. Wars and civil disturbances that destroy homelands and displace people are included among the causes of disasters. Other causes can be: building collapse, blizzard, drought, epidemic, earthquake, explosion, fire, flood, hazardous material or transportation incident (such as a chemical spill), hurricane, nuclear incident, tornado, or volcano.

Industrial accident: disaster type term used in EM-DAT to describe technological accidents of an industrial nature/ involving industrial buildings (e.g. factories). It comprises of a number of disaster subsets: "chemical spill/leak"; "explosions"; "radiation leakages"; "collapses"; "gas leaks" from industrial sites; "poisoning"; "fires"; and other technological accidents involving industrial sites.

- "Chemical spill": accident release occurring during the production, transportation or handling of hazardous chemical substances.

- "Explosions": explosions involving buildings or structures. Can either involve industrial structures or domestic/non-industrial structures.

- "Collapse": accident involving the collapse of building or structure. Can either involve industrial structures or domestic/non-industrial structures.

- "Poisoning": poisoning of atmosphere or water courses due to industrial sources.

- "Urban fire": urban fire involving buildings or structures. Can either involve industrial structures or domestic/non-industrial structures.

Industrial Accident - Affected People

Units: number of people

Data Source: EM-DAT: the OFDA/CRED International Disaster Database (data as of July 2001)

Data Provider: The OFDA/CRED International Disaster Database - www.cred.be/emdat - Université Catholique de Louvain - Brussels - Belgium

Years: 1980-2000

	1980	1990	1991	1992	1993	1994	1995	1996	1997	1998	1999	2000
AFRICA	----	----	----	----	----	----	----	----	----	----	----	0
Central Africa	----	----	----	----	----	----	----	----	----	----	----	----
Eastern Africa	----	----	----	----	----	----	----	----	----	----	----	----
Northern Africa	----	----	----	----	----	----	----	----	----	----	----	0
Southern Africa	----	----	----	----	----	----	----	----	----	----	----	0
Western Africa	----	----	----	----	----	----	----	----	----	----	----	0
Western Indian Ocean	----	----	----	----	----	----	----	----	----	----	----	----
ASIA AND PACIFIC	4500	2000	18644	600	1600	200	25500	2033	101000	0	500	14359
Australia and New Zealand	1500	2000	----	----	----	----	----	----	----	----	----	----
Central Asia	----	----	----	----	1600	----	----	----	----	----	----	----
North West Pacific and East Asia	----	0	3644	----	----	200	0	----	----	0	0	14159
South Asia	----	----	0	600	----	0	25000	----	101000	----	500	0
South East Asia	3000	----	15000	----	----	0	500	2033	----	----	0	200
South Pacific	----	----	----	----	----	----	----	----	----	----	----	----
EUROPE	10000	----	3000	150	----	5175	200	400	1000	46000	809	----
Central Europe	----	----	----	----	----	175	----	200	----	----	----	----
Eastern Europe	----	----	----	150	----	----	----	----	----	----	109	----
Western Europe	10000	----	3000	----	----	5000	200	200	1000	46000	700	----
LATIN AMERICA AND CARIBBEAN	----	----	30500	----	----	0	200	----	----	10000	1166	----
Caribbean	----	----	----	----	----	----	----	----	----	----	----	----
Meso-America	----	----	30500	----	----	----	----	----	----	10000	1166	----
South America	----	----	----	----	----	0	200	----	----	0	----	----
NORTH AMERICA	42100	4000	16000	----	----	9230	200	1200	6000	4000	----	----
North America	42100	4000	16000	----	----	9230	200	1200	6000	4000	----	----
POLAR	----	----	----	----	----	----	----	----	----	----	----	----
Antarctic	----	----	----	----	----	----	----	----	----	----	----	----
Arctic	----	----	----	----	----	----	----	----	----	----	----	----
WEST ASIA	----	----	----	----	----	----	----	----	----	----	0	----
Arabian Peninsula	----	----	----	----	----	----	----	----	----	----	----	----
Mashriq	----	----	----	----	----	----	----	----	----	----	0	----
GLOBAL TOTALS	56600	6000	68144	750	1600	14605	26100	3633	108000	60000	2475	14359
REPORTED GLOBAL TOTALS FROM ORIGINAL DATA SOURCES	56600	56600	56600	56600	56600	56600	56600	56600	56600	56600	56600	56600

Comments:

Affected: people requiring immediate assistance during a period of emergency, for example requiring basic survival needs such as food, water, shelter, sanitation and immediate medical assistance (definition considered in EM-DAT; included in the field "total affected").

Appearance of a significant number of cases of an infectious disease introduced in a region or a population that is usually free from that disease. (100 or more people affected).

Industrial Accident - Homeless People

Units: number of people

Data Source: EM-DAT: the OFDA/CRED International Disaster Database (data as of July 2001)

Data Provider: The OFDA/CRED International Disaster Database - www.cred.be/emdat -
Université Catholique de Louvain - Brussels - Belgium

Years: 1980-2000

	1980	1990	1991	1992	1993	1994	1995	1996	1997	1998	1999	2000
AFRICA	----	----	----	----	----	----	----	----	----	----	----	100
Central Africa	----	----	----	----	----	----	----	----	----	----	----	----
Eastern Africa	----	----	----	----	----	----	----	----	----	----	----	----
Northern Africa	----	----	----	----	----	----	----	----	----	----	----	0
Southern Africa	----	----	----	----	----	----	----	----	----	----	----	0
Western Africa	----	----	----	----	----	----	----	----	----	----	----	100
Western Indian Ocean	----	----	----	----	----	----	----	----	----	----	----	----
ASIA AND PACIFIC	----	0	5000	----	----	0	0	0	----	0	0	0
Australia and New Zealand	----	0	----	----	----	----	----	----	----	----	----	----
Central Asia	----	----	----	----	----	----	----	----	----	----	----	----
North West Pacific and East Asia	----	0	0	----	----	----	0	----	----	0	0	0
South Asia	----	----	0	----	----	0	----	----	----	----	0	0
South East Asia	----	----	5000	----	----	0	----	0	----	----	0	0
South Pacific	----	----	----	----	----	0	----	----	----	----	----	----
EUROPE	0	----	1400	----	----	3000	----	1	----	----	0	----
Central Europe	----	----	----	----	----	----	----	----	----	----	----	----
Eastern Europe	----	----	----	----	----	3000	----	----	----	----	0	----
Western Europe	0	----	1400	----	----	0	----	1	----	----	0	----
LATIN AMERICA AND CARIBBEAN	----	----	1600	15000	----	0	----	----	----	900	0	----
Caribbean	----	----	----	----	----	----	----	----	----	----	----	----
Meso-America	----	----	1600	15000	----	----	----	----	----	----	0	----
South America	----	----	----	----	----	0	----	----	----	900	----	----
NORTH AMERICA	----	----	----	----	----	300	----	----	----	----	----	----
North America	----	----	----	----	----	300	----	----	----	----	----	----
POLAR	----	----	----	----	----	----	----	----	----	----	----	----
Antarctic	----	----	----	----	----	----	----	----	----	----	----	----
Arctic	----	----	----	----	----	----	----	----	----	----	----	----
WEST ASIA	----	----	----	----	----	----	----	----	----	----	0	----
Arabian Peninsula	----	----	----	----	----	----	----	----	----	----	----	----
Mashriq	----	----	----	----	----	----	----	----	----	----	0	----
GLOBAL TOTALS	0	0	8000	15000	----	3300	0	1	----	900	0	100
REPORTED GLOBAL TOTALS FROM ORIGINAL DATA SOURCES	0	0	0	0	0	0	0	0	0	0	0	0

Comments:

Homeless: people needing immediate assistance in the form of shelter (included in the field "total affected").

They are always part of the affected population. Reporting from the field should give the number of individuals that are homeless; if only the number of families or houses is reported, the figure is multiplied by the average family size for the affected area (x5 for the developing countries, x3 for the industrialised countries, according to UNDP country list). (100 or more people homeless).

AGGREGATIONS

Units: number of people

Data Source: EM-DAT: the OFDA/CRED International Disaster Database (data as of July 2001)

Data Provider: The OFDA/CRED International Disaster Database - www.cred.be/emdat -
Université Catholique de Louvain - Brussels - Belgium

Years: 1980-2000

	1980	1990	1991	1992	1993	1994	1995	1996	1997	1998	1999	2000
AFRICA	----	----	233	400	----	----	----	----	2000	200	----	92
Central Africa	----	----	----	----	----	----	----	----	----	----	----	----
Eastern Africa	----	----	200	----	----	----	----	----	----	----	----	----
Northern Africa	----	----	----	----	----	----	----	----	2000	----	----	38
Southern Africa	----	----	----	----	----	----	----	----	----	----	----	0
Western Africa	----	----	33	400	----	----	----	----	----	200	----	54
Western Indian Ocean	----	----	----	----	----	----	----	----	----	----	----	----
ASIA AND PACIFIC	550	297	1179	295	2972	686	277	10547	297	1419	347	2060
Australia and New Zealand	----	0	----	----	----	----	----	----	----	----	----	----
Central Asia	----	----	----	----	----	----	----	----	----	600	----	----
North West Pacific and East Asia	----	100	726	51	1737	259	17	10370	217	763	83	1503
South Asia	----	100	260	124	633	342	210	160	80	56	157	458
South East Asia	550	97	193	120	602	25	50	17	----	----	107	99
South Pacific	----	----	----	----	----	60	----	----	----	----	----	----
EUROPE	12	110	7	64	37	144	287	210	933	185	10	----
Central Europe	----	----	----	64	3	----	222	----	----	137	----	----
Eastern Europe	----	110	----	----	----	17	----	130	233	48	10	----
Western Europe	12	----	7	----	34	127	65	80	700	----	0	----
LATIN AMERICA AND CARIBBEAN	----	----	344	1754	81	401	300	867	187	165	340	----
Caribbean	----	----	----	154	----	----	----	----	----	----	----	----
Meso-America	----	----	300	1600	----	----	----	797	----	45	340	----
South America	----	----	44	----	81	401	300	70	187	120	----	----
NORTH AMERICA	352	----	96	----	3000	137	105	92	8	12	----	----
North America	352	----	96	----	3000	137	105	92	8	12	----	----
POLAR	----	----	----	----	----	----	----	----	----	----	----	----
Antarctic	----	----	----	----	----	----	----	----	----	----	----	----
Arctic	----	----	----	----	----	----	----	----	----	----	----	----
WEST ASIA	----	----	----	----	100	----	----	100	----	----	20	----
Arabian Peninsula	----	----	----	----	100	----	----	100	----	----	----	----
Mashriq	----	----	----	----	----	----	----	----	----	----	20	----
GLOBAL TOTALS	914	407	1859	2513	6190	1368	969	11816	3425	1981	717	2152
REPORTED GLOBAL TOTALS FROM ORIGINAL DATA SOURCES	914	914	914	914	914	914	914	914	914	914	914	914

Comments:

Injured: people suffering from physical injuries, trauma or an illness requiring medical treatment as a direct result of a disaster (included in the field "total affected").

The number of injured is entered when the term "injured" is written in the source. Injured people are always part of the affected population. Any related word like "hospitalised" is considered as injured. If there is no precise number like "hundreds of injured", 200 injured will be entered (although it is probably underestimated). (100 or more people injured).

Industrial Accident - Killed People

Units: number of people

Data Source: EM-DAT: the OFDA/CRED International Disaster Database (data as of July 2001)

Data Provider: The OFDA/CRED International Disaster Database - www.cred.be/emdat -
Université Catholique de Louvain - Brussels - Belgium

Years: 1980-2000

	1980	1990	1991	1992	1993	1994	1995	1996	1997	1998	1999	2000
AFRICA	----	109	114	225	73	----	118	----	0	1162	34	700
Central Africa	----	----	----	----	----	----	----	----	----	----	----	----
Eastern Africa	----	----	100	----	20	----	0	----	----	----	----	----
Northern Africa	----	----	----	25	----	----	----	----	0	----	----	19
Southern Africa	----	----	----	----	53	----	104	----	----	----	19	11
Western Africa	----	109	14	200	----	----	14	----	----	1162	15	670
Western Indian Ocean	----	----	----	----	----	----	----	----	----	----	----	----
ASIA AND PACIFIC	91	462	1203	490	1007	630	285	584	797	543	569	820
Australia and New Zealand	0	0	----	----	----	11	----	----	----	----	----	----
Central Asia	----	----	----	----	30	----	----	----	4	----	----	----
North West Pacific and East Asia	----	164	474	124	666	259	124	381	584	490	371	761
South Asia	50	235	283	341	46	234	127	191	185	24	94	58
South East Asia	41	63	446	25	265	115	34	12	28	25	104	1
South Pacific	----	----	----	----	----	11	----	----	----	----	----	----
EUROPE	271	281	148	431	70	71	97	20	114	165	49	93
Central Europe	97	281	80	304	50	18	58	0	----	36	0	----
Eastern Europe	----	----	62	127	----	53	25	20	91	118	49	93
Western Europe	174	----	6	----	20	0	14	0	23	11	0	----
LATIN AMERICA AND CARIBBEAN	----	33	113	213	94	73	13	18	19	55	63	----
Caribbean	----	----	----	13	----	----	----	----	----	0	----	----
Meso-America	----	33	21	200	----	----	0	6	----	16	63	----
South America	----	----	92	----	94	73	13	12	19	39	----	----
NORTH AMERICA	0	17	25	26	0	5	0	0	25	0	----	----
North America	0	17	25	26	0	5	0	0	25	0	----	----
POLAR	----	----	----	----	----	----	----	----	----	----	----	----
Antarctic	----	----	----	----	----	----	----	----	----	----	----	----
Arctic	----	----	----	----	----	----	----	----	----	----	----	----
WEST ASIA	----	----	0	----	0	----	0	38	----	----	14	----
Arabian Peninsula	----	----	0	----	0	----	----	38	----	----	----	----
Mashriq	----	----	----	----	----	----	0	----	----	----	14	----
GLOBAL TOTALS	362	902	1603	1385	1244	779	513	660	955	1925	729	1613
REPORTED GLOBAL TOTALS FROM ORIGINAL DATA SOURCES	****	****	****	****	****	****	****	****	****	****	****	****

Comments:

Killed: persons confirmed as dead and persons missing and presumed dead (10 or more people killed).

Units: number of people

Data Source: EM-DAT: the OFDA/CRED International Disaster Database (data as of July 2001)

Data Provider: The OFDA/CRED International Disaster Database - www.cred.be/emdat -
Université Catholique de Louvain - Brussels - Belgium

Years: 1980-2000

	1980	1990	1991	1992	1993	1994	1995	1996	1997	1998	1999	2000
AFRICA	----	----	233	400	----	----	----	----	2000	200	----	192
Central Africa	----	----	----	----	----	----	----	----	----	----	----	----
Eastern Africa	----	----	200	----	----	----	----	----	----	----	----	----
Northern Africa	----	----	----	----	----	----	----	----	2000	----	----	38
Southern Africa	----	----	----	----	----	----	----	----	----	----	----	0
Western Africa	----	----	33	400	----	----	----	----	----	200	----	154
Western Indian Ocean	----	----	----	----	----	----	----	----	----	----	----	----
ASIA AND PACIFIC	5050	2297	24823	895	4572	886	25777	12580	101297	1419	847	16419
Australia and New Zealand	1500	2000	----	----	----	----	----	----	----	----	----	----
Central Asia	----	----	----	----	1600	----	----	----	----	600	----	----
North West Pacific and East Asia	----	100	4370	51	1737	459	17	10370	217	763	83	15662
South Asia	----	100	260	724	633	342	25210	160	101080	56	657	458
South East Asia	3550	97	20193	120	602	25	550	2050	----	----	107	299
South Pacific	----	----	----	----	----	60	----	----	----	----	----	----
EUROPE	10012	110	4207	214	37	8319	487	611	1933	46185	819	----
Central Europe	----	0	----	64	3	175	222	200	----	137	----	----
Eastern Europe	----	110	----	150	----	3017	----	130	233	48	119	----
Western Europe	10012	0	4207	----	34	5127	265	281	1700	46000	700	----
LATIN AMERICA AND CARIBBEAN	----	0	32444	16754	81	401	500	867	187	11065	1506	----
Caribbean	----	----	----	154	----	----	----	----	----	----	----	----
Meso-America	----	0	32400	16600	----	----	----	797	----	10045	1506	----
South America	----	----	44	----	81	401	500	70	187	1020	----	----
NORTH AMERICA	42452	4000	16096	----	3000	9667	305	1292	6008	4012	----	----
North America	42452	4000	16096	----	3000	9667	305	1292	6008	4012	----	----
POLAR	----	----	----	----	----	----	----	----	----	----	----	----
Antarctic	----	----	----	----	----	----	----	----	----	----	----	----
Arctic	----	----	----	----	----	----	----	----	----	----	----	----
WEST ASIA	----	----	----	----	100	----	----	100	----	----	20	----
Arabian Peninsula	----	----	----	----	100	----	----	100	----	----	----	----
Mashriq	----	----	----	----	----	----	----	----	----	----	20	----
GLOBAL TOTALS	57514	6407	77803	18263	7790	19273	27069	15450	111425	62881	3192	16611
REPORTED GLOBAL TOTALS FROM ORIGINAL DATA SOURCES	****	****	****	****	****	****	****	****	****	****	****	****

Comments:

Total affected: in EM-DAT, people that have been injured, affected and left homeless after a disaster are included in this category.

Units: number of people

Data Source: EM-DAT: the OFDA/CRED International Disaster Database (data as of July 2001)

Data Provider: The OFDA/CRED International Disaster Database - www.cred.be/emdat -
    Université Catholique de Louvain - Brussels - Belgium

Years: 1980-2000

	1980	1990	1991	1992	1993	1994	1995	1996	1997	1998	1999	2000
AFRICA	----	----	----	0	----	71	0	----	----	----	0	6
Central Africa	----	----	----	----	----	----	----	----	----	----	0	6
Eastern Africa	----	----	----	----	----	----	----	----	----	----	----	0
Northern Africa	----	----	----	0	----	0	----	----	----	----	----	----
Southern Africa	----	----	----	----	----	0	0	----	----	----	----	0
Western Africa	----	----	----	----	----	0	----	----	----	----	0	0
Western Indian Ocean	----	----	----	----	----	71	----	----	----	----	0	0
ASIA AND PACIFIC	----	----	631	50000	0	0	0	28	0	550	0	106
Australia and New Zealand	----	----	----	----	----	----	----	----	----	----	----	----
Central Asia	----	----	30	----	----	0	----	----	----	0	0	0
North West Pacific and East Asia	----	----	----	----	0	----	0	----	----	550	0	0
South Asia	----	----	580	50000	----	0	----	----	----	0	0	106
South East Asia	----	----	----	----	0	0	0	28	0	0	0	0
South Pacific	----	----	21	----	----	----	0	----	----	----	----	0
EUROPE	----	281	0	0	----	140	----	----	0	0	0	0
Central Europe	----	----	----	0	----	140	----	----	----	----	0	0
Eastern Europe	----	----	0	----	----	0	----	----	----	0	0	0
Western Europe	----	281	----	0	----	----	----	----	0	0	0	0
LATIN AMERICA AND CARIBBEAN	----	----	0	----	----	0	----	100	----	----	0	54
Caribbean	----	----	----	----	----	0	----	----	----	----	----	----
Meso-America	----	----	----	----	----	0	----	----	----	----	0	0
South America	----	----	0	----	----	0	----	100	----	----	0	54
NORTH AMERICA	----	185	----	----	----	0	0	----	----	----	0	2500
North America	----	185	----	----	----	0	0	----	----	----	0	2500
POLAR	----	----	----	----	----	----	----	----	----	----	----	0
Antarctic	----	----	----	----	----	----	----	----	----	----	----	----
Arctic	----	----	----	----	----	----	----	----	----	----	----	0
WEST ASIA	----	----	24	----	----	----	----	50	----	----	0	----
Arabian Peninsula	----	----	24	----	----	----	----	----	----	----	----	----
Mashriq	----	----	----	----	----	----	----	50	----	----	0	----
GLOBAL TOTALS	----	466	655	50000	0	211	0	178	0	550	0	2666
REPORTED GLOBAL TOTALS FROM ORIGINAL DATA SOURCES	****	****	****	****	****	****	****	****	****	****	****	****

Comments:

Transport accident: disaster type term used in EM-DAT to describe technological transport accidents involving mechanised modes of transport. It comprises of four disaster subsets: accidents involving aeroplanes, helicopters, airships and balloons "transport air"; accidents involving sailing boats, ferries, cruise ships, other boats "transport boat"; accidents involving trains "transport rail"; and accidents involving motor vehicles on roads and tracks "transport road".

Affected: people requiring immediate assistance during a period of emergency, i.e. requiring basic survival needs such as food, water, shelter, sanitation and immediate medical assistance (definition considered in EM-DAT; included in the field "total affected"); appearance of a significant number of cases of an infectious disease introduced in a region or a population that is usually free from that disease. (100 or more people affected).

AGGREGATIONS

Transport Accident - Injured People

Units: number of people

Data Source: EM-DAT: the OFDA/CRED International Disaster Database (data as of July 2001)

Data Provider: The OFDA/CRED International Disaster Database - www.cred.be/emdat -
Université Catholique de Louvain - Brussels - Belgium

Years: 1980-2000

	1980	1990	1991	1992	1993	1994	1995	1996	1997	1998	1999	2000
AFRICA	----	1	151	189	339	354	300	780	146	339	991	1205
Central Africa	----	----	----	57	----	8	----	----	51	3	24	458
Eastern Africa	----	----	50	----	----	----	4	500	----	61	----	128
Northern Africa	----	1	43	36	----	16	----	18	----	----	----	----
Southern Africa	----	----	----	42	----	117	13	----	----	----	----	68
Western Africa	----	----	18	40	231	201	168	172	95	174	491	246
Western Indian Ocean	----	----	40	14	108	12	115	90	----	101	476	305
ASIA AND PACIFIC	1122	902	1533	688	1428	1438	1922	1161	1988	1958	2711	1576
Australia and New Zealand	----	----	----	----	----	----	----	----	----	----	----	----
Central Asia	----	88	36	32	206	101	225	53	366	389	463	368
North West Pacific and East Asia	1122	----	88	39	221	213	105	18	339	71	64	107
South Asia	----	----	111	83	123	528	79	199	79	579	296	298
South East Asia	----	800	543	384	660	561	1512	683	1184	909	1888	706
South Pacific	----	14	755	150	218	35	1	208	20	10	----	97
EUROPE	227	244	578	497	332	351	260	255	540	418	459	260
Central Europe	127	39	373	25	19	140	24	20	172	----	245	12
Eastern Europe	----	----	59	34	116	211	4	50	61	103	74	75
Western Europe	100	205	146	438	197	----	232	185	307	315	140	173
LATIN AMERICA AND CARIBBEAN	----	20	345	145	195	290	175	173	248	256	411	444
Caribbean	----	----	200	----	----	156	1	----	71	133	----	----
Meso-America	----	----	80	----	30	60	65	28	----	----	166	58
South America	----	20	65	145	165	74	109	145	177	123	245	386
NORTH AMERICA	11	336	155	24	156	512	216	311	121	----	484	0
North America	11	336	155	24	156	512	216	311	121	----	484	0
POLAR	----	----	----	----	----	----	----	----	----	----	----	4
Antarctic	----	----	----	----	----	----	----	----	----	----	----	----
Arctic	----	----	----	----	----	----	----	----	----	----	----	4
WEST ASIA	----	----	----	----	----	39	----	10	----	----	51	----
Arabian Peninsula	----	----	----	----	----	----	----	----	----	----	----	----
Mashriq	----	----	----	----	----	39	----	10	----	----	51	----
GLOBAL TOTALS	1360	1503	2762	1543	2450	2984	2873	2690	3043	2971	5115	3491
REPORTED GLOBAL TOTALS FROM ORIGINAL DATA SOURCES	****	****	****	****	****	****	****	****	****	****	****	****

Comments:

Transport accident: disaster type term used in EM-DAT to describe technological transport accidents involving mechanised modes of transport. It comprises of four disaster subsets: accidents involving aeroplanes, helicopters, airships and balloons "transport air"; accidents involving sailing boats, ferries, cruise ships, other boats "transport boat"; accidents involving trains "transport rail"; and accidents involving motor vehicles on roads and tracks "transport road".

Injured: people suffering from physical injuries, trauma or an illness requiring medical treatment as a direct result of a disaster (included in the field "total affected"). The number of injured is entered when the term "injured" is written in the source. Injured people are always part of the affected population. Any related word like "hospitalized" is considered as injured. If there is no precise number like "hundreds of injured", 200 injured will be entered (although it is probably underestimated). (100 or more people injured).

Transport Accident - Killed People

Units: number of people

Data Source: EM-DAT: the OFDA/CRED International Disaster Database (data as of July 2001)

Data Provider: The OFDA/CRED International Disaster Database - www.cred.be/emdat -
Université Catholique de Louvain - Brussels - Belgium

Years: 1980-2000

	1980	1990	1991	1992	1993	1994	1995	1996	1997	1998	1999	2000
AFRICA	----	435	1625	1450	1225	819	1267	1678	587	1348	1328	2729
Central Africa	----	23	176	541	170	145	96	654	363	587	566	1521
Eastern Africa	----	67	100	58	367	----	437	547	46	400	55	134
Northern Africa	----	162	95	221	71	24	----	22	----	----	39	----
Southern Africa	----	93	139	101	30	204	401	249	----	55	78	261
Western Africa	----	80	641	333	103	134	141	125	142	182	443	532
Western Indian Ocean	----	10	474	196	484	312	192	81	36	124	147	281
ASIA AND PACIFIC	455	3440	2938	2190	3145	3148	2195	3327	3285	3187	3797	2702
Australia and New Zealand	----	----	----	----	----	----	----	----	----	----	----	----
Central Asia	----	628	517	552	411	639	250	169	445	476	663	668
North West Pacific and East Asia	164	573	376	278	824	553	211	486	624	361	860	512
South Asia	----	76	329	99	150	223	65	1148	100	265	372	228
South East Asia	128	1991	1515	1237	1190	1399	1588	1446	1858	1762	1885	1170
South Pacific	163	172	201	24	570	334	81	78	258	323	17	124
EUROPE	254	517	483	473	254	1324	264	412	523	291	209	560
Central Europe	80	16	23	88	27	1053	154	68	53	10	31	42
Eastern Europe	----	20	124	74	135	196	25	41	111	112	63	90
Western Europe	174	481	336	311	92	75	85	303	359	169	115	428
LATIN AMERICA AND CARIBBEAN	----	753	837	787	2216	414	632	970	852	508	717	595
Caribbean	----	60	190	483	1800	108	71	339	593	115	13	45
Meso-America	----	206	104	47	54	57	141	90	58	16	340	149
South America	----	487	543	257	362	249	420	541	201	377	364	401
NORTH AMERICA	22	112	156	59	101	321	43	364	72	262	350	107
North America	22	112	156	59	101	321	43	364	72	262	350	107
POLAR	----	----	----	147	----	----	131	----	----	180	----	14
Antarctic	----	10	----	----	----	----	----	----	----	----	----	143
Arctic	----	----	----	147	----	----	131	----	----	180	----	14
WEST ASIA	----	47	60	----	----	40	15	37	----	----	33	163
Arabian Peninsula	----	----	33	----	----	28	15	19	----	----	21	----
Mashriq	----	47	27	----	----	12	----	18	----	----	12	163
GLOBAL TOTALS	1032	5314	6458	5106	6941	6066	4547	6818	5319	5776	6458	7061
REPORTED GLOBAL TOTALS FROM ORIGINAL DATA SOURCES	****	****	****	****	****	****	****	****	****	****	****	****

Comments:

Transport accident: disaster type term used in EM-DAT to describe technological transport accidents involving mechanised modes of transport. It comprises of four disaster subsets: accidents involving aeroplanes, helicopters, airships and balloons "transport air"; accidents involving sailing boats, ferries, cruise ships, other boats "transport boat"; accidents involving trains "transport rail"; and accidents involving motor vehicles on roads and tracks "transport toad".

Killed: persons confirmed as dead and persons missing and presumed dead (10 or more people killed).

AGGREGATIONS

Transport Accident - Total Affected People

Units: number of people

Data Source: EM-DAT: the OFDA/CRED International Disaster Database (data as of July 2001)

Data Provider: The OFDA/CRED International Disaster Database - www.cred.be/emdat - Université Catholique de Louvain - Brussels - Belgium

Years: 1980-2000

	1980	1990	1991	1992	1993	1994	1995	1996	1997	1998	1999	2000
AFRICA	----	1	151	190	339	425	300	780	146	339	991	1211
Central Africa	----	----	----	58	----	8	----	----	51	3	24	464
Eastern Africa	----	0	50	----	----	----	4	500	----	61	----	128
Northern Africa	----	1	43	36	----	16	----	18	----	----	----	----
Southern Africa	----	0	----	42	----	117	13	----	----	----	----	68
Western Africa	----	0	18	40	231	201	168	172	95	174	491	246
Western Indian Ocean	----	0	40	14	108	83	115	90	----	101	476	305
ASIA AND PACIFIC	1122	902	2164	688	1428	1438	1922	1189	1988	2508	2711	1682
Australia and New Zealand	----	----	----	----	----	----	----	----	----	----	----	----
Central Asia	----	88	66	32	206	101	225	53	366	389	463	368
North West Pacific and East Asia	1122	0	88	39	221	213	105	18	339	621	64	107
South Asia	----	0	691	83	123	528	79	199	79	579	296	404
South East Asia	----	800	543	384	660	561	1512	711	1184	909	1888	706
South Pacific	----	14	776	150	218	35	1	208	20	10	----	97
EUROPE	227	525	580	497	332	491	260	255	540	418	459	260
Central Europe	127	39	375	25	19	280	24	20	172	----	245	12
Eastern Europe	----	----	59	34	116	211	4	50	61	103	74	75
Western Europe	100	486	146	438	197	----	232	185	307	315	140	173
LATIN AMERICA AND CARIBBEAN	----	20	345	145	195	290	175	283	248	256	411	498
Caribbean	----	0	200	----	----	156	1	----	71	133	----	----
Meso-America	----	0	80	----	30	60	65	28	----	----	166	58
South America	----	20	65	145	165	74	109	255	177	123	245	440
NORTH AMERICA	11	521	155	24	156	512	216	311	121	----	484	2500
North America	11	521	155	24	156	512	216	311	121	----	484	2500
POLAR	----	----	----	----	----	----	----	----	----	----	----	4
Antarctic	----	0	----	----	----	----	----	----	----	----	----	----
Arctic	----	----	----	----	----	----	----	----	----	----	----	4
WEST ASIA	----	0	24	----	----	39	----	60	----	----	51	----
Arabian Peninsula	----	----	24	----	----	----	----	----	----	----	----	----
Mashriq	----	0	----	----	----	39	----	60	----	----	51	----
GLOBAL TOTALS	1360	1969	3419	1544	2450	3195	2873	2878	3043	3521	5115	6157
REPORTED GLOBAL TOTALS FROM ORIGINAL DATA SOURCES	****	****	****	****	****	****	****	****	****	****	****	****

Comments:

Transport accident: disaster type term used in EM-DAT to describe technological transport accidents involving mechanised modes of transport. It comprises of four disaster subsets: accidents involving aeroplanes, helicopters, airships and balloons "transport air"; accidents involving sailing boats, ferries, cruise ships, other boats "transport boat"; accidents involving trains "transport rail"; and accidents involving motor vehicles on roads and tracks "transport toad".

Total affected: in EM-DAT, people that have been injured, affected and left homeless after a disaster are included in this category.

1.1.9 Tourism

International Tourism – Arrivals

Units: thousand people

Data Source: World Development Indicators 2001

Data Provider: The World Bank

Years: 1980-1999

	1980	1990	1991	1992	1993	1994	1995	1996	1997	1998	1999
AFRICA	8701.68	17309.71	18228.43	20853.14	20590.86	21064.57	22796.29	25043.00	26584.00	27841.00	----
Central Africa	189.00	305.00	310.00	280.00	309.00	314.00	336.00	358.00	387.00	428.00	----
Eastern Africa	561.00	1166.00	1163.00	1265.00	1333.00	1469.00	1568.00	1714.00	1701.00	1422.00	----
Northern Africa	5352.00	10905.00	10797.00	12131.00	11202.00	10572.00	10233.00	10856.00	11707.00	11922.00	13979.00
Southern Africa	1578.00	3081.00	4159.00	5234.00	5862.00	6727.00	8106.00	9125.00	9807.00	10799.00	11534.68
Western Africa	819.68	1395.71	1356.43	1436.14	1314.86	1378.57	1912.29	2265.00	2189.00	2436.00	----
Western Indian Ocean	202.00	457.00	443.00	507.00	570.00	604.00	641.00	725.00	793.00	834.00	865.00
ASIA AND PACIFIC	22323.51	55778.00	57795.00	65894.00	72807.00	78561.00	84017.00	91994.00	91760.00	91427.00	100548.33
Australia and New Zealand	1370.00	3191.00	3333.00	3659.00	4153.00	4685.00	5135.00	5694.00	5815.00	5652.00	6066.00
Central Asia	----	----	----	----	----	----	----	----	----	----	----
North West Pacific and East Asia	9407.29	26035.00	29298.00	34764.00	38827.00	40265.00	41770.00	46877.00	46348.00	47848.00	52813.33
South Asia	2242.00	3158.00	3258.00	3582.00	3511.00	3866.00	4200.00	4434.00	4834.00	5219.00	5627.00
South East Asia	8369.00	21469.00	20076.00	21812.00	24205.00	27135.00	29173.00	31042.00	30611.00	28951.00	33293.00
South Pacific	935.22	1925.00	1830.00	2077.00	2061.00	2504.00	2882.00	3003.00	3044.00	2615.00	2680.00
EUROPE	----	----	----	302703.50	310297.13	334456.00	337889.00	353156.00	370386.00	382478.00	395172.00
Central Europe	----	51880.00	54339.00	65422.50	70054.13	78199.00	76778.00	79768.00	79069.00	76125.00	70362.00
Eastern Europe	----	----	----	----	----	9724.00	13361.00	18924.00	23742.00	22853.00	26858.00
Western Europe	145900.00	223098.00	221400.00	233954.00	234212.00	246533.00	247750.00	254464.00	267575.00	283500.00	297952.00
LATIN AMERICA AND CARIBBEAN	25131.67	36500.00	36140.00	38486.00	39262.00	41837.00	46451.00	49172.00	48865.00	51825.00	52196.00
Caribbean	5868.00	9684.00	9594.00	9995.00	10855.00	11420.00	11812.00	12194.00	12992.00	13529.00	13765.00
Meso-America	13454.67	18924.00	17992.00	19413.00	18790.00	19580.00	22846.00	24042.00	22353.00	22825.00	22978.00
South America	5809.00	7892.00	8554.00	9078.00	9617.00	10837.00	11793.00	12936.00	13520.00	15471.00	15453.00
NORTH AMERICA	35376.00	54572.00	57586.00	62002.00	60884.00	60725.00	60250.00	63774.00	65388.00	65262.00	68048.00
North America	35376.00	54572.00	57586.00	62002.00	60884.00	60725.00	60250.00	63774.00	65388.00	65262.00	68048.00
POLAR	----	----	----	----	----	----	----	----	----	----	----
Antarctic	----	----	----	----	----	----	----	----	----	----	----
Arctic	----	----	----	----	----	----	----	----	----	----	----
WEST ASIA	6088.00	6452.00	6164.00	7474.00	8180.00	8707.00	9425.00	9658.00	10551.00	11847.00	----
Arabian Peninsula	3234.00	4570.00	4837.00	5415.00	6277.00	6662.00	7025.00	7250.00	7924.00	8673.00	----
Mashriq	2854.00	1882.00	1327.00	2059.00	1903.00	2045.00	2400.00	2408.00	2627.00	3174.00	----
GLOBAL TOTALS	----	445589.71	451652.43	497412.65	512020.99	545350.57	560828.29	592797.00	613534.00	630680.00	653189.01
REPORTED GLOBAL TOTALS FROM ORIGINAL DATA SOURCES	286130.00	461480.00	467490.00	508960.00	523980.00	554480.00	569860.00	601550.00	622590.00	639910.00	668480.00

Comments:

International tourism arrivals are the number of visitors who travel to a country other than that where they have their usual residence for a period not exceeding 12 months and whose main purpose in visiting is other than an activity remunerated from within the country visited.

AGGREGATIONS

International Tourism – Departures

Units: thousand people

Data Source: World Development Indicators 2001

Data Provider: The World Bank

Years: 1980-1998

	1980	1990	1991	1992	1993	1994	1995	1996	1997	1998
AFRICA	----	----	----	----	----	----	----	----	----	----
Central Africa	----	----	----	----	----	----	----	----	----	----
Eastern Africa	----	----	----	----	----	----	----	----	----	----
Northern Africa	3190.32	9358.02	8945.22	8873.00	9099.00	8637.00	8244.00	7645.00	7533.00	7575.80
Southern Africa	----	----	----	----	----	----	----	----	----	----
Western Africa	----	----	----	----	----	----	----	----	----	----
Western Indian Ocean	----	141.00	137.00	157.00	162.00	171.00	177.00	187.00	197.00	209.00
ASIA AND PACIFIC	----	----	43519.00	44802.00	49651.00	55128.00	62015.00	69892.00	71275.00	72100.00
Australia and New Zealand	1601.00	2887.00	2878.00	3027.00	3067.00	3178.00	3439.00	3825.00	4065.00	4327.00
Central Asia	----	----	----	----	----	----	----	----	----	----
North West Pacific and East Asia	----	----	14657.00	16800.00	18125.00	20504.00	23683.00	26473.00	26752.00	27404.00
South Asia	2007.54	3857.00	4035.00	4064.00	4739.00	4777.00	5522.00	6238.00	6642.00	6932.00
South East Asia	----	----	----	----	----	----	----	----	----	----
South Pacific	----	----	----	----	----	----	----	----	----	----
EUROPE	----	----	----	----	302292.60	325560.96	340437.73	356467.32	365897.35	377784.91
Central Europe	----	----	----	----	70192.60	77066.96	73079.73	81955.32	87745.35	89238.74
Eastern Europe	----	----	----	----	----	15799.00	29008.00	28380.00	22933.00	21832.50
Western Europe	120973.70	185541.00	194639.00	210354.00	222886.00	232695.00	238350.00	246132.00	255219.00	266713.67
LATIN AMERICA AND CARIBBEAN	----	16640.00	18089.50	22170.00	22067.00	25082.00	22409.00	24113.00	26429.00	28491.00
Caribbean	----	----	----	----	----	----	----	----	----	----
Meso-America	4319.00	8882.00	9226.00	12866.00	11674.00	13451.00	10009.00	10741.00	10863.00	12061.00
South America	----	6251.00	7284.00	7649.00	8696.00	9858.00	10576.00	11427.00	13555.00	14421.00
NORTH AMERICA	35554.00	65038.00	63503.00	65599.00	64972.00	64795.00	68969.00	71284.00	72055.00	73927.00
North America	35554.00	65038.00	63503.00	65599.00	64972.00	64795.00	68969.00	71284.00	72055.00	73927.00
POLAR	----	----	----	----	----	----	----	----	----	----
Antarctic	----	----	----	----	----	----	----	----	----	----
Arctic	----	----	----	----	----	----	----	----	----	----
WEST ASIA	----	----	----	----	----	----	----	----	----	----
Arabian Peninsula	----	----	----	----	----	----	----	----	----	----
Mashriq	2273.40	2423.00	2075.00	----	----	----	----	----	----	----
GLOBAL TOTALS	----	----	----	----	----	----	----	----	----	----
REPORTED GLOBAL TOTALS FROM ORIGINAL DATA SOURCES	----	458110.00	468080.00	508760.00	538400.00	569770.00	597170.00	629240.00	648920.00	670820.00

Comments:

International tourism departures are the number of departures that people make from their country of usual residence to any other country for any purpose other than a remunerated activity in the country visited.

1.1.10 Trade Balance

Exports of Goods and Services

Units: million constant 1995 US$

Data Source: World Development Indicators 2001

Data Provider: The World Bank

Years: 1960-1999

	1960	1970	1980	1990	1995	1996	1997	1998	1999
AFRICA	----	----	----	104922.63	126383.19	136326.78	143414.62	146205.06	149439.29
Central Africa	----	----	----	----	----	----	----	----	----
Eastern Africa	----	----	----	----	4780.37	5141.56	5441.65	4907.98	5098.29
Northern Africa	12822.23	14216.18	21034.77	35036.19	41911.85	43098.26	45748.73	45806.31	49006.90
Southern Africa	----	----	----	36479.26	46147.16	50302.28	52644.35	54316.46	54587.95
Western Africa	5295.75	13407.24	20915.51	19997.81	23412.03	26695.26	27421.45	28714.46	27603.81
Western Indian Ocean	695.25	1518.82	1699.59	2877.17	3449.25	3845.34	4060.56	4393.27	4596.41
ASIA AND PACIFIC	----	174724.77	419495.79	788360.82	1233172.00	1325406.84	1495537.66	1510294.96	1496221.79
Australia and New Zealand	12796.91	25054.05	35407.87	63141.27	91451.87	99790.34	103580.11	105440.91	----
Central Asia	----	----	----	----	15466.50	16902.85	18151.53	18054.70	20161.76
North West Pacific and East Asia	----	110181.64	292156.92	545903.76	803427.53	861664.11	1005010.23	1021323.75	1088257.49
South Asia	----	10993.87	24480.62	50885.77	77594.17	86311.38	86769.09	90132.22	92414.82
South East Asia	----	----	----	126989.92	241257.11	256840.60	278630.66	----	----
South Pacific	----	----	----	----	----	----	----	----	----
EUROPE	----	----	----	1986513.44	3036201.82	3195248.65	3512840.43	3718383.18	3683744.24
Central Europe	----	----	----	115020.31	159245.64	177144.90	204018.78	225950.85	226716.46
Eastern Europe	----	----	----	231980.23	130630.20	141845.72	149070.02	141607.44	138213.35
Western Europe	290038.50	622675.57	1068907.63	1639512.91	2746325.98	2876258.02	3159751.64	3350824.89	3318814.44
LATIN AMERICA AND CARIBBEAN	26584.25	43328.54	106643.48	179948.06	268542.31	297035.56	320982.18	343391.31	369104.04
Caribbean	----	----	----	----	----	----	----	----	----
Meso-America	6426.99	12374.64	33745.63	59290.01	101928.30	118826.43	131468.04	146452.47	166187.64
South America	17419.69	26771.83	66288.26	109879.62	154855.16	166363.32	176903.95	183511.05	189352.34
NORTH AMERICA	89727.35	213350.07	423046.72	728360.03	1038211.94	1118194.52	1247247.65	1291967.23	----
North America	89727.35	213350.07	423046.72	728360.03	1038211.94	1118194.52	1247247.65	1291967.23	----
POLAR	----	----	----	----	----	----	----	----	----
Antarctic	----	----	----	----	----	----	----	----	----
Arctic	----	----	----	----	----	----	----	----	----
WEST ASIA	----	----	----	----	----	----	----	----	----
Arabian Peninsula	----	----	----	----	----	----	----	----	----
Mashriq	----	----	----	----	----	----	----	----	----
GLOBAL TOTALS	497856.93	1117578.80	2135777.02	3795214.36	5713762.04	6083364.16	6730932.43	7021443.57	----
REPORTED GLOBAL TOTALS FROM ORIGINAL DATA SOURCES	750180.00	1582700.00	2898800.00	4811700.00	6368800.00	6779000.00	7485500.00	7796600.00	8157000.00

Comments:

Exports of goods and services represent the value of all goods and other market services provided to the rest of the world. They include the value of merchandise, freight, insurance, transport, travel, royalties, license fees, and other services, such as communication, construction, financial, information, business, personal, and government services. They exclude labor and property income (formerly called factor services) as well as transfer payments.

Data sources: the WTO publishes data on world trade in its Annual Report. Estimates of total exports of goods are also published in the IMF's International Financial Statistics and Direction of Trade Statistics and in the United Nations Statistics Division's Monthly Bulletin of Statistics. The United Nations Conference on Trade and Development (UNCTAD) publishes data on the structure of exports and imports in its Handbook of International Trade and Development Statistics. Tariff line records of exports and imports are compiled in the United Nations Statistics Division's COMTRADE database.

Imports of Goods and Services

Units: million constant 1995 US$

Data Source: World Development Indicators 2001

Data Provider: The World Bank

Years: 1960-1999

	1960	1970	1980	1990	1995	1996	1997	1998	1999
AFRICA	----	----	----	115606.87	137990.95	143077.37	151669.91	157471.57	162583.43
Central Africa	----	----	----	----	----	----	----	----	----
Eastern Africa	----	----	----	----	7096.49	7596.82	7924.49	8149.50	8172.56
Northern Africa	18295.66	19258.18	46293.38	45679.41	49413.82	47146.27	49603.97	53458.92	57404.94
Southern Africa	----	----	----	33482.38	47592.22	50759.69	52966.32	53642.89	51733.97
Western Africa	9135.03	15263.53	41127.52	22828.60	24909.20	26771.94	29660.69	30075.80	33229.85
Western Indian Ocean	1576.32	2370.99	2945.97	3564.64	3871.57	4239.08	4640.95	4827.79	5157.10
ASIA AND PACIFIC	----	206201.04	417359.02	762452.18	1188956.15	1312286.88	1374067.36	1234627.37	1226894.17
Australia and New Zealand	16401.35	27204.51	39853.53	65146.56	92098.91	100692.70	109578.96	114731.94	----
Central Asia	----	----	----	----	16438.54	21466.14	21982.52	19681.22	18775.08
North West Pacific and East Asia	----	130432.35	251871.67	484015.16	716884.30	796274.02	824306.27	751111.60	850093.31
South Asia	----	22428.35	58459.60	77317.88	93659.69	100589.64	105606.16	103125.75	99960.14
South East Asia	----	----	----	134232.07	266728.62	289872.45	309147.87	----	----
South Pacific	----	----	----	----	----	----	----	----	----
EUROPE	----	----	----	2120694.06	2895378.49	3041771.96	3332347.40	3603916.46	3581606.77
Central Europe	----	----	----	123786.22	173971.33	201365.71	235953.19	260081.76	261645.58
Eastern Europe	----	----	----	320243.04	124130.59	136571.85	151792.22	131078.56	101174.02
Western Europe	300517.18	676233.49	1067580.45	1676664.80	2597276.57	2703834.40	2944601.99	3212756.15	3218787.17
LATIN AMERICA AND CARIBBEAN	41055.02	67094.36	161728.61	159647.95	283863.76	312291.83	367249.79	403770.14	388994.66
Caribbean	----	----	----	----	----	----	----	----	----
Meso-America	14110.96	24117.81	58713.31	67175.96	97894.58	116060.42	140722.55	163605.01	182717.88
South America	24477.36	37949.10	91311.00	79478.20	173413.82	182677.32	211853.56	224408.17	190379.56
NORTH AMERICA	115917.24	268338.73	409396.29	792097.90	1103807.55	1193389.10	1359680.45	1506643.56	----
North America	115917.24	268338.73	409396.29	792097.90	1103807.55	1193389.10	1359680.45	1506643.56	----
POLAR	----	----	----	----	----	----	----	----	----
Antarctic	----	----	----	----	----	----	----	----	----
Arctic	----	----	----	----	----	----	----	----	----
WEST ASIA	----	----	----	----	----	----	----	----	----
Arabian Peninsula	----	----	----	----	----	----	----	----	----
Mashriq	----	----	----	----	----	----	----	----	----
GLOBAL TOTALS	----	----	----	----	5630036.38	6023665.42	6606464.43	6928635.99	----
REPORTED GLOBAL TOTALS FROM ORIGINAL DATA SOURCES	852290.00	1828500.00	3078900.00	4970500.00	6268400.00	6699200.00	7346300.00	7696600.00	8037100.00

Comments:

Imports of goods and services represent the value of all goods and other market services received from the rest of the world. They include the value of merchandise, freight, insurance, transport, travel, royalties, license fees, and other services, such as communication, construction, financial, information, business, personal, and government services. They exclude labor and property income (formerly called factor services) as well as transfer payments.

Data sources: the WTO publishes data on world trade in its Annual Report. Estimates of total exports of goods are also published in the IMF's International Financial Statistics and Direction of Trade Statistics and in the United Nations Statistics Division's Monthly Bulletin of Statistics. The United Nations Conference on Trade and Development (UNCTAD) publishes data on the structure of exports and imports in its Handbook of International Trade and Development Statistics. Tariff line records of exports and imports are compiled in the United Nations Statistics Division's COMTRADE database.

Units: percent of GDP

Data Source: World Development Indicators 2001

Data Provider: The World Bank

Years: 1960-1999

	1960	1970	1980	1990	1991	1992	1993	1994	1995	1996	1997	1998	1999
AFRICA	----	45.04	63.15	53.73	53.33	54.93	55.79	57.77	59.73	58.22	59.10	57.52	57.55
Central Africa	46.54	49.54	59.24	57.16	51.87	49.64	47.54	67.29	64.06	64.71	67.99	69.42	72.42
Eastern Africa	----	----	----	37.11	40.42	39.88	52.84	51.30	51.65	51.73	49.67	45.02	44.60
Northern Africa	65.47	42.09	65.35	57.31	60.38	59.00	58.17	57.33	59.93	55.18	55.93	51.96	53.80
Southern Africa	----	----	64.56	48.28	45.96	48.86	47.86	51.75	53.94	54.86	54.76	56.41	53.51
Western Africa	35.98	37.85	55.33	62.19	59.48	65.85	72.17	71.49	73.43	71.02	74.86	69.67	72.68
Western Indian Ocean	39.25	52.40	72.80	91.38	89.35	85.98	87.19	91.83	93.19	96.24	99.41	102.14	103.24
ASIA AND PACIFIC	----	22.07	33.75	31.19	30.21	30.43	29.85	32.04	33.88	34.67	34.64	36.10	34.53
Australia and New Zealand	30.28	28.03	33.07	36.47	37.34	39.82	40.44	42.01	42.03	41.58	43.24	42.76	----
Central Asia	----	----	----	----	----	114.99	71.32	66.73	75.05	78.42	72.92	64.48	72.28
North West Pacific and East Asia	----	20.13	29.56	23.99	22.66	22.24	20.93	22.85	24.02	25.22	27.48	27.81	27.10
South Asia	----	11.60	22.85	25.69	25.45	26.51	30.34	30.74	30.78	30.94	30.63	29.52	31.16
South East Asia	34.01	58.88	103.76	112.40	112.98	111.18	113.90	117.22	123.89	120.82	97.39	118.67	----
South Pacific	----	----	99.11	86.57	65.53	63.27	63.63	66.16	68.87	67.70	66.18	68.12	65.94
EUROPE	----	----	58.14	56.74	54.23	57.17	54.67	56.25	58.98	59.64	63.27	64.60	65.66
Central Europe	----	----	----	61.45	61.89	63.55	62.07	64.05	67.38	70.78	76.12	76.56	76.14
Eastern Europe	----	----	----	42.53	33.12	97.88	69.52	60.27	60.98	54.63	55.47	66.35	81.18
Western Europe	42.26	46.93	58.51	57.95	55.43	54.20	53.39	55.63	58.42	59.22	62.85	63.82	64.44
LATIN AMERICA AND CARIBBEAN	24.80	22.68	28.58	30.26	29.52	30.86	30.99	31.35	34.57	35.46	36.06	35.18	36.65
Caribbean	----	----	----	----	----	----	----	----	----	----	----	----	----
Meso-America	23.29	22.09	29.59	41.39	38.57	38.90	38.30	41.92	59.43	62.95	62.08	64.87	64.48
South America	21.27	18.78	23.83	23.57	23.87	25.35	25.71	25.48	25.17	25.49	26.35	25.68	27.35
NORTH AMERICA	9.59	13.47	23.34	23.00	22.94	23.32	23.86	25.34	27.05	27.32	28.42	28.09	----
North America	9.59	13.47	23.34	23.00	22.94	23.32	23.86	25.34	27.05	27.32	28.42	28.09	----
POLAR	----	----	----	----	----	----	----	----	----	----	----	----	----
Antarctic	----	----	----	----	----	----	----	----	----	----	----	----	----
Arctic	----	----	----	----	----	----	----	----	----	----	----	----	----
WEST ASIA	----	----	----	90.89	93.28	93.96	90.44	86.61	90.16	92.41	91.80	84.33	70.34
Arabian Peninsula	----	----	----	90.51	94.28	95.21	91.24	86.91	91.44	93.89	94.12	85.95	68.29
Mashriq	----	----	----	----	----	----	----	----	----	----	----	----	----
GLOBAL TOTALS	----	28.58	39.57	37.68	38.09	39.31	38.17	39.65	41.83	42.23	43.70	44.37	----
REPORTED GLOBAL TOTALS FROM ORIGINAL DATA SOURCES	25.37	27.95	39.83	38.74	39.31	40.13	39.26	40.72	42.76	43.10	44.32	44.86	51.99

Comments:

Trade is the sum of exports and imports of goods and services measured as a share of gross domestic product.

The weighting factor, GDP Constant US$ 1995, is used in the World Development Indicators 2001.

Data sources: the WTO publishes data on world trade in its Annual Report. Estimates of total exports of goods are also published in the IMF's International Financial Statistics and Direction of Trade Statistics and in the United Nations Statistics Division's Monthly Bulletin of Statistics. The United Nations Conference on Trade and Development (UNCTAD) publishes data on the structure of exports and imports in its Handbook of International Trade and Development Statistics. Tariff line records of exports and imports are compiled in the United Nations Statistics Division's COMTRADE database.

1.1.11 Transport and Freight

Passenger Cars

Units: number per thousand people

Data Source: World Development Indicators 2001

Data Provider: The World Bank

Years: 1978-1999

	1978	1980	1990	1991	1992	1993	1994	1995	1996	1997	1998	1999
AFRICA	----	----	16.12	16.23	15.97	16.04	16.08	16.54	16.86	----	----	----
Central Africa	----	----	----	----	----	----	----	----	----	----	----	----
Eastern Africa	----	2.37	2.95	3.01	3.01	3.05	3.09	3.16	3.22	----	----	----
Northern Africa	----	----	23.40	23.61	23.82	24.16	25.34	27.67	28.17	----	----	----
Southern Africa	----	----	36.23	36.29	36.02	34.98	36.21	35.56	36.18	----	----	----
Western Africa	----	----	9.29	9.39	8.42	9.02	7.25	7.43	7.75	----	----	----
Western Indian Ocean	7.69	----	7.14	7.53	7.90	7.90	8.78	9.02	9.38	----	----	----
ASIA AND PACIFIC	----	----	19.86	20.98	22.07	23.02	24.13	25.38	26.61	----	----	----
Australia and New Zealand	----	404.72	447.32	447.09	450.59	453.53	458.07	473.20	477.48	----	----	----
Central Asia	----	----	----	----	----	----	----	----	----	----	----	----
North West Pacific and East Asia	----	----	29.16	31.02	33.02	34.89	36.85	38.82	40.94	----	----	----
South Asia	----	----	3.55	4.32	4.55	4.69	4.93	5.15	5.43	5.79	----	----
South East Asia	----	----	12.96	13.13	13.95	14.88	16.28	17.78	19.81	21.61	----	----
South Pacific	----	----	----	----	----	12.92	12.67	12.41	12.18	----	----	----
EUROPE	----	----	218.63	226.49	232.49	237.62	252.40	258.42	266.76	275.30	276.69	----
Central Europe	----	----	102.29	109.86	115.99	124.53	129.36	136.23	144.39	152.50	159.63	162.83
Eastern Europe	----	----	----	64.37	68.96	74.42	81.65	88.30	97.08	105.91	----	----
Western Europe	----	285.77	372.54	381.64	388.06	390.79	414.36	419.01	426.65	434.77	443.63	----
LATIN AMERICA AND CARIBBEAN	----	----	----	----	----	72.83	----	----	----	----	----	----
Caribbean	----	----	----	45.98	45.20	47.43	46.74	47.32	47.89	----	----	----
Meso-America	----	59.67	69.37	72.03	73.56	72.27	74.10	79.39	80.66	82.30	84.90	87.69
South America	----	----	----	----	----	75.93	----	----	----	----	----	----
NORTH AMERICA	----	524.28	562.94	556.63	492.19	488.99	487.17	481.32	479.97	475.06	----	----
North America	----	524.28	562.94	556.63	492.19	488.99	487.17	481.32	479.97	475.06	----	----
POLAR	----	----	----	----	----	----	----	----	----	----	----	----
Antarctic	----	----	----	----	----	----	----	----	----	----	----	----
Arctic	----	----	----	----	----	----	----	----	----	----	----	----
WEST ASIA	----	----	59.17	57.58	56.33	55.24	63.68	62.42	61.53	----	----	----
Arabian Peninsula	----	39.94	91.13	87.89	85.32	82.80	80.51	78.22	76.25	----	----	----
Mashriq	----	----	30.51	30.40	30.42	30.74	48.78	48.42	48.42	----	----	----
GLOBAL TOTALS	----	----	84.38	85.65	83.58	83.10	84.49	85.89	87.38	----	----	----
REPORTED GLOBAL TOTALS FROM ORIGINAL DATA SOURCES	----	----	90.68	90.72	88.60	87.18	87.77	88.93	90.24	----	----	----

Comments:

Passenger cars refer to road motor vehicles, other than two-wheelers, intended for the carriage of passengers and designed to seat no more than nine people (including the driver).

Data source: the data are from the International Road Federation's (IRF) electronic files and its annual World Road Statistics.

Vehicles per 1000 People

Units: number of vehicles

Data Source: World Development Indicators 2001

Data Provider: The World Bank

Years: 1990-1999

	1990	1991	1992	1993	1994	1995	1996	1997	1998	1999
AFRICA	25.86	25.96	26.00	25.90	25.56	26.50	26.57	----	----	----
Central Africa	----	----	----	----	----	----	----	----	----	----
Eastern Africa	3.98	4.18	4.24	4.34	4.45	4.59	4.74	----	----	----
Northern Africa	37.16	37.45	37.71	37.65	39.15	41.33	41.96	----	----	----
Southern Africa	51.86	52.06	51.78	50.83	51.92	51.43	51.81	----	----	----
Western Africa	20.06	19.94	19.93	20.10	16.60	18.40	17.87	----	----	----
Western Indian Ocean	10.99	10.12	10.60	11.04	13.39	13.76	14.12	----	----	----
ASIA AND PACIFIC	32.53	33.71	35.18	36.56	38.38	39.86	41.38	----	----	----
Australia and New Zealand	529.35	533.53	562.77	567.56	589.89	583.13	590.15	----	----	----
Central Asia	----	----	----	----	----	----	----	----	----	----
North West Pacific and East Asia	50.30	52.26	54.32	56.38	58.53	60.61	62.80	----	----	----
South Asia	5.82	6.56	6.97	7.18	7.61	8.04	8.55	9.21	----	----
South East Asia	23.69	24.14	25.89	29.12	33.91	37.57	40.93	44.27	46.28	----
South Pacific	----	----	----	35.97	35.26	34.29	33.62	----	----	----
EUROPE	251.97	258.88	267.96	271.66	286.56	294.81	304.38	314.02	323.69	----
Central Europe	122.38	130.93	138.23	146.97	151.20	160.62	169.47	177.00	186.88	191.46
Eastern Europe	74.12	79.41	84.29	90.20	97.80	105.05	119.13	129.18	135.76	141.93
Western Europe	423.09	430.07	442.14	441.46	465.27	472.90	479.60	489.52	500.39	----
LATIN AMERICA AND CARIBBEAN	90.80	94.91	92.95	99.23	88.91	91.72	90.89	91.85	104.85	----
Caribbean	----	74.05	71.85	74.81	64.20	66.71	68.09	64.42	----	----
Meso America	97.94	106.54	105.38	109.28	111.54	115.66	118.04	120.64	124.56	128.38
South America	90.01	92.90	90.62	98.16	83.03	85.34	82.97	83.75	84.92	----
NORTH AMERICA	742.68	731.47	726.88	729.93	737.22	737.49	745.60	741.59	745.96	----
North America	742.68	731.47	726.88	729.93	737.22	737.49	745.60	741.59	745.96	----
POLAR	----	----	----	----	----	----	----	----	----	----
Antarctic	----	----	----	----	----	----	----	----	----	----
Arctic	----	----	----	----	----	----	----	----	----	----
WEST ASIA	91.38	88.76	87.14	85.62	94.10	92.77	91.77	----	----	----
Arabian Peninsula	142.14	136.95	133.65	130.15	126.85	123.60	120.88	----	----	----
Mashriq	45.86	45.52	45.57	46.03	65.10	65.46	65.84	----	----	----
GLOBAL TOTALS	109.00	109.88	111.11	112.59	114.87	116.73	118.76	----	----	----
REPORTED GLOBAL TOTALS FROM ORIGINAL DATA SOURCES	117.58	115.55	117.26	116.29	118.12	120.26	122.27	----	----	----

Comments:

Vehicles include cars, buses, and freight vehicles but do not include two-wheelers.

1.2 Land

1.2.1 Agriculture Production

Agricultural Production Index – Total

Units: net production index number (PIN) base 1989-91

Data Source: FAOSTAT (data as of May 2001)

Data Provider: Food and Agriculture Organization of the United Nations (FAO)

Years: 1961-2000

	1961	1970	1980	1990	1991	1992	1993	1994	1995	1996	1997	1998	1999	2000
AFRICA	47.07	63.59	73.18	98.45	104.71	104.44	108.45	110.20	113.62	123.98	124.79	128.07	131.94	132.54
Central Africa	51.04	63.80	76.19	100.15	102.99	104.04	104.47	107.66	110.62	106.63	105.01	109.37	108.04	107.06
Eastern Africa	49.10	70.09	75.07	99.99	102.51	100.75	100.28	100.18	105.41	109.68	110.01	110.37	113.64	115.73
Northern Africa	41.42	59.14	72.45	98.59	106.82	109.35	111.17	112.60	118.28	135.33	135.29	137.56	142.90	143.49
Southern Africa	53.40	66.17	87.22	98.56	100.18	87.17	97.38	98.71	92.53	105.53	105.08	105.04	109.22	110.52
Western Africa	45.45	62.94	61.99	96.70	108.42	113.19	118.74	121.68	127.91	135.97	138.95	145.79	149.85	150.13
Western Indian Ocean	64.10	74.52	81.54	100.15	102.48	104.38	105.37	102.29	105.71	107.02	111.17	108.74	105.77	106.99
ASIA AND PACIFIC	40.15	54.18	69.42	102.51	103.20	107.43	111.77	114.41	120.29	126.21	130.90	134.63	139.44	142.02
Australia and New Zealand	55.19	74.17	80.05	102.47	101.40	105.36	107.90	99.23	108.00	117.14	118.34	123.34	130.80	129.31
Central Asia	77.16	98.58	103.81	127.17	110.74	104.54	96.64	86.68	77.23	71.16	73.24	71.12	79.70	75.74
North West Pacific and East Asia	31.51	45.23	61.68	101.06	104.13	109.77	118.18	125.11	132.89	140.26	149.17	154.64	159.40	164.98
South Asia	44.20	54.60	67.08	99.79	102.11	106.13	109.35	112.90	117.02	121.02	123.03	125.92	130.20	129.94
South East Asia	36.23	49.29	71.79	99.67	102.04	108.00	110.76	113.90	118.42	122.87	123.77	121.75	126.89	129.59
South Pacific	53.30	71.66	86.78	102.08	97.56	101.11	103.17	103.59	105.04	109.76	106.35	103.66	107.46	108.62
EUROPE	61.19	75.84	87.31	100.56	94.53	91.80	89.69	84.05	83.91	86.53	87.16	86.02	86.62	85.75
Central Europe	57.55	69.93	88.35	99.92	97.17	89.58	92.11	85.43	88.01	89.20	89.91	90.62	88.81	85.02
Eastern Europe	61.30	78.32	82.48	101.04	87.99	83.16	79.55	66.82	63.53	63.47	64.07	56.19	57.17	57.66
Western Europe	63.07	76.40	92.78	100.33	100.11	100.91	97.96	97.00	97.39	101.76	102.46	102.61	104.03	103.78
LATIN AMERICA AND CARIBBEAN	44.01	58.11	79.95	100.24	102.44	105.76	104.20	111.47	119.83	118.07	123.33	126.58	132.71	135.47
Caribbean	78.53	101.57	102.08	101.36	98.50	107.81	89.25	104.62	125.71	95.70	112.07	115.84	110.91	111.02
Meso-America	40.90	59.07	86.25	101.48	104.26	104.71	106.49	111.16	118.28	116.76	121.04	122.86	127.32	130.21
South America	41.34	53.83	75.97	99.81	102.80	105.61	105.80	112.48	119.29	121.16	125.25	128.63	136.31	139.34
NORTH AMERICA	58.94	69.26	88.79	101.52	100.97	108.63	100.53	115.03	109.49	114.94	118.37	120.12	122.71	125.09
North America	58.94	69.26	88.79	101.52	100.97	108.63	100.53	115.03	109.49	114.94	118.37	120.12	122.71	125.09
POLAR	51.70	90.00	91.80	100.20	100.00	105.30	105.70	105.70	105.60	105.70	105.90	105.80	105.80	105.80
Antarctic	----	----	----	----	----	----	----	----	----	----	----	----	----	----
Arctic	51.70	90.00	91.80	100.20	100.00	105.30	105.70	105.70	105.60	105.70	105.90	105.80	105.80	105.80
WEST ASIA	27.46	34.29	51.52	107.40	98.57	111.51	108.48	111.92	115.92	120.95	123.51	136.41	123.70	130.37
Arabian Peninsula	18.14	23.16	33.02	103.94	103.29	109.63	103.82	105.54	103.18	105.18	124.79	129.26	127.86	131.82
Mashriq	42.31	50.62	82.62	109.72	95.40	112.76	111.37	115.81	122.64	128.63	122.71	140.46	121.04	129.44
GLOBAL TOTALS	48.89	63.05	78.76	101.23	100.09	102.82	102.55	106.20	108.90	113.31	116.75	119.39	123.16	124.99
REPORTED GLOBAL TOTALS FROM ORIGINAL DATA SOURCES	49.80	62.90	78.50	100.70	101.20	103.50	104.10	107.10	109.30	113.70	116.60	118.20	121.40	122.80

Comments:

Index of agricultural production: The FAO index of agricultural production shows the relative level of the aggregate volume of agricultural production for each year in comparison with the base period 1989-91. They are based on the sum of price-weighted quantities of different agricultural commodities produced after deductions of quantities used as seed and feed weighted in a similar manner. The resulting aggregate represents, therefore, disposable production for any use except as seed and feed.

Net production index number (PIN) base 1989-91: presents net production (production - feed - seed) indices. All indices are calculated by the Laspeyres formula. Net production quantities of each commodity are weighted by 1989-91 average international commodity prices and summed for each year. To obtain the index, the aggregate for a given year is divided by the average aggregate for the base period 1989-91. Indices are calculated from net production data presented on a calendar year basis.

Cereals - Area Harvested

Figures from 1995 to 2000 do not have visible decimals due to space constraint. However no information has been lost since all decimals were nil.

Units: hectares

Data Source: FAOSTAT (data as of May 2001)

Data Provider: Food and Agriculture Organization of the United Nations (FAO)

Years: 1961-2000

	1961	1970	1980	1990	1995	1996	1997	1998	1999	2000
AFRICA	57253706.00	66857867.00	64443679.00	79371400.00	88803961	94304495	90848041	94210007	91125119	918658
Central Africa	2946622.00	2942891.00	3405050.00	3732444.00	4737815	4674495	4867489	5166617	4779865	54984
Eastern Africa	8794266.00	10263178.00	8600986.00	9050154.00	10802419	11993638	12104982	10885733	11848713	111760
Northern Africa	11662399.00	14266491.00	15743278.00	15822035.00	17852546	22977997	18723476	23082618	18597543	181714
Southern Africa	12569616.00	14410911.00	15368378.00	15738829.00	15591628	16380283	16661047	15181511	15399966	166600
Western Africa	20378684.00	23852019.00	19975389.00	33683563.00	38458928	36922766	37101961	38483238	39045747	389450
Western Indian Ocean	902119.00	1122377.00	1350598.00	1344375.00	1360625	1355316	1389086	1410290	1453285	14148
ASIA AND PACIFIC	289144766.30	311161851.91	329440566.65	325522142.57	316885315	322239798	319663474	319194090	319417641	3151355
Australia and New Zealand	8565998.00	10793758.00	15778944.00	13600810.00	14771300	16879705	16148600	16794400	16742875	180228
Central Asia	27973479.30	27077849.91	28877030.65	24639167.57	19520050	18874738	17813956	14729577	14147891	152327
North West Pacific and East Asia	99237950.00	101246183.00	101589635.00	99778968.00	95154825	97758126	97501225	97464897	96782600	909267
South Asia	119436056.00	133672995.00	139758942.00	141271863.00	138320604	139053320	138856080	140404256	139413032	1396627
South East Asia	33915649.00	38355930.00	43417876.00	46214607.00	49106682	49660228	49330486	49787896	52316543	512756
South Pacific	15634.00	15136.00	18139.00	16727.00	11854	13681	13127	13064	14700	147
EUROPE	176428867.70	171972122.09	176852933.35	159363641.43	146265428	143564221	148681816	133194440	129662905	1345739
Central Europe	45808421.40	43109946.91	42432097.65	41963339.80	41794257	41040445	42714011	42226551	39174438	395753
Eastern Europe	87396455.30	84598275.18	90219385.70	76979194.62	67766841	64569649	66844490	52487012	53078080	562678
Western Europe	43223991.00	44263900.00	44201450.00	40421107.00	36704330	37954127	39123315	38480877	37410387	387307
LATIN AMERICA AND CARIBBEAN	37265137.00	46549063.00	49038472.00	47409566.00	48593078	49743385	48841139	46007897	47248467	496707
Caribbean	853341.00	911802.00	799846.00	714651.00	761471	816774	876725	806006	845760	8265
Meso-America	9549463.00	11793115.00	11664814.00	12774862.00	12848514	13709048	12639212	13308900	12228371	142103
South America	26862333.00	33844146.00	36573812.00	33920053.00	34983093	35217563	35325202	31892991	34174336	346338
NORTH AMERICA	81937984.00	72815430.00	90959594.00	87246800.00	78027010	85394230	82927280	79852860	75986530	769564
North America	81937984.00	72815430.00	90959594.00	87246800.00	78027010	85394230	82927280	79852860	75986530	769564
POLAR	----	----	----	----	----	----	----	----	----	
Antarctic	----	----	----	----	----	----	----	----	----	
Arctic	----	----	----	----	----	----	----	----	----	
WEST ASIA	6099095.00	6403883.00	6553056.00	9374794.00	8462238	7986348	7711936	7833772	7572977	70157
Arabian Peninsula	1475377.00	1424925.00	1307425.00	1824657.00	1443186	1272586	1381963	1398183	1329074	13173
Mashriq	4623718.00	4978958.00	5245631.00	7550137.00	7019052	6713762	6329973	6435589	6243903	56984
GLOBAL TOTALS	648129556.00	675760217.00	717288301.00	708288344.00	687037030	703232477	698673686	680293066	671013639	6752183
REPORTED GLOBAL TOTALS FROM ORIGINAL DATA SOURCES	648129670.00	675760311.00	717288429.00	708288469.00	687177039	703406612	698865086	680480865	671186038	6754054

Comments:

Cereals are generally of the gramineous family and, in the FAO concept, refer to crops harvested for dry grain only. Crops harvested green for forage, silage or grazing are classified as fodder crops. Also excluded are industrial crops, for example broom sorghum (crude organic materials nes) and sweet sorghum when grown for syrup (sugar crops nes). For international trade classifications, fresh cereals (other than sweet corn), whether or not suitable for use as fresh vegetables, are classified as cereals. Cereals are identified according to their genus; this category includes: heat rice, paddy, barley, maize, pop corn, rye, oats, millet, sorghum, buckwheat, quinoa, fonio, triticale, canary seed, mixed grain and cereals nes.

Area harvested: data refer to the area from which a crop is gathered. Area harvested, therefore, excludes the area from which, although sown or planted, there was no harvest due to damage, failure, etc. It is usually net for temporary crops and gross for permanent crops. Net area differs from gross area insofar as the latter includes uncultivated patches, footpaths, ditches, headlands, shoulders, shelterbelts, etc.

If the crop under consideration is harvested more than once during the year as a consequence of successive cropping (that is to say the same crop is sown or planted more than once in the same field during the year), the area is counted as many times as harvested. On the contrary, area harvested will be recorded only once in the case of successive gathering of the crop during the year from the same standing crops.

AGGREGATIONS

Cereals - Area Harvested

Figures from 1995 to 2000 do not have visible decimals due to space constraint. However no information has been lost since all decimals were nil.

Units: metric tonnes

Data Source: FAOSTAT (data as of May 2001)

Data Provider: Food and Agriculture Organization of the United Nations (FAO)

Years: 1961-2000

	1961	1970	1980	1990	1995	2000
AFRICA	46281779.00	60507494.00	72602167.00	92762230.00	97671027	112404578
Central Africa	2025062.00	2189565.00	2476429.00	3035842.00	3848141	4571090
Eastern Africa	6992708.00	9442804.00	10470271.00	11584967.00	12864182	13132000
Northern Africa	9833843.00	16712760.00	19147501.00	24539684.00	24096139	27905145
Southern Africa	12165815.00	13841052.00	22072562.00	22055927.00	17455374	25619651
Western Africa	13652951.00	16241733.00	16168236.00	28933245.00	36727413	38678492
Western Indian Ocean	1611400.00	2079580.00	2267168.00	2612565.00	2679778	2498200
ASIA AND PACIFIC	339429601.46	499042381.75	642866703.26	887356287.02	929332541	982444286
Australia and New Zealand	9538587.00	13486512.00	17158140.00	23913758.00	28083492	31459000
Central Asia	17913640.46	25188031.75	25693427.26	31537602.02	15090351	18402790
North West Pacific and East Asia	140315798.00	231182256.00	306194631.00	436086909.00	443711323	432005915
South Asia	120845426.00	158198391.00	198037944.00	267543655.00	291693639	331896854
South East Asia	50789260.00	70959004.00	95741896.00	128239802.00	150724039	168642408
South Pacific	26890.00	28187.00	40665.00	34561.00	29697	37319
EUROPE	258788381.54	341815891.25	427551169.74	493013948.98	406047799	412917327
Central Europe	68709546.93	81812379.52	116202449.64	130095388.48	119607056	96350085
Eastern Europe	97321105.61	136841369.73	139587079.10	171337272.50	103597115	96545411
Western Europe	92757729.00	123162142.00	171761641.00	191581288.00	182843628	220021831
LATIN AMERICA AND CARIBBEAN	47399044.00	71364892.00	88248642.00	99021897.00	123902967	138613613
Caribbean	1003682.00	1288301.00	1476029.00	1413982.00	1286992	1560334
Meso-America	10220089.00	17598189.00	24182124.00	29641820.00	30897490	33637136
South America	36175273.00	52478402.00	62590489.00	67966095.00	91718485	103416143
NORTH AMERICA	180531264.00	215542656.00	311499202.00	369202900.00	326914232	395180600
North America	180531264.00	215542656.00	311499202.00	369202900.00	326914232	395180600
POLAR	----	----	----	----	----	----
Antarctic	----	----	----	----	----	----
Arctic	----	----	----	----	----	----
WEST ASIA	4596610.00	4417119.00	7172442.00	11727345.00	12383376	7638517
Arabian Peninsula	1302100.00	1282323.00	1135463.00	4916627.00	3491919	3161745
Mashriq	3294510.00	3134796.00	6036979.00	6810718.00	8891457	4476772
GLOBAL TOTALS	877026680.00	1192690434.00	1549940326.00	1953084608.00	1896251942	2049198921
REPORTED GLOBAL TOTALS FROM ORIGINAL DATA SOURCES	877026930.00	1192690540.00	1549940470.00	1953084760.00	1896509240	2049414990

Comments:

Cereals are generally of the gramineous family and, in the FAO concept, refer to crops harvested for dry grain only. Crops harvested green for forage, silage or grazing are classified as fodder crops. Also excluded are industrial crops, for example broom sorghum (crude organic materials nes) and sweet sorghum when grown for syrup (sugar crops nes). For international trade classifications, fresh cereals (other than sweet corn), whether or not suitable for use as fresh vegetables, are classified as cereals. Cereals are identified according to their genus; this category includes: heat rice, paddy, barley, maize, pop corn, rye, oats, millet, sorghum, buckwheat, quinoa, fonio, triticale, canary seed, mixed grain and cereals nes.

Production: data refer to the actual harvested production from the field or orchard and gardens, excluding harvesting and threshing losses and that part of crop not harvested for any reason. Production therefore includes the quantities of the commodity sold in the market (marketed production) and the quantities consumed or used by the producers (auto-consumption).When the production data available refers to a production period falling into two successive calendar years and it is not possible to allocate the relative production to each of them, it is usual to refer production data to that year into which the bulk of the production falls.

Production data are reported in terms of clean, dry weight of grains (12-14 percent moisture) in the form usually marketed. Rice, however, is reported in terms of paddy.

AGGREGATIONS

Cereals - Yield

Units: hectograms per hectare

Data Source: FAOSTAT (data as of May 2001)

Data Provider: Food and Agriculture Organization of the United Nations (FAO)

Years: 1961-2000

	1961	1970	1980	1990	1995	1996	1997	1998	1999	2000
AFRICA	8083.77	9050.19	11266.04	11687.04	10998.64	13219.61	12178.69	12286.57	12282.30	12235.85
Central Africa	6872.46	7440.36	7272.86	8133.80	8122.17	8370.83	8405.59	9149.19	8507.25	8313.21
Eastern Africa	7951.75	9200.44	12173.11	12801.03	11908.85	12146.80	12203.49	12223.07	11685.31	11750.22
Northern Africa	8432.17	11714.58	12162.35	15509.74	13497.38	17312.73	15256.83	15216.78	16088.90	15356.67
Southern Africa	9678.88	9604.60	14362.52	14013.53	11195.52	16801.72	14607.27	14357.43	14578.89	15377.94
Western Africa	6699.77	6809.54	8094.15	8589.61	9549.93	9793.44	9728.29	9913.33	9929.53	9931.80
Western Indian Ocean	17862.19	18528.81	16786.83	19433.30	19695.48	20087.72	20014.04	18778.33	19729.37	17657.07
ASIA AND PACIFIC	11707.79	15997.41	19474.30	27210.26	29327.01	30687.40	30843.51	31415.00	32172.52	31175.34
Australia and New Zealand	11135.40	12494.49	10874.12	17582.83	19012.00	21193.40	19870.13	19888.18	21450.34	17454.58
Central Asia	6078.58	8829.87	8445.89	12150.51	7730.62	9136.59	10688.23	9423.60	15690.97	12081.10
North West Pacific and East Asia	14139.28	22833.49	30140.21	43705.26	46629.99	48871.99	48207.44	49461.87	49525.30	47511.75
South Asia	10118.13	11834.63	14169.99	18938.05	21088.34	21682.65	22302.30	22667.89	23206.23	23764.11
South East Asia	14976.23	18504.87	22051.49	27748.77	30693.35	31524.79	31438.21	31152.12	31854.68	32889.40
South Pacific	17199.69	18622.13	22418.61	20661.99	25052.44	22410.09	23777.65	15516.78	25386.92	25386.92
EUROPE	14719.49	19950.29	24249.10	31037.03	27761.05	29406.46	31799.46	31631.66	31238.49	30683.26
Central Europe	14980.38	18948.96	27355.83	30965.76	28618.31	26884.48	30319.10	29294.86	28972.28	24346.13
Eastern Europe	11249.07	16340.65	15630.04	22485.86	15287.11	15513.24	19839.15	15444.87	16038.16	17157.95
Western Europe	21459.98	27824.35	38859.03	47396.13	49815.43	55769.43	53850.59	56274.32	55177.90	56808.12
LATIN AMERICA AND CARIBBEAN	12719.50	15331.30	17995.61	20886.64	25498.21	25759.87	27164.88	28305.58	28420.19	27906.39
Caribbean	11762.02	14129.03	18454.20	19785.30	16900.98	17773.97	18465.42	16680.70	19336.31	18877.59
Meso-America	10702.10	14922.73	20730.79	23203.36	24047.23	24413.30	25087.24	24814.29	26392.63	23670.76
South America	13467.09	15506.06	17113.23	20037.32	26218.26	26469.26	28124.16	30056.27	29370.51	29859.75
NORTH AMERICA	22032.30	29601.46	34246.11	42316.83	41897.93	46162.25	46554.02	50135.36	51254.36	51350.83
North America	22032.30	29601.46	34246.11	42316.83	41897.93	46162.25	46554.02	50135.36	51254.36	51350.83
POLAR	----	----	----	----	----	----	----	----	----	----
Antarctic	----	----	----	----	----	----	----	----	----	----
Arctic	----	----	----	----	----	----	----	----	----	----
WEST ASIA	7536.63	7015.27	11053.14	12461.44	14598.07	14770.76	12604.37	14072.07	10773.18	10845.03
Arabian Peninsula	8825.78	8999.08	8685.05	26945.46	24195.89	20511.25	21695.97	21811.41	23823.71	24000.84
Mashriq	7125.27	6447.52	11643.37	8961.05	12624.67	13682.66	10619.49	12390.64	7995.25	7803.67
GLOBAL TOTALS	13531.64	17650.91	21609.30	27574.07	27600.03	29433.26	30026.21	30595.83	30946.03	30348.23
REPORTED GLOBAL TOTALS FROM ORIGINAL DATA SOURCES	13532.00	17650.00	21608.00	27575.00	27599.00	29431.00	30022.00	30593.00	30943.00	30343.00

Comments:

Cereals are generally of the gramineous family and, in the FAO concept, refer to crops harvested for dry grain only. Crops harvested green for forage, silage or grazing are classified as fodder crops. Also excluded are industrial crops, for example broom sorghum (crude organic materials nes) and sweet sorghum when grown for syrup (sugar crops nes). For international trade classifications, fresh cereals (other than sweet corn), whether or not suitable for use as fresh vegetables, are classified as cereals. Cereals are identified according to their genus; this category includes: heat rice, paddy, barley, maize, pop corn, rye, oats, millet, sorghum, buckwheat, quinoa, fonio, triticale, canary seed, mixed grain and cereals nes.

Yield represents the harvested production per unit of harvested area for crop products. In most of the cases yield data are not recorded but obtained by dividing the data stored under production element by those recorded under element: area harvested.

Meat Production – Per Capita

Units: kilograms per person

Data Source: FAOSTAT (data as of May 2001)

Data Provider: Food and Agriculture Organization of the United Nations (FAO)

Years: 1961-2000

	1961	1970	1980	1990	1991	1992	1993	1994	1995	1996	1997	1998	1999	2000
AFRICA	13.78	14.13	14.21	13.93	13.92	13.78	13.41	13.15	13.29	13.34	13.61	13.59	13.48	13.30
Central Africa	11.71	10.78	9.51	10.06	9.79	9.79	9.51	9.15	9.18	9.26	9.04	9.09	8.85	8.75
Eastern Africa	17.34	16.16	14.20	13.11	12.74	12.14	11.54	11.42	11.42	11.67	11.52	11.24	11.12	10.73
Northern Africa	12.19	12.72	14.00	16.06	16.61	16.83	16.95	17.26	18.01	18.40	20.14	20.33	20.06	20.04
Southern Africa	18.69	19.83	19.92	18.93	18.91	19.10	17.83	16.53	16.25	15.85	15.39	15.51	15.85	16.00
Western Africa	9.47	10.59	11.43	10.05	9.93	9.52	9.50	9.47	9.55	9.56	9.64	9.59	9.42	9.18
Western Indian Ocean	22.68	21.79	21.24	20.60	20.60	20.51	20.48	20.34	20.47	20.31	20.32	19.89	19.77	17.87
ASIA AND PACIFIC	6.97	10.11	13.14	18.48	19.38	20.14	21.49	22.62	23.43	23.13	24.98	26.46	26.92	27.53
Australia and New Zealand	165.82	199.96	214.18	206.80	214.69	216.59	213.15	212.60	213.49	204.85	207.71	217.73	215.00	217.48
Central Asia	54.99	59.27	58.43	63.30	57.65	40.81	41.94	39.05	34.07	30.44	27.72	56.88	55.45	54.48
North West Pacific and East Asia	4.48	9.94	16.01	26.35	28.22	30.26	32.92	35.53	37.66	37.40	41.67	44.31	45.49	47.18
South Asia	4.56	4.58	4.89	5.96	6.02	6.22	6.45	6.58	6.64	6.47	6.59	6.48	6.50	6.45
South East Asia	7.81	9.21	10.16	13.91	14.89	15.61	16.67	17.07	17.08	17.57	17.97	17.59	18.13	18.17
South Pacific	13.48	14.72	16.15	16.23	16.07	16.11	16.28	16.95	16.99	17.06	17.13	16.79	16.62	16.41
EUROPE	44.85	56.29	70.41	78.12	75.74	72.73	70.22	68.10	66.90	66.95	65.76	67.09	66.99	65.43
Central Europe	42.07	48.59	63.63	64.90	60.57	58.14	54.57	50.58	51.76	52.33	51.07	51.12	51.29	49.47
Eastern Europe	37.91	49.26	56.69	70.67	65.18	57.25	50.56	46.40	39.92	37.05	34.04	32.83	31.08	30.76
Western Europe	49.90	63.59	81.55	89.09	89.56	89.20	89.65	89.60	90.30	91.73	91.53	94.87	95.50	93.22
LATIN AMERICA AND CARIBBEAN	36.16	38.65	43.88	45.47	47.15	47.90	49.17	50.92	53.95	54.46	54.82	55.33	58.06	59.83
Caribbean	17.60	21.27	22.15	25.19	24.06	21.24	21.55	21.49	22.12	22.61	22.77	23.34	24.01	26.21
Meso America	24.11	24.26	33.78	30.91	32.42	32.78	33.82	35.38	36.95	35.49	36.56	37.81	38.56	39.20
South America	42.73	45.96	50.25	53.31	55.38	56.69	58.15	60.19	64.05	65.34	65.46	65.68	69.41	71.59
NORTH AMERICA	86.76	100.94	105.96	111.28	113.12	116.92	116.67	121.58	124.02	124.79	125.36	128.27	132.20	132.71
North America	86.76	100.94	105.96	111.28	113.12	116.92	116.67	121.58	124.02	124.79	125.36	128.27	132.20	132.71
POLAR	7.04	13.57	10.96	10.44	10.40	10.84	10.90	10.90	10.90	10.89	10.88	10.86	10.83	10.81
Antarctic	----	----	----	----	----	----	----	----	----	----	----	----	----	----
Arctic	7.04	13.57	10.96	10.44	10.40	10.84	10.90	10.90	10.90	10.89	10.88	10.86	10.83	10.81
WEST ASIA	12.04	11.39	14.68	19.49	16.38	17.33	17.17	17.74	17.63	17.55	18.84	17.92	18.57	18.17
Arabian Peninsula	10.27	8.95	12.96	20.25	19.29	20.27	20.18	21.25	20.45	20.86	23.05	21.05	21.13	20.49
Mashriq	13.26	13.04	16.03	18.80	13.77	14.70	14.50	14.63	15.13	14.60	15.05	15.07	16.22	16.02
GLOBAL TOTALS	23.19	27.30	30.92	34.29	34.52	34.65	34.99	35.62	36.21	36.03	36.94	38.07	38.65	38.84
REPORTED GLOBAL TOTALS FROM ORIGINAL DATA	****	****	****	****	****	****	****	****	****	****	****	****	****	****

Comments:

Meat production includes meat from animals slaughtered in countries, irrespective of their origin and comprises horse meat, poultry meat and meat from all other domestic or wild animals such as camels, rabbits, reindeer and game animals.

AGGREGATIONS

Meat Production – Total

Figures from 1995 to 2000 do not have visible decimals due to space constraint. However no information has been lost since all decimals were nil.

Units: kilograms per person

Data Source: FAOSTAT (data as of May 2001)

Data Provider: Food and Agriculture Organization of the United Nations (FAO)

Years: 1961-2000

	1961	1970	1980	1990	1995	1996	1997	1998	1999	2000
AFRICA	3880202.00	5006230.00	6599098.00	8587839.00	9348990	9617925	10051088	10272759	10441541	10544
Central Africa	323552.00	373804.00	429059.00	613680.00	664681	689058	689902	710373	709252	719
Eastern Africa	825316.00	981735.00	1137965.00	1420685.00	1426617	1496784	1520950	1527875	1555700	1543
Northern Africa	839545.00	1093656.00	1551339.00	2298675.00	2858206	2976352	3319808	3405096	3421879	3483
Southern Africa	1006128.00	1348644.00	1788287.00	2241517.00	2192624	2190356	2176436	2240689	2338778	2407
Western Africa	731081.00	1021971.00	1459458.00	1720548.00	1874212	1926122	1995045	2037759	2057327	2056
Western Indian Ocean	154580.00	186420.00	232990.00	292734.00	332650	339253	348947	350967	358605	333
ASIA AND PACIFIC	11728394.32	20924612.75	33283241.04	56019481.42	76657647	76771040	84025024	88585852	91427478	94813
Australia and New Zealand	2142625.00	3070321.00	3787167.00	4187372.00	4627619	4494507	4610213	4886163	4876257	4983
Central Asia	1393411.32	1966471.75	2414013.04	3204237.42	1836623	1658983	1527531	1503525	1519379	1538
North West Pacific and East Asia	3595222.00	9766598.00	18780114.00	35417300.00	53224443	53326725	59930331	64277309	66509905	69503
South Asia	2776694.00	3439006.00	4593883.00	6997678.00	8628549	8570028	8891040	8903759	9088146	9191
South East Asia	1779603.00	2627269.00	3631846.00	6117407.00	8228391	8605783	8947879	8896962	9314383	9474
South Pacific	40839.00	54947.00	76218.00	95487.00	112022	115014	118030	118134	119408	120
EUROPE	29163715.68	39687178.25	53042510.96	62133502.58	54170115	54327869	53450213	54596877	54572127	53319
Central Europe	5695940.50	7321045.58	10776577.37	12124004.29	9900769	10056246	9863886	9923435	10004965	9693
Eastern Europe	6989369.18	9863847.68	12108720.58	16072496.28	9076272	8401114	7693635	7390384	6962851	6854
Western Europe	16478406.00	22502285.00	30157213.00	33937002.00	35193074	35870509	35892692	37283058	37604311	36770
LATIN AMERICA AND CARIBBEAN	8111259.00	11007064.00	15855025.00	20024458.00	25882429	26558188	27160996	27845229	29669979	31041
Caribbean	366515.00	527952.00	646934.00	852697.00	793700	820334	835007	865039	899196	991
Meso-America	1227629.00	1637410.00	3034817.00	3443380.00	4553580	4458519	4679518	4929839	5119757	5297
South America	6517115.00	8841702.00	12173274.00	15728381.00	20535149	21279335	21646471	22050351	23651026	24752
NORTH AMERICA	17991542.00	23402358.00	27011675.00	31434598.00	36970125	37601751	38181550	39477473	41104641	41669
North America	17991542.00	23402358.00	27011675.00	31434598.00	36970125	37601751	38181550	39477473	41104641	41669
POLAR	----	----	----	----	----	----	----	----	----	
Antarctic	----	----	----	----	----	----	----	----	----	
Arctic	242.00	630.00	550.00	580.00	608	608	608	608	608	
WEST ASIA	303965.00	397390.00	733443.00	1396267.00	1473908	1513259	1675813	1645021	1759076	1775
Arabian Peninsula	110411.00	126120.00	284095.00	685872.00	803177	847215	970365	919487	957628	962
Mashriq	193554.00	271270.00	449348.00	710395.00	670731	666044	705448	725534	801448	813
GLOBAL TOTALS	71179320.00	100425463.00	136525543.00	179596726.00	204503822	206390640	214545292	222423819	228975450	233163
REPORTED GLOBAL TOTALS FROM ORIGINAL DATA SOURCES	71186853.00	100435844.00	136526557.00	179597944.00	204552750	206439660	214595116	222475885	229025372	233217

Comments:

Meat production includes meat from animals slaughtered in countries, irrespective of their origin and comprises horse meat, poultry meat and meat from all other domestic or wild animals such as camels, rabbits, reindeer and game animals.

AGGREGATIONS

1.2.2 Boundaries

Land Area

Units: thousand hectares

Data Source: FAOSTAT (data as of July 2001)

Data Provider: Food and Agriculture Organization of the United Nations (FAO)

Years: 1961-1999

	1961	1970	1980	1990	1995	1996	1997	1998	1999
AFRICA	2963389.00	2963389.00	2963305.00	2963305.00	2963305.00	2963305.00	2963305.00	2963305.00	2963305.00
Central Africa	524361.00	524361.00	524281.00	524281.00	524281.00	524281.00	524281.00	524281.00	524281.00
Eastern Africa	256811.00	256811.00	256811.00	256811.00	256811.00	256811.00	256811.00	256811.00	256811.00
Northern Africa	838043.00	838043.00	838039.00	838039.00	838039.00	838039.00	838039.00	838039.00	838039.00
Southern Africa	679762.00	679762.00	679762.00	679762.00	679762.00	679762.00	679762.00	679762.00	679762.00
Western Africa	605537.00	605537.00	605537.00	605537.00	605537.00	605537.00	605537.00	605537.00	605537.00
Western Indian Ocean	58875.00	58875.00	58875.00	58875.00	58875.00	58875.00	58875.00	58875.00	58875.00
ASIA AND PACIFIC	3463407.00	3463396.00	3463345.00	3463174.00	3463220.00	3463210.00	3463210.00	3463210.00	3463210.00
Australia and New Zealand	795029.00	795029.00	795029.00	795029.00	795029.00	795029.00	795029.00	795029.00	795029.00
Central Asia	391627.00	391627.00	391627.00	391627.00	391627.00	391627.00	391627.00	391627.00	391627.00
North West Pacific and East Asia	1147981.00	1147981.00	1147939.00	1147768.00	1147768.00	1147758.00	1147758.00	1147758.00	1147758.00
South Asia	640337.00	640326.00	640326.00	640326.00	640326.00	640326.00	640326.00	640326.00	640326.00
South East Asia	434551.00	434551.00	434542.00	434542.00	434542.00	434542.00	434542.00	434542.00	434542.00
South Pacific	53882.00	53882.00	53882.00	53882.00	53928.00	53928.00	53928.00	53928.00	53928.00
EUROPE	2359435.00	2359435.00	2359393.00	2359379.00	2359383.00	2359383.00	2359383.00	2359383.00	2359383.00
Central Europe	209313.00	209313.00	209299.00	209287.00	209287.00	209287.00	209287.00	209287.00	209287.00
Eastern Europe	1789274.00	1789274.00	1789274.00	1789274.00	1789274.00	1789274.00	1789274.00	1789274.00	1789274.00
Western Europe	360848.00	360848.00	360820.00	360818.00	360822.00	360822.00	360822.00	360822.00	360822.00
LATIN AMERICA AND CARIBBEAN	2017767.00	2017767.00	2017768.00	2017768.00	2017768.00	2017768.00	2017768.00	2017768.00	2017768.00
Caribbean	22880.00	22880.00	22880.00	22880.00	22880.00	22880.00	22880.00	22880.00	22880.00
Meso-America	241955.00	241955.00	241942.00	241942.00	241942.00	241942.00	241942.00	241942.00	241942.00
South America	1752932.00	1752932.00	1752946.00	1752946.00	1752946.00	1752946.00	1752946.00	1752946.00	1752946.00
NORTH AMERICA	1838016.00	1838016.00	1838016.00	1838016.00	1838016.00	1838016.00	1838016.00	1838016.00	1838016.00
North America	1838016.00	1838016.00	1838016.00	1838016.00	1838016.00	1838016.00	1838016.00	1838016.00	1838016.00
POLAR	62298.27	62298.27	62298.27	62298.27	62298.27	62298.27	62298.27	62298.27	62298.27
Antarctic	28128.27	28128.27	28128.27	28128.27	28128.27	28128.27	28128.27	28128.27	28128.27
Arctic	34170.00	34170.00	34170.00	34170.00	34170.00	34170.00	34170.00	34170.00	34170.00
WEST ASIA	372392.00	372492.00	372392.00	372392.00	372392.00	372392.00	372392.00	372392.00	372392.00
Arabian Peninsula	300323.00	300323.00	300323.00	300323.00	300323.00	300323.00	300323.00	300323.00	300323.00
Mashriq	72069.00	72169.00	72069.00	72069.00	72069.00	72069.00	72069.00	72069.00	72069.00
GLOBAL TOTALS	13076704.27	13076793.27	13076517.27	13076332.27	13076382.27	13076372.27	13076372.27	13076372.27	13076372.27
REPORTED GLOBAL TOTALS FROM ORIGINAL DATA SOURCES	13043270.00	13043360.00	13043070.00	13042870.00	13050530.00	13050520.00	13050520.00	13050520.00	13050520.00

Comments:

Land area: total area excluding area under inland water bodies. The definition of inland water bodies generally includes major rivers and lakes.

Total Area

Units: thousand hectares

Data Source: FAOSTAT (data as of July 2001)

Data Provider: Food and Agriculture Organization of the United Nations (FAO)

Years: 1961-1999

	1961	1970	1980	1990	1995	1996	1997	1998	1999
AFRICA	3030973.00	3031198.90	3031198.90	3031198.90	3031198.90	3031198.90	3031198.90	3031198.90	3031198.90
Central Africa	536596.00	536596.00	536596.00	536596.00	536596.00	536596.00	536596.00	536596.00	536596.00
Eastern Africa	275834.00	275834.00	275834.00	275834.00	275834.00	275834.00	275834.00	275834.00	275834.00
Northern Africa	852470.00	852470.00	852470.00	852470.00	852470.00	852470.00	852470.00	852470.00	852470.00
Southern Africa	693031.00	693031.00	693031.00	693031.00	693031.00	693031.00	693031.00	693031.00	693031.00
Western Africa	613803.00	613803.00	613803.00	613803.00	613803.00	613803.00	613803.00	613803.00	613803.00
Western Indian Ocean	59239.00	59464.90	59464.90	59464.90	59464.90	59464.90	59464.90	59464.90	59464.90
ASIA AND PACIFIC	3558234.20	3558234.20	3558234.20	3558234.20	3558280.20	3558280.20	3558280.20	3558280.20	3558280.20
Australia and New Zealand	801175.00	801175.00	801175.00	801175.00	801175.00	801175.00	801175.00	801175.00	801175.00
Central Asia	400340.00	400340.00	400340.00	400340.00	400340.00	400340.00	400340.00	400340.00	400340.00
North West Pacific and East Asia	1176217.00	1176217.00	1176217.00	1176217.00	1176217.00	1176217.00	1176217.00	1176217.00	1176217.00
South Asia	677279.00	677279.00	677279.00	677279.00	677279.00	677279.00	677279.00	677279.00	677279.00
South East Asia	448008.50	448008.50	448008.50	448008.50	448008.50	448008.50	448008.50	448008.50	448008.50
South Pacific	55214.70	55214.70	55214.70	55214.70	55260.70	55260.70	55260.70	55260.70	55260.70
EUROPE	2404083.24	2404082.24	2404082.24	2404082.24	2404082.24	2404082.24	2404082.24	2404082.24	2404082.24
Central Europe	213698.00	213698.00	213698.00	213698.00	213698.00	213698.00	213698.00	213698.00	213698.00
Eastern Europe	1810665.00	1810665.00	1810665.00	1810665.00	1810665.00	1810665.00	1810665.00	1810665.00	1810665.00
Western Europe	379720.24	379719.24	379719.24	379719.24	379719.24	379719.24	379719.24	379719.24	379719.24
LATIN AMERICA AND CARIBBEAN	2058068.00	2058068.00	2058068.00	2058068.00	2058068.00	2058068.00	2058068.00	2058068.00	2058068.00
Caribbean	23475.00	23475.00	23475.00	23475.00	23475.00	23475.00	23475.00	23475.00	23475.00
Meso-America	247980.00	247980.00	247980.00	247980.00	247980.00	247980.00	247980.00	247980.00	247980.00
South America	1786613.00	1786613.00	1786613.00	1786613.00	1786613.00	1786613.00	1786613.00	1786613.00	1786613.00
NORTH AMERICA	1959994.00	1959994.00	1959994.00	1959994.00	1959994.00	1959994.00	1959994.00	1959994.00	1959994.00
North America	1959994.00	1959994.00	1959994.00	1959994.00	1959994.00	1959994.00	1959994.00	1959994.00	1959994.00
POLAR	1435385.00	1435385.00	1435385.00	1435385.00	1435385.00	1435385.00	1435385.00	1435385.00	1435385.00
Antarctic	1401215.00	1401215.00	1401215.00	1401215.00	1401215.00	1401215.00	1401215.00	1401215.00	1401215.00
Arctic	34170.00	34170.00	34170.00	34170.00	34170.00	34170.00	34170.00	34170.00	34170.00
WEST ASIA	373256.00	373256.00	373256.00	373256.00	373256.00	373256.00	373256.00	373256.00	373256.00
Arabian Peninsula	300323.00	300323.00	300323.00	300323.00	300323.00	300323.00	300323.00	300323.00	300323.00
Mashriq	72933.00	72933.00	72933.00	72933.00	72933.00	72933.00	72933.00	72933.00	72933.00
GLOBAL TOTALS	14819993.44	14820218.34	14820218.34	14820218.34	14820264.34	14820264.34	14820264.34	14820264.34	14820264.34
REPORTED GLOBAL TOTALS FROM ORIGINAL DATA SOURCES	14819993.44	14819993.44	14819993.44	14819993.44	14819993.44	14819993.44	14819993.44	14819993.44	14819993.44

Comments:

Total area is the whole area of the country, including area under inland water bodies. The definition of inland water bodies generally includes major rivers and lakes.

The World Factbook can be accessed on the Internet at: http://www.cia.gov/cia/publications/factbook/index.html.

1.2.3 Fertilizer and Pesticides Consumption and Production

The Statistics Division of the Food and Agriculture Organization of the United Nations started the collection of data on consumption of major individual pesticides products about three decades ago. However, the response to the related Pesticides Consumption Annual Questionnaire sent to all member countries was not very encouraging. Therefore, in 1986 in co-operation with the Commission of the European Union, a study was undertaken to find ways to improve the country coverage of the data.

The present work of collecting data on groups of pesticides is a result of the recommendations of this study. Data collected earlier have been published in various issues of the production yearbook. The present database refers to the quantity of pesticides used in or sold to the agricultural sector expressed in metric tonnes of active ingredients. A strict inter-country comparison on the basis of the database is not feasible because the country coverage and time series are incomplete due to a high rate of non-response.

Fertilizer – Consumption

Units: metric tonnes

Data Source: FAOSTAT (data as of December 2001)

Data Provider: Food and Agriculture Organization of the United Nations (FAO)

Years: 1961-1999

	1961	1970	1980	1990	1995	1996	1997	1998	1999
AFRICA	716141.00	1615122.00	3241614.00	3654952.00	3449489.00	3753605.00	3675659.00	3795243.00	3960153.00
Central Africa	5398.00	38127.00	42990.00	36536.00	50258.00	56400.00	----	65204.00	73925.00
Eastern Africa	15400.00	65155.00	109200.00	236181.00	218614.00	347542.00	274684.00	305612.00	341250.00
Northern Africa	364609.00	651260.00	1301917.00	1658795.00	1659495.00	1740300.00	1738690.00	1799777.00	1927550.00
Southern Africa	283565.00	752160.00	1455787.00	1141647.00	1068202.00	1140461.00	1113786.00	1088138.00	1134819.00
Western Africa	19506.00	59424.00	291948.00	531105.00	401970.00	405042.00	446689.00	488834.00	439203.00
Western Indian Ocean	27663.00	48996.00	39772.00	50688.00	50950.00	63860.00	50256.00	47678.00	43406.00
ASIA AND PACIFIC	5003488.33	14126996.98	33460187.48	57483332.97	67725221.00	69198918.00	71429821.00	72844528.00	75996704.00
Australia and New Zealand	962313.00	1419690.00	1626643.00	1506800.00	2550000.00	2675900.00	2831800.00	2911500.00	2972100.00
Central Asia	394404.33	1500773.98	2729685.48	3148539.97	798000.00	811100.00	1214427.00	1001623.00	1002151.00
North West Pacific and East Asia	2740284.00	7235502.00	18691150.00	30914803.00	38306529.00	38550452.00	38324300.00	38535000.00	39185100.00
South Asia	501631.00	2907840.00	7881166.00	16194072.00	18895912.00	19379531.00	21484120.00	21811283.00	23976316.00
South East Asia	403122.00	1055244.00	2507540.00	5681118.00	7141180.00	7747335.00	7540274.00	8556622.00	8835193.00
South Pacific	----	----	24003.00	38000.00	33600.00	34600.00	34900.00	28500.00	25844.00
EUROPE	16398940.67	34209977.02	48772553.52	46912979.03	26301119.00	27180155.00	25670124.00	26538773.00	26222368.00
Central Europe	2580918.30	7379569.02	11654244.34	8936795.30	5174267.00	5657232.00	5764385.00	5881797.00	5654618.00
Eastern Europe	2231107.37	8489734.01	15441568.18	17810987.73	3348000.00	3046800.00	3150500.00	2664042.00	2763207.00
Western Europe	11586915.00	18340674.00	21676741.00	20165196.00	17778852.00	18476123.00	16755239.00	17992934.00	17804543.00
LATIN AMERICA AND CARIBBEAN	979127.00	2908835.00	7508539.00	7926786.00	8522113.00	10252730.00	11225068.00	11349345.00	11427316.00
Caribbean	160264.00	503573.00	645317.00	747380.00	415057.00	412391.00	429014.00	361727.00	327259.00
Meso-America	260404.00	772555.00	1572460.00	2198063.00	1816487.00	2231280.00	2403825.00	2487025.00	2487938.00
South America	558459.00	1632707.00	5290762.00	4981343.00	6290569.00	7609059.00	8392229.00	8500593.00	8612119.00
NORTH AMERICA	8048099.00	16337600.00	23418446.00	20660788.00	22605781.00	23006639.00	22891250.00	22423126.00	22450549.00
North America	8048099.00	16337600.00	23418446.00	20660788.00	22605781.00	23006639.00	22891250.00	22423126.00	22450549.00
POLAR	----	----	----	----	----	----	----	----	---
Antarctic	----	----	----	----	----	----	----	----	---
Arctic	----	----	----	----	----	----	----	----	---
WEST ASIA	----	108972.00	318270.00	1086890.00	1080243.00	1153178.00	1187721.00	1206217.00	1296939.00
Arabian Peninsula	----	4800.00	57638.00	535420.00	344500.00	366379.00	379595.00	410053.00	432702.00
Mashriq	32448.00	104172.00	260632.00	551470.00	735743.00	786799.00	808126.00	796164.00	864237.00
GLOBAL TOTALS	31182244.00	69307503.00	116719610.00	137725728.00	129683966.00	134545225.00	136079643.00	138157232.00	141354029.00
REPORTED GLOBAL TOTALS FROM ORIGINAL DATA SOURCES	31182244.00	69307503.00	116719610.00	137725728.00	129690966.00	134555125.00	136087643.00	138166852.00	141360187.00

Comments:

Fertilizer consumption refers to the application of nutrients in terms of nitrogen (N), phosphate (P_2O_5), and potash (K_2O) consumed in agriculture by a country. The fertilizer year is July 1 to June 30. For countries that report their fertilizer statistics on a calendar year basis, data are shown under the fertilizer year that begins in that calendar year, for example, 1991 data are under 1991/92.

Fertilizer – Production

Units: metric tonnes

Data Source: FAOSTAT (data as of December 2001)

Data Provider: Food and Agriculture Organization of the United Nations (FAO)

Years: 1961-1999

	1961	1970	1980	1990	1995	1996	1997	1998	1999
AFRICA	----	----	----	----	----	----	----	----	----
Central Africa	----	----	----	----	----	----	----	----	----
Eastern Africa	----	----	----	----	----	----	----	----	----
Northern Africa	225924.00	532335.00	1229572.00	3470799.00	3688598.00	3887660.00	3600400.00	3976580.00	4191800.00
Southern Africa	----	----	1069165.00	----	----	----	----	----	----
Western Africa	----	----	----	----	----	----	----	----	----
Western Indian Ocean	----	----	----	----	----	----	----	----	----
ASIA AND PACIFIC	3562576.38	10095783.50	25343712.54	41272216.88	48898477.00	50926673.00	52103261.00	54040691.00	56554648.00
Australia and New Zealand	807628.00	1151602.00	1389000.00	724700.00	1012100.00	922100.00	985200.00	1008500.00	997720.00
Central Asia	317956.38	1149507.50	2136335.54	2730617.88	1267700.00	1337400.00	1126400.00	954400.00	941700.00
North West Pacific and East Asia	2106473.00	6237500.00	16385200.00	22469217.00	27685100.00	29429600.00	29122400.00	30577000.00	32241600.00
South Asia	242815.00	1305006.00	4014422.00	11527982.00	14772130.00	14813557.00	16632146.00	17300738.00	18051305.00
South East Asia	----	252168.00	1418755.00	3819700.00	4161447.00	4424016.00	4237115.00	4200053.00	4322323.00
South Pacific	----	----	----	----	----	----	----	----	----
EUROPE	20304983.62	39807225.50	59072838.46	59632803.12	41328065.00	41589309.00	41906933.00	40945398.00	42177534.00
Central Europe	2305951.18	5373677.01	9868389.90	8712648.11	7520629.00	7589376.00	7099133.00	6203498.00	5869402.00
Eastern Europe	3224710.43	11658293.49	21666693.55	27693899.01	15096700.00	14855900.00	15621400.00	15826600.00	17602600.00
Western Europe	14774322.00	22775255.00	27537755.00	23226256.00	18710736.00	19144033.00	19186400.00	18915300.00	18705532.00
LATIN AMERICA AND CARIBBEAN	585006.00	1316711.00	3650207.00	4911152.00	5262743.00	5695936.00	5645242.00	5450058.00	5438440.00
Caribbean	----	----	----	----	----	----	----	----	----
Meso-America	----	602042.00	997587.00	1797679.00	1781000.00	1977400.00	1794700.00	1701235.00	1498300.00
South America	445533.00	566632.00	2494491.00	2734673.00	3163643.00	3385636.00	3511042.00	3458123.00	3614140.00
NORTH AMERICA	8658193.00	20253955.00	33193000.00	34556632.00	37953140.00	39358600.00	37788500.00	36103627.00	34488506.00
North America	8658193.00	20253955.00	33193000.00	34556632.00	37953140.00	39358600.00	37788500.00	36103627.00	34488506.00
POLAR	----	----	----	----	----	----	----	----	----
Antarctic	----	----	----	----	----	----	----	----	----
Arctic	----	----	----	----	----	----	----	----	----
WEST ASIA	----	----	1139624.00	3216655.00	4242200.00	4280020.00	4408231.00	4850586.00	4934471.00
Arabian Peninsula	----	----	----	----	----	----	----	----	----
Mashriq	----	59400.00	483924.00	1867029.00	2000400.00	2071620.00	1986931.00	2032018.00	2133549.00
GLOBAL TOTALS	33510747.00	72935228.00	124751936.00	148436999.00	142513274.00	146914488.00	146567588.00	146451940.00	148923199.00
REPORTED GLOBAL TOTALS FROM ORIGINAL DATA SOURCES	33510747.00	72935228.00	124751936.00	148436999.00	142513274.00	146914488.00	146567588.00	146451940.00	148923199.00

Comments:

Fertilizer production refers to the production of plant nutrient for domestic use in agriculture in terms of nitrogen (N), phosphate (P_2O_5), and potash (K_2O).

Production based on imported ammonia, phosphoric acid or rock phosphate is considered national production, while that based on imported finished fertilizers (ammonium phosphate, potassium chloride, etc.) is excluded from national production to avoid double counting at the World level.

The fertilizer year is July 1 to June 30. For countries that report their fertilizer statistics on a calendar year basis, data are shown under the fertilizer year that begins in that calendar year; for example, 1991 data are under 1991/92.

Units: metric tonnes

Data Source: FAOSTAT (data as of December 2001)

Data Provider: Food and Agriculture Organization of the United Nations (FAO)

Years: 1990-1998

	1990	1991	1992	1993	1994	1995	1996	1997	1998
AFRICA	----	----	----	----	----	----	----	----	----
Central Africa	----	----	----	----	----	----	----	----	----
Eastern Africa	----	----	----	----	----	----	----	----	----
Northern Africa	----	----	----	----	----	----	----	----	----
Southern Africa	----	----	----	----	----	6370.00	----	----	----
Western Africa	----	----	----	----	----	----	----	----	----
Western Indian Ocean	153.00	28.00	11.00	24.00	24.00	12.00	26.00	5.00	16.00
ASIA AND PACIFIC	----	----	----	----	----	----	----	----	----
Australia and New Zealand	----	----	94193.00	----	----	----	----	----	----
Central Asia	----	----	----	----	----	----	----	----	----
North West Pacific and East Asia	----	----	----	----	----	----	----	----	----
South Asia	12923.00	13166.00	13236.00	----	13453.00	16740.00	12146.00	----	----
South East Asia	----	----	----	----	----	----	----	----	----
South Pacific	----	----	39.00	40.00	57.00	----	----	----	----
EUROPE	177882.19	159198.93	----	----	----	----	----	104609.00	----
Central Europe	----	21996.73	21259.00	23550.00	23586.00	23198.00	----	----	----
Eastern Europe	17776.14	15383.20	----	----	----	----	----	12110.00	----
Western Europe	----	----	----	----	----	166015.00	132419.00	----	----
LATIN AMERICA AND CARIBBEAN	----	----	----	----	----	----	----	----	----
Caribbean	----	----	----	----	----	----	----	----	----
Meso-America	----	----	----	----	----	----	----	----	----
South America	----	----	----	19732.00	20845.00	23920.00	----	----	----
NORTH AMERICA	24650.00	21319.00	20412.00	21319.00	25552.00	22226.00	24040.00	24040.00	----
North America	24650.00	21319.00	20412.00	21319.00	25552.00	22226.00	24040.00	24040.00	----
POLAR	----	----	----	----	----	----	----	----	----
Antarctic	----	----	----	----	----	----	----	----	----
Arctic	----	----	----	----	----	----	----	----	----
WEST ASIA	----	----	----	----	----	----	----	----	----
Arabian Peninsula	----	----	----	----	----	----	234.00	578.00	----
Mashriq	----	----	----	----	----	----	----	----	----
GLOBAL TOTALS	----	----	----	----	----	----	----	----	----
REPORTED GLOBAL TOTALS FROM ORIGINAL DATA SOURCES	****	****	****	****	****	****	****	****	****

Comments:

Pesticides consumption corresponds to the quantities of pesticides used in (or sold to) the agricultural sector. Figures are generally expressed in terms of active ingredients data.

Fungicides, bactericides and seed treatments include: inorganics, dithiocarbamates, benzimidazoles, triazoles, diazoles, diazines, morpholines and others not elsewhere classified.

Pesticides Consumption – Herbicides

Units: metric tonnes

Data Source: FAOSTAT (data as of December 2001)

Data Provider: Food and Agriculture Organization of the United Nations (FAO)

Years: 1990-1998

	1990	1991	1992	1993	1994	1995	1996	1997	1998
AFRICA	----	----	----	----	----	----	----	----	----
Central Africa	----	----	----	----	----	----	----	----	----
Eastern Africa	----	----	----	----	----	----	----	----	----
Northern Africa	----	----	----	----	----	----	----	----	----
Southern Africa	----	----	----	----	8834.00	8652.00	9055.00	----	----
Western Africa	----	----	----	----	----	----	----	----	----
Western Indian Ocean	657.00	9.00	24.00	27.00	946.00	907.00	1039.00	1010.00	31.00
ASIA AND PACIFIC	----	----	----	----	----	----	----	----	----
Australia and New Zealand	----	----	18031.00	----	----	----	----	----	----
Central Asia	----	----	----	----	----	----	----	----	----
North West Pacific and East Asia	----	----	----	----	----	----	----	----	----
South Asia	8846.00	8750.00	11816.00	----	11060.00	9299.00	12210.00	----	----
South East Asia	----	----	----	----	----	----	----	----	----
South Pacific	----	----	163.00	116.00	150.00	110.00	----	----	----
EUROPE	----	----	----	----	----	----	----	94567.00	----
Central Europe	----	24104.46	27788.00	41415.00	37113.00	34455.00	----	----	----
Eastern Europe	32839.10	41769.85	----	----	----	----	----	12152.00	----
Western Europe	111919.00	109527.00	97200.00	101215.00	112673.00	114123.00	106775.00	----	----
LATIN AMERICA AND CARIBBEAN	----	----	----	----	----	----	----	----	----
Caribbean	----	----	----	----	----	----	----	----	----
Meso-America	----	----	----	----	----	----	----	----	----
South America	----	23997.00	29619.00	39634.00	54153.00	66366.00	----	----	----
NORTH AMERICA	226869.00	195580.00	199762.00	188777.00	237902.00	205106.00	218177.00	213187.00	----
North America	226869.00	195580.00	199762.00	188777.00	237902.00	205106.00	218177.00	213187.00	----
POLAR	----	----	----	----	----	----	----	----	----
Antarctic	----	----	----	----	----	----	----	----	----
Arctic	----	----	----	----	----	----	----	----	----
WEST ASIA	----	----	----	----	----	----	1204.00	2035.00	----
Arabian Peninsula	----	----	----	----	----	----	353.00	966.00	----
Mashriq	----	----	----	----	----	----	----	----	----
GLOBAL TOTALS	----	----	----	----	----	----	----	----	----
REPORTED GLOBAL TOTALS FROM ORIGINAL DATA SOURCES	****	****	****	****	****	****	****	****	****

Comments:

Pesticides consumption corresponds to the quantities of pesticides used in (or sold to) the agricultural sector. Figures are generally expressed in terms of active ingredients data.

Units: metric tonnes

Data Source: FAOSTAT (data as of December 2001)

Data Provider: Food and Agriculture Organization of the United Nations (FAO)

Years: 1990-1998

	1990	1991	1992	1993	1994	1995	1996	1997	1998
AFRICA	----	----	----	----	----	----	----	----	----
Central Africa	----	----	----	----	----	----	----	----	----
Eastern Africa	----	----	----	----	----	----	----	----	----
Northern Africa	----	----	----	----	----	----	----	----	----
Southern Africa	----	----	----	----	5552.00	5644.00	----	----	----
Western Africa	----	----	----	----	----	----	----	----	----
Western Indian Ocean	166.00	101.00	66.00	77.00	412.00	655.00	438.00	741.00	49.00
ASIA AND PACIFIC	----	----	----	----	----	----	----	----	----
Australia and New Zealand	----	----	7430.00	----	----	----	----	----	----
Central Asia	----	----	----	----	----	----	----	----	----
North West Pacific and East Asia	----	----	----	----	----	----	----	----	----
South Asia	65335.00	62622.00	61672.00	----	49599.00	49728.00	48513.00	----	----
South East Asia	----	----	----	----	----	----	----	----	----
South Pacific	----	----	452.00	117.00	17.00	1.00	----	----	----
EUROPE	----	----	----	----	----	----	----	27643.00	----
Central Europe	----	10904.06	10701.00	13674.00	11103.00	10712.00	----	----	----
Eastern Europe	----	----	----	----	----	----	----	1686.00	----
Western Europe	----	56026.00	50171.00	52219.00	51652.00	57362.00	26483.00	----	----
LATIN AMERICA AND CARIBBEAN	----	----	----	----	----	----	----	----	----
Caribbean	----	----	----	----	----	----	----	----	----
Meso-America	----	----	----	----	----	----	----	----	----
South America	----	----	----	17775.00	23475.00	30065.00	----	----	----
NORTH AMERICA	92724.00	----	----	----	106845.00	----	----	----	----
North America	92724.00	----	----	----	106845.00	----	----	----	----
POLAR	----	----	----	----	----	----	----	----	----
Antarctic	----	----	----	----	----	----	----	----	----
Arctic	----	----	----	----	----	----	----	----	----
WEST ASIA	----	----	----	----	----	----	1121.00	2413.00	----
Arabian Peninsula	----	----	----	----	----	----	572.00	1662.00	----
Mashriq	----	----	----	----	----	----	----	----	----
GLOBAL TOTALS	----	----	----	----	----	----	----	----	----
REPORTED GLOBAL TOTALS FROM ORIGINAL DATA SOURCES	****	****	****	****	****	****	****	****	****

Comments:

Pesticides consumption corresponds to the quantities of pesticides used in (or sold to) the agricultural sector. Figures are generally expressed in terms of active ingredients data.

Insecticides include: chlorinated hydrocarbons, organo-phosphates, carbamates-insecticides, pyrethroids, botanical and biological products and others not elsewhere classified.

Pesticides Consumption – Mineral Oils

Units: metric tonnes

Data Source: FAOSTAT (data as of December 2001)

Data Provider: Food and Agriculture Organization of the United Nations (FAO)

Years: 1990-1998

	1990	1991	1992	1993	1994	1995	1996	1997	1998
AFRICA	----	----	----	----	----	----	----	----	----
Central Africa	----	----	----	----	----	----	----	----	----
Eastern Africa	----	----	----	----	----	----	----	----	----
Northern Africa	----	----	----	----	----	----	----	----	----
Southern Africa	----	----	----	----	----	----	----	----	----
Western Africa	----	----	----	----	----	----	----	----	----
Western Indian Ocean	----	----	----	----	----	----	----	----	----
ASIA AND PACIFIC	----	----	----	----	----	----	----	----	----
Australia and New Zealand	----	----	----	----	----	----	----	----	----
Central Asia	----	----	----	----	----	----	----	----	----
North West Pacific and East Asia	----	----	----	----	----	----	----	----	----
South Asia	----	----	----	----	----	----	----	----	----
South East Asia	----	----	----	----	----	----	----	----	----
South Pacific	----	----	----	----	----	----	----	----	----
EUROPE	----	----	----	----	----	----	----	----	----
Central Europe	----	----	----	----	----	----	----	----	----
Eastern Europe	----	----	----	----	----	----	----	----	----
Western Europe	----	----	----	----	----	11690.00	11496.00	----	----
LATIN AMERICA AND CARIBBEAN	----	----	----	----	----	----	----	----	----
Caribbean	----	----	----	----	----	----	----	----	----
Meso-America	----	----	----	----	----	----	----	----	----
South America	----	----	----	----	----	----	----	----	----
NORTH AMERICA	74389.00	63503.00	75296.00	73028.00	73935.00	76203.00	68946.00	78925.00	----
North America	74389.00	63503.00	75296.00	73028.00	73935.00	76203.00	68946.00	78925.00	----
POLAR	----	----	----	----	----	----	----	----	----
Antarctic	----	----	----	----	----	----	----	----	----
Arctic	----	----	----	----	----	----	----	----	----
WEST ASIA	----	----	----	----	----	----	----	----	----
Arabian Peninsula	----	----	----	----	----	----	----	----	----
Mashriq	----	----	----	----	----	----	----	----	----
GLOBAL TOTALS	----	----	----	----	----	----	----	----	----
REPORTED GLOBAL TOTALS FROM ORIGINAL DATA SOURCES	****	****	****	****	****	****	****	****	****

Comments:

Pesticides consumption corresponds to the quantities of pesticides used in (or sold to) the agricultural sector. Figures are generally expressed in terms of active ingredients data.

AGGREGATIONS

Units: metric tonnes

Data Source: FAOSTAT (data as of December 2001)

Data Provider: Food and Agriculture Organization of the United Nations (FAO)

Years: 1990-1998

	1990	1991	1992	1993	1994	1995	1996	1997	1998
AFRICA	----	----	----	----	----	----	----	----	----
Central Africa	----	----	----	----	----	----	----	----	----
Eastern Africa	----	----	----	----	----	----	----	----	----
Northern Africa	----	----	----	----	----	----	----	----	----
Southern Africa	----	----	----	----	----	----	----	----	----
Western Africa	----	----	----	----	----	----	----	----	----
Western Indian Ocean	----	----	----	----	----	----	9.00	----	----
ASIA AND PACIFIC	----	----	----	----	----	----	----	----	----
Australia and New Zealand	----	----	----	----	----	----	----	----	----
Central Asia	----	----	----	----	----	----	----	----	----
North West Pacific and East Asia	----	----	----	----	----	----	----	----	----
South Asia	----	----	46.00	----	62.00	67.00	----	----	----
South East Asia	----	----	----	----	----	----	----	----	----
South Pacific	----	----	----	----	----	----	----	----	----
EUROPE	----	----	----	----	----	----	----	----	----
Central Europe	----	----	----	----	----	----	----	----	----
Eastern Europe	----	----	----	----	----	----	----	----	----
Western Europe	22225.00	17243.00	18328.00	----	14923.00	15728.00	15072.00	----	----
LATIN AMERICA AND CARIBBEAN	----	----	----	----	----	----	----	----	----
Caribbean	----	----	----	----	----	----	----	----	----
Meso-America	----	----	----	----	----	----	----	----	----
South America	----	----	----	----	----	1111.00	----	----	----
NORTH AMERICA	4200.00	4000.00	4354.00	4000.00	4090.00	4000.00	----	----	----
North America	4200.00	4000.00	4354.00	4000.00	4090.00	4000.00	----	----	----
POLAR	----	----	----	----	----	----	----	----	----
Antarctic	----	----	----	----	----	----	----	----	----
Arctic	----	----	----	----	----	----	----	----	----
WEST ASIA	----	----	----	----	----	----	----	----	----
Arabian Peninsula	----	----	----	----	----	----	----	5.00	----
Mashriq	----	----	----	----	----	----	----	----	----
GLOBAL TOTALS	----	----	----	----	----	----	----	----	----
REPORTED GLOBAL TOTALS FROM ORIGINAL DATA SOURCES	****	****	****	****	****	****	****	****	****

Comments:

Pesticides consumption corresponds to the quantities of pesticides used in (or sold to) the agricultural sector. Figures are generally expressed in terms of active ingredients data.

AGGREGATIONS

Pesticides Consumption – Rodenticides

Units: metric tonnes

Data Source: FAOSTAT (data as of December 2001)

Data Provider: Food and Agriculture Organization of the United Nations (FAO)

Years: 1990-1998

	1990	1991	1992	1993	1994	1995	1996	1997	1998
AFRICA	----	----	----	----	----	----	----	----	----
Central Africa	----	----	----	----	----	----	----	----	----
Eastern Africa	----	----	----	----	----	----	----	----	----
Northern Africa	----	----	----	----	----	----	----	----	----
Southern Africa	----	----	----	----	----	----	----	----	----
Western Africa	----	----	----	----	----	----	----	----	----
Western Indian Ocean	----	----	----	----	----	----	----	----	----
ASIA AND PACIFIC	----	----	----	----	----	----	----	----	----
Australia and New Zealand	----	----	----	----	----	----	----	----	----
Central Asia	----	----	----	----	----	----	----	----	----
North West Pacific and East Asia	----	----	----	----	----	----	----	----	----
South Asia	----	611.00	705.00	----	336.00	285.00	329.00	----	----
South East Asia	----	----	----	----	----	----	----	----	----
South Pacific	----	----	----	53.00	----	----	----	----	----
EUROPE	----	----	----	----	----	----	----	139.00	----
Central Europe	----	----	----	----	----	----	----	----	----
Eastern Europe	----	----	----	----	----	----	----	13.00	----
Western Europe	----	----	----	----	----	----	----	----	----
LATIN AMERICA AND CARIBBEAN	----	----	----	----	----	----	----	----	----
Caribbean	----	----	----	----	----	----	----	----	----
Meso-America	----	----	----	----	----	----	----	----	----
South America	----	----	----	----	----	----	----	----	----
NORTH AMERICA	11340.00	11340.00	11340.00	11340.00	11340.00	11340.00	----	----	----
North America	11340.00	11340.00	11340.00	11340.00	11340.00	11340.00	----	----	----
POLAR	----	----	----	----	----	----	----	----	----
Antarctic	----	----	----	----	----	----	----	----	----
Arctic	----	----	----	----	----	----	----	----	----
WEST ASIA	----	----	----	----	----	----	----	----	----
Arabian Peninsula	----	----	----	----	----	----	----	----	----
Mashriq	----	----	----	----	----	----	----	----	----
GLOBAL TOTALS	----	----	----	----	----	----	----	----	----
REPORTED GLOBAL TOTALS FROM ORIGINAL DATA SOURCES	****	****	****	****	****	****	****	****	****

Comments:

Rodenticides include anti-coagulants and others not elsewhere classified.

AGGREGATIONS

1.2.4 Land Use

Arable and Permanent Crops - Percent of Land Area

Units: percent of land area

Data Source: FAOSTAT (data as of July 2001)

Data Provider: Food and Agriculture Organization of the United Nations (FAO)

Years: 1961-1999

	1961	1970	1980	1990	1991	1992	1993	1994	1995	1996	1997	1998	1999
AFRICA	12.28	12.74	12.86	12.72	12.70	12.61	12.49	12.32	12.22	12.23	12.23	12.23	12.23
Central Africa	3.36	3.56	3.91	4.06	4.06	4.08	4.08	4.09	4.10	4.11	4.11	4.12	4.12
Eastern Africa	8.58	9.83	10.62	11.21	11.22	11.23	10.50	9.95	9.96	9.97	10.03	10.10	10.15
Northern Africa	4.15	4.36	4.59	4.93	4.95	5.04	5.28	5.33	5.49	5.62	5.63	5.62	5.56
Southern Africa	4.55	4.83	5.09	5.47	5.54	5.61	5.68	5.74	5.81	5.82	5.82	5.82	5.81
Western Africa	8.01	8.46	9.13	10.05	10.13	10.22	10.54	10.72	10.93	10.91	10.97	10.97	10.96
Western Indian Ocean	4.03	4.46	5.54	5.74	5.75	5.73	5.74	5.73	5.73	5.73	5.74	5.74	5.73
ASIA AND PACIFIC	5.08	5.80	6.87	7.42	7.49	7.56	7.63	7.66	7.92	7.89	7.89	7.90	7.91
Australia and New Zealand	4.26	5.70	6.00	6.53	6.26	6.39	6.27	7.32	6.74	6.98	6.73	7.18	6.48
Central Asia	11.36	11.04	10.97	10.84	10.84	11.23	11.27	11.21	10.42	10.22	9.96	9.95	9.95
North West Pacific and East Asia	10.08	9.86	9.67	12.38	12.38	12.41	12.49	12.54	12.64	12.66	12.68	12.68	12.67
South Asia	33.29	34.54	34.95	35.72	35.90	35.87	36.06	36.11	36.16	36.11	36.12	36.20	36.29
South East Asia	15.66	16.53	18.00	20.82	20.43	20.44	20.64	20.64	20.64	20.75	20.48	20.49	20.46
South Pacific	1.67	1.90	2.08	2.42	2.46	2.48	2.48	2.50	2.57	2.61	2.61	2.61	2.61
EUROPE	15.79	15.41	15.19	14.98	14.94	14.87	14.73	14.66	14.54	14.57	14.63	14.46	14.37
Central Europe	39.56	40.07	39.89	39.01	38.98	38.47	38.37	38.43	38.04	38.95	39.03	38.06	37.77
Eastern Europe	10.56	10.26	10.20	10.08	10.08	10.07	9.94	9.89	9.83	9.74	9.81	9.72	9.64
Western Europe	27.92	26.63	25.63	25.31	25.10	24.99	24.80	24.54	24.26	24.40	24.37	24.28	24.24
LATIN AMERICA AND CARIBBEAN	5.27	5.64	6.05	6.51	6.55	6.61	6.70	6.71	6.82	6.86	6.88	6.88	6.87
Caribbean	18.28	23.33	27.66	31.03	31.53	32.15	32.66	33.31	33.22	33.06	32.88	33.12	33.21
Meso-America	12.26	12.05	12.93	13.95	14.05	14.19	14.54	14.71	14.89	14.87	14.83	14.83	14.81
South America	3.92	4.71	5.77	6.21	6.27	6.32	6.35	6.35	6.63	6.60	6.61	6.61	6.63
NORTH AMERICA	13.75	14.31	14.57	16.09	16.00	16.09	16.15	16.41	16.23	16.28	16.16	16.28	16.13
North America	12.28	12.74	12.86	12.72	12.70	12.61	12.49	12.32	12.22	12.23	12.23	12.23	12.23
POLAR	3.81	3.88	4.10	4.61	4.65	4.63	4.67	4.68	4.69	4.61	4.70	4.69	4.69
Antarctic	----	----	----	----	----	----	----	----	----	----	----	----	----
Arctic	----	----	----	----	----	----	----	----	----	----	----	----	----
WEST ASIA	----	----	----	----	----	----	----	----	----	----	----	----	----
Arabian Peninsula	0.85	0.96	1.17	1.75	1.81	1.79	1.89	1.89	1.90	1.81	1.91	1.90	1.90
Mashriq	16.17	16.02	16.35	16.55	16.48	16.46	16.24	16.28	16.31	16.27	16.34	16.29	16.32
GLOBAL TOTALS	10.35	10.70	11.01	11.55	11.55	11.57	11.57	11.61	11.59	11.62	11.60	11.60	11.55
REPORTED GLOBAL TOTALS FROM ORIGINAL DATA SOURCES	10.35	10.35	10.35	10.35	10.35	10.35	10.35	10.35	10.35	10.35	10.35	10.35	10.35

Comments:

Arable and permanent crops shows the total of data from arable land and land under permanent crops:

Arable land: land under temporary crops (double-cropped areas are counted only once), temporary meadows for mowing or pasture, land under market and kitchen gardens and land temporarily fallow (less than five years). The abandoned land resulting from shifting cultivation is not included in this category. Data for "arable land" are not meant to indicate the amount of land that is potentially cultivable.

Permanent crops: land cultivated with crops that occupy the land for long periods and need not be replanted after each harvest, such as cocoa, coffee and rubber; this category includes land under flowering shrubs, fruit trees, nut trees and vines, but excludes land under trees grown for wood or timber.

AGGREGATIONS

Arable and Permanent Crops - Total

Units: thousand hectares

Data Source: FAOSTAT (data as of July 2001)

Data Provider: Food and Agriculture Organization of the United Nations (FAO)

Years: 1961-1999

	1961	1970	1980	1990	1995	1996	1997	1998	1999
AFRICA	154948.00	165752.00	177954.00	191207.00	200406.00	201451.00	202200.00	202250.00	201782.00
Central Africa	17627.00	18646.00	20518.00	21261.00	21506.00	21548.00	21571.00	21594.00	21596.00
Eastern Africa	21829.00	25029.00	27035.00	28517.00	25356.00	25362.00	25518.00	25715.00	25839.00
Northern Africa	33669.00	35398.00	37224.00	39967.00	44521.00	45568.00	45699.00	45600.00	45110.00
Southern Africa	30930.00	32820.00	34611.00	37202.00	39482.00	39546.00	39585.00	39554.00	39516.00
Western Africa	48518.00	51235.00	55307.00	60881.00	66168.00	66054.00	66450.00	66410.00	66347.00
Western Indian Ocean	2375.00	2624.00	3259.00	3379.00	3373.00	3373.00	3377.00	3377.00	3374.00
ASIA AND PACIFIC	476106.57	495683.30	504721.23	557061.53	562046.00	563675.00	559734.00	563828.00	558629.00
Australia and New Zealand	33845.00	45287.00	47686.00	51946.00	53576.00	55499.00	53480.00	57066.00	51509.00
Central Asia	44478.57	43217.30	42942.23	42460.53	40823.00	40019.00	39000.00	38972.00	38975.00
North West Pacific and East Asia	115667.00	113146.00	110958.00	142120.00	145045.00	145340.00	145566.00	145502.00	145448.00
South Asia	213183.00	221176.00	223800.00	228754.00	231527.00	231248.00	231276.00	231829.00	232370.00
South East Asia	68031.00	71829.00	78210.00	90472.00	89676.00	90148.00	88990.00	89037.00	88905.00
South Pacific	902.00	1028.00	1125.00	1309.00	1399.00	1421.00	1422.00	1422.00	1422.00
EUROPE	372450.43	363506.70	358360.77	353292.47	343011.00	343808.00	345085.00	341089.00	338981.00
Central Europe	82807.14	83871.09	83499.83	81650.07	79622.00	81521.00	81685.00	79662.00	79057.00
Eastern Europe	188972.29	183613.61	182444.94	180398.40	175916.00	174327.00	175526.00	173867.00	172528.00
Western Europe	100671.00	96022.00	92416.00	91244.00	87473.00	87960.00	87874.00	87560.00	87396.00
LATIN AMERICA AND CARIBBEAN	102471.00	116945.00	138620.00	149642.00	159761.00	159166.00	159116.00	159295.00	159559.00
Caribbean	4179.00	5332.00	6322.00	7092.00	7593.00	7557.00	7515.00	7569.00	7591.00
Meso-America	29675.00	29144.00	31279.00	33761.00	36016.00	35975.00	35891.00	35885.00	35837.00
South America	68617.00	82469.00	101019.00	108789.00	116152.00	115634.00	115710.00	115841.00	116131.00
NORTH AMERICA	225712.00	234203.00	236327.00	233729.00	224603.00	224703.00	224703.00	224703.00	224703.00
North America	225712.00	234203.00	236327.00	233729.00	224603.00	224703.00	224703.00	224703.00	224703.00
POLAR	----	----	----	----	----	----	----	----	----
Antarctic	----	----	----	----	----	----	----	----	----
Arctic	----	----	----	----	----	----	----	----	----
WEST ASIA	14200.00	14449.00	15284.00	17168.00	17451.00	17173.00	17513.00	17449.00	17460.00
Arabian Peninsula	2543.00	2887.00	3501.00	5243.00	5695.00	5444.00	5737.00	5708.00	5698.00
Mashriq	11657.00	11562.00	11783.00	11925.00	11756.00	11729.00	11776.00	11741.00	11762.00
GLOBAL TOTALS	1345888.00	1390539.00	1431267.00	1502100.00	1507278.00	1509976.00	1508351.00	1508614.00	1501114.00
REPORTED GLOBAL TOTALS FROM ORIGINAL DATA SOURCES	1346202.00	1390865.00	1431603.00	1502448.00	1507618.00	1510316.00	1508690.00	1508953.00	1501452.00

Comments:

Arable and permanent crops shows the total of data from arable land and land under permanent crops:

Arable land: land under temporary crops (double-cropped areas are counted only once), temporary meadows for mowing or pasture, land under market and kitchen gardens and land temporarily fallow (less than five years). The abandoned land resulting from shifting cultivation is not included in this category. Data for "arable land" are not meant to indicate the amount of land that is potentially cultivable.

Permanent crops: land cultivated with crops that occupy the land for long periods and need not be replanted after each harvest, such as cocoa, coffee and rubber; this category includes land under flowering shrubs, fruit trees, nut trees and vines, but excludes land under trees grown for wood or timber.

Arable Land

Units: thousand hectares

Data Source: FAOSTAT (data as of July 2001)

Data Provider: Food and Agriculture Organization of the United Nations (FAO)

Years: 1961-1999

	1961	1970	1980	1990	1995	1996	1997	1998	1999
AFRICA	140353.00	149101.00	158210.00	168440.00	176576.00	177246.00	177718.00	177684.00	177249.00
Central Africa	16298.00	16933.00	18096.00	18384.00	18632.00	18684.00	18717.00	18740.00	18742.00
Eastern Africa	19721.00	22228.00	23694.00	24912.00	21853.00	21838.00	21964.00	22141.00	22237.00
Northern Africa	31136.00	32689.00	33994.00	35786.00	40126.00	41037.00	41037.00	40893.00	40399.00
Southern Africa	28465.00	30271.00	31963.00	34441.00	36649.00	36698.00	36722.00	36672.00	36634.00
Western Africa	42599.00	44691.00	47736.00	52189.00	56539.00	56212.00	56501.00	56461.00	56461.00
Western Indian Ocean	2134.00	2289.00	2727.00	2728.00	2777.00	2777.00	2777.00	2777.00	2776.00
ASIA AND PACIFIC	449205.74	465714.93	470613.47	509529.99	506802.00	507633.00	503066.00	506921.00	501640.00
Australia and New Zealand	32685.00	44114.00	46531.00	50461.00	51717.00	53559.00	51525.00	55096.00	49534.00
Central Asia	43821.74	42406.93	42149.47	41773.99	40036.00	39236.00	38225.00	38199.00	38203.00
North West Pacific and East Asia	113134.00	109660.00	106650.00	133469.00	133493.00	133526.00	133462.00	133408.00	133363.00
South Asia	206255.00	214438.00	215977.00	219134.00	220250.00	219688.00	219372.00	219727.00	220213.00
South East Asia	53066.00	54858.00	59041.00	64331.00	60870.00	61186.00	60043.00	60052.00	59888.00
South Pacific	244.00	238.00	265.00	361.00	436.00	438.00	439.00	439.00	439.00
EUROPE	352727.26	341896.07	336341.53	332319.01	322708.00	323593.00	324787.00	320871.00	318702.00
Central Europe	78075.53	78344.21	77566.39	76037.70	74457.00	76385.00	76471.00	74501.00	73918.00
Eastern Europe	185311.74	179328.86	178240.13	176652.31	171700.00	170161.00	171452.00	169969.00	168647.00
Western Europe	89340.00	84223.00	80535.00	79629.00	76551.00	77047.00	76864.00	76401.00	76137.00
LATIN AMERICA AND CARIBBEAN	86612.00	99079.00	117147.00	125753.00	133679.00	133408.00	133179.00	133348.00	133216.00
Caribbean	3248.00	4116.00	4642.00	5218.00	5660.00	5635.00	5585.00	5639.00	5640.00
Meso-America	27142.00	26414.00	28302.00	30232.00	31979.00	31921.00	31591.00	31586.00	31434.00
South America	56222.00	68549.00	84203.00	90303.00	96040.00	95852.00	96003.00	96123.00	96142.00
NORTH AMERICA	223753.00	232348.00	234333.00	231565.00	222413.00	222513.00	222513.00	222513.00	222513.00
North America	223753.00	232348.00	234333.00	231565.00	222413.00	222513.00	222513.00	222513.00	222513.00
POLAR	----	----	----	----	----	----	----	----	----
Antarctic	----	----	----	----	----	----	----	----	----
Arctic	----	----	----	----	----	----	----	----	----
WEST ASIA	13614.00	13740.00	14289.00	15647.00	15810.00	15467.00	15774.00	15669.00	15599.00
Arabian Peninsula	2418.00	2714.00	3292.00	4980.00	5366.00	5088.00	5369.00	5326.00	5264.00
Mashriq	11196.00	11026.00	10997.00	10667.00	10444.00	10379.00	10405.00	10343.00	10335.00
GLOBAL TOTALS	1266265.00	1301879.00	1330934.00	1383254.00	1377988.00	1379860.00	1377037.00	1377006.00	1368919.00
REPORTED GLOBAL TOTALS FROM ORIGINAL DATA SOURCES	1266452.00	1302073.00	1331130.00	1383456.00	1378180.00	1380052.00	1377229.00	1377198.00	1369110.00

Comments:

Arable land: land under temporary crops (double-cropped areas are counted only once), temporary meadows for mowing or pasture, land under market and kitchen gardens and land temporarily fallow (less than five years). The abandoned land resulting from shifting cultivation is not included in this category. Data for "arable land" are not meant to indicate the amount of land that is potentially cultivable.

AGGREGATIONS

Forests and Woodland - Percent of Land Area

Units: percent of land area

Data Source: FAOSTAT (data as of July 2001)

Data Provider: Food and Agriculture Organization of the United Nations (FAO)

Years: 1961-1994

	1961	1970	1980	1981	1982	1983	1984	1985	1986	1987	1988	1989	1990	1991	1992	1993	1994
AFRICA	40.85	40.70	39.78	39.70	39.63	39.60	39.57	39.54	40.43	40.40	40.44	40.58	40.77	40.77	40.77	40.77	40.77
Central Africa	64.96	64.42	63.79	63.72	61.57	61.56	61.56	61.56	61.56	61.55	61.55	61.56	61.58	61.55	61.55	61.55	61.55
Eastern Africa	20.66	20.64	20.54	20.52	20.52	20.52	20.52	20.52	20.52	20.52	20.52	20.52	21.06	21.03	21.03	20.96	20.92
Northern Africa	6.68	6.82	6.94	6.95	6.95	6.96	6.96	6.99	6.97	7.01	7.05	7.09	7.04	6.97	6.91	7.01	6.96
Southern Africa	24.48	24.39	24.28	24.34	24.34	24.34	24.34	24.34	24.30	24.27	24.27	24.27	24.64	24.52	24.48	24.40	24.35
Western Africa	16.16	16.08	15.86	15.78	15.77	15.76	15.75	15.70	15.69	15.68	15.67	15.66	15.38	15.28	15.21	15.15	15.15
Western Indian Ocean	39.75	39.74	39.74	39.72	39.72	39.73	39.73	39.73	39.73	39.73	39.73	39.73	39.75	39.71	39.71	39.71	39.71
ASIA AND PACIFIC	51.62	50.01	49.73	49.73	49.73	49.73	49.79	49.80	50.31	50.37	50.41	50.43	50.56	50.42	50.20	50.07	49.90
Australia and New Zealand	19.25	19.25	19.25	19.25	19.25	19.25	19.25	19.25	19.25	19.25	19.25	19.25	19.28	19.20	19.20	19.20	19.20
Central Asia	18.85	18.85	19.17	19.17	19.17	19.17	19.17	19.17	19.17	19.17	19.17	19.17	19.44	19.44	16.76	16.75	16.75
North West Pacific and East Asia	18.29	17.41	16.48	16.38	16.28	16.19	16.09	15.99	15.90	15.69	15.60	15.60	15.57	15.92	15.92	15.92	15.92
South Asia	13.37	14.84	14.95	14.96	14.95	14.94	14.98	14.99	14.99	14.89	14.98	15.01	15.21	15.25	15.22	15.21	15.29
South East Asia	61.62	59.02	55.02	54.67	54.05	53.54	53.13	53.38	53.18	53.03	52.79	52.55	52.85	52.72	52.70	52.86	52.79
South Pacific	88.51	88.51	88.49	88.49	88.49	88.38	88.38	88.38	88.38	88.38	88.38	88.38	88.51	88.28	88.28	88.28	88.28
EUROPE	44.83	45.17	46.03	46.04	46.04	46.05	46.05	46.06	46.08	46.09	46.09	46.09	46.66	46.67	41.23	41.27	41.21
Central Europe	29.38	29.85	30.19	30.19	30.20	30.22	30.26	30.27	30.28	30.29	30.31	30.25	30.30	30.35	29.32	29.27	29.62
Eastern Europe	49.58	49.58	50.43	50.43	50.43	50.43	50.43	50.43	50.43	50.43	50.43	50.43	51.13	51.13	44.08	44.12	43.99
Western Europe	30.26	32.20	33.45	33.48	33.48	33.55	33.54	33.57	33.71	33.72	33.75	33.78	34.00	34.01	34.02	34.11	34.14
LATIN AMERICA AND CARIBBEAN	24.88	24.79	24.63	24.61	24.23	24.23	24.23	24.23	24.21	24.21	24.22	24.23	24.30	24.22	24.18	24.17	24.14
Caribbean	17.90	18.63	19.48	19.97	20.32	20.05	20.22	20.43	20.45	19.54	19.39	18.64	18.37	18.32	18.31	19.62	19.62
Meso-America	35.48	32.56	29.40	29.00	28.67	28.35	28.59	28.47	28.38	28.71	28.83	28.76	29.10	29.02	28.99	28.99	28.99
South America	54.29	52.82	52.93	52.97	53.02	53.06	53.11	53.12	53.73	53.76	53.79	53.83	53.94	53.79	53.54	53.38	53.18
NORTH AMERICA	24.19	23.85	23.09	23.02	22.91	22.81	22.73	22.73	22.68	22.57	22.53	22.50	22.61	22.69	22.38	22.40	22.40
North America	40.85	40.70	39.78	39.70	39.63	39.60	39.57	39.54	40.43	40.40	40.44	40.58	40.77	40.77	40.77	40.77	40.77
POLAR	1.87	1.87	1.75	1.62	1.47	1.35	1.36	1.23	1.25	1.28	1.28	1.30	1.33	1.33	1.33	1.33	1.33
Antarctic	----	----	----	----	----	----	----	----	----	----	----	----	----	----	----	----	----
Arctic	0.03	0.03	0.03	0.03	0.03	0.03	0.03	0.03	0.03	0.03	0.03	0.03	0.03	0.03	0.03	0.03	0.03
WEST ASIA	0.03	0.03	0.03	0.03	0.03	0.03	0.03	0.03	0.03	0.03	0.03	0.03	0.03	0.03	0.03	0.03	0.03
Arabian Peninsula	2.07	2.05	1.89	1.73	1.55	1.39	1.41	1.24	1.26	1.30	1.30	1.33	1.37	1.37	1.37	1.37	1.37
Mashriq	1.03	1.13	1.12	1.15	1.16	1.17	1.17	1.19	1.21	1.22	1.20	1.17	1.15	1.15	1.15	1.15	1.15
GLOBAL TOTALS	33.89	33.56	33.31	33.27	33.14	33.11	33.10	33.09	33.28	33.26	33.26	33.28	33.47	33.46	32.35	32.34	32.30
REPORTED GLOBAL TOTALS FROM ORIGINAL DATA SOURCES	33.89	33.89	33.89	33.89	33.89	33.89	33.89	33.89	33.89	33.89	33.89	33.89	33.89	33.89	33.89	33.89	33.89

Comments:

Forests and woodland: land under natural or planted stands of trees, whether productive or not. This category includes land from which forests have been cleared but that will be reforested in the foreseeable future, but it excludes woodland or forest used only for recreation purposes. The question of shrub land, savannah, etc. raises the same problem as in the category "permanent meadows and pastures".

The forest and woodlands reporting was done from an agricultural perspective and handled forests as a side issue. Partly for this reason, the time series was discontinued in mid-1990s as it did not correspond to FAO figures on forest developments.

Irrigated Land

Units: thousand hectares

Data Source: FAOSTAT (data as of July 2001)

Data Provider: Food and Agriculture Organization of the United Nations (FAO)

Years: 1961-1999

	1961	1970	1980	1990	1995	1996	1997	1998	1999
AFRICA	7410.00	8483.00	9491.00	11235.00	12380.00	12451.00	12458.00	12520.00	12538.00
Central Africa	27.00	35.00	55.00	75.00	90.00	90.00	90.00	90.00	90.00
Eastern Africa	276.00	315.00	389.00	480.00	567.00	567.00	567.00	567.00	568.00
Northern Africa	5373.00	6001.00	6183.00	7006.00	7873.00	7914.00	7911.00	7951.00	7965.00
Southern Africa	978.00	1251.00	1570.00	1838.00	1901.00	1931.00	1936.00	1956.00	1960.00
Western Africa	445.00	531.00	628.00	808.00	832.00	832.00	834.00	834.00	833.00
Western Indian Ocean	311.00	350.00	666.00	1028.00	1117.00	1117.00	1120.00	1122.00	1122.00
ASIA AND PACIFIC	91867.27	111928.87	136125.86	156786.90	169443.00	171534.00	174509.00	178351.00	181347.00
Australia and New Zealand	1078.00	1587.00	1683.00	2112.00	2685.00	2675.00	2665.00	2650.00	2536.00
Central Asia	4598.27	5429.87	8413.86	10174.90	10207.00	10087.00	10021.00	10204.00	10222.00
North West Pacific and East Asia	35006.00	43230.00	50987.00	53656.00	55354.00	56407.00	57229.00	58260.00	59102.00
South Asia	43135.00	52588.00	63251.00	76029.00	85023.00	86063.00	88156.00	90624.00	92720.00
South East Asia	8049.00	9093.00	11790.00	14814.00	16171.00	16299.00	16435.00	16610.00	16764.00
South Pacific	----	----	----	----	----	----	----	----	----
EUROPE	14601.73	18027.13	25698.14	31411.10	31842.00	31744.00	31778.00	31458.00	31357.00
Central Europe	3094.09	4441.82	7128.62	9420.40	9194.00	9166.00	9122.00	9099.00	8979.00
Eastern Europe	4786.63	5652.30	8758.52	10591.70	10580.00	10258.00	10080.00	9743.00	9668.00
Western Europe	6721.00	7933.00	9811.00	11399.00	12068.00	12320.00	12576.00	12616.00	12710.00
LATIN AMERICA AND CARIBBEAN	8260.00	10190.00	13807.00	16786.00	18123.00	18263.00	18411.00	18553.00	18601.00
Caribbean	441.00	705.00	1072.00	1267.00	1282.00	1282.00	1284.00	1289.00	1293.00
Meso-America	3158.00	3812.00	5343.00	6020.00	6855.00	6963.00	6973.00	6975.00	6982.00
South America	4661.00	5673.00	7392.00	9499.00	9986.00	10018.00	10154.00	10289.00	10326.00
NORTH AMERICA	14350.00	16421.00	21178.00	21618.00	22520.00	22720.00	23002.00	23020.00	23120.00
North America	14350.00	16421.00	21178.00	21618.00	22520.00	22720.00	23002.00	23020.00	23120.00
POLAR	----	----	----	----	----	----	----	----	----
Antarctic	----	----	----	----	----	----	----	----	----
Arctic	----	----	----	----	----	----	----	----	----
WEST ASIA	2490.00	2744.00	3407.00	6458.00	7062.00	7110.00	7165.00	7215.00	7191.00
Arabian Peninsula	602.00	702.00	985.00	2080.00	2257.00	2262.00	2268.00	2270.00	2273.00
Mashriq	1888.00	2042.00	2422.00	4378.00	4805.00	4848.00	4897.00	4945.00	4918.00
GLOBAL TOTALS	138979.00	167794.00	209707.00	244295.00	261370.00	263822.00	267323.00	271117.00	274154.00
REPORTED GLOBAL TOTALS FROM ORIGINAL DATA SOURCES	138989.00	167803.00	209716.00	244305.00	261380.00	263833.00	267335.00	271129.00	274166.00

Comments:

Irrigated land: data on irrigation relates to areas equipped to provide water to the crops. These include areas equipped for full and partial control irrigation, spate irrigation areas and equipped wetland or inland valley bottoms.

Land in Permanent Crops

Units: thousand hectares

Data Source: FAOSTAT (data as of July 2001)

Data Provider: Food and Agriculture Organization of the United Nations (FAO)

Years: 1961-1999

	1961	1970	1980	1990	1992	1993	1994	1995	1996	1997	1998	1999
AFRICA	14595.00	16650.00	19744.00	22767.00	23108.00	23125.00	23452.00	23830.00	24191.00	24482.00	24566.00	24596.00
Central Africa	1329.00	1713.00	2422.00	2877.00	2866.00	2869.00	2869.00	2874.00	2864.00	2854.00	2854.00	2854.00
Eastern Africa	2108.00	2801.00	3341.00	3605.00	3637.00	3532.00	3502.00	3503.00	3524.00	3554.00	3574.00	3602.00
Northern Africa	2533.00	2709.00	3230.00	4181.00	4264.00	4309.00	4342.00	4395.00	4517.00	4662.00	4707.00	4711.00
Southern Africa	2465.00	2549.00	2648.00	2761.00	2790.00	2805.00	2820.00	2833.00	2848.00	2863.00	2882.00	2882.00
Western Africa	5919.00	6543.00	7571.00	8692.00	8914.00	9002.00	9321.00	9629.00	9842.00	9949.00	9949.00	9949.00
Western Indian Ocean	241.00	335.00	532.00	651.00	637.00	608.00	598.00	596.00	596.00	600.00	600.00	598.00
ASIA AND PACIFIC	26937.20	29977.96	34118.11	47556.21	49308.00	51401.00	53617.00	55239.00	56037.00	56663.00	56902.00	56984.00
Australia and New Zealand	1160.00	1173.00	1155.00	1485.00	1513.00	1586.00	1737.00	1859.00	1940.00	1955.00	1970.00	1975.00
Central Asia	697.20	823.96	808.11	716.21	839.00	799.00	798.00	787.00	783.00	775.00	773.00	772.00
North West Pacific and East Asia	2533.00	3486.00	4308.00	8651.00	8948.00	9839.00	10431.00	11552.00	11814.00	12104.00	12094.00	12085.00
South Asia	6928.00	6738.00	7823.00	9620.00	10454.00	10759.00	11047.00	11277.00	11560.00	11904.00	12102.00	12157.00
South East Asia	14965.00	16971.00	19169.00	26141.00	26602.00	27466.00	28653.00	28806.00	28962.00	28947.00	28985.00	29017.00
South Pacific	654.00	786.00	855.00	943.00	952.00	952.00	951.00	958.00	978.00	978.00	978.00	978.00
EUROPE	19682.80	21597.04	22003.89	20943.79	21297.00	21232.00	21305.00	20303.00	20215.00	20298.00	20218.00	20279.00
Central Europe	4731.60	5519.62	5926.74	5609.85	5768.00	5924.00	5841.00	5165.00	5136.00	5214.00	5161.00	5139.00
Eastern Europe	3620.21	4278.42	4196.15	3718.94	4277.00	4227.00	4388.00	4216.00	4166.00	4074.00	3898.00	3881.00
Western Europe	11331.00	11799.00	11881.00	11615.00	11252.00	11081.00	11076.00	10922.00	10913.00	11010.00	11159.00	11259.00
LATIN AMERICA AND CARIBBEAN	15857.00	17864.00	21471.00	23887.00	24995.00	25435.00	25837.00	26080.00	25756.00	25935.00	25945.00	26341.00
Caribbean	929.00	1214.00	1678.00	1872.00	1918.00	1918.00	1918.00	1931.00	1920.00	1928.00	1928.00	1949.00
Meso-America	2533.00	2730.00	2977.00	3529.00	3670.00	3878.00	3974.00	4037.00	4054.00	4300.00	4299.00	4403.00
South America	12395.00	13920.00	16816.00	18486.00	19407.00	19639.00	19945.00	20112.00	19782.00	19707.00	19718.00	19989.00
NORTH AMERICA	1959.00	1855.00	1994.00	2164.00	2180.00	2180.00	2185.00	2190.00	2190.00	2190.00	2190.00	2190.00
North America	1959.00	1855.00	1994.00	2164.00	2180.00	2180.00	2185.00	2190.00	2190.00	2190.00	2190.00	2190.00
POLAR	----	----	----	----	----	----	----	----	----	----	----	---
Antarctic	----	----	----	----	----	----	----	----	----	----	----	---
Arctic	----	----	----	----	----	----	----	----	----	----	----	---
WEST ASIA	586.00	709.00	995.00	1521.00	1618.00	1526.00	1581.00	1641.00	1706.00	1739.00	1780.00	1861.00
Arabian Peninsula	125.00	173.00	209.00	263.00	283.00	304.00	314.00	329.00	356.00	368.00	382.00	434.00
Mashriq	461.00	536.00	786.00	1258.00	1335.00	1222.00	1267.00	1312.00	1350.00	1371.00	1398.00	1427.00
GLOBAL TOTALS	79617.00	88653.00	100326.00	118839.00	122506.00	124899.00	127977.00	129283.00	130095.00	131307.00	131601.00	132251.00
REPORTED GLOBAL TOTALS FROM ORIGINAL DATA SOURCES	79750.00	88791.00	100473.00	118992.00	122660.00	125053.00	128131.00	129438.00	130250.00	131461.00	131755.00	132405.00

Comments:

Permanent crops: land cultivated with crops that occupy the land for long periods and need not be replanted after each harvest, such as cocoa, coffee and rubber; this category includes land under flowering shrubs, fruit trees, nut trees and vines, but excludes land under trees grown for wood or timber.

Permanent Pasture - Percent of Land Area

Units: percent of land area

Data Source: FAOSTAT (data as of July 2001)

Data Provider: Food and Agriculture Organization of the United Nations (FAO)

Years: 1961-1999

	1961	1970	1980	1990	1991	1992	1993	1994	1995	1996	1997	1998	1999
AFRICA	31.31	31.17	31.21	31.55	31.55	31.58	30.84	30.89	30.91	30.91	30.90	31.02	31.02
Central Africa	15.29	15.26	15.22	15.22	15.22	15.22	15.22	15.22	15.21	15.22	15.23	15.24	15.24
Eastern Africa	46.66	46.48	46.41	46.22	46.20	46.18	37.37	37.38	37.39	37.39	37.39	37.39	37.39
Northern Africa	23.21	23.59	24.10	25.09	25.07	25.15	25.15	25.28	25.28	25.26	25.25	25.66	25.66
Southern Africa	49.76	48.83	48.56	48.72	48.77	48.81	48.84	48.88	48.92	48.94	48.94	48.95	48.95
Western Africa	28.24	28.19	28.05	28.23	28.22	28.22	28.30	28.30	28.37	28.36	28.31	28.31	28.31
Western Indian Ocean	40.85	40.85	40.85	40.85	40.85	40.85	40.85	40.85	40.86	40.86	40.86	40.86	40.86
ASIA AND PACIFIC	33.13	34.63	36.03	37.47	37.46	38.08	37.80	37.79	37.60	37.60	37.57	37.51	37.39
Australia and New Zealand	55.85	56.94	56.97	54.07	54.19	54.42	53.80	53.78	53.63	53.62	53.49	53.24	52.68
Central Asia	54.20	56.47	57.76	58.74	58.60	64.31	64.30	64.45	63.18	63.21	63.22	63.22	63.23
North West Pacific and East Asia	33.03	36.06	39.91	45.73	45.77	45.55	45.16	45.11	45.10	45.10	45.10	45.10	45.10
South Asia	14.94	14.83	14.73	14.64	14.66	14.64	14.58	14.57	14.54	14.54	14.55	14.55	14.55
South East Asia	3.67	3.67	3.67	4.25	3.93	3.95	3.95	3.94	3.94	3.89	3.89	3.89	3.89
South Pacific	0.90	0.95	1.07	1.07	1.07	1.08	1.08	1.07	1.09	1.09	1.09	1.09	1.09
EUROPE	8.07	8.17	8.12	8.15	8.06	8.37	8.35	8.36	8.34	8.42	8.43	8.48	8.52
Central Europe	15.65	15.49	15.22	16.18	16.38	15.92	15.93	15.94	15.94	16.21	16.13	16.08	15.96
Eastern Europe	4.91	5.11	5.23	5.32	5.30	5.80	5.78	5.76	5.75	5.82	5.87	5.91	5.95
Western Europe	19.38	19.08	18.37	17.55	16.90	16.74	16.70	16.84	16.82	16.75	16.69	16.81	16.97
LATIN AMERICA AND CARIBBEAN	25.12	26.91	28.32	29.58	29.69	29.79	29.74	29.77	29.79	29.82	29.81	29.83	29.83
Caribbean	22.89	25.47	25.87	27.03	26.84	26.02	25.16	24.73	24.25	23.81	23.42	23.28	23.32
Meso-America	34.56	35.01	35.58	37.53	37.74	37.96	38.24	38.45	38.63	38.65	38.64	38.64	38.64
South America	23.85	25.81	27.35	28.52	28.62	28.71	28.63	28.63	28.64	28.67	28.67	28.69	28.70
NORTH AMERICA	15.45	14.50	14.44	14.59	14.59	14.59	14.59	14.59	14.59	14.59	14.59	14.59	14.59
North America	15.45	14.50	14.44	14.59	14.59	14.59	14.59	14.59	14.59	14.59	14.59	14.59	14.59
POLAR	0.69	0.69	0.69	0.69	0.69	0.69	0.69	0.69	0.69	0.69	0.69	0.69	0.69
Antarctic	----	----	----	----	----	----	----	----	----	----	----	----	----
Arctic	0.69	0.69	0.69	0.69	0.69	0.69	0.69	0.69	0.69	0.69	0.69	0.69	0.69
WEST ASIA	31.13	30.82	31.05	40.32	40.35	40.39	53.86	53.88	53.88	53.89	53.88	53.88	53.88
Arabian Peninsula	34.11	34.11	34.11	45.78	45.79	45.79	62.44	62.45	62.45	62.45	62.45	62.45	62.45
Mashriq	18.68	17.12	18.29	17.59	17.69	17.86	18.08	18.19	18.18	18.22	18.17	18.15	18.15
GLOBAL TOTALS	24.18	24.69	25.28	26.21	26.21	26.45	26.58	26.60	26.55	26.57	26.56	26.59	26.56
REPORTED GLOBAL TOTALS FROM ORIGINAL DATA SOURCES	****	****	****	****	****	****	****	****	****	****	****	****	****

Comments:

Permanent pasture: land used permanently (five years or more) for herbaceous forage crops, either cultivated or growing wild (wild prairie or grazing land). The dividing line between this category and the category "forests and woodland"; is rather indefinite, especially in the case of shrubs, savannah, etc., which may have been reported under either of these two categories.

1.3 Forests

The Forest Stewardship Council (FSC) is an international non-profit organisation founded in 1993 to support environmentally appropriate, socially beneficial, and economically viable management of the world's forests.

It is the only independent third-party certification body for forests and forest products. Membership in the FSC consists of representatives from social and environmental groups, timber trade and forestry professionals, indigenous peoples' organizations, community forestry groups, and forest product certification organizations such as the Certified Forest Products Council. The nine-member board has representatives from Sweden, USA, Cameroon, Indonesia, Malaysia, Brazil, Ecuador and Canada. These representatives are mostly from non-governmental organizations (NGOs).

The FSC is introducing an international labelling scheme for forest products, which provides a credible guarantee that the product comes from a well-managed forest. According to the FSC all forest products carrying their logo have been independently certified as coming from forests that meet the internationally recognised FSC Principles and Criteria of Forest Stewardship. For more detailed information on FSC Principles and Criteria please refer to Forest Stewardship Council (FSC) On-line Data Service, available at: http://www.fscoax.org (FSC, Oaxaca, Mexico, February 2000).

1.3.1   Vegetation and Land Cover

Forest Change

Forest plantation and natural forests are included in the term forest, a term that refers to land with a tree cover of more than 10 percent and area of more than 0.5 ha. Forests are determined both by the presence of trees and the absence of other predominant land uses. The trees should be able to reach a minimum height of 5 m. Young stands that have not yet reached, but are expected to reach, a crown density of 10 percent and tree height of 5 m are included under forest, as are temporarily unstocked areas. The term includes forests used for purposes of production, protection, multiple use or conservation (i.e. forest in national parks, nature reserves and other protected areas), as well as forest stands on agricultural lands (for example windbreaks and shelterbelts of trees with a width of more than 20 m) and rubberwood plantations and cork oak stands. The term specifically excludes stands of trees established primarily for agricultural production, for example fruit tree plantations. It also excludes trees planted in agroforestry systems.

Units: thousand hectares

Data Source: Global Forest Resources Assessment 2000

Data Provider: Food and Agriculture Organization of the United Nations (FAO)

Years: 2000

Annual Natural Forest Change

	2000
AFRICA	----
Central Africa	----
Eastern Africa	----
Northern Africa	----
Southern Africa	----
Western Africa	----
Western Indian Ocean	----
ASIA AND PACIFIC	----
Australia and New Zealand	----
Central Asia	----
North West Pacific and East Asia	652.20
South Asia	-1731.50
South East Asia	-3011.40
South Pacific	-137.36
EUROPE	----
Central Europe	----
Eastern Europe	----
Western Europe	----
LATIN AMERICA AND CARIBBEAN	-5234.42
Caribbean	-8.92
Meso-America	-1054.10
South America	-4171.40
NORTH AMERICA	----
North America	----
POLAR	----
Antarctic	----
Arctic	----
WEST ASIA	-17.46
Arabian Peninsula	7.47
Mashriq	-24.93
GLOBAL TOTALS	----
REPORTED GLOBAL TOTALS FROM ORIGINAL DATA SOURCES	-12500.00

Annual Forest Plantation Change

	2000
AFRICA	193.94
Central Africa	6.30
Eastern Africa	12.40
Northern Africa	89.58
Southern Africa	----
Western Africa	59.40
Western Indian Ocean	5.68
ASIA AND PACIFIC	----
Australia and New Zealand	----
Central Asia	----
North West Pacific and East Asia	1153.80
South Asia	1634.50
South East Asia	686.40
South Pacific	15.36
EUROPE	----
Central Europe	----
Eastern Europe	----
Western Europe	----
LATIN AMERICA AND CARIBBEAN	622.42
Caribbean	30.92
Meso-America	82.10
South America	509.40
NORTH AMERICA	----
North America	----
POLAR	----
Antarctic	----
Arctic	----
WEST ASIA	25.46
Arabian Peninsula	0.53
Mashriq	24.93
GLOBAL TOTALS	4457.88
REPORTED GLOBAL TOTALS FROM ORIGINAL DATA SOURCES	4457.88

Comments:

Annual forest plantation change is the annual change of a forest established by planting and/or seeding in the process of afforestation or reforestation. It consists of introduced species or, in some cases, indigenous species.

AGGREGATIONS

Units: thousand hectares

Data Source: Global Forest Resources Assessment 2000

Data Provider: Food and Agriculture Organization of the United Nations (FAO)

Years: 1990-00

Forest Average Annual Change – Total

	1990-00
AFRICA	-5262.00
Central Africa	-904.00
Eastern Africa	-336.00
Northern Africa	-925.00
Southern Africa	-1714.00
Western Africa	-1265.00
Western Indian Ocean	-118.00
ASIA AND PACIFIC	-774.00
Australia and New Zealand	-243.00
Central Asia	269.00
North West Pacific and East Asia	1744.00
South Asia	-97.00
South East Asia	-2325.00
South Pacific	-122.00
EUROPE	930.00
Central Europe	132.00
Eastern Europe	440.00
Western Europe	358.00
LATIN AMERICA AND CARIBBEAN	-4669.00
Caribbean	14.00
Meso-America	-972.00
South America	-3711.00
NORTH AMERICA	----
North America	388.00
POLAR	----
Antarctic	----
Arctic	----
WEST ASIA	-1.00
Arabian Peninsula	-1.00
Mashriq	0.00
GLOBAL TOTALS	-9388.00
REPORTED GLOBAL TOTALS FROM ORIGINAL DATA SOURCES	-9391.00

Forest Average Annual Change – Rate

	1990-00
AFRICA	-0.74
Central Africa	-0.37
Eastern Africa	-0.87
Northern Africa	-1.22
Southern Africa	-0.70
Western Africa	-1.53
Western Indian Ocean	-0.90
ASIA AND PACIFIC	-0.06
Australia and New Zealand	-0.15
Central Asia	1.55
North West Pacific and East Asia	0.89
South Asia	-0.06
South East Asia	-1.05
South Pacific	-0.37
EUROPE	0.09
Central Europe	0.26
Eastern Europe	0.04
Western Europe	0.37
LATIN AMERICA AND CARIBBEAN	-0.46
Caribbean	0.33
Meso-America	-1.24
South America	-0.40
NORTH AMERICA	0.10
North America	0.10
POLAR	----
Antarctic	----
Arctic	----
WEST ASIA	0.02
Arabian Peninsula	0.03
Mashriq	-0.01
GLOBAL TOTALS	-0.21
REPORTED GLOBAL TOTALS FROM ORIGINAL DATA SOURCES	-0.22

Comments:

Forest average annual change – total is the net change in forests and includes expansion of forest plantations and losses and gains in the area of natural forests.

Comments:

Forest average annual change – rate is the net change in forests and includes expansion of forest plantations and losses and gains in the area of natural forests.

Total forest includes natural forests and forest plantations. The term is used to refer to land with a tree cover of more than 10 percent and area of more than 0.5 ha. Forests are determined both by the presence of trees and the absence of other predominant land uses. The trees should be able to reach a minimum height of 5 m. Young stands that have not yet reached, but are expected to reach, a crown density of 10 percent and tree height of 5 m are included under forest, as are temporarily unstocked areas. The term includes forests used for purposes of production, protection, multiple use or conservation (that is to say forest in national parks, nature reserves and other protected areas), as well as forest stands on agricultural lands (for example windbreaks and shelterbelts of trees with a width of more than 20 m) and rubberwood plantations and cork oak stands. The term specifically excludes stands of trees established primarily for agricultural production, for example fruit tree plantations. It also excludes trees planted in agroforestry systems.

Units: hectares

Data Source: Forest Stewardship Council On-line Data Service (data as of January 2002)

Data Provider: Forest Stewardship Council (FSC)

Years: 1996-2001

	1996	1997	1998	1999	2000	2001
AFRICA	26308	89626	254558	804513	855700	980077
Central Africa	0	0	0	0	0	0
Eastern Africa	0	0	0	0	0	0
Northern Africa	0	0	0	0	0	0
Southern Africa	26308	89626	250248	800203	851390	975767
Western Africa	0	0	4310	4310	4310	4310
Western Indian Ocean	0	0	0	0	0	0
ASIA AND PACIFIC	12726	67809	70114	70114	565380	793096
Australia and New Zealand	0	0	2305	2305	378701	502404
Central Asia	0	0	0	0	0	0
North West Pacific and East Asia	0	0	0	0	3319	6390
South Asia	12726	12726	12726	12726	12726	18000
South East Asia	0	55083	55083	55083	131232	226900
South Pacific	0	0	0	0	39402	39402
EUROPE	3133871	3809188	5561549	9258239	15032370	17030126
Central Europe	1555563	1555563	2240069	2853006	3555605	4498226
Eastern Europe	0	0	0	0	184515	418715
Western Europe	1578308	2253625	3321480	6405233	11292250	12113185
LATIN AMERICA AND CARIBBEAN	223213	425605	826452	1598880	2179431	3600873
Caribbean	0	0	0	0	0	0
Meso-America	122251	244072	262389	413980	662270	1263816
South America	100962	181533	564063	1184900	1517161	2337057
NORTH AMERICA	598491	873349	1442287	1851719	3031225	3153267
North America	598491	873349	1442287	1851719	3031225	3153267
POLAR	0	0	0	0	0	0
Antarctic	0	0	0	0	0	0
Arctic	0	0	0	0	0	0
WEST ASIA	0	0	0	0	0	0
Arabian Peninsula	0	0	0	0	0	0
Mashriq	0	0	0	0	0	0
GLOBAL TOTALS	3994609	5265577	8154960	13583465	21664106	25557439
REPORTED GLOBAL TOTALS FROM ORIGINAL DATA SOURCES	3994609	3994609	3994609	3994609	3994609	3994609

Forests Extent

Units: thousand hectares

Data Source: Global Forest Resources Assessment 2000

Data Provider: Food and Agriculture Organization of the United Nations (FAO)

Total Forest Extent

Years: 1990, 2000

	1990	2000
AFRICA	702502	649866
Central Africa	249370	240329
Eastern Africa	38753	35386
Northern Africa	77146	67889
Southern Africa	239085	221938
Western Africa	85112	72472
Western Indian Ocean	13036	11852
ASIA AND PACIFIC	734036	726284
Australia and New Zealand	164915	162485
Central Asia	16591	19275
North West Pacific and East Asia	195218	212664
South Asia	86294	85315
South East Asia	234662	211407
South Pacific	36356	35138
EUROPE	1042041	1051326
Central Europe	48948	50268
Eastern Europe	870732	875136
Western Europe	122361	125922
LATIN AMERICA AND CARIBBEAN	1011049	964358
Caribbean	5580	5711
Meso-America	82738	73029
South America	922731	885618
NORTH AMERICA	466684	470564
North America	466684	470564
POLAR	----	----
Antarctic	----	----
Arctic	----	----
WEST ASIA	3675	3663
Arabian Peninsula	2292	2281
Mashriq	1383	1382
GLOBAL TOTALS	3959987	3866061
REPORTED GLOBAL TOTALS FROM ORIGINAL DATA SOURCES	3959987	3959987

Total Forest Extent

Years: 1990, 2000

	2000
AFRICA	8035
Central Africa	314.
Eastern Africa	849
Northern Africa	2333
Southern Africa	2385
Western Africa	1781
Western Indian Ocean	371.
ASIA AND PACIFIC	115905.
Australia and New Zealand	2585.
Central Asia	384.
North West Pacific and East Asia	55765.
South Asia	36936.
South East Asia	19971
South Pacific	262
EUROPE	34192.
Central Europe	4111.
Eastern Europe	22194.
Western Europe	7887
LATIN AMERICA AND CARIBBEAN	11750.
Caribbean	566.
Meso-America	728.
South America	10455.
NORTH AMERICA	-
North America	-
POLAR	-
Antarctic	-
Arctic	-
WEST ASIA	610.
Arabian Peninsula	324.
Mashriq	286.
GLOBAL TOTALS	186733.
REPORTED GLOBAL TOTALS FROM ORIGINAL DATA SOURCES	186733.

Comments:

Forest plantation is a forest established by planting and/or seeding in the process of afforestation or re-forestation. It consists of introduced species or, in some cases, indigenous species.

Total forest includes natural forests and forest plantations. The term is used to refer to land with a tree cover of more than 10 percent and area of more than 0.5 ha. Forests are determined both by the presence of trees and the absence of other predominant land uses. The trees should be able to reach a minimum height of 5 m. Young stands that have not yet reached, but are expected to reach, a crown density of 10 percent and tree height of 5 m are included under forest, as are temporarily unstocked areas. The term includes forests used for purposes of production, protection, multiple use or conservation (that is to say forest in national parks, nature reserves and other protected areas), as well as forest stands on agricultural lands (for example windbreaks and shelterbelts of trees with a width of more than 20 m) and rubberwood plantations and cork oak stands. The term specifically excludes stands of trees established primarily for agricultural production, for example fruit tree plantations. It also excludes trees planted in agroforestry systems.

Natural Forests Extent

Units: thousand hectares

Data Source: Global Forest Resources Assessment 2000

Data Provider: Food and Agriculture Organization of the United Nations (FAO)

Years: 2000

	2000
AFRICA	639893.19
Central Africa	238235.50
Eastern Africa	34530.80
Northern Africa	65403.83
Southern Africa	219552.06
Western Africa	70690.80
Western Indian Ocean	11480.20
ASIA AND PACIFIC	583879.26
Australia and New Zealand	159900.00
Central Asia	18890.60
North West Pacific and East Asia	131796.00
South Asia	47026.40
South East Asia	191433.20
South Pacific	34833.06
EUROPE	980092.60
Central Europe	46411.40
Eastern Europe	852941.90
Western Europe	80739.30
LATIN AMERICA AND CARIBBEAN	951718.81
Caribbean	4255.81
Meso-America	72300.10
South America	875162.90
NORTH AMERICA	----
North America	----
POLAR	----
Antarctic	----
Arctic	----
WEST ASIA	2603.67
Arabian Peninsula	1508.23
Mashriq	1095.45
GLOBAL TOTALS	3367942.50
REPORTED GLOBAL TOTALS FROM ORIGINAL DATA SOURCES	3367942.50

Comments:

Natural forest is a forest composed of indigenous trees, and not classified as forest plantation.

Roundwood – Production

Units: thousand cubic meters

Data Source: FAOSTAT (data as of August 2001)

Data Provider: Food and Agriculture Organization of the United Nations (FAO)

Years: 1961-2000

	1961	1970	1980	1990	1995	1996	1997	1998	1999	2000
AFRICA	195095.00	259910.60	337967.50	438708.70	542100.69	552355.04	565759.12	572471.94	582485.74	584171.6
Central Africa	29097.00	37105.00	46861.00	63808.00	75281.00	76754.00	79026.00	79534.00	81010.00	81010.0
Eastern Africa	41458.00	54106.60	71112.00	90408.70	141172.60	144117.88	149079.86	151865.74	155394.39	157080.3
Northern Africa	8279.00	10255.00	13469.70	17132.00	18958.40	19279.80	18921.80	20253.80	19976.80	19976.8
Southern Africa	41201.00	61867.00	79422.00	103457.20	122329.10	124271.20	126708.10	125013.00	126636.60	126636.6
Western Africa	71099.00	90919.00	120712.00	155652.80	174548.70	177732.20	181791.20	185670.80	189039.20	189039.2
Western Indian Ocean	3961.00	5658.00	6390.80	8250.00	9810.89	10199.96	10232.16	10134.60	10428.75	10428.7
ASIA AND PACIFIC	593134.12	712831.04	902238.29	1073243.57	1152186.99	1174504.01	1182931.59	1150275.41	1156469.60	1156469.6
Australia and New Zealand	19530.00	20739.00	26892.00	31910.00	39171.00	38433.00	39717.00	39186.00	40891.00	40891.0
Central Asia	----	----	----	----	----	----	----	----	----	---
North West Pacific and East Asia	219049.00	229697.60	280891.02	318582.02	336164.00	343519.00	341469.00	327194.00	319714.00	319714.0
South Asia	184426.00	231004.99	289610.41	353058.41	386229.69	393372.70	402044.69	406076.00	412988.99	412988.9
South East Asia	165675.00	225403.00	296242.61	359806.52	379832.50	388413.51	388999.50	367626.51	372717.51	372717.5
South Pacific	3806.50	5276.10	7944.30	9173.70	10450.80	10450.80	10348.80	10192.90	10158.10	10158.1
EUROPE	667388.38	756163.35	710691.05	789811.87	505754.00	485745.10	540366.00	486528.34	553349.22	588971.1
Central Europe	101082.46	140233.84	121993.35	106773.17	100288.00	101507.10	106109.00	108167.34	118331.62	124603.0
Eastern Europe	334250.93	366628.51	339583.70	367961.70	126225.00	123325.00	162707.00	111327.00	160217.60	174302.1
Western Europe	232055.00	249301.00	249114.00	315077.00	279241.00	260913.00	271550.00	267034.00	274800.00	290066.0
LATIN AMERICA AND CARIBBEAN	160777.30	202305.90	296295.39	345385.02	362072.00	360031.00	360791.00	361384.00	359914.60	359914.6
Caribbean	6229.00	6584.40	8082.90	9683.40	9393.10	9447.20	9571.30	9553.00	9456.00	9456.0
Meso-America	22601.30	31632.50	44153.20	52580.90	55967.60	56455.30	58677.30	59861.00	60947.60	60947.6
South America	131947.00	164089.00	244059.29	283120.72	296711.30	294128.50	292542.40	291970.00	289511.00	289511.0
NORTH AMERICA	383339.02	449570.01	570619.02	672227.00	687742.00	680349.00	677058.01	679971.33	683299.83	686092.8
North America	383339.02	449570.01	570619.02	672227.00	687742.00	680349.00	677058.01	679971.33	683299.83	686092.8
POLAR	----	----	----	----	----	----	----	----	----	---
Antarctic	----	----	----	----	----	----	----	----	----	---
Arctic	----	----	----	----	----	----	----	----	----	---
WEST ASIA	----	----	----	----	----	----	----	----	----	---
Arabian Peninsula	----	----	----	----	----	----	----	----	----	---
Mashriq	379.00	450.00	479.60	450.60	508.55	518.55	622.55	645.55	650.55	650.5
GLOBAL TOTALS	2000112.82	2381230.90	2818290.84	3319826.76	3250364.23	3253502.69	3327528.27	3251276.57	3336169.54	3376270.3
REPORTED GLOBAL TOTALS FROM ORIGINAL DATA SOURCES	2000112.82	2381230.90	2818290.84	3319826.76	3250364.23	3253502.69	3327590.47	3251276.57	3336169.54	3376270.3

Comments:

Roundwood comprises wood in the rough, wood in its natural state as felled, or otherwise harvested, with or without bark, round, split, roughly squared or other forms (for example roots, stumps, burls, etc.). It may also be impregnated (for example telegraph poles) or roughly shaped or pointed. It comprises all wood obtained from removals, that is to say the quantities removed from forests and from trees outside the forest, including wood recovered from natural, felling and logging losses during the period - calendar year or forest year. Commodities included are sawlogs and veneer logs, pulpwood, other industrial roundwood (including pitprops) and fuelwood. The statistics include recorded volumes, as well as estimated unrecorded volumes.

Figures are given in solid volume of roundwood (or roundwood equivalent) without bark.

Production of primary products is reported, even though a portion may immediately be consumed in the production of another commodity (for example wood pulp, which may immediately be converted into paper as part of a continuous process). An exception is made in the case of veneer production, which excludes veneer sheets used for plywood production within the country.

Forest Fire Extent – Annual Average

Units: thousand hectares

Data Source: Global Forest Resources Assessment 2000

Data Provider: Food and Agriculture Organization of the United Nations (FAO)

Years: 1990-00

	1990-00
AFRICA	----
Central Africa	----
Eastern Africa	----
Northern Africa	----
Southern Africa	----
Western Africa	----
Western Indian Ocean	----
ASIA AND PACIFIC	4.46
Australia and New Zealand	0.33
Central Asia	1.80
North West Pacific and East Asia	2.33
South Asia	----
South East Asia	----
South Pacific	----
EUROPE	1050.67
Central Europe	33.45
Eastern Europe	825.71
Western Europe	191.50
LATIN AMERICA AND CARIBBEAN	556.49
Caribbean	0.42
Meso-America	66.78
South America	489.30
NORTH AMERICA	501.63
North America	501.63
POLAR	----
Antarctic	----
Arctic	----
WEST ASIA	----
Arabian Peninsula	----
Mashriq	----
GLOBAL TOTALS	----
REPORTED GLOBAL TOTALS FROM ORIGINAL DATA SOURCES	----

Comments:

Forest fire extent - annual average comprises the reported forest areas exposed to fire.

Mangroves Forest

Units: thousand hectares

Data Source: Statistical Analysis of Global Forest Conservation, 1997

Data Provider: World Conservation Monitoring Centre (UNEP-WCMC) and Centre for International Forestry Research (CIFOR)

Years: Various

Mangroves Forest Extent - Protected Area

	VARIOUS
AFRICA	54.70
Central Africa	23.10
Eastern Africa	0.00
Northern Africa	0.00
Southern Africa	22.50
Western Africa	8.50
Western Indian Ocean	0.60
ASIA AND PACIFIC	1431.80
Australia and New Zealand	254.40
Central Asia	0.00
North West Pacific and East Asia	0.00
South Asia	217.10
South East Asia	853.40
South Pacific	106.90
EUROPE	0.00
Central Europe	0.00
Eastern Europe	0.00
Western Europe	0.00
LATIN AMERICA AND CARIBBEAN	1154.50
Caribbean	93.00
Meso-America	121.80
South America	939.70
NORTH AMERICA	119.50
North America	119.50
POLAR	0.00
Antarctic	0.00
Arctic	0.00
WEST ASIA	----
Arabian Peninsula	----
Mashriq	----
GLOBAL TOTALS	2760.50
REPORTED GLOBAL TOTALS FROM ORIGINAL DATA SOURCES	2261.70

Mangroves Forest Extent - Total Area

	VARIOUS
AFRICA	3801.40
Central Africa	440.90
Eastern Africa	0.00
Northern Africa	0.00
Southern Africa	887.90
Western Africa	2162.90
Western Indian Ocean	309.70
ASIA AND PACIFIC	9499.20
Australia and New Zealand	4902.40
Central Asia	0.00
North West Pacific and East Asia	0.00
South Asia	825.40
South East Asia	3207.30
South Pacific	564.10
EUROPE	0.00
Central Europe	0.00
Eastern Europe	0.00
Western Europe	0.00
LATIN AMERICA AND CARIBBEAN	4441.10
Caribbean	862.40
Meso-America	649.20
South America	2929.50
NORTH AMERICA	199.00
North America	199.00
POLAR	0.00
Antarctic	0.00
Arctic	0.00
WEST ASIA	----
Arabian Peninsula	----
Mashriq	----
GLOBAL TOTALS	17940.70
REPORTED GLOBAL TOTALS FROM ORIGINAL DATA SOURCES	16945.20

Comments:

Mangrove trees and shrubs, including ferns and palms, are found along river banks and coastlines in tropical and subtropical countries. Their main characteristic is that they can tolerate salt and brackish water environments. Globally, there are seventy known species of mangroves.

The data for forest cover by forest type are for 1996. The WCMC analysis was carried out by overlaying forest and protected areas coverages in a Geographic Information System. The forest GIS coverages were created at WCMC by compiling many national and regional data showing forest extent. Many of the map sources where forest, land cover and/or vegetation maps.

Data source: Iremonger, S., C. Ravilious, T. Quinton, "A statistical analysis of global forest conservation." In: S. Iremonger, C. Ravilious, and T. Quinton (Eds). "A Global Overview of Forest Conservation CD-ROM" (World Conservation Monitoring Centre (WCMC) and Centre for International Forestry Research, Cambridge, U.K., 1997).

1.4 Biodiversity

1.4.1   Protected Areas and Environmental Protection

A **protected area** is defined by The World Conservation Union (IUCN) as: an area of land and/or sea especially dedicated to the protection and maintenance of biological diversity, and of natural and associated cultural resources, and managed through legal or other effective means. Although all protected areas meet the general purposes contained in this definition, in practice the precise purposes for which protected areas are managed differ greatly.

Following are the definitions of IUCN Management categories I-VI:

Category Ia. Strict Nature Reserve: protected area managed mainly for science. Area of land and/or sea possessing some outstanding or representative ecosystems, geological or physiological features and/or species, available primarily for scientific research and/or environmental monitoring.

Category Ib. Wilderness Area: protected area managed mainly for wilderness protection.
Large area of unmodified or slightly modified land, and/or sea, retaining its natural character and influence, without permanent or significant habitation, which is protected and managed so as to preserve its natural condition.

Category II.   National Park: protected area managed mainly for ecosystem protection and recreation. Natural area of land and/or sea, designated to a) protect the ecological integrity of one or more ecosystems for present and future generations; b) exclude exploitation or occupation inimical to the purposes of designation of the area and c) provide a foundation for spiritual, scientific, educational, recreational and visitor opportunities, all of which must be environmentally and culturally compatible.

Category III. Natural Monument: protected area managed mainly for conservation of specific natural features. Area containing one, or more, specific natural or natural/cultural feature which is of outstanding or unique value because of its inherent rarity, representative or aesthetic qualities or cultural significance.

Category IV.   Habitat/Species Management Area: protected area managed mainly for conservation through management intervention. Area of land and/or sea subject to active intervention for management purposes so as to ensure the maintenance of habitats and/or to meet the requirements of specific species.

Category V.   Protected Landscape/Seascape: protected area managed mainly for landscape/seascape conservation and recreation. Area of land, with coast and sea as appropriate, where the interaction of people and nature over time has produced an area of distinct character with significant aesthetic, ecological and/or cultural value, and often with high biological diversity. Safeguarding the integrity of this traditional interaction is vital to the protection, maintenance and evolution of such an area.

Category VI.   Managed Resource Protected Area: protected area managed mainly for the sustainable use of natural ecosystems. Area containing predominantly unmodified natural systems, managed to ensure long term protection and maintenance of biological diversity, while providing at the same time a sustainable flow of natural products and services to meet community needs.

Please refer to the original source for the most recent data (http://quin.unep-wcmc.org/wdbpa/GEO3.cfm). For more information on the variables and collection methodologies see original source or the IUCN web site: http://wcpa.iucn.org.

Protected Areas (IUCN Categories I-VI)

Units: hectares

Data Source: World Database of Protected Areas

Data Provider: World Conservation Monitoring Centre (UNEP-WCMC)

Years: 1970, 1980, 1990, 2002

Protected Areas (IUCN Categories I-VI) - Area

	1970	1980	1990	2002
AFRICA	95072911	174810658	200611691	206274903
Central Africa	29453375	31029169	31760369	33519586
Eastern Africa	10149148	28614104	29332798	29514399
Northern Africa	7113060	8700004	19345347	19665372
Southern Africa	----	----	----	----
Western Africa	12056765	18556298	28466074	30699160
Western Indian Ocean	1085799	1109024	1213705	1273065
ASIA AND PACIFIC	48105467	135196921	249609663	295851606
Australia and New Zealand	16808387	66391288	97005388	108424536
Central Asia	5054196	8300025	9948026	10450307
North West Pacific and East Asia	8274030	13573238	73452715	89784764
South Asia	5470474	22099703	31256144	33596650
South East Asia	12436599	23786868	36765800	52345560
South Pacific	61781	1045799	1181590	1249789
EUROPE	24481271	55512523	91861814	126368040
Central Europe	1857486	4305729	8140654	10930863
Eastern Europe	8591098	18620177	37887630	62167901
Western Europe	14032687	32586617	45833530	53269276
LATIN AMERICA AND CARIBBEAN	44171363	92359073	161401928	213104123
Caribbean	1399502	1368022	2746964	2720245
Meso-America	5192900	7530232	16191416	26877929
South America	37578961	83460819	142463548	183505949
NORTH AMERICA	75095863	190137768	206883435	267391191
North America	75095863	190137768	206883435	267391191
POLAR	44470	98370164	98466435	98649045
Antarctic	44470	120164	216435	399045
Arctic	----	98250000	98250000	98250000
WEST ASIA	210786	217106	16039211	86341355
Arabian Peninsula	420	3430	15736435	85943079
Mashriq	210366	213676	302776	398276
GLOBAL TOTALS	287182131	746604213	1024874177	1293980263
REPORTED GLOBAL TOTALS FROM ORIGINAL DATA SOURCES	285190725	752614639	1032212740	1303488421

Protected Areas (IUCN Categories I-VI) - Number

	1970	1980	1990	2002
AFRICA	366	694	844	909
Central Africa	58	68	73	75
Eastern Africa	82	139	175	183
Northern Africa	30	44	79	92
Southern Africa	----	----	----	----
Western Africa	68	110	131	141
Western Indian Ocean	51	78	110	131
ASIA AND PACIFIC	1913	3581	5852	6961
Australia and New Zealand	749	1601	2543	3242
Central Asia	41	158	184	195
North West Pacific and East Asia	65	197	849	1041
South Asia	167	430	743	779
South East Asia	859	1101	1406	1565
South Pacific	32	94	127	139
EUROPE	4444	9133	17087	21900
Central Europe	1122	1683	2808	3348
Eastern Europe	431	2274	3905	5381
Western Europe	2891	5176	10374	13171
LATIN AMERICA AND CARIBBEAN	592	1012	1902	2649
Caribbean	264	286	461	499
Meso-America	72	159	338	532
South America	256	567	1103	1618
NORTH AMERICA	1906	3071	4176	4616
North America	1906	3071	4176	4616
POLAR	13	30	57	65
Antarctic	13	28	55	63
Arctic	----	2	2	2
WEST ASIA	10	16	39	54
Arabian Peninsula	1	6	22	35
Mashriq	9	10	17	19
GLOBAL TOTALS	9244	17537	29957	37154
REPORTED GLOBAL TOTALS FROM ORIGINAL DATA SOURCES	9394	17924	30623	37957

1.4.2   Total and Threatened Species

The assessments for the different species were assessed in different years; 1996, 1998, 1999 and 2000 (hence "various" at the top of the "year" column).

The knowledge about the number of species is constantly changing with new assessments and according to the data provider the current numbers of threaten species are underestimated. Changes are going to be made by the provider in 2003, for the latest data see http://www.redlist.org.

Critically endangered: a specie is "critically endangered" when it is facing an extremely high risk of extinction in the wild in the immediate future, as defined by any of the criteria (A to E) as described below.

Endangered: a species is "endangered" when it is not "critically endangered" but is facing a very high risk of extinction in the wild in the near future.

Extinct: a species is "extinct" when there is no reasonable doubt that the last individual has died.

Extinct in the wild: a species is "extinct in the wild" when it is known only to survive in cultivation, in captivity or as a naturalised population (or populations) well outside the past range. A taxon is presumed extinct in the wild when exhaustive surveys in known and/or expected habitat, at appropriate times (diurnal, seasonal, annual), throughout its historic range have failed to record an individual. Surveys should be over a time frame appropriate to the taxon's life cycle and life form.

Vulnerable: a species is "vulnerable" when it is not "critically endangered" or "endangered" but is facing a high risk of extinction in the wild in the medium-term future, as defined by any of the criteria (A to E) as described below.

A) Population reduction in the form of either of the following:
    1) An observed, estimated, inferred or suspected reduction of at least 20 percent over the last 10 years or three generations, whichever is the longer, based on (and specifying) any of the following:
        a) direct observation
        b) an index of abundance appropriate for the taxon
        c) a decline in area of occupancy, extent of occurrence and/or quality of habitat
        d) actual or potential levels of exploitation
        e) the effects of introduced taxa, hybridisation, pathogens, pollutants, competitors or parasites.

    2) A reduction of at least 20 percent, projected or suspected to be met within the next ten years or three generations, whichever is the longer, based on (and specifying) any of (b), (c), (d) or (e) above.

B) Extent of occurrence estimated to be less than 20000 km^2 or area of occupancy estimated to be less than 2000 km^2, and estimates indicating any two of the following:
    1) Severely fragmented or known to exist at no more than ten locations.

    2) Continuing decline, inferred, observed or projected, in any of the following:
        a) extent of occurrence
        b) area of occupancy
        c) area, extent and/or quality of habitat
        d) number of locations or subpopulations
        e) number of mature individuals

    3) Extreme fluctuations in any of the following:
        a) extent of occurrence
        b) area of occupancy
        c) number of locations or subpopulations
        d) number of mature individuals

C) Population estimated to number less than 10000 mature individuals and either:
    1) An estimated continuing decline of at least 10 percent within 10 years or three generations, whichever is longer, or
    2) A continuing decline, observed, projected, or inferred, in numbers of mature individuals and population structure in the form of either:
        a) severely fragmented (that is to say no subpopulation estimated to contain more than 1000 mature individuals)
        b) all individuals are in a single subpopulation

D) Population very small or restricted in the form of either of the following:

    1) Population estimated to number less than 1000 mature individuals.
    2) Population is characterised by an acute restriction in its area of occupancy (typically less than 100 km^2) or in the number of locations (typically less than five). Such a taxon would thus be prone to the effects of human activities (or stochastic events whose impact is increased by human activities) within a very short period of time in an unforeseeable future, and is thus capable of becoming Critically Endangered or even Extinct in a very short period.

E) Quantitative analysis showing the probability of extinction in the wild is at least 10 percent within 100 years.

The knowledge about the number of species is constantly changing with new assessments and according to the data provider the current numbers of threaten amphibian species are underestimated. Changes are going to be made by the provider in 2003, for the latest data see http://www.redlist.org.

Amphibian Species

Units: number of species

Data Source: 2000 IUCN Red List of Threatened Species On-line Data Service (data as of February 2002)

Data Provider: International Union for the Conservation of Nature
and Natural Resources (IUCN) and Species Survival Commission (SSC)

Years: various

	Critically Endangered	Endangered	Extinct in the Wild	Vulnerable
AFRICA	0	3	0	13
Central Africa	0	0	0	1
Eastern Africa	0	0	0	0
Northern Africa	0	0	0	0
Southern Africa	0	2	0	7
Western Africa	0	1	0	0
Western Indian Ocean	0	1	0	5
ASIA AND PACIFIC	15	23	0	32
Australia and New Zealand	8	10	0	9
Central Asia	0	1	0	0
North West Pacific and East Asia	0	4	0	7
South Asia	0	2	0	4
South East Asia	7	6	0	12
South Pacific	0	1	0	0
EUROPE	3	2	0	8
Central Europe	0	1	0	3
Eastern Europe	0	0	0	1
Western Europe	3	1	0	7
LATIN AMERICA AND CARIBBEAN	7	4	0	17
Caribbean	1	0	0	7
Meso-America	2	1	0	2
South America	4	3	0	8
NORTH AMERICA	1	8	0	16
North America	1	8	0	16
POLAR	0	0	0	0
Antarctic	0	0	0	0
Arctic	0	0	0	0
WEST ASIA	0	0	0	0
Arabian Peninsula	0	0	0	0
Mashriq	0	0	0	0
GLOBAL TOTALS	26	40	0	85
REPORTED GLOBAL TOTALS FROM ORIGINAL DATA SOURCES	26	40	0	85

Bird Species

Units: number of species

Data Source: 2000 IUCN Red List of Threatened Species On-line Data Service (Data as of February 2002)

Data Provider: International Union for the Conservation of Nature
and Natural Resources (IUCN) and Species Survival Commission (SSC)

Years: various

	Critically Endangered	Endangered	Extinct	Extinct in the Wild	Vulnerable
AFRICA	28	57	37	0	118
Central Africa	3	15	0	0	35
Eastern Africa	6	17	0	0	31
Northern Africa	2	3	0	0	8
Southern Africa	3	20	9	0	55
Western Africa	3	5	0	0	17
Western Indian Ocean	15	15	28	0	26
ASIA AND PACIFIC	70	118	46	1	335
Australia and New Zealand	8	27	22	0	57
Central Asia	1	1	0	0	13
North West Pacific and East Asia	6	13	4	0	71
South Asia	8	11	0	0	66
South East Asia	36	53	2	0	146
South Pacific	18	28	19	1	82
EUROPE	5	10	3	0	30
Central Europe	2	1	0	0	10
Eastern Europe	3	9	1	0	27
Western Europe	3	2	2	0	14
LATIN AMERICA AND CARIBBEAN	70	135	17	2	212
Caribbean	12	21	12	0	24
Meso-America	10	16	4	1	38
South America	48	100	2	1	160
NORTH AMERICA	15	13	25	0	32
North America	15	13	25	0	32
POLAR	1	0	2	0	10
Antarctic	1	0	1	0	10
Arctic	0	0	1	0	0
WEST ASIA	2	1	0	0	17
Arabian Peninsula	2	1	0	0	15
Mashriq	2	1	0	0	12
GLOBAL TOTALS	182	321	128	3	680
REPORTED GLOBAL TOTALS FROM ORIGINAL DATA SOURCES	182	321	128	3	680

AGGREGATIONS

Mammal Species

Units: number of species

Data Source: 2000 IUCN Red List of Threatened Species On-line Data Service (Data as of February 2002)

Data Provider: International Union for the Conservation of Nature and Natural Resources (IUCN) and Species Survival Commission (SSC)

Years: various

	Critically Endangered	Endangered	Extinct	Extinct in the Wild	Vulnerable
AFRICA	79	108	9	2	173
Central Africa	22	47	0	1	48
Eastern Africa	22	31	1	0	61
Northern Africa	15	18	2	2	24
Southern Africa	13	26	5	0	58
Western Africa	10	37	0	1	29
Western Indian Ocean	18	23	1	0	38
ASIA AND PACIFIC	114	236	45	2	356
Australia and New Zealand	13	25	26	1	57
Central Asia	2	10	0	1	28
North West Pacific and East Asia	28	64	3	1	8C
South Asia	18	69	3	0	89
South East Asia	47	96	11	0	152
South Pacific	13	20	4	0	56
EUROPE	15	25	7	1	71
Central Europe	5	8	0	0	29
Eastern Europe	9	16	1	0	45
Western Europe	4	14	6	1	45
LATIN AMERICA AND CARIBBEAN	48	102	38	1	178
Caribbean	7	11	31	0	15
Meso-America	19	51	5	1	49
South America	22	46	4	0	133
NORTH AMERICA	17	31	12	1	62
North America	17	31	12	1	62
POLAR	1	4	0	0	4
Antarctic	0	2	0	0	0
Arctic	1	4	0	0	4
WEST ASIA	5	8	3	1	19
Arabian Peninsula	3	5	3	1	10
Mashriq	2	5	1	1	17
GLOBAL TOTALS	276	485	109	7	784
REPORTED GLOBAL TOTALS FROM ORIGINAL DATA SOURCES	276	485	109	7	784

AGGREGATIONS

Plant Species

Units: number of species

Data Source: 2000 IUCN Red List of Threatened Species On-line Data Service (Data as of February 2002)

Data Provider: International Union for the Conservation of Nature
and Natural Resources (IUCN) and Species Survival Commission (SSC)

Years: various

	Critically Endangered	Endangered	Extinct	Extinct in the Wild	Vulnerable
AFRICA	184	195	8	3	695
Central Africa	47	34	1	0	200
Eastern Africa	7	21	0	0	159
Northern Africa	2	4	0	0	19
Southern Africa	20	62	5	3	341
Western Africa	19	39	2	0	172
Western Indian Ocean	113	88	2	0	118
ASIA AND PACIFIC	555	564	26	7	1471
Australia and New Zealand	7	15	0	0	39
Central Asia	1	0	0	0	1
North West Pacific and East Asia	43	77	2	3	149
South Asia	130	206	9	2	232
South East Asia	288	189	3	2	812
South Pacific	100	103	12	0	336
EUROPE	16	19	1	1	42
Central Europe	1	2	0	1	11
Eastern Europe	0	2	0	0	7
Western Europe	15	16	1	0	31
LATIN AMERICA AND CARIBBEAN	273	506	20	4	1360
Caribbean	97	138	6	1	237
Meso-America	72	161	0	1	279
South America	104	222	14	2	887
NORTH AMERICA	60	81	22	2	72
North America	60	81	22	2	72
POLAR	0	0	0	0	1
Antarctic	0	0	0	0	0
Arctic	0	0	0	0	1
WEST ASIA	0	6	0	0	52
Arabian Peninsula	0	6	0	0	51
Mashriq	0	0	0	0	1
GLOBAL TOTALS	1095	1389	79	17	3765
REPORTED GLOBAL TOTALS FROM ORIGINAL DATA SOURCES	1095	1389	79	17	3765

AGGREGATIONS

Reptile Species

Units: number of species

Data Source: 2000 IUCN Red List of Threatened Species On-line Data Service (Data as of February 2002)

Data Provider: International Union for the Conservation of Nature
and Natural Resources (IUCN) and Species Survival Commission (SSC)

Years: various

	Critically Endangered	Endangered	Extinct	Extinct in the Wild	Vulnerable
AFRICA	4	12	12	1	34
Central Africa	2	1	0	0	2
Eastern Africa	2	3	0	0	4
Northern Africa	1	3	0	0	3
Southern Africa	4	5	1	0	17
Western Africa	2	3	1	0	2
Western Indian Ocean	1	9	10	1	14
ASIA AND PACIFIC	27	47	3	0	74
Australia and New Zealand	4	12	1	0	42
Central Asia	0	1	0	0	2
North West Pacific and East Asia	11	20	1	0	7
South Asia	5	17	0	0	13
South East Asia	16	21	0	0	18
South Pacific	3	6	1	0	7
EUROPE	13	12	0	0	12
Central Europe	6	10	0	0	3
Eastern Europe	3	3	0	0	4
Western Europe	7	6	0	0	9
LATIN AMERICA AND CARIBBEAN	32	29	6	2	49
Caribbean	22	18	5	0	13
Meso-America	5	6	0	0	10
South America	10	11	1	2	29
NORTH AMERICA	3	9	1	0	17
North America	3	9	1	0	17
POLAR	0	0	0	0	0
Antarctic	0	0	0	0	0
Arctic	0	0	0	0	0
WEST ASIA	2	4	0	0	0
Arabian Peninsula	1	3	0	0	0
Mashriq	1	1	0	0	2
GLOBAL TOTALS	71	93	22	3	175
REPORTED GLOBAL TOTALS FROM ORIGINAL DATA SOURCES	71	93	22	3	175

AGGREGATIONS

1.5 Freshwater

1.5.1   Waste Production and Management

Reused Treated Wastewater

Units: million cubic meters

Data Source: AQUASTAT Information System on Water in Agriculture (data as of January 2001)

Data Provider: Food and Agriculture Organization of the United Nations (FAO)

Years: various

	VARIOUS
AFRICA	320.00
Central Africa	0.00
Eastern Africa	0.00
Northern Africa	320.00
Southern Africa	0.00
Western Africa	0.00
Western Indian Ocean	0.00
ASIA AND PACIFIC	299.10
Australia and New Zealand	----
Central Asia	299.10
North West Pacific and East Asia	0.00
South Asia	0.00
South East Asia	0.00
South Pacific	----
EUROPE	----
Central Europe	----
Eastern Europe	0.10
Western Europe	----
LATIN AMERICA AND CARIBBEAN	0.00
Caribbean	0.00
Meso-America	0.00
South America	0.00
NORTH AMERICA	----
North America	----
POLAR	----
Antarctic	----
Arctic	----
WEST ASIA	858.53
Arabian Peninsula	436.23
Mashriq	422.30
GLOBAL TOTALS	----
REPORTED GLOBAL TOTALS FROM ORIGINAL DATA SOURCES	----

Comments:

Reused treated wastewater: annual quantity of treated wastewater that is reused. The treated wastewater is wastewater discharged from treatment plant (effluent).

1.5.2 Water Consumption and Resources

Desalinated Water

Units: million cubic meters

Data Source: AQUASTAT Information System on Water in Agriculture (data as of January 2001)

Data Provider: Food and Agriculture Organization of the United Nations (FAO)

Years: various

	VARIOUS
AFRICA	**173.00**
Central Africa	0.00
Eastern Africa	0.20
Northern Africa	171.10
Southern Africa	0.00
Western Africa	1.70
Western Indian Ocean	0.00
ASIA AND PACIFIC	**1330.90**
Australia and New Zealand	----
Central Asia	1328.00
North West Pacific and East Asia	0.00
South Asia	2.90
South East Asia	0.00
South Pacific	----
EUROPE	**----**
Central Europe	----
Eastern Europe	0.00
Western Europe	----
LATIN AMERICA AND CARIBBEAN	**3.30**
Caribbean	3.30
Meso-America	0.00
South America	0.00
NORTH AMERICA	**----**
North America	----
POLAR	**----**
Antarctic	----
Arctic	----
WEST ASIA	**1518.70**
Arabian Peninsula	1516.70
Mashriq	2.00
GLOBAL TOTALS	----
REPORTED GLOBAL TOTALS FROM ORIGINAL DATA SOURCES	****

Comments:

Desalinated water corresponds to the annual amount of fresh water generated by desalination of sea or brackish waters (annually estimated on the basis of the total capacity of water desalination installations).

Groundwater Recharge

Units: cubic kilometres

Data Source: various

Data Provider: World Resources Institute

Groundwater Recharge – Total

Years: various

	VARIOUS
AFRICA	1586.60
Central Africa	858.50
Eastern Africa	106.00
Northern Africa	23.80
Southern Africa	181.60
Western Africa	361.02
Western Indian Ocean	55.68
ASIA AND PACIFIC	2818.64
Australia and New Zealand	270.00
Central Asia	78.51
North West Pacific and East Asia	896.20
South Asia	573.33
South East Asia	1000.60
South Pacific	----
EUROPE	1416.71
Central Europe	113.18
Eastern Europe	854.34
Western Europe	449.19
LATIN AMERICA AND CARIBBEAN	4084.50
Caribbean	----
Meso-America	331.00
South America	3740.00
NORTH AMERICA	1884.00
North America	1884.00
POLAR	----
Antarctic	----
Arctic	----
WEST ASIA	28.58
Arabian Peninsula	3.60
Mashriq	24.98
GLOBAL TOTALS	11819.03
REPORTED GLOBAL TOTALS FROM ORIGINAL DATA SOURCES	****

Groundwater Recharge – per Capita

Year: 2000

	2000
AFRICA	2023.85
Central Africa	10454.20
Eastern Africa	766.82
Northern Africa	140.89
Southern Africa	1273.44
Western Africa	1601.02
Western Indian Ocean	3255.97
ASIA AND PACIFIC	809.60
Australia and New Zealand	11636.53
Central Asia	1401.58
North West Pacific and East Asia	606.02
South Asia	405.90
South East Asia	1950.21
South Pacific	----
EUROPE	1744.14
Central Europe	656.72
Eastern Europe	3740.67
Western Europe	1138.30
LATIN AMERICA AND CARIBBEAN	7882.41
Caribbean	----
Meso-America	2569.81
South America	10771.97
NORTH AMERICA	6069.91
North America	6069.91
POLAR	----
Antarctic	----
Arctic	----
WEST ASIA	312.50
Arabian Peninsula	80.37
Mashriq	527.35
GLOBAL TOTALS	1965.67
REPORTED GLOBAL TOTALS FROM ORIGINAL DATA SOURCES	****

Comments:

Groundwater recharge is the average annual amount of water that is estimated to infiltrate soils, including water from rivers and streams that lose it to underlying strata. In general, this figure would represent the maximum amount of water that could be withdrawn annually without ultimately depleting the groundwater resource.

Groundwater Withdrawal

Data Source: various

Data Provider: World Resources Institute

Years: various

Groundwater Withdrawal – Total

Units: cubic kilometres

	VARIOUS
AFRICA	----
Central Africa	----
Eastern Africa	----
Northern Africa	16.39
Southern Africa	----
Western Africa	----
Western Indian Ocean	4.99
ASIA AND PACIFIC	380.15
Australia and New Zealand	2.24
Central Asia	13.05
North West Pacific and East Asia	69.26
South Asia	289.69
South East Asia	----
South Pacific	----
EUROPE	77.48
Central Europe	21.44
Eastern Europe	----
Western Europe	47.84
LATIN AMERICA AND CARIBBEAN	----
Caribbean	----
Meso-America	----
South America	----
NORTH AMERICA	110.87
North America	110.87
POLAR	----
Antarctic	----
Arctic	----
WEST ASIA	21.15
Arabian Peninsula	18.22
Mashriq	2.93
GLOBAL TOTALS	----
REPORTED GLOBAL TOTALS FROM ORIGINAL DATA SOURCES	600.00

Comments:

Total groundwater withdrawal refers to the annual abstractions from all groundwater sources-even nonrenewable sources.

Groundwater Withdrawals – from Agriculture

Units: percent

	VARIOUS
AFRICA	----
Central Africa	----
Eastern Africa	----
Northern Africa	64.66
Southern Africa	----
Western Africa	----
Western Indian Ocean	----
ASIA AND PACIFIC	78.76
Australia and New Zealand	66.67
Central Asia	43.60
North West Pacific and East Asia	47.93
South Asia	89.20
South East Asia	----
South Pacific	----
EUROPE	----
Central Europe	----
Eastern Europe	----
Western Europe	38.94
LATIN AMERICA AND CARIBBEAN	----
Caribbean	----
Meso-America	----
South America	----
NORTH AMERICA	62.03
North America	62.03
POLAR	----
Antarctic	----
Arctic	----
WEST ASIA	87.23
Arabian Peninsula	89.37
Mashriq	73.95
GLOBAL TOTALS	----
REPORTED GLOBAL TOTALS FROM ORIGINAL DATA SOURCES	****

Comments:

Groundwater withdrawal - agricultural refers to the annual sector share of water abstractions from all groundwater sources (even nonrenewable) for agricultural use (irrigation and livestock).

AGGREGATIONS

Units: percent

Data Source: various

Data Provider: World Resources Institute

Years: various

Groundwater Withdrawal – from Industry	
	VARIOUS
AFRICA	----
Central Africa	----
Eastern Africa	----
Northern Africa	2.21
Southern Africa	----
Western Africa	----
Western Indian Ocean	----
ASIA AND PACIFIC	----
Australia and New Zealand	19.51
Central Asia	25.05
North West Pacific and East Asia	----
South Asia	1.94
South East Asia	----
South Pacific	----
EUROPE	----
Central Europe	----
Eastern Europe	----
Western Europe	20.48
LATIN AMERICA AND CARIBBEAN	----
Caribbean	----
Meso-America	----
South America	----
NORTH AMERICA	5.42
North America	5.42
POLAR	----
Antarctic	----
Arctic	----
WEST ASIA	2.62
Arabian Peninsula	1.87
Mashriq	7.32
GLOBAL TOTALS	----
REPORTED GLOBAL TOTALS FROM ORIGINAL DATA SOURCES	----

Comments:

Groundwater withdrawal - industrial refers to the annual sector share of water abstractions from all groundwater sources (even nonrenewable) for industrial use, including water withdrawn to cool thermoelectric plants.

Groundwater Withdrawals – from Domestic	
	VARIOUS
AFRICA	----
Central Africa	----
Eastern Africa	----
Northern Africa	33.11
Southern Africa	----
Western Africa	----
Western Indian Ocean	2.51
ASIA AND PACIFIC	----
Australia and New Zealand	13.00
Central Asia	32.03
North West Pacific and East Asia	----
South Asia	9.23
South East Asia	----
South Pacific	----
EUROPE	----
Central Europe	----
Eastern Europe	----
Western Europe	42.00
LATIN AMERICA AND CARIBBEAN	----
Caribbean	----
Meso-America	----
South America	----
NORTH AMERICA	20.05
North America	20.05
POLAR	----
Antarctic	----
Arctic	----
WEST ASIA	10.15
Arabian Peninsula	8.79
Mashriq	18.60
GLOBAL TOTALS	----
REPORTED GLOBAL TOTALS FROM ORIGINAL DATA SOURCES	----

Comments:

Groundwater withdrawal - domestic refers to the annual sector share of water abstractions from all groundwater sources (even nonrenewable) for domestic use. It includes drinking water, homes, commercial establishments, public services (for example hospitals) and municipal use or provision.

Internal Renewable Water Resources

Units: cubic meters per person

Data Source: AQUASTAT Information System on Water in Agriculture (data as of February 2002)

Data Provider: Food and Agriculture Organization of the United Nations (FAO)

Internal Renewable Water Resources - Total

Years: various

	VARIOUS
AFRICA	3990.82
Central Africa	1777.20
Eastern Africa	188.20
Northern Africa	84.82
Southern Africa	537.50
Western Africa	1062.90
Western Indian Ocean	340.20
ASIA AND PACIFIC	12973.63
Australia and New Zealand	670.00
Central Asia	205.87
North West Pacific and East Asia	3409.05
South Asia	2140.27
South East Asia	5674.19
South Pacific	874.25
EUROPE	6835.34
Central Europe	651.21
Eastern Europe	4479.32
Western Europe	1704.82
LATIN AMERICA AND CARIBBEAN	13428.97
Caribbean	85.53
Meso-America	1097.46
South America	12245.98
NORTH AMERICA	5308.60
North America	5308.60
POLAR	----
Antarctic	----
Arctic	----
WEST ASIA	55.37
Arabian Peninsula	7.69
Mashriq	47.68
GLOBAL TOTALS	42592.73
REPORTED GLOBAL TOTALS FROM ORIGINAL DATA SOURCES	41022.00

Comments:

Internal renewable water resources is that part of the water resources (surface water and groundwater) generated from endogenous precipitation. It is computed by adding up average annual surface runoff and groundwater recharge occurring within the countries borders.

Internal Renewable Water Resources - per Capita

Year: 2000

	2000
AFRICA	5048.03
Central Africa	21602.24
Eastern Africa	1307.76
Northern Africa	487.76
Southern Africa	3572.15
Western Africa	4741.22
Western Indian Ocean	19072.00
ASIA AND PACIFIC	3755.33
Australia and New Zealand	29237.66
Central Asia	3624.59
North West Pacific and East Asia	2302.43
South Asia	1502.92
South East Asia	10882.93
South Pacific	144022.07
EUROPE	8362.84
Central Europe	3323.67
Eastern Europe	19764.70
Western Europe	4322.40
LATIN AMERICA AND CARIBBEAN	25923.09
Caribbean	2668.02
Meso-America	8121.56
South America	35436.93
NORTH AMERICA	16907.07
North America	16907.07
POLAR	----
Antarctic	----
Arctic	----
WEST ASIA	602.87
Arabian Peninsula	170.72
Mashriq	1002.85
GLOBAL TOTALS	7077.37
REPORTED GLOBAL TOTALS FROM ORIGINAL DATA SOURCES	7077.37

Comments:

Internal renewable water resources - per capita is the amount of water resources on a per person basis, using 2000 population estimates.

Water Withdrawal

Data Source: AQUASTAT Information System on Water in Agriculture (data as of February 2002)

Data Provider: Food and Agriculture Organization of the United Nations (FAO)

Years: various

Water Withdrawal – Total

Units: cubic kilometres

	VARIOUS
AFRICA	151.92
Central Africa	1.12
Eastern Africa	6.14
Northern Africa	96.12
Southern Africa	20.49
Western Africa	11.40
Western Indian Ocean	16.66
ASIA AND PACIFIC	1849.99
Australia and New Zealand	16.60
Central Asia	137.46
North West Pacific and East Asia	655.15
South Asia	805.13
South East Asia	235.52
South Pacific	0.13
EUROPE	455.92
Central Europe	97.48
Eastern Europe	131.71
Western Europe	226.73
LATIN AMERICA AND CARIBBEAN	262.82
Caribbean	15.85
Meso-America	90.01
South America	156.95
NORTH AMERICA	512.44
North America	512.44
POLAR	----
Antarctic	----
Arctic	----
WEST ASIA	83.83
Arabian Peninsula	24.34
Mashriq	59.49
GLOBAL TOTALS	3316.92
REPORTED GLOBAL TOTALS FROM ORIGINAL DATA SOURCES	3253.30

Comments:

Total water withdrawal: annual quantity of water extracted from the resources for agricultural, industrial and domestic purposes. It does not include other withdrawals.

Water Withdrawal – from Agriculture

Units: percent

	VARIOUS
AFRICA	85.43
Central Africa	38.51
Eastern Africa	83.91
Northern Africa	87.10
Southern Africa	75.58
Western Africa	75.36
Western Indian Ocean	98.52
ASIA AND PACIFIC	85.79
Australia and New Zealand	34.33
Central Asia	91.33
North West Pacific and East Asia	75.38
South Asia	93.38
South East Asia	89.20
South Pacific	51.54
EUROPE	36.42
Central Europe	46.69
Eastern Europe	30.74
Western Europe	35.31
LATIN AMERICA AND CARIBBEAN	73.46
Caribbean	74.00
Meso-America	77.89
South America	70.86
NORTH AMERICA	39.36
North America	39.36
POLAR	----
Antarctic	----
Arctic	----
WEST ASIA	90.40
Arabian Peninsula	87.27
Mashriq	91.68
GLOBAL TOTALS	70.95
REPORTED GLOBAL TOTALS FROM ORIGINAL DATA SOURCES	71.00

Comments:

Agricultural water withdrawal: annual quantity of water withdrawn for agricultural purposes including irrigation and livestock watering

Units: percent

Data Source: AQUASTAT Information System on Water in Agriculture (data as of February 2002)

Data Provider: Food and Agriculture Organization of the United Nations (FAO)

Years: various

Water Withdrawal – from Industry

	VARIOUS
AFRICA	5.81
Central Africa	14.82
Eastern Africa	2.92
Northern Africa	6.01
Southern Africa	8.95
Western Africa	7.37
Western Indian Ocean	0.15
ASIA AND PACIFIC	8.39
Australia and New Zealand	2.96
Central Asia	5.75
North West Pacific and East Asia	17.57
South Asia	2.48
South East Asia	4.93
South Pacific	21.54
EUROPE	48.98
Central Europe	39.87
Eastern Europe	52.66
Western Europe	50.76
LATIN AMERICA AND CARIBBEAN	8.72
Caribbean	0.95
Meso-America	5.39
South America	11.41
NORTH AMERICA	47.20
North America	47.20
POLAR	----
Antarctic	----
Arctic	----
WEST ASIA	3.52
Arabian Peninsula	1.82
Mashriq	4.22
GLOBAL TOTALS	19.75
REPORTED GLOBAL TOTALS FROM ORIGINAL DATA SOURCES	19.75

Comments:

Industrial water withdrawal: annual quantity of water withdrawn for industrial purposes. This sector usually refers to self-supplied industries not connected to any distribution network.

Water Withdrawal – from Domestic

	VARIOUS
AFRICA	8.66
Central Africa	46.68
Eastern Africa	13.29
Northern Africa	6.74
Southern Africa	15.42
Western Africa	17.27
Western Indian Ocean	1.32
ASIA AND PACIFIC	6.22
Australia and New Zealand	62.71
Central Asia	2.83
North West Pacific and East Asia	7.85
South Asia	4.33
South East Asia	6.16
South Pacific	26.92
EUROPE	14.74
Central Europe	13.06
Eastern Europe	17.18
Western Europe	14.05
LATIN AMERICA AND CARIBBEAN	17.82
Caribbean	24.99
Meso-America	16.71
South America	17.73
NORTH AMERICA	13.44
North America	13.44
POLAR	----
Antarctic	----
Arctic	----
WEST ASIA	6.08
Arabian Peninsula	10.93
Mashriq	4.10
GLOBAL TOTALS	9.00
REPORTED GLOBAL TOTALS FROM ORIGINAL DATA SOURCES	9.00

Comments:

Domestic water withdrawal: annual quantity of water withdrawn for domestic purposes (drinking water, homes, commercial establishments, public services, and municipal use). It is usually computed as the total amount of water withdrawn by public distribution networks, and usually includes the withdrawal by those industries connected to public networks.

1.6 Coastal and Marine Areas

All data reported as "0.5" values are according to the provider considered as greater than zero, but less than 0.5 metric tonnes. Database downloadable with Fishstat-Plus software (Version 2.3 2000) at: http://www.fao.org/fi/statist/fisoft/fishplus.asp.

1.6.1 Aquaculture Production

Aquaculture Production – Total

Units: metric tonnes

Data Source: FAO Fisheries Department, Fishery Information, Data and Statistics Units

Data Provider: Food and Agriculture Organization of the United Nations (FAO)

Years: 1970-1999

	1970	1980	1990	1995	1996	1997	1998	1999
AFRICA	----	----	82480.50	104621.00	120660.50	121636.50	192790.50	284127.00
Central Africa	----	----	1153.00	1045.00	976.00	853.00	945.00	1346.00
Eastern Africa	----	----	1518.00	1680.00	977.00	751.00	710.00	1015.50
Northern Africa	----	19154.00	64065.00	66316.00	80836.00	79041.00	144729.00	231473.00
Southern Africa	----	----	7367.00	12693.00	11961.00	12792.50	15167.50	16618.50
Western Africa	----	----	7972.50	17800.00	19860.50	18779.00	21950.00	23811.00
Western Indian Ocean	----	69.00	405.00	5087.00	6050.00	9420.00	9289.00	9863.00
ASIA AND PACIFIC	2760115.45	6200333.73	14500809.00	28544080.50	30915017.50	32660627.00	35673995.50	38922834.50
Australia and New Zealand	8421.00	12214.00	41001.00	92786.00	100123.00	103487.00	121913.00	125379.00
Central Asia	8350.45	16733.23	38407.00	14277.00	7249.00	10237.00	8862.00	7591.00
North West Pacific and East Asia	2159909.00	4811813.00	11355028.00	23450124.00	25509750.00	27169817.00	29895765.00	32880816.00
South Asia	188955.00	473214.00	1256706.00	2063200.00	2222894.00	2358613.00	2618195.00	2728841.00
South East Asia	394479.00	886349.00	1803004.00	2916993.00	3063946.00	3009280.00	3020909.00	3165285.00
South Pacific	----	----	6663.00	6700.50	11055.50	9193.00	8351.50	14922.50
EUROPE	581563.05	912758.77	1638940.00	1647949.00	1743471.00	1832290.00	2037024.00	2181641.00
Central Europe	82546.59	114424.79	140820.00	112631.00	117626.00	130869.00	147438.00	166923.00
Eastern Europe	68222.46	136708.98	372033.00	112603.00	94293.00	96385.00	101162.00	112685.00
Western Europe	430794.00	661625.00	1126087.00	1422715.00	1531552.00	1605036.00	1788424.00	1902033.00
LATIN AMERICA AND CARIBBEAN	1221.50	28374.00	230656.00	489949.00	657812.00	781248.00	807825.00	769388.50
Caribbean	----	----	13170.50	27926.00	40871.00	48402.00	51356.00	60530.50
Meso-America	540.50	8919.00	31470.00	52533.00	60595.00	72598.00	77951.00	81600.00
South America	331.00	17126.00	186015.50	409490.00	556346.00	660248.00	678518.00	627258.00
NORTH AMERICA	172272.00	171931.00	356664.00	478618.00	464709.00	519248.00	536279.00	592279.00
North America	172272.00	171931.00	356664.00	478618.00	464709.00	519248.00	536279.00	592279.00
POLAR	----	----	----	----	----	----	----	----
Antarctic	----	----	----	----	----	----	----	----
Arctic	----	----	----	----	----	----	----	----
WEST ASIA	----	----	----	----	----	----	----	----
Arabian Peninsula	----	----	----	----	----	----	----	----
Mashriq	----	4171.00	4469.00	8927.00	9386.00	9496.00	15426.00	9077.00
GLOBAL TOTALS	----	----	16816007.50	31276934.00	33914985.50	35929496.00	39268705.50	42764678.50
REPORTED GLOBAL TOTALS FROM ORIGINAL DATA SOURCES	****	****	****	****	****	****	****	****

Comments:

Aquaculture total production includes all fish, molluscs, crustaceans, aquatic plants, aquatic animals and animal products, cultivated in freshwater, marine and brackish environments. Aquaculture is defined by FAO as "the farming of aquatic organisms". Farming implies some form of intervention in the rearing process to enhance production, such as regular stocking, feeding, protection from predators, etc. Farming also implies individual or corporate ownership of the stock being cultivated. For statistical purposes, aquatic organisms which are harvested by an individual or corporate body which has owned them throughout their rearing period contribute to aquaculture while aquatic organisms which are exploitable by the public as a common property resource, with or without appropriate licenses, are the harvest of fisheries. Production of fish, crustaceans and molluscs is expressed in live weight, that is the nominal weight of the aquatic organisms at the time of capture. The harvest of aquatic plants is given in wet weight. Fish, crustaceans, molluscs and all other aquatic organisms included in the database have been classified according to approximately 350 commercial species items, further arranged within the 50 groups of species constituting the nine divisions of the FAO International Standard Statistical Classification of Aquatic Animals and Plants (ISSCAAP). The taxonomic code descriptors are taken from FAO's Aquatic Science and Fisheries Information System.

Crustaceans and Molluscs Catch – Marine

Units: metric tonnes

Data Source: FAO Fisheries Department, Fishery Information, Data and Statistics Units

Data Provider: Food and Agriculture Organization of the United Nations (FAO)

Years: 1970-1999

	1970	1980	1990	1995	1996	1997	1998	1999
AFRICA	44504.00	102307.00	234811.50	251857.00	240313.00	232347.00	247498.00	336683.00
Central Africa	1601.00	2595.00	2888.00	2231.00	3103.00	3751.00	4763.00	4384.00
Eastern Africa	2200.00	2251.00	1749.00	2202.00	2344.00	2577.00	2177.00	2008.00
Northern Africa	7400.50	35040.00	103465.50	130942.00	122908.00	102475.00	108616.00	164079.00
Southern Africa	17200.50	26713.00	28632.00	27358.00	28496.00	28304.00	33119.00	31537.00
Western Africa	10801.50	29103.00	86413.00	75965.00	69942.00	80428.00	83836.00	120845.00
Western Indian Ocean	5300.50	6605.00	11664.00	13159.00	13520.00	14812.00	14987.00	13830.00
ASIA AND PACIFIC	2588222.50	3301374.00	5110972.00	7294013.00	7222472.00	8206841.00	7948440.00	8470597.00
Australia and New Zealand	54600.00	71230.00	88425.00	162741.00	105698.00	132936.00	131455.00	122508.00
Central Asia	----	----	----	----	----	----	----	----
North West Pacific and East Asia	1807912.00	2196024.00	3576889.00	5378519.00	5293662.00	6126553.00	5953548.00	6468173.00
South Asia	162080.00	310856.00	336208.00	397105.00	399842.00	488841.00	458222.00	427988.00
South East Asia	562857.00	720979.00	1098625.00	1340372.00	1409157.00	1442272.00	1393540.00	1439768.00
South Pacific	773.50	2285.00	10825.00	15276.00	14113.00	16239.00	11675.00	12160.00
EUROPE	603663.50	1224554.00	1540922.00	1190354.00	1140782.00	1119099.00	1192213.00	1144161.00
Central Europe	17449.45	89931.03	159948.00	46030.00	63184.00	56618.00	52539.00	62249.00
Eastern Europe	97157.55	475565.97	526831.00	202443.00	188963.00	170612.00	155975.00	177471.00
Western Europe	489056.50	659057.00	854143.00	941881.00	888635.00	891869.00	983699.00	904441.00
LATIN AMERICA AND CARIBBEAN	423318.50	514218.00	670501.50	861821.00	1024363.00	1182868.00	959834.00	1127892.00
Caribbean	21253.50	28664.00	42446.50	36145.00	35114.00	39040.00	38005.00	37939.00
Meso-America	160400.00	217050.00	229162.00	252888.00	350018.00	343207.00	226811.00	285978.00
South America	241665.00	268504.00	398893.00	572788.00	639231.00	800621.00	695018.00	803975.00
NORTH AMERICA	899673.00	1256480.00	1364124.00	1361969.00	1334343.00	1308819.00	1276350.00	1371427.00
North America	899673.00	1256480.00	1364124.00	1361969.00	1334343.00	1308819.00	1276350.00	1371427.00
POLAR	----	----	----	----	----	----	----	----
Antarctic	----	----	----	----	----	----	----	----
Arctic	8429.00	35942.00	69319.00	88211.00	74176.00	68375.00	73728.00	82074.00
WEST ASIA	----	----	----	----	----	----	----	----
Arabian Peninsula	11080.00	16165.00	16135.00	18543.00	25266.00	31860.00	26502.00	22858.00
Mashriq	----	----	----	----	----	----	----	----
GLOBAL TOTALS	4579490.50	6451300.00	9007194.00	11067505.00	11062492.00	12151232.00	11725596.00	12556817.00
REPORTED GLOBAL TOTALS FROM ORIGINAL DATA SOURCES	4579490.50	4579490.50	4579490.50	4579490.50	4579490.50	4579490.50	4579490.50	4579490.50

Comments:

Crustaceans and molluscs catch data relate to nominal catch of crustaceans and molluscs, taken for commercial, industrial, recreational and subsistence purposes from marine waters. The harvest from mariculture, aquaculture and other kinds of fish farming is excluded.

Data include all quantities caught and landed for both food and feed purposes, but exclude discards. Catches of fish are expressed in live weight, that is the nominal weight of the aquatic organisms at the time of capture. To assign nationality to catches, the flag of the fishing vessel is used, unless the wording of chartering and joint operation contracts indicates otherwise.

AGGREGATIONS

Fish Catch

Units: metric tonnes

Data Source: FAO Fisheries Department, Fishery Information, Data and Statistics Units

Data Provider: Food and Agriculture Organization of the United Nations (FAO)

Years: 1970-1999

	1970	1980	1990	1995	1996	1997	1998	1999
AFRICA	3577558.50	3503907.00	4737127.50	5201994.00	5002643.00	5377880.50	5449378.00	5588027.50
Central Africa	309701.00	309878.00	387169.50	443309.50	466022.50	441596.50	473770.50	505143.50
Eastern Africa	186501.00	235985.00	477375.50	443604.00	416616.00	419204.00	440639.00	481413.00
Northern Africa	391972.50	564741.50	949900.00	1290679.50	1079659.50	1284604.50	1221445.00	1237213.50
Southern Africa	1848972.00	1279613.00	1440942.50	1410727.50	1249277.50	1362583.00	1444892.00	1435006.00
Western Africa	782510.00	1047267.50	1355341.50	1474107.50	1656244.50	1723175.50	1712017.50	1742727.50
Western Indian Ocean	57902.00	66422.00	126398.50	139566.00	134823.00	146717.00	156614.00	186524.00
ASIA AND PACIFIC	17448450.70	23104453.94	29375433.50	30392490.00	32125644.00	32869889.50	33832999.50	34512507.50
Australia and New Zealand	95500.00	203809.00	460136.00	581627.00	515090.00	656495.00	704904.00	682394.00
Central Asia	59254.70	43699.94	52756.00	42411.00	36542.50	28306.50	21518.50	23374.50
North West Pacific and East Asia	10613499.50	14396317.50	16789989.50	15445410.50	16920931.50	17282607.50	17971926.50	17683985.50
South Asia	2405214.50	2825732.50	3820611.50	4438287.50	4704121.50	4726475.50	4503305.50	4856218.50
South East Asia	4221927.50	5506647.00	8084145.00	9649394.50	9733467.50	9926916.00	10340500.00	10926348.00
South Pacific	53054.50	128248.00	167795.50	235359.50	215491.00	249089.00	290845.00	340187.00
EUROPE	17389470.30	19830850.56	18521713.50	16258499.00	16535187.00	16849199.00	15983616.00	15022309.00
Central Europe	1608647.83	2407007.77	2053745.00	1442752.50	1209288.50	1094603.50	1041564.50	1073466.50
Eastern Europe	5493302.47	7143765.29	7788501.00	4197567.00	4631856.00	4537235.00	4388108.00	3955604.00
Western Europe	10287520.00	10280077.50	8679467.50	10618179.50	10694042.50	11217360.50	10553943.50	9993238.50
LATIN AMERICA AND CARIBBEAN	14908984.50	8982930.50	15381384.00	20460399.50	20525419.00	17881332.00	11308251.50	16841294.00
Caribbean	127686.00	221194.50	222292.50	131632.50	136472.00	136783.00	140735.50	124028.00
Meso-America	278616.00	1287634.00	1306068.00	1350281.50	1333882.50	1389486.50	1235222.50	1150278.50
South America	14502682.50	7474102.00	13853023.50	18978485.50	19055064.50	16355062.50	9932293.50	15566987.50
NORTH AMERICA	2775855.50	3232230.50	5331783.50	4060423.50	4061427.50	4256283.50	4044538.50	3943024.50
North America	2775855.50	3232230.50	5331783.50	4060423.50	4061427.50	4256283.50	4044538.50	3943024.50
POLAR	----	----	----	----	----	----	----	----
Antarctic	----	----	----	----	----	----	----	----
Arctic	29653.50	66375.50	73657.50	40554.50	41717.50	52099.50	54786.50	78155.50
WEST ASIA	303506.50	296331.50	365247.50	436029.50	420309.50	431479.50	432385.50	432891.50
Arabian Peninsula	267423.50	241549.50	337907.00	396612.00	377083.00	386748.00	395705.00	393370.00
Mashriq	36083.00	54782.00	27340.50	39417.50	43226.50	44731.50	36680.50	39521.50
GLOBAL TOTALS	56433679.50	59017379.50	73786412.00	76850465.50	78712423.00	77718239.00	71106031.00	76418285.00
REPORTED GLOBAL TOTALS FROM ORIGINAL DATA SOURCES	56433679.50	56433679.50	56433679.50	56433679.50	56433679.50	56433679.50	56433679.50	56433679.50

Comments:

Fish catch relates to nominal catch of fish, taken for commercial, industrial, recreational and subsistence purposes from freshwater, marine and brackish environments. The harvest from mariculture, aquaculture and other kinds of fish farming is excluded.

Data include all quantities caught and landed for both food and feed purposes, but exclude discards. Catches of fish are expressed in live weight, that is the nominal weight of the aquatic organisms at the time of capture. To assign nationality to catches, the flag of the fishing vessel is used, unless the wording of chartering and joint operation contracts indicates otherwise.

Fish Catch Marine

Units: metric tonnes

Data Source: FAO Fisheries Department, Fishery Information, Data and Statistics Units

Data Provider: Food and Agriculture Organization of the United Nations (FAO)

Years: 1970-1999

	1970	1980	1990	1995	1996	1997	1998	1999
AFRICA	2540310.50	2276370.50	2991340.50	3374495.50	3239288.50	3559851.50	3588042.50	371427
Central Africa	74000.00	102277.00	94452.00	122929.00	127217.00	123225.00	132789.00	13190
Eastern Africa	27800.00	18279.00	31897.00	34976.00	33354.00	28786.00	28390.00	3200
Northern Africa	322571.50	452584.50	761683.50	1045944.50	826543.50	1019989.50	950513.50	98959
Southern Africa	1614020.00	983614.00	981844.00	1019334.00	896604.00	976106.00	1076867.00	109982
Western Africa	479119.00	691160.00	1025027.00	1041748.00	1250749.00	1295030.00	1272871.00	130443
Western Indian Ocean	22800.00	28456.00	96437.00	109564.00	104821.00	116715.00	126612.00	15652
ASIA AND PACIFIC	15183723.64	20937318.57	26709542.00	27363383.00	28929091.00	29666046.00	30467344.00	3062943
Australia and New Zealand	94500.00	201908.00	457056.00	580825.00	514237.00	655668.00	704290.00	68106
Central Asia	69.64	92.57	93.00	----	----	----	----	2
North West Pacific and East Asia	10227908.00	13995710.00	16080028.00	14499313.00	15876177.00	16201330.00	16706980.00	1624494
South Asia	1282041.50	1799265.50	2744812.50	3278384.50	3495349.50	3486512.50	3270353.50	346184
South East Asia	3526161.50	4819621.50	7271365.50	8780301.50	8838668.50	9084340.50	9506028.50	991252
South Pacific	53043.00	120721.00	156187.00	224528.00	204629.00	238175.00	279677.00	32902
EUROPE	16983168.86	19422475.43	18116009.00	15956421.50	16228504.50	16562956.50	15681980.50	1474309
Central Europe	1546604.54	2327940.70	1955525.00	1352182.50	1123609.50	1024791.50	966768.50	100170
Eastern Europe	5225151.82	6946192.72	7579620.00	4096692.00	4519141.00	4429735.00	4268021.00	385473
Western Europe	10211412.50	10148342.00	8580864.00	10507547.00	10585754.00	11108430.00	10447191.00	988666
LATIN AMERICA AND CARIBBEAN	14760702.50	8674590.50	14944898.00	19956854.00	20049190.00	17433114.00	10871211.00	1641301
Caribbean	126977.00	213576.00	204859.00	119971.00	124424.00	127228.00	133559.00	11778
Meso-America	270845.00	1260467.00	1187877.00	1224820.00	1209273.00	1271823.00	1128014.00	105170
South America	14362880.50	7200547.50	13552162.00	18612063.00	18715493.00	16034063.00	9609638.00	1524352
NORTH AMERICA	2703205.00	3157621.00	5294388.00	4026116.00	4031917.00	4224620.00	4013539.00	390454
North America	2703205.00	3157621.00	5294388.00	4026116.00	4031917.00	4224620.00	4013539.00	390454
POLAR	----	----	----	----	----	----	----	
Antarctic	----	----	----	----	----	----	----	
Arctic	29653.00	66375.00	73657.00	40554.00	41717.00	52099.00	54786.00	7815
WEST ASIA	272220.00	280023.00	344564.00	408869.00	397784.00	407030.00	418554.00	41566
Arabian Peninsula	267420.00	241546.00	337904.00	396609.00	377080.00	386745.00	395702.00	39336
Mashriq	4800.00	38477.00	6660.00	12260.00	20704.00	20285.00	22852.00	2229
GLOBAL TOTALS	52473183.50	54815074.00	68474463.50	71126768.00	72917567.00	71905792.00	65095532.00	6989824
REPORTED GLOBAL TOTALS FROM ORIGINAL DATA SOURCES	52473183.50	52473183.50	52473183.50	52473183.50	52473183.50	52473183.50	52473183.50	5247318

Comments:

Fish catch data relate to nominal catch of fish, taken for commercial, industrial, recreational and subsistence purposes from marine waters. The harvest from mariculture, aquaculture and other kinds of fish farming is excluded.

Data include all quantities caught and landed for both food and feed purposes but exclude discards. Catches of fish are expressed in live weight, that is the nominal weight of the aquatic organisms at the time of capture. To assign nationality to catches, the flag of the fishing vessel is used, unless the wording of chartering and joint operation contracts indicates otherwise.

AGGREGATIONS

Fishery Production – Marine

Units: metric tonnes

Data Source: FAO Fisheries Department, Fishery Information, Data and Statistics Units

Data Provider: Food and Agriculture Organization of the United Nations (FAO)

Years: 1960-1999

	1960	1970	1980	1990	1995	1996	1997	1998	1999
AFRICA	1637800.00	2586738.50	2383351.50	3219197.50	3623428.50	3486842.50	3795438.50	3843337.50	4057219.50
Central Africa	42600.00	75600.00	104872.00	97340.00	125160.00	130320.00	126976.00	137552.00	136285.00
Eastern Africa	17700.00	30000.00	20530.00	33873.00	37181.00	35700.00	31367.00	30571.00	34011.00
Northern Africa	230200.00	330199.50	486054.50	849041.50	1163993.50	943095.50	1114084.50	1056220.50	1152710.50
Southern Africa	1137500.00	1631220.00	1010340.00	1012233.00	1049419.00	927275.00	1007480.00	1113262.00	1134253.00
Western Africa	198900.00	491619.00	726494.00	1118556.00	1123161.00	1329376.00	1380931.00	1360980.00	1425886.00
Western Indian Ocean	10900.00	28100.00	35061.00	108154.00	124514.00	121076.00	134600.00	144752.00	174074.00
ASIA AND PACIFIC	11631561.00	18709443.00	26071970.00	36270713.00	44140036.00	45932394.00	47830719.00	48949500.00	50623271.00
Australia and New Zealand	103100.00	158721.00	287221.00	587934.00	836529.00	719677.00	891210.00	956493.00	928437.00
Central Asia	----	----	----	----	----	----	----	----	----
North West Pacific and East Asia	8703556.00	12827862.00	17730316.00	23368922.00	28093494.00	29705203.00	31070515.00	31872647.00	32828809.00
South Asia	1077600.00	1467391.50	2149496.50	3268862.50	3972733.50	4213695.50	4269671.50	4044129.50	4268026.50
South East Asia	1716100.50	4201255.50	5781493.50	8878513.50	10997909.50	11075418.50	11345085.50	11784937.50	12255016.50
South Pacific	31204.50	54213.00	123443.00	166481.00	239370.00	218400.00	254237.00	291293.00	342982.00
EUROPE	9931136.05	18181059.50	21357611.00	20677587.00	18480892.50	18782992.50	19213693.50	18623076.50	17742848.50
Central Europe	585680.00	1581674.75	2419293.73	2116940.00	1396769.50	1191527.50	1079832.50	1026709.50	1074044.50
Eastern Europe	2008106.55	5511519.25	7579865.27	8221794.00	4469831.00	4848640.00	4798793.00	4637322.00	4232736.00
Western Europe	7337349.50	11087865.50	11358452.00	10338853.00	12614292.00	12742825.00	13335068.00	12959045.00	12436068.00
LATIN AMERICA AND CARIBBEAN	4546701.00	15185417.50	9200773.50	15745325.00	21140964.00	21459789.00	19088692.00	12354184.00	18033828.00
Caribbean	67300.50	148577.00	243350.00	249629.00	158014.00	161218.00	168379.00	173525.00	157135.00
Meso-America	202100.00	432045.00	1477981.00	1431846.00	1517241.00	1597360.00	1660951.00	1409186.00	1390446.00
South America	4277300.50	14604795.50	7479442.50	14063850.00	19465709.00	19701211.00	17259362.00	10771473.00	16486247.00
NORTH AMERICA	3476717.00	4030505.00	4865146.00	7215726.00	6158823.00	5979714.00	6052691.00	5829573.00	5908857.00
North America	3476717.00	4030505.00	4865146.00	7215726.00	6158823.00	5979714.00	6052691.00	5829573.00	5908857.00
POLAR	----	----	----	----	----	----	----	----	----
Antarctic	----	----	----	----	----	----	----	----	----
Arctic	34786.00	39461.00	103738.00	143331.00	128890.00	116018.00	120596.00	128590.00	160253.00
WEST ASIA	127200.00	283600.00	296338.00	361302.00	429403.00	425026.00	441635.00	448639.00	441846.00
Arabian Peninsula	123100.00	278800.00	257801.00	354535.00	416831.00	403970.00	420752.00	425181.00	418853.00
Mashriq	4100.00	4800.00	38537.00	6767.00	12572.00	21056.00	20883.00	23458.00	22993.00
GLOBAL TOTALS	31386501.05	59017024.50	64279428.00	83633548.50	94102937.00	96183276.00	96543965.00	90177400.00	96968623.00
REPORTED GLOBAL TOTALS FROM ORIGINAL DATA SOURCES	31386501.05	31386501.05	31386501.05	31386501.05	31386501.05	31386501.05	31386501.05	31386501.05	31386501.05

Comments:

Fishery production - marine includes: marine fish, marine, crustaceans, marine molluscs and diadromous fish caught in marine areas.

Fishery production relates to catch of aquatic organisms, taken for commercial, industrial, recreational and subsistence purposes from inland, brackish and marine waters. The harvest from mariculture, aquaculture and other kinds of fish farming is also included. Data include all quantities caught for both food and feed purposes but exclude discards. Catches of fish, crustaceans and molluscs are expressed in live weight, that is the nominal weight of the aquatic organisms at the time of capture. To assign nationality to catches, the flag of the fishing vessel is used, unless the wording of chartering and joint operation contracts indicates otherwise. According to the inland or marine area where caught, capture production is also classified into seven major inland fishing areas and 18 major marine fishing areas, internationally established for fishery statistical purposes.

1.6.2 Boundaries

The United Nations Convention on the Law of the Sea (UNCLOS) was opened for signature on 10 December 1982 in Montego Bay, Jamaica, and it entered into force on 16 November 1994. As of January 2000, there were 132 countries that have ratified UNCLOS. Under UNCLOS, coastal States can claim sovereign rights in a 200-nautical mile exclusive economic zone (EEZ). This allows for sovereign rights over the EEZ in terms of exploration, exploitation, conservation and management of all natural resources in the seabed, its subsoil and overlaying waters. UNCLOS allows other states to navigate and fly over the EEZ, as well as to lay submarine cables and pipelines. The inner limit of the EEZ starts at the outer boundary of the Territorial Sea (that is to say 12 nautical miles from the low-water line along the coast).

In cases where a country's low-water lines is within 400 nautical miles of each other the EEZ boundaries are generally established by treaty, though there are many cases where these are in dispute. Under UNCLOS, "land-locked and geographically disadvantaged States have the right to participate on an equitable basis in exploitation of an appropriate part of the surplus of the living resources of the EEZ's of coastal States of the same region or sub-region". Some States have not ratified UNCLOS and many have not yet claimed their EEZ. These areas of unclaimed EEZ are the areas that a State has the right to claim under UNCLOS, but has not done so yet. Given the uncertainties surrounding much of the delimitation of the EEZ, these figures should be used with caution. Further information on the Web site: http://www.maritimeboundaries.com. Data source: exclusive economic zone area calculations are from L. Pruett and J. Cimino, unpublished data, Global Maritime Boundaries Database (GMBD), Veridian - MRJ Technology Solutions, (Fairfax, Virginia, January, 2000).

Units: square kilometres

Data Source: The Global Maritime Boundaries Database (GMBD)

Data Provider: Viridian - MRJ Technology Solutions

Years: 2000

Claimed Exclusive Economic Zone (EEZ)

	2000
AFRICA	----
Central Africa	----
Eastern Africa	106544
Northern Africa	----
Southern Africa	----
Western Africa	1784273
Western Indian Ocean	4114902
ASIA AND PACIFIC	----
Australia and New Zealand	10551548
Central Asia	----
North West Pacific and East Asia	----
South Asia	3845876
South East Asia	4185408
South Pacific	19603432
EUROPE	11642594
Central Europe	270650
Eastern Europe	6361067
Western Europe	----
LATIN AMERICA AND CARIBBEAN	15847750
Caribbean	2356050
Meso-America	4132911
South America	9358789
NORTH AMERICA	11084405
North America	11084405
POLAR	----
Antarctic	----
Arctic	----
WEST ASIA	----
Arabian Peninsula	----
Mashriq	----
GLOBAL TOTALS	----
REPORTED GLOBAL TOTALS FROM ORIGINAL DATA SOURCES	102108403

Exclusive Fishing Zone (EFZ)

	2000
AFRICA	----
Central Africa	41483
Eastern Africa	759253
Northern Africa	----
Southern Africa	----
Western Africa	----
Western Indian Ocean	----
ASIA AND PACIFIC	----
Australia and New Zealand	----
Central Asia	----
North West Pacific and East Asia	----
South Asia	----
South East Asia	----
South Pacific	----
EUROPE	----
Central Europe	----
Eastern Europe	----
Western Europe	----
LATIN AMERICA AND CARIBBEAN	----
Caribbean	----
Meso-America	----
South America	----
NORTH AMERICA	----
North America	----
POLAR	----
Antarctic	----
Arctic	----
WEST ASIA	----
Arabian Peninsula	----
Mashriq	----
GLOBAL TOTALS	----
REPORTED GLOBAL TOTALS FROM ORIGINAL DATA SOURCES	12885241

Comments: The exclusive fishing zone or fishery zone refers to an area beyond the outer limit of the territorial sea (12 nautical miles from the coast) in which the coastal State has the right to fish, subject to any concessions which may be granted to foreign fishermen. Some countries have made no claim beyond the territorial sea. Some states have claimed an exclusive fishing zone instead of the more encompassing 200 nautical mile EEZ.

AGGREGATIONS

Units: square kilometres

Data Source: The Global Maritime Boundaries Database (GMBD)

Data Provider: Viridian - MRJ Technology Solutions

Year: 2000

Territorial Sea

	2000
AFRICA	**1069124**
Central Africa	61038
Eastern Africa	125325
Northern Africa	229874
Southern Africa	249572
Western Africa	197739
Western Indian Ocean	205576
ASIA AND PACIFIC	**8631507**
Australia and New Zealand	949726
Central Asia	----
North West Pacific and East Asia	851989
South Asia	498281
South East Asia	4450878
South Pacific	1880633
EUROPE	**2694369**
Central Europe	194138
Eastern Europe	1378150
Western Europe	1122081
LATIN AMERICA AND CARIBBEAN	**2079913**
Caribbean	575449
Meso-America	474504
South America	1029960
NORTH AMERICA	**3484108**
North America	3484108
POLAR	**----**
Antarctic	----
Arctic	----
WEST ASIA	**277302**
Arabian Peninsula	267931
Mashriq	9371
GLOBAL TOTALS	18236323
REPORTED GLOBAL TOTALS FROM ORIGINAL DATA SOURCES	18816919

Comments:

World total ("reported global totals from original data") excludes 2,867,050 square kilometres of disputed territorial seas.

Territorial sea is defined under UNCLOS as the 12-nautical mile zone from the baseline or low-water line along the coast. The coastal State's sovereignty extends to the territorial sea, including its sea-bed, subsoil, and air space above it. Foreign vessels are allowed "innocent passage" through those waters. Even though the established limit for a territorial sea is 12 nautical miles, some countries claim larger areas. Territorial seas with overlapping claims from different countries are shown separately as disputed territorial seas.

Continental Shelf Area

	2000
AFRICA	**1246082.90**
Central Africa	67391.10
Eastern Africa	99726.90
Northern Africa	274922.00
Southern Africa	391344.30
Western Africa	254813.50
Western Indian Ocean	157885.10
ASIA AND PACIFIC	**8158697.10**
Australia and New Zealand	2312994.10
Central Asia	237621.00
North West Pacific and East Asia	1446521.30
South Asia	684836.10
South East Asia	3224698.90
South Pacific	252025.70
EUROPE	**6456161.30**
Central Europe	239877.80
Eastern Europe	4295697.00
Western Europe	1920586.50
LATIN AMERICA AND CARIBBEAN	**3009595.90**
Caribbean	187606.90
Meso-America	619006.50
South America	2202982.50
NORTH AMERICA	**5107516.90**
North America	5107516.90
POLAR	**----**
Antarctic	----
Arctic	----
WEST ASIA	**307772.00**
Arabian Peninsula	304635.30
Mashriq	3136.70
GLOBAL TOTALS	24285826.10
REPORTED GLOBAL TOTALS FROM ORIGINAL DATA SOURCES	24285959.12

Comments:

According to UNCLOS, the continental shelf is the area of the seabed and subsoil which extends beyond the territorial sea to a distance of 200 nautical miles from the territorial sea baseline and beyond that distance to the outer edge of the continental margin.

Areas of continental shelf that are disputed by overlapping claims by one or more nations have been excluded from this table. Areas that are of cooperative joint development between two or more nations have also been excluded. Coastal States have sovereign rights over the continental shelf (the national area of the seabed) for exploring and exploiting it; the shelf can extend at least 200 nautical miles from the shore, and more under specified circumstances.

Length of Coastline

Units: kilometres

Data Source: World Vector Shoreline, United States Defence Mapping Agency

Data Provider: Viridian - MRJ Technology Solutions

Year: 2000

	2000
AFRICA	76540.50
Central Africa	5071.30
Eastern Africa	9373.30
Northern Africa	15660.80
Southern Africa	18158.70
Western Africa	16411.20
Western Indian Ocean	11865.20
ASIA AND PACIFIC	429920.30
Australia and New Zealand	83738.90
Central Asia	7524.40
North West Pacific and East Asia	77567.30
South Asia	33803.90
South East Asia	173252.20
South Pacific	54033.60
EUROPE	336155.50
Central Europe	21151.10
Eastern Europe	116510.30
Western Europe	198494.10
LATIN AMERICA AND CARIBBEAN	218270.00
Caribbean	35246.30
Meso-America	38456.90
South America	144566.80
NORTH AMERICA	398835.20
North America	398835.20
POLAR	----
Antarctic	----
Arctic	----
WEST ASIA	18959.80
Arabian Peninsula	18321.20
Mashriq	638.60
GLOBAL TOTALS	1478681.30
REPORTED GLOBAL TOTALS FROM ORIGINAL DATA SOURCES	1634700.69

Comments:

The measurement of an irregular and curving feature such as a nation's coastal length is scale-dependent and very difficult to measure. Maps of individual islands for example, frequently show great detail, whereas regional maps summarize complex coastlines into a few simple lines. In addition, coastal features are constantly changing due to erosion, etc. The only way to derive comparable statistics on coastline length is to use a single source which uses a constant scale. This is what has been attempted with the data presented in this table, however, highly complex coastlines will appear longer at higher resolutions. Estimates may differ from other published sources.

Because of the difficulty in trying to measure coastline length, these figures should be interpreted as approximations and should be used with caution. Coastline length was derived from the World Vector Shoreline database at 1:250,000 kilometres. The estimates presented here were calculated using a geographic information system (GIS) and an underlying database consistent for the entire world.

The methodology used to estimate length is based on the following: 1) A country's coastline is made up of individual lines, and an individual line has two or more vertices and/or nodes. 2) The length between two vertices is calculated on the surface of a sphere. 3) The sum of the lengths of the pairs of vertices is aggregated for each individual line, and 4) the sum of the lengths of individual lines was aggregated for a country. In general, the coastline length of islands that are part of a country, but are not overseas territories, are included in the coastline estimate for that country (that is to say Canary Islands are included in Spain). Coastline length for overseas territories and dependencies are listed separately. Disputed areas are not included in country or regional totals. Data source: coastal length data are based on the World Vector Shoreline, United States Defence Mapping Agency, 1989. Figures were calculated by L. Pruett and J. Cimino, unpublished data, Global Maritime Boundaries Database (GMBD), Viridian - MRJ Technology Solutions, (Fairfax, Virginia, January, 2000).

AGGREGATIONS

1.6.3 Population

Population Within 100 Kilometres of Coast

Units: percent

Data Source: Gridded Population of the World (GPW), Version 2 alpha

Data Provider: Consortium for International Earth Science Information Network (CIESIN)

Year: 1995

	1995
AFRICA	28.89
Central Africa	8.08
Eastern Africa	7.84
Northern Africa	51.54
Southern Africa	26.20
Western Africa	30.76
Western Indian Ocean	61.90
ASIA AND PACIFIC	38.14
Australia and New Zealand	91.50
Central Asia	2.84
North West Pacific and East Asia	33.82
South Asia	27.45
South East Asia	80.75
South Pacific	73.39
EUROPE	41.30
Central Europe	30.32
Eastern Europe	17.13
Western Europe	60.98
LATIN AMERICA AND CARIBBEAN	50.71
Caribbean	99.91
Meso-America	41.47
South America	48.73
NORTH AMERICA	41.39
North America	41.39
POLAR	----
Antarctic	----
Arctic	----
WEST ASIA	38.79
Arabian Peninsula	54.28
Mashriq	25.07
GLOBAL TOTALS	38.69
REPORTED GLOBAL TOTALS FROM ORIGINAL DATA SOURCES	39.00

Comments:

Globally 2.2 billion people or 39 percent of the population live within 100 km from the coast. Estimates were based on 1995 population figures. The Gridded Population of the World is a raster data set that provides information on the spatial distribution of the world's human population. The grid cells are approximately 4.6 km on each side. Populations are distributed according to administrative districts which vary in scale, level and size from country to country. A 100 km coastal buffer with a 10 km "safe area" falling into the sea were used in the geographic information system to calculate the number of people in the coastal zone for each country individually. The percentage population in the coastal zone were calculated from 1995 United Nations (UN) Population Division totals for each country. Please refer to the original source for further information on the variables and collection methodologies.

Data source: percent of the population living within 100 kilometres from the coast is derived from the Centre for International Earth Science Information Network (CIESIN), World Resources Institute, and International Food Policy Research Institute, "Gridded Population of the World, Version 2 alpha" (Columbia University, Palisades, NY, 2000), available at http://sedac.ciesin.columbia.edu/plue/gpw/index.html.

1.7 Atmosphere

1.7.1 Climate

Precipitation

Data Source: Mitchell, T.D. and Hulme, M. (2001), Climate Data for Political Areas

Data Provider: Intergovernmental Panel on Climate Change (IPCC) Data Distribution Centre (DDC)

Years: 1961-90

Precipitation - Cubic Kilometres

Units: cubic kilometres per year

	1961-90
AFRICA	20445.51
Central Africa	6745.23
Eastern Africa	1918.30
Northern Africa	1624.50
Southern Africa	5394.42
Western Africa	3860.44
Western Indian Ocean	902.63
ASIA AND PACIFIC	29929.43
Australia and New Zealand	4606.10
Central Asia	997.59
North West Pacific and East Asia	7267.55
South Asia	5156.64
South East Asia	10227.74
South Pacific	1673.81
EUROPE	12930.33
Central Europe	1403.27
Eastern Europe	8471.30
Western Europe	3055.76
LATIN AMERICA AND CARIBBEAN	30796.12
Caribbean	341.94
Meso-America	2738.57
South America	27715.61
NORTH AMERICA	12435.42
North America	12435.42
POLAR	----
Antarctic	----
Arctic	199.72
WEST ASIA	421.53
Arabian Peninsula	251.28
Mashriq	170.26
GLOBAL TOTALS	107163.02
REPORTED GLOBAL TOTALS FROM ORIGINAL DATA SOURCES	107163.02

Precipitation - Cubic Kilometres

Units: millimetres per year

	1961-90
AFRICA	674.50
Central Africa	1257.04
Eastern Africa	695.45
Northern Africa	190.56
Southern Africa	778.38
Western Africa	628.94
Western Indian Ocean	1517.93
ASIA AND PACIFIC	844.50
Australia and New Zealand	574.92
Central Asia	249.19
North West Pacific and East Asia	619.77
South Asia	761.38
South East Asia	2335.05
South Pacific	3055.25
EUROPE	537.88
Central Europe	656.66
Eastern Europe	467.86
Western Europe	804.93
LATIN AMERICA AND CARIBBEAN	1496.47
Caribbean	1466.40
Meso-America	1104.35
South America	1551.29
NORTH AMERICA	634.47
North America	634.47
POLAR	----
Antarctic	----
Arctic	584.50
WEST ASIA	113.33
Arabian Peninsula	83.67
Mashriq	235.45
GLOBAL TOTALS	832.52
REPORTED GLOBAL TOTALS FROM ORIGINAL DATA SOURCES	832.52

Comments:

The Global totals do not include values for Antarctica.

Precipitation is the annual average (within 1961-90) of water falling on the country. The original data took the form of a value for each month and each box on a 0.5 degree latitude / longitude grid. The seasonal and annual values are the means of their constituent months. Original Data Station observations were first collected by national meteorological, hydrological and related services, and were acquired through the free and unrestricted exchange of meteorological and related data. These observations were gridded by collaborators at the Climatic Research Unit (www.cru.uea.ac.uk).

The gridded data-set is publicly available, and has been published in a peer-reviewed scientific journal: New, M., Hulme, M., and Jones, P., 1999: Representing twentieth-century space-time climate variability. Part I: Development of a 1961-1990 mean monthly terrestrial climatology. Journal of Climate 12: 829-856.

New, M.G., Hulme,M., and Jones,P.D., 2000: Representing twentieth-century space-time climate variability. Part II: Development of 1901-1996 monthly grids of terrestrial surface climate. Journal of Climate 13:2217-2238.).

AGGREGATIONS

1.7.2 Emissions of GHG, ODS, Dust, Metals

Some of the figures may be negative since the figures are for each calendar year, it is quite possible that in some years the feedstock figure may exceed the production figure of that year, if the feedstock use is from a carry-over stock. The production could be negative in such cases. For the same reason, the consumption could also be negative. The data formats used by the Parties, before 1997, to report data were replaced by a revised format after the Ninth Meeting of the Parties (for more information see Article 7 of the Protocol).

Carbon to the Atmosphere from Land-Use Change - Annual Net Flux

Units: million tonnes of carbon

Data Source: Houghton and Hackler (2001), CDIAC NPD-050/R1

Data Provider: Carbon Dioxide Information Analysis Centre (CDIAC)

Years: 1960-1990

	1960	1970	1980	1981	1982	1983	1984	1985	1986	1987	1988	1989	1990
AFRICA	95.34	149.73	238.69	256.24	261.24	265.19	267.79	269.92	318.45	326.58	332.71	337.54	341.50
Central Africa	----	----	----	----	----	----	----	----	----	----	----	----	----
Eastern Africa	----	----	----	----	----	----	----	----	----	----	----	----	----
Northern Africa	----	----	----	----	----	----	----	----	----	----	----	----	----
Southern Africa	----	----	----	----	----	----	----	----	----	----	----	----	----
Western Africa	----	----	----	----	----	----	----	----	----	----	----	----	----
Western Indian Ocean	----	----	----	----	----	----	----	----	----	----	----	----	----
ASIA AND PACIFIC	527.77	635.99	808.66	1068.87	1125.17	1166.90	1196.46	1219.02	1237.49	1215.52	1224.47	1235.19	1147.00
Australia and New Zealand	----	----	----	----	----	----	----	----	----	----	----	----	----
Central Asia	----	----	----	----	----	----	----	----	----	----	----	----	----
North West Pacific and East Asia	----	----	----	----	----	----	----	----	----	----	----	----	----
South Asia	----	----	----	----	----	----	----	----	----	----	----	----	----
South East Asia	----	----	----	----	----	----	----	----	----	----	----	----	----
South Pacific	----	----	----	----	----	----	----	----	----	----	----	----	----
EUROPE	230.33	128.03	21.35	18.07	14.34	11.59	9.28	7.43	4.74	3.85	3.20	2.77	2.03
Central Europe	----	----	----	----	----	----	----	----	----	----	----	----	----
Eastern Europe	----	----	----	----	----	----	----	----	----	----	----	----	----
Western Europe	----	----	----	----	----	----	----	----	----	----	----	----	----
LATIN AMERICA AND CARIBBEAN	366.65	638.01	511.59	516.39	517.03	516.74	552.39	563.87	567.57	592.06	596.17	579.12	577.16
Caribbean	----	----	----	----	----	----	----	----	----	----	----	----	----
Meso-America	----	----	----	----	----	----	----	----	----	----	----	----	----
South America	----	----	----	----	----	----	----	----	----	----	----	----	----
NORTH AMERICA	-2.61	-9.26	-7.09	-7.01	-5.80	-4.30	-2.55	-0.55	1.67	4.08	6.69	9.47	12.42
North America	----	----	----	----	----	----	----	----	----	----	----	----	----
POLAR	----	----	----	----	----	----	----	----	----	----	----	----	----
Antarctic	----	----	----	----	----	----	----	----	----	----	----	----	----
Arctic	----	----	----	----	----	----	----	----	----	----	----	----	----
WEST ASIA	----	----	----	----	----	----	----	----	----	----	----	----	----
Arabian Peninsula	----	----	----	----	----	----	----	----	----	----	----	----	----
Mashriq	----	----	----	----	----	----	----	----	----	----	----	----	----
GLOBAL TOTALS	----	----	----	----	----	----	----	----	----	----	----	----	----
REPORTED GLOBAL TOTALS FROM ORIGINAL DATA SOURCES	1240.68	1563.98	1595.05	1869.54	1928.54	1972.56	2041.41	2079.48	2150.28	2163.10	2184.95	2186.55	2103.34

Comments:

Carbon to the atmosphere from land-use change - annual net flux is a numeric database with annual estimations, from 1850 through 1990, of the net flux of carbon between terrestrial ecosystems and the atmosphere. The data is the result of deliberate changes in land cover and land use, especially forest clearing for agriculture and the harvest of wood for wood products or energy.

The database is an attempt to calculate the flux between the atmosphere and the terrestrial ecosystems, for both temperate and tropical ecosystems. The database is a revision to Houghton and Hackler (1995) and it provides and documents the data corresponding to the analysis reported by Houghton (1999).

AGGREGATIONS

Consumption of ODS: Carbon Tetrachloride

Units: ozone depleting potential (ODP) tonnes

Data Source: Production and Consumption of Ozone Depleting Substances 1986-2000

Data Provider: Secretariat for the Vienna Convention and the Montreal Protocol (the Ozone Secretariat)

Years: 1989-2000

	1989	1990	1991	1992	1993	1994	1995	1996	1997	1998	1999	200
AFRICA	----	----	----	----	----	----	898.99	518.36	489.87	348.20	----	--
Central Africa	----	----	----	----	----	----	----	----	----	----	----	--
Eastern Africa	7.37	----	----	----	----	----	40.72	24.78	71.88	69.70	----	--
Northern Africa	262.35	----	180.51	----	185.68	94.95	250.25	245.30	138.53	83.60	59.40	--
Southern Africa	----	----	----	----	----	----	----	----	----	----	----	--
Western Africa	----	----	----	----	590.02	656.87	607.37	230.52	244.42	160.78	151.58	--
Western Indian Ocean	----	----	----	----	----	----	0.09	0.05	0.11	0.03	0.03	--
ASIA AND PACIFIC	110148.86	----	----	41696.31	50877.84	9642.79	-11604.51	7394.70	8850.30	92126.54	16942.24	--
Australia and New Zealand	9.12	----	----	12.32	1.51	0.86	0.24	0.43	0.52	0.24	0.01	--
Central Asia	136.62	----	----	----	86.24	----	----	44.53	35.43	20.90	----	--
North West Pacific and East Asia	104820.10	----	14567.30	35884.20	39486.70	-18.70	-15579.30	-2160.40	149.60	85661.40	147.40	--
South Asia	4868.38	----	----	5529.71	11059.96	9386.85	3752.62	9381.35	8627.88	6432.78	16786.50	--
South East Asia	314.43	----	----	267.77	----	263.11	213.79	128.68	36.87	11.22	8.25	--
South Pacific	0.21	----	----	----	0.00	0.00	0.00	0.11	0.00	0.00	----	--
EUROPE	185103.09	----	----	14365.62	20619.15	14714.93	-493.51	-3656.14	-1943.93	7854.07	2270.53	--
Central Europe	8237.45	----	----	----	1122.13	7195.65	----	636.65	-98.66	2135.27	-685.76	--
Eastern Europe	103299.90	1989.90	2209.90	3000.03	-3011.69	35.31	16.25	548.78	-2177.52	4956.17	2038.10	--
Western Europe	73565.74	----	----	10156.91	22508.71	7483.97	693.91	-4841.57	332.25	762.63	----	--
LATIN AMERICA AND CARIBBEAN	37547.20	37390.14	----	22136.50	23170.70	1196.85	5671.46	-617.83	-235.48	15.23	4024.38	--
Caribbean	----	----	----	----	----	----	----	----	----	----	----	--
Meso-America	18409.60	6551.60	2966.04	612.04	602.91	16.17	13.97	13.53	16.06	19.51	12.32	--
South America	19046.48	30641.13	----	21524.38	22529.51	1180.68	5561.79	-638.18	-317.54	-50.16	4011.99	--
NORTH AMERICA	18091.70	----	----	3671.80	37921.40	405.90	-18829.80	11.00	-27239.80	-32361.67	-12486.77	--
North America	18091.70	----	----	3671.80	37921.40	405.90	-18829.80	11.00	-27239.80	-32361.67	-12486.77	--
POLAR	----	----	----	----	----	----	----	----	----	----	----	--
Antarctic	----	----	----	----	----	----	----	----	----	----	----	--
Arctic	----	----	----	----	----	----	----	----	----	----	----	--
WEST ASIA	----	----	----	----	----	----	----	----	----	----	----	--
Arabian Peninsula	----	----	----	----	----	----	990.13	568.84	337.21	683.21	----	--
Mashriq	----	----	----	----	----	----	----	----	----	----	----	--
GLOBAL TOTALS	----	----	----	----	----	----	----	----	----	----	----	--
REPORTED GLOBAL TOTALS FROM ORIGINAL DATA SOURCES	****	****	****	****	****	****	****	****	****	****	****	**

Comments:

Consumption means production plus imports minus exports of controlled substances, as reported to the Secretariat by 183 parties to the Montreal Protocol on Substances that Deplete the Ozone Layer. Annex B, Group II: carbon tetrachloride comprises CCl_4.

AGGREGATIONS

Units: ozone depleting potential (ODP) tonnes

Data Source: Production and Consumption of Ozone Depleting Substances 1986-2000

Data Provider: Secretariat for the Vienna Convention and the Montreal Protocol (the Ozone Secretariat)

Years: 1986, 1989-2000

	1986	1989	1990	1995	1996	1997	1998	1999	2000
AFRICA	----	20632.54	----	12327.30	13630.08	13262.02	12076.61	----	----
Central Africa	----	----	----	----	----	----	----	----	----
Eastern Africa	305.96	317.11	----	401.52	272.29	361.48	359.38	----	----
Northern Africa	10258.20	7728.00	----	6804.84	6879.52	6215.88	5097.90	3736.30	----
Southern Africa	----	----	----	----	----	----	----	----	----
Western Africa	----	1158.84	1322.98	2436.32	5454.13	5554.93	5540.44	4826.04	----
Western Indian Ocean	111.57	----	----	49.55	61.20	136.25	68.50	48.43	----
ASIA AND PACIFIC	181146.86	220271.64	----	135675.69	79031.13	79638.05	79191.05	65940.73	----
Australia and New Zealand	16378.40	15477.28	7974.80	2773.65	235.86	183.93	195.06	274.14	----
Central Asia	3157.76	4019.86	----	----	1115.54	757.41	1170.21	----	----
North West Pacific and East Asia	147378.40	181398.88	139552.20	98361.80	46486.76	50975.90	55226.54	43027.20	----
South Asia	3296.96	5516.96	----	9338.17	9761.31	9019.10	7575.21	10683.79	----
South East Asia	10864.58	13813.89	----	24775.22	21334.81	18643.99	14961.54	11871.92	----
South Pacific	70.76	----	----	77.10	96.85	57.72	62.49	----	----
EUROPE	499785.38	452814.26	333708.39	47825.07	42645.09	34786.97	37829.55	31555.06	----
Central Europe	46102.04	39865.44	----	11607.08	7014.04	6826.52	5884.56	2662.98	----
Eastern Europe	110986.00	107546.20	104500.02	22328.36	14749.78	12994.74	13356.14	16099.65	----
Western Europe	342697.34	305402.62	207651.75	13889.63	20881.27	14965.71	18588.85	----	----
LATIN AMERICA AND CARIBBEAN	35496.73	32605.90	32225.41	33096.52	30608.92	27778.85	24789.96	25360.89	----
Caribbean	----	----	----	----	----	----	----	----	----
Meso-America	10193.82	10701.97	12733.05	6261.53	6875.84	5518.03	4227.84	3979.24	----
South America	23813.14	20032.31	18131.33	25387.62	22234.83	20838.46	19273.05	20500.59	----
NORTH AMERICA	325921.80	336386.00	211481.80	40346.00	2026.00	1053.57	2563.24	2898.96	----
North America	325921.80	336386.00	211481.80	40346.00	2026.00	1053.57	2563.24	2898.96	----
POLAR	----	----	----	----	----	----	----	----	----
Antarctic	----	----	----	----	----	----	----	----	----
Arctic	----	----	----	----	----	----	----	----	----
WEST ASIA	----	----	----	----	----	----	----	----	----
Arabian Peninsula	8508.46	----	----	3575.81	3484.38	3867.57	4043.04	----	----
Mashriq	----	----	----	----	----	----	----	----	----
GLOBAL TOTALS	----	----	----	----	----	----	----	----	----
REPORTED GLOBAL TOTALS FROM ORIGINAL DATA SOURCES	****	****	****	****	****	****	****	****	****

Comments:

Consumption means production plus imports minus exports of controlled substances, as reported to the Secretariat by 183 parties to the Montreal Protocol on Substances that Deplete the Ozone Layer. The Annex A, Group I: chlorofluorocarbons (CFCs) comprise $CFCl_3$, CF_2Cl_2, $C_2F_3Cl_3$, $C_2F_4Cl_2$, C_2F_5Cl.

Consumption of ODS: Hydrochlorofluorocarbons

Units: ozone depleting potential (ODP) tonnes

Data Source: Production and Consumption of Ozone Depleting Substances 1986-2000

Data Provider: Secretariat for the Vienna Convention and the Montreal Protocol (the Ozone Secretariat)

Years: 1989-2000

	1989	1990	1991	1992	1993	1994	1995	1996	1997	1998	1999	2000
AFRICA	----	----	----	----	----	----	235.91	367.92	448.52	503.38	----	----
Central Africa	----	----	----	----	----	----	----	----	----	----	----	----
Eastern Africa	2.75	----	----	----	----	----	6.73	13.35	13.36	18.94	----	----
Northern Africa	5.72	----	21.77	----	20.85	22.16	9.52	86.63	96.14	125.13	117.94	----
Southern Africa	----	----	----	----	----	----	----	----	----	----	----	----
Western Africa	----	----	28.19	29.85	12.99	8.11	26.51	35.86	49.47	74.83	60.82	----
Western Indian Ocean	----	----	----	----	----	----	3.73	1.23	11.09	6.96	5.62	----
ASIA AND PACIFIC	2781.55	----	----	3254.07	6071.78	5271.45	7764.54	6604.73	7183.61	6621.02	9359.00	----
Australia and New Zealand	171.37	----	----	148.04	223.22	142.42	196.69	254.74	248.87	284.94	286.58	----
Central Asia	7.34	----	----	----	7.22	----	----	3.95	3.30	1.78	----	----
North West Pacific and East Asia	2080.50	----	2399.03	2568.74	5038.42	4236.55	6429.13	5059.71	5680.21	5010.30	7717.79	----
South Asia	124.28	----	----	147.79	181.56	176.40	187.99	214.39	38.13	42.72	49.16	----
South East Asia	396.99	----	----	387.18	----	709.75	943.74	1068.45	1210.32	1273.58	1301.45	----
South Pacific	1.07	----	----	----	----	----	3.81	3.49	2.78	7.70	----	----
EUROPE	4258.26	----	----	4978.73	5791.65	7748.75	9681.56	9260.77	9214.35	11074.52	9552.91	
Central Europe	124.91	----	----	----	114.46	155.21	----	264.31	351.86	393.61	408.31	----
Eastern Europe	1243.23	474.20	356.30	289.53	181.49	121.03	115.26	105.33	114.74	102.23	157.95	----
Western Europe	2890.12	----	----	4625.69	5495.70	7472.51	9328.11	8891.13	8747.75	10578.68	----	----
LATIN AMERICA AND CARIBBEAN	507.83	427.52	----	740.67	782.40	870.40	1031.88	948.00	1168.27	1297.16	1970.71	----
Caribbean	----	----	----	----	----	----	----	----	----	----	----	----
Meso-America	150.16	132.58	127.38	333.99	346.83	335.34	322.23	362.72	463.47	513.31	1171.03	----
South America	338.39	290.50	----	388.77	398.94	517.26	665.45	545.03	682.05	746.52	765.87	----
NORTH AMERICA	6728.00	----	----	5956.59	7272.65	11478.34	14663.15	12113.29	11656.65	14706.60	14794.52	----
North America	6728.00	----	----	5956.59	7272.65	11478.34	14663.15	12113.29	11656.65	14706.60	14794.52	----
POLAR	----	----	----	----	----	----	----	----	----	----	----	----
Antarctic	----	----	----	----	----	----	----	----	----	----	----	----
Arctic	----	----	----	----	----	----	----	----	----	----	----	----
WEST ASIA	----	----	----	----	----	----	----	----	----	----	----	----
Arabian Peninsula	----	----	----	----	----	----	214.96	280.10	312.36	322.85	----	----
Mashriq	----	----	----	----	----	----	----	----	----	----	----	----
GLOBAL TOTALS	----	----	----	----	----	----	----	----	----	----	----	----
REPORTED GLOBAL TOTALS FROM ORIGINAL DATA SOURCES	****	****	****	****	****	****	****	****	****	****	****	****

Comments:

Consumption means production plus imports minus exports of controlled substances, as reported to the Secretariat by 183 parties to the Montreal Protocol on Substances that Deplete the Ozone Layer. The Annex C, Group I: hydrochlorofluorocarbons (HCFCs) comprise $CHFCl_2$, CHF_2Cl, CH_2FCl, C_2HFCl_4, $C_2HF_2Cl_3$, $C_2HF_3Cl_2$, $CHCl_2CF_3$, C_2HF_4Cl, $CHFClCF_3$, $C_2H_2FCl_3$, $C_2H_2F_2Cl_2$, $C_2H_2F_3Cl$, $C_2H_3FCl_2$, CH_3CFCl_2, $C_2H_3F_2Cl$, CH_3CF_2Cl, C_2H_4FCl, C_3HFCl_6, $C_3HF_2Cl_5$, $C_3HF_3Cl_4$, $C_3HF_4Cl_3$, $C_3HF_5Cl_2$, $CF_3CF_2CHCl_2$, CF_2ClCF_2CHClF, C_3HF_6Cl, $C_3H_2FCl_5$, $C_3H_2F_2Cl_4$, $C_3H_2F_3Cl_3$, $C_3H_2F_4Cl_2$, $C_3H_2F_5Cl$, $C_3H_3FCl_4$, $C_3H_3F_2Cl_3$, $C_3H_3F_3Cl_2$, $C_3H_3F_4Cl$, $C_3H_4FCl_3$, $C_3H_4F_2Cl_2$, $C_3H_4F_3Cl$, $C_3H_5FCl_2$, $C_3H_5F_2Cl$, C_3H_6FCl.

AGGREGATIONS

Consumption of ODS: Methyl Chloroform

Units: ozone depleting potential (ODP) tonnes

Data Source: Production and Consumption of Ozone Depleting Substances 1986-2000

Data Provider: Secretariat for the Vienna Convention and the Montreal Protocol (the Ozone Secretariat)

Years: 1989-2000

	1989	1990	1991	1992	1993	1994	1995	1996	1997	1998	1999	2000
AFRICA	----	----	----	----	----	----	0.00	0.00	0.00	0.00	----	----
Central Africa	----	----	----	----	----	----	----	----	----	----	----	----
Eastern Africa	0.00	----	----	----	----	----	0.00	0.00	0.00	0.00	----	----
Northern Africa	0.00	----	0.00	----	0.00	0.00	0.00	0.00	0.00	0.00	0.00	----
Southern Africa	----	----	----	----	----	----	----	----	----	----	----	----
Western Africa	----	----	----	----	0.00	0.00	0.00	0.00	0.00	0.00	0.00	----
Western Indian Ocean	----	----	----	----	----	----	0.00	0.00	0.00	0.00	0.00	----
ASIA AND PACIFIC	15683.10	----	----	15773.70	7221.35	388.90	5149.10	967.00	1183.10	1033.70	1170.30	----
Australia and New Zealand	0.00	----	0.00	0.00	0.00	0.00	0.00	0.00	0.00	0.00	0.00	----
Central Asia	0.00	----	----	----	0.00	----	----	0.00	0.00	0.00	----	----
North West Pacific and East Asia	15636.40	44.00	17119.10	15725.40	7165.45	388.90	5149.10	967.00	1183.10	1033.70	1170.30	----
South Asia	46.70	----	----	48.30	55.90	0.00	0.00	0.00	0.00	0.00	0.00	----
South East Asia	0.00	----	----	----	----	0.00	0.00	0.00	0.00	0.00	0.00	----
South Pacific	0.00	----	----	----	0.00	0.00	0.00	0.00	0.00	0.00	----	----
EUROPE	13421.90	----	----	11361.40	5534.70	3044.47	1641.95	79.86	225.91	184.30	259.40	----
Central Europe	27.20	----	----	----	9.90	7.80	----	12.90	2.91	0.00	0.00	----
Eastern Europe	330.00	310.00	310.00	400.00	100.00	196.57	196.90	0.00	0.00	0.00	0.00	----
Western Europe	13064.70	----	----	10961.40	5424.80	2840.10	1439.00	66.96	223.00	184.30	----	----
LATIN AMERICA AND CARIBBEAN	1130.00	660.00	----	750.00	764.00	39.20	0.00	0.00	0.00	0.00	97.20	----
Caribbean	----	----	----	----	----	----	----	----	----	----	----	----
Meso-America	0.00	0.00	0.00	0.00	0.00	0.00	0.00	0.00	0.00	0.00	0.00	----
South America	1130.00	660.00	----	750.00	764.00	39.20	0.00	0.00	0.00	0.00	97.20	----
NORTH AMERICA	32649.10	----	----	25722.60	20637.10	5794.60	4598.60	447.50	437.30	262.30	245.80	----
North America	32649.10	----	----	25722.60	20637.10	5794.60	4598.60	447.50	437.30	262.30	245.80	----
POLAR	----	----	----	----	----	----	----	----	----	----	----	----
Antarctic	----	----	----	----	----	----	----	----	----	----	----	----
Arctic	----	----	----	----	----	----	----	----	----	----	----	----
WEST ASIA	----	----	----	----	----	----	----	----	----	----	----	----
Arabian Peninsula	----	----	----	----	----	----	0.00	0.00	0.00	0.00	----	----
Mashriq	----	----	----	----	----	----	----	----	----	----	----	----
GLOBAL TOTALS	----	----	----	----	----	----	----	----	----	----	----	----
REPORTED GLOBAL TOTALS FROM ORIGINAL DATA SOURCES	****	****	****	****	****	****	****	****	****	****	****	****

Comments:

Production means the amount of controlled substances produced, minus the amount destroyed by technologies approved by the parties (to the Montreal Protocol on Substances that Deplete the Ozone Layer) and minus the amount entirely used as feedstock in the manufacture of other chemicals. The amount recycled and reused is not to be considered as production. The data forms prescribe reporting of feedstock use and of quantities destroyed separately, and reporting of total production without deduction. The Secretariat would make the necessary deduction.

AGGREGATIONS

Units: ozone depleting potential (ODP) tonnes

Data Source: Production and Consumption of Ozone Depleting Substances 1986-2000

Data Provider: Secretariat for the Vienna Convention and the Montreal Protocol (the Ozone Secretariat)

Years: 1989-2000

	1989	1990	1991	1992	1993	1994	1995	1996	1997	1998	1999	2000
AFRICA	----	----	----	----	----	----	0.00	0.00	0.00	0.00	----	----
Central Africa	----	----	----	----	----	----	----	----	----	----	----	----
Eastern Africa	0.00	----	----	----	----	----	0.00	0.00	0.00	0.00	----	----
Northern Africa	0.00	----	0.00	----	0.00	0.00	0.00	0.00	0.00	0.00	0.00	----
Southern Africa	----	----	----	----	----	----	----	----	----	----	----	----
Western Africa	----	----	----	----	0.00	0.00	0.00	0.00	0.00	0.00	0.00	----
Western Indian Ocean	----	----	----	----	----	----	0.00	0.00	0.00	0.00	0.00	----
ASIA AND PACIFIC	2342.00	----	----	1600.00	808.00	136.00	35.00	17.00	27.00	26.00	27.00	----
Australia and New Zealand	0.00	----	----	0.00	0.00	0.00	0.00	0.00	0.00	0.00	0.00	----
Central Asia	0.00	----	----	----	0.00	----	----	0.00	0.00	0.00	----	----
North West Pacific and East Asia	2342.00	----	1585.00	1600.00	808.00	136.00	35.00	17.00	27.00	26.00	27.00	----
South Asia	0.00	----	----	0.00	0.00	0.00	0.00	0.00	0.00	0.00	0.00	----
South East Asia	0.00	----	----	----	----	0.00	0.00	0.00	0.00	0.00	0.00	----
South Pacific	0.00	----	----	----	0.00	0.00	0.00	0.00	0.00	0.00	----	----
EUROPE	384.00	----	----	----	51.60	-145.50	-257.77	-126.00	75.00	13.05	16.50	----
Central Europe	0.00	----	----	----	0.00	----	----	0.00	0.00	0.00	0.00	----
Eastern Europe	300.00	300.00	250.00	17.00	0.60	25.00	24.78	20.00	75.00	13.05	16.50	----
Western Europe	84.00	----	----	20.00	51.00	-170.50	-282.00	-146.00	0.00	0.00	----	----
LATIN AMERICA AND CARIBBEAN	0.00	0.00	----	10.80	0.00	0.00	0.00	0.00	0.00	0.00	0.00	----
Caribbean	----	----	----	----	----	----	----	----	----	----	----	----
Meso-America	0.00	0.00	0.00	0.00	0.00	0.00	0.00	0.00	0.00	0.00	0.00	----
South America	0.00	0.00	----	10.80	0.00	0.00	0.00	0.00	0.00	0.00	0.00	----
NORTH AMERICA	577.00	----	----	75.00	106.00	101.00	38.00	0.00	0.00	-1.00	0.00	----
North America	577.00	----	----	75.00	106.00	101.00	38.00	0.00	0.00	-1.00	0.00	----
POLAR	----	----	----	----	----	----	----	----	----	----	----	----
Antarctic	----	----	----	----	----	----	----	----	----	----	----	----
Arctic	----	----	----	----	----	----	----	----	----	----	----	----
WEST ASIA	----	----	----	----	----	----	----	----	----	----	----	----
Arabian Peninsula	----	----	----	----	----	----	0.00	0.00	0.00	0.00	----	----
Mashriq	----	----	----	----	----	----	----	----	----	----	----	----
GLOBAL TOTALS	----	----	----	----	----	----	----	----	----	----	----	----
REPORTED GLOBAL TOTALS FROM ORIGINAL DATA SOURCES	****	****	****	****	****	****	****	****	****	****	****	****

Comments:

Production means the amount of controlled substances produced, minus the amount destroyed by technologies approved by the Parties (to the Montreal Protocol on Substances that Deplete the Ozone Layer) and minus the amount entirely used as feedstock in the manufacture of other chemicals. The amount recycled and reused is not to be considered as production. The data forms prescribe reporting of feedstock use and of quantities destroyed separately, and reporting of total production without deduction. The Secretariat would make the necessary deduction.

The data is reported to the Secretariat by the parties to the Montreal Protocol on Substances that Deplete the Ozone Layer. Annex B, Group I: other fully halogenated CFCs comprise CF_3Cl, C_2FCl_5, C_2F2Cl_4, C_3FCl_7, $C_3F_2Cl_6$, $C_3F_3Cl_5$, $C_3F_4Cl_4$, $C_3F_5Cl_3$, $C_3F_6Cl_2$, C_3F_7Cl.

Emissions of CO₂ - from Cement Production (CDIAC)

Units: thousand metric tonnes of CO$_2$

Data Source: Energy Statistics Published by the United Nations and the U.S. Bureau of Mines and Energy Information Administration

Data Provider: Carbon Dioxide Information Analysis Centre (CDIAC)

Years: 1960-1998

	1960	1970	1980	1990	1991	1992	1993	1994	1995	1996	1997	1998
RICA	4136.37	8886.90	15897.20	24981.73	26786.10	27480.57	27280.00	28732.73	30627.30	31525.63	31428.10	32714.37
ntral Africa	99.73	224.03	545.60	643.87	494.27	511.13	506.37	427.53	517.37	495.00	478.50	382.07
stern Africa	225.87	579.70	694.10	985.97	914.10	969.10	948.57	1013.47	1170.77	1357.03	1263.53	1122.00
rthern Africa	2101.73	3396.80	7929.17	15362.23	17485.23	18096.10	17737.13	18488.80	19661.03	20286.93	19762.23	20990.20
uthern Africa	1540.00	3731.93	4337.30	4949.27	4816.53	4615.97	4808.47	4992.53	5530.07	5589.47	5971.53	6051.10
estern Africa	169.03	916.30	2361.33	3010.33	3045.90	3258.20	3249.40	3630.37	3572.07	3608.00	3754.30	3959.63
estern Indian Ocean	0.00	38.13	29.70	30.07	30.07	30.07	30.07	180.03	176.00	189.20	198.00	209.37
IA AND PACIFIC	30935.85	63941.31	140693.89	255070.17	283324.83	317884.60	355851.47	390703.50	426102.60	448582.20	462728.93	444907.47
stralia and New Zealand	1699.50	2662.37	3045.53	3899.50	3333.00	2987.23	3141.60	3690.13	3715.07	3727.17	3727.90	3727.53
ntral Asia	3220.98	6739.84	8848.62	9694.27	8986.66	7430.13	5684.80	4039.20	3422.47	2725.07	2545.77	2582.07
rth West Pacific and East Asia	19989.93	40805.97	103302.83	181889.40	206685.97	240026.60	272274.57	301827.90	330623.33	341479.60	351332.67	338996.17
uth Asia	4948.17	9528.93	14971.00	35128.87	37300.63	37176.33	39684.33	41317.47	44202.77	51144.87	53128.90	56572.27
uth East Asia	1077.27	4174.87	10455.87	24387.00	26934.60	30177.77	34981.47	39737.13	44043.63	49413.83	51902.03	42939.60
uth Pacific	0.00	29.33	70.03	71.13	83.97	86.53	84.70	91.67	95.33	91.67	91.67	89.83
ROPE	81902.52	154842.09	198751.81	196275.96	183453.94	169979.33	156751.10	151022.30	150404.10	144763.30	149699.00	151526.47
ntral Europe	11719.74	23810.67	40857.31	39049.16	34989.66	33271.33	34044.27	34381.60	36163.60	37336.93	38440.97	39691.30
stern Europe	18953.21	39659.19	52067.90	57043.94	52880.15	43134.67	33909.33	25329.33	22893.20	17410.43	17203.27	17101.70
estern Europe	51229.57	91372.23	105826.60	100182.87	95584.13	93573.33	88797.50	91311.37	91347.30	90015.93	94054.77	94733.47
TIN AMERICA AND CARIBBEAN	8704.30	17901.03	37731.83	42431.40	43553.03	43876.07	45530.83	48804.07	48495.70	52486.13	56905.20	60681.50
ribbean	962.50	2227.50	3292.30	3645.40	3212.00	3293.40	2603.70	2651.73	2963.40	3180.83	3534.67	3606.90
eso-America	1673.47	4006.20	9500.33	14408.17	14536.87	15247.47	15712.77	17068.33	14266.63	15056.43	16345.27	16873.63
uth America	6068.33	11667.33	24939.20	24377.83	25804.17	25335.20	27214.37	29084.00	31265.67	34248.87	37025.27	40200.97
RTH AMERICA	30574.50	36540.53	39264.50	40693.77	38127.83	37474.07	40132.40	44147.77	43556.33	45305.33	47172.40	47869.43
rth America	30574.50	36540.53	39264.50	40693.77	38127.83	37474.07	40132.40	44147.77	43556.33	45305.33	47172.40	47869.43
LAR	0.00	0.00	0.00	0.00	0.00	0.00	0.00	0.00	0.00	0.00	0.00	0.00
tarctic	0.00	0.00	0.00	0.00	0.00	0.00	0.00	0.00	0.00	0.00	0.00	0.00
ctic	0.00	0.00	0.00	0.00	0.00	0.00	0.00	0.00	0.00	0.00	0.00	0.00
ST ASIA	1104.77	2478.67	8038.07	17265.23	14075.97	16219.13	17644.37	18771.87	20044.20	20611.80	18958.50	18708.07
abian Peninsula	44.73	463.10	3105.67	9176.93	8708.70	11066.00	11154.73	11815.47	13253.53	13772.73	12968.63	12930.50
shriq	1060.03	2015.57	4932.40	8088.30	5367.27	5153.13	6489.63	6956.40	6790.67	6839.07	5989.87	5777.57
OBAL TOTALS	157358.30	284590.53	440377.30	576718.27	589321.70	612913.77	643190.17	682182.23	719230.23	743274.40	766892.13	756407.30
PORTED GLOBAL TOTALS FROM IGINAL DATA SOURCES	157666.67	286000.00	440000.00	575666.67	590333.33	612333.33	645333.33	682000.00	718666.67	744333.33	766333.33	759000.00

Comments:

Emissions of CO$_2$ - from cement production (CDIAC) is the amount of CO$_2$ created by the conversion of calcium carbonate to calcium oxide inside the kilns, and by burning large quantities of fossil fuels to heat the kilns to the 1450°C necessary for roasting limestone.

The annual estimates of CO$_2$ emissions from cement production have been derived from energy statistics published by the United Nations and the cement production estimates by the U.S. Department of Interior's Bureau of Mines. These estimations were calculated using the methods of G. Marland and R. M. Rotty (1984). For information about the data collection methodology used by G. Marland and Rotty see http://cdiac.esd.ornl.gov.

Emissions of CO$_2$ - Total (CDIAC)

Units: thousand metric tonnes of CO$_2$

Data Source: Energy Statistics Published by the United Nations U.S. and the Bureau of Mines and Energy Information Administrat

Data Provider: Carbon Dioxide Information Analysis Centre (CDIAC)

Years: 1960-1998

	1960	1970	1980	1990	1995	1996	1997	1998
AFRICA	145797.67	298939.67	516743.33	700498.33	804961.67	797797.00	803733.33	819001.33
Central Africa	3153.33	5470.67	14325.67	14831.67	10189.67	9654.33	8884.33	9529.67
Eastern Africa	3322.00	6193.00	9995.33	10699.33	11421.67	11304.33	10992.67	13515.33
Northern Africa	29788.00	84139.00	167119.33	234219.33	282091.33	290264.33	304901.67	307241.00
Southern Africa	100903.00	169040.67	237127.00	321548.33	377076.33	360374.67	367568.67	374036.67
Western Africa	7967.67	32365.67	85140.00	115690.67	119280.33	120978.00	106069.33	109200.67
Western Indian Ocean	663.67	1730.67	3036.00	3509.00	4902.33	5221.33	5316.67	5478.00
ASIA AND PACIFIC	1559932.53	2579303.27	4175394.02	6294189.83	7781502.67	8185191.67	8234299.33	7946572.33
Australia and New Zealand	99733.33	156783.00	220513.33	289784.00	323205.67	344648.33	348135.33	361779.00
Central Asia	189134.53	300290.27	441037.02	484257.49	309129.33	284837.67	278659.33	271718.33
North West Pacific and East Asia	1063010.67	1664897.67	2738717.67	4096645.67	5136123.67	5358690.33	5333562.67	5053073.67
South Asia	160728.33	308854.33	508728.00	978937.67	1294777.00	1402522.00	1441462.00	1485414.33
South East Asia	45811.33	142945.00	258489.00	435882.33	707498.00	783709.67	821674.33	763887.67
South Pacific	1514.33	5533.00	7909.00	8682.67	10769.00	10783.67	10805.67	10699.33
EUROPE	3965554.47	5988013.73	7633248.98	7608750.51	6272071.67	6392961.67	6196802.33	6148255.67
Central Europe	530635.11	905166.14	1296123.30	1201114.34	1047075.33	1088215.33	1069896.67	1029394.67
Eastern Europe	1234204.36	1959555.26	2878003.35	3160040.17	2090205.33	2033515.00	1950879.33	1907227.67
Western Europe	2200715.00	3123292.33	3459122.33	3247596.00	3134791.00	3271231.33	3176026.33	3211633.33
LATIN AMERICA AND CARIBBEAN	295097.00	508555.67	874786.00	994755.67	1183116.00	1218840.33	1281998.67	1339286.67
Caribbean	27665.00	61468.00	103194.67	98446.33	111763.67	115008.67	116226.00	127629.33
Meso-America	67030.33	114150.67	269518.33	324848.33	347042.67	357635.67	386738.00	409856.33
South America	200401.67	332937.00	502073.00	571461.00	724309.67	746196.00	779034.67	801801.00
NORTH AMERICA	3113938.67	4551796.33	5051409.00	5248686.67	5668259.67	5816503.00	5955297.33	5919221.00
North America	3113938.67	4551796.33	5051409.00	5248686.67	5668259.67	5816503.00	5955297.33	5919221.00
POLAR	223.67	381.33	557.33	564.67	520.67	535.33	531.67	535.33
Antarctic	----	----	----	----	----	----	----	----
Arctic	223.67	381.33	557.33	553.67	502.33	517.00	520.67	528.00
WEST ASIA	29700.00	123156.00	296057.67	432171.67	606980.00	640915.00	650598.67	683965.33
Arabian Peninsula	14879.33	88337.33	221826.00	327712.00	455748.33	489170.00	494541.67	520567.67
Mashriq	14820.67	34818.67	74231.67	104459.67	151231.67	151745.00	156057.00	163397.67
GLOBAL TOTALS	9110244.00	14050146.00	18548196.33	21279617.33	22317412.33	23052744.00	23123261.33	22856837.67
REPORTED GLOBAL TOTALS FROM ORIGINAL DATA SOURCES	9452666.67	14941666.67	19422333.33	22352000.00	23386000.00	23943333.33	24302666.67	24229333.33

Comments:

Emissions of CO$_2$ - total (CDIAC) are the sum of CO$_2$ produced during the consumption of solid, liquid, and gaseous fuels, and from gas flaring, and cement manufacturing. The data is primarily derived from U.N. data. Thereafter supplemented with data from the U.S. Department of Energy's Energy Information Administration (R. M. Rotty) and with a few national estimates provided by G. Marland. For information about the data collection methodology used by R. M. Rotty (1974) and G. Marland see http://cdiac.esd.ornl.gov.

The sum of emissions estimates for all countries is not equal to the estimate of global total emissions because:

1. The global total includes emissions from bunker fuels (that is to say fuels used by ships and aircraft during international trade) whereas these are not included in any national totals.

2. The global total includes estimates for the oxidation of non-fuel hydrocarbon products (for example asphalt) whereas national totals do not.

3. National totals include annual changes in fuel stocks whereas the global total does not.

4. Due to statistical differences in the international statistics, the sum of exports from all exporters is not identical to the sum of imports by all importers.

AGGREGATIONS

Emissions of CO₂ - Total per Capita (CDIAC)

Units: kilograms of CO_2 per capita

Data Source: Energy Statistics Published by the United Nations and the U.S. Bureau of Mines and Energy Information Administration

Data Provider: Carbon Dioxide Information Analysis Centre (CDIAC)

Years: 1960-1998

	1960	1970	1980	1990	1991	1992	1993	1994	1995	1996	1997	1998
RICA	641.55	866.78	1144.26	1179.13	1170.16	1236.07	1173.28	1160.47	1185.84	1146.76	1127.43	1135.97
ntral Africa	116.72	158.26	317.91	242.93	176.33	177.04	191.72	158.14	141.00	129.90	116.48	121.91
stern Africa	81.61	100.76	123.77	98.59	86.65	90.39	95.61	97.16	93.98	96.64	91.67	110.00
rthern Africa	443.55	981.11	1509.01	1632.21	1642.02	1631.48	1716.31	1658.74	1763.82	1777.99	1827.18	1867.20
uthern Africa	2513.28	2636.20	2844.36	2971.70	2955.97	2955.35	2758.20	2909.26	3018.02	2818.53	2813.03	2802.97
stern Africa	105.35	328.08	658.21	674.74	678.29	924.65	757.53	668.25	611.88	606.61	519.70	522.50
stern Indian Ocean	100.80	203.52	279.61	248.74	272.23	276.46	282.71	292.10	303.79	315.13	312.73	314.10
IA AND PACIFIC	992.04	1275.28	1637.59	2042.16	2083.85	2121.50	2151.66	2237.98	2314.36	2400.17	2378.60	2359.80
stralia and New Zealand	7885.01	10210.60	12470.96	14311.50	13690.52	14377.04	14509.47	14797.89	14983.98	15802.51	15795.37	16256.42
ntral Asia	9872.92	11008.40	12428.53	10537.97	9980.69	8207.92	7301.24	6702.53	5798.52	5292.96	5127.96	11119.62
rth West Pacific and East Asia	1329.38	1658.83	2252.35	2937.92	3031.29	3117.16	3193.01	3333.40	3484.23	3600.86	3542.56	3311.09
uth Asia	326.00	451.55	538.53	832.10	861.55	902.33	901.92	984.25	994.35	1055.96	1065.68	1078.33
uth East Asia	211.86	499.22	719.17	987.43	1036.55	1134.07	1247.90	1291.65	1474.50	1610.13	1665.66	1529.79
uth Pacific	518.82	1443.02	1650.30	1469.47	1494.58	1493.17	1629.15	1710.68	1569.37	1531.87	1499.26	1478.13
ROPE	6271.76	8534.46	10072.25	9356.58	9028.06	8638.60	8255.75	7744.39	7724.92	7856.82	7600.85	7528.60
ntral Europe	3404.17	5348.85	7086.14	6195.17	5650.21	5493.94	5595.66	5315.81	5474.02	5662.36	5539.19	5301.36
stern Europe	7552.95	10380.30	13652.66	13307.58	12761.99	11945.30	10830.63	9279.73	9050.02	8817.52	8473.90	8300.99
stern Europe	6722.58	8832.93	9342.93	8512.42	8427.92	8189.79	8016.92	8019.49	8043.25	8368.05	8103.82	8177.69
TIN AMERICA AND CARIBBEAN	1353.77	1787.85	2423.18	2259.01	2326.20	2426.22	2438.59	2580.67	2469.02	2503.40	2591.85	2665.76
ribbean	1368.51	2495.18	3556.66	2905.19	3007.01	3137.76	3167.37	3404.55	3148.59	3214.56	3221.37	3498.79
so-America	1357.52	1690.76	2999.85	2915.63	3009.09	3178.75	2960.03	3078.94	2815.88	2846.53	3020.64	3142.60
uth America	1350.48	1730.34	2072.23	1936.67	1989.55	2059.09	2156.98	2296.84	2259.33	2291.48	2355.77	2388.20
RTH AMERICA	15259.00	19659.93	19809.86	18622.63	18508.24	18414.90	19041.34	19327.59	19112.85	19434.02	19722.89	19442.84
rth America	15259.00	19659.93	19809.86	18622.63	18508.24	18414.90	19041.34	19327.59	19112.85	19434.02	19722.89	19442.84
LAR	6839.96	8289.86	11146.67	9886.90	9755.95	8577.38	8904.76	8970.24	8970.24	9232.14	9297.62	9428.57
tarctic	----	----	----	----	----	----	----	----	----	----	----	----
ctic	6839.96	8289.86	11146.67	9886.90	9755.95	8577.38	8904.76	8970.24	8970.24	9232.14	9297.62	9428.57
ST ASIA	1217.88	3571.28	5982.25	5979.75	12623.10	6672.57	7021.05	7113.99	7178.91	7335.98	7230.86	7374.11
abian Peninsula	1504.32	6266.26	10120.08	9524.00	23647.85	10615.61	11190.05	11249.21	11266.14	11672.66	11405.70	11579.86
shriq	1021.77	1744.87	2746.37	2801.73	2731.80	3148.71	3315.84	3452.13	3557.82	3474.43	3481.70	3557.51
OBAL TOTALS	3080.42	3844.47	4185.35	4016.61	4064.49	3941.35	3919.54	3908.45	3920.62	3993.98	3951.73	3913.85
PORTED GLOBAL TOTALS FROM ORIGINAL TA SOURCES	3116.67	4033.33	4363.33	4216.67	4216.67	4106.67	4033.33	4070.00	4106.67	4143.33	4143.33	4143.33

Comments:

Emissions of CO_2 - total per capita (CDIAC) are the sum of CO_2 produced during the consumption of solid, liquid, and gaseous fuels, and from gas flaring, and cement manufacturing weighted with the population. The data is primarily derived from U.N. data. Thereafter supplemented with data from the U.S. Department of Energy's Energy Information Administration (R. M. Rotty) and with a few national estimates provided by G. Marland. For information about the data collection methodology used by R. M. Rotty (1974) and G. Marland see http://cdiac.esd.ornl.gov.

The sum of emissions estimates for all countries is not equal to the estimate of global total emissions because:

1. The global total includes emissions from bunker fuels (that is to say fuels used by ships and aircraft during international trade) whereas these are not included in any national totals.

2. The global total includes estimates for the oxidation of non-fuel hydrocarbon products (for example asphalt) whereas national totals do not.

3. National totals include annual changes in fuel stocks whereas the global total does not.

4. Due to statistical differences in the international statistics, the sum of exports from all exporters is not identical to the sum of imports by all importers.

AGGREGATIONS

Emissions of CO$_2$ - from Gas Flaring (CDIAC)

Units: thousand metric tonnes of CO$_2$

Data Source: Energy Statistics Published by the United Nations and the U.S. Bureau of Mines and Energy Information Administrati*

Data Provider: Carbon Dioxide Information Analysis Centre (CDIAC)

Years: 1971-1998

	1971	1980	1990	1991	1992	1993	1994	1995	1996	1997	1998
AFRICA	56018.60	76304.07	64380.80	65157.40	67229.07	68045.63	70663.27	73128.00	72668.57	62582.30	59263.23
Central Africa	536.07	3234.73	3556.30	168.67	167.20	167.20	166.83	166.83	166.83	166.83	167.20
Eastern Africa	0.00	0.00	1.83	1.83	1.83	1.83	1.83	1.83	1.83	1.83	1.83
Northern Africa	29460.93	27615.13	15924.33	15828.63	18040.37	16535.20	16886.47	19277.50	18148.17	17151.57	17122.97
Southern Africa	1389.67	2417.07	1760.00	1689.97	1713.80	1735.07	1807.67	1735.07	1762.20	1758.17	1739.47
Western Africa	24631.93	43037.13	43138.33	47468.30	47305.87	49606.33	51800.47	51946.77	52589.53	43503.90	40231.77
Western Indian Ocean	0.00	0.00	0.00	0.00	0.00	0.00	0.00	0.00	0.00	0.00	0.00
ASIA AND PACIFIC	56982.20	50221.60	43726.10	47210.17	47531.00	45592.43	43891.83	42641.13	42875.07	41938.23	41679.00
Australia and New Zealand	24.57	382.80	0.00	0.00	0.00	0.00	0.00	0.00	0.00	0.00	0.00
Central Asia	0.00	0.00	0.00	0.00	0.00	0.00	0.00	0.00	0.00	0.00	0.00
North West Pacific and East Asia	830.87	3092.83	0.00	0.00	0.00	0.00	0.00	0.00	0.00	0.00	0.00
South Asia	44199.83	21133.93	31818.23	31070.23	30654.43	26710.93	26404.40	25503.50	25672.17	24363.53	22530.57
South East Asia	11926.93	25612.03	11907.87	16139.93	16876.57	18881.50	17487.43	17137.63	17202.90	17574.70	19148.43
South Pacific	0.00	0.00	0.00	0.00	0.00	0.00	0.00	0.00	0.00	0.00	0.00
EUROPE	15265.33	24934.23	14011.02	13463.86	22777.76	18309.13	17861.43	17080.07	17792.87	17787.73	17998.93
Central Europe	282.70	401.50	0.00	0.00	0.00	0.00	0.00	0.00	0.00	0.00	0.00
Eastern Europe	5935.50	5786.90	6769.90	6297.60	15621.83	10803.83	8993.23	9915.77	9866.27	9866.27	9866.27
Western Europe	9047.13	18745.83	7241.12	7166.26	7155.93	7505.30	8868.20	7164.30	7926.60	7921.47	8132.67
LATIN AMERICA AND CARIBBEAN	53967.10	31594.93	24131.80	22588.87	23737.63	29379.53	31401.33	31331.30	24678.13	24525.23	24138.03
Caribbean	1784.57	3881.17	2490.77	2718.10	2933.70	2560.07	2169.57	1951.77	2756.60	1959.47	2129.23
Meso-America	10125.50	9893.40	2560.80	2426.23	2522.30	3269.20	3453.27	3546.77	3551.90	3561.80	3564.37
South America	42057.03	17820.37	19080.23	17444.53	18281.63	23550.27	25778.50	25832.77	18369.63	19003.97	18444.43
NORTH AMERICA	19820.53	9831.07	12436.97	14023.53	13622.03	16883.90	16953.93	20024.03	19093.07	17734.57	16610.73
North America	19820.53	9831.07	12436.97	14023.53	13622.03	16883.90	16953.93	20024.03	19093.07	17734.57	16610.73
POLAR	0.00	0.00	0.00	0.00	0.00	0.00	0.00	0.00	0.00	0.00	0.00
Antarctic	----	----	----	----	----	----	----	----	----	----	----
Arctic	0.00	0.00	0.00	0.00	0.00	0.00	0.00	0.00	0.00	0.00	0.00
WEST ASIA	103120.60	117135.33	31301.60	62075.57	32293.80	28416.30	27302.37	31429.20	30940.43	12583.27	13885.67
Arabian Peninsula	89101.47	97950.23	19244.50	56988.80	26294.77	21430.20	22223.67	26794.53	26404.40	7674.33	8892.40
Mashriq	14019.13	19185.10	12057.10	5086.77	5999.03	6986.10	5078.70	4634.67	4536.03	4908.93	4993.27
GLOBAL TOTALS	305174.37	310021.23	189988.28	224519.39	207191.29	206626.93	208074.17	215633.73	208048.13	177151.33	173575.60
REPORTED GLOBAL TOTALS FROM ORIGINAL DATA SOURCES	322666.67	326333.33	205333.33	238333.33	201666.67	205333.33	209000.00	216333.33	209000.00	176000.00	172333.33

Comments:

Emissions of CO$_2$ - from gas flaring (CDIAC): annual estimations of CO$_2$ emissions from gas flaring, primarily derived from U.N. data. Thereafter supplemented with data from the U.S. Department of Energy's Energy Information Administration, R. M. Rotty (1974), and with a few national estimates provided by G. Marland. For information about the data collection methodology used by R. M. Rotty (1974) and G. Marland see http://cdiac.esd.ornl.gov.

The sum of emissions estimates for all countries is not equal to the estimate of global total emissions because:

1. The global total includes emissions from bunker fuels (that is to say fuels used by ships and aircraft during international trade) whereas these are not included in any national totals.

2. The global total includes estimates for the oxidation of non-fuel hydrocarbon products (for example asphalt) whereas national totals do not.

3. National totals include annual changes in fuel stocks whereas the global total does not.

4. Due to statistical differences in the international statistics, the sum of exports from all exporters is not identical to the sum of imports by all importers.

Emissions of CO₂ - from Gas Fuel Consumption (CDIAC)

Units: thousand metric tonnes of CO_2

Data Source: Energy Statistics Published by the United Nations and the U.S. Bureau of Mines and Energy Information Administration

Data Provider: Carbon Dioxide Information Analysis Centre (CDIAC)

Years: 1960-1998

	1960	1970	1980	1990	1995	1996	1997	1998
AFRICA	47.67	817.67	34657.33	76457.33	101783.00	109446.33	112379.67	110583.00
Central Africa	14.67	117.33	33.00	194.33	1591.33	1485.00	1415.33	1265.00
Eastern Africa	0.00	0.00	3.67	0.00	0.00	0.00	0.00	0.00
Northern Africa	33.00	403.33	31929.33	64100.67	86456.33	92887.67	95502.00	94240.67
Southern Africa	0.00	84.33	146.67	4458.67	4583.33	4583.33	4224.00	3743.67
Western Africa	0.00	212.67	2544.67	7703.67	9152.00	10490.33	11238.33	11333.67
Western Indian Ocean	0.00	0.00	0.00	0.00	0.00	0.00	0.00	0.00
ASIA AND PACIFIC	21206.44	82017.96	209405.42	486304.99	628441.00	694235.67	720712.67	716653.67
Australia and New Zealand	0.00	2962.67	19506.67	42720.33	47043.33	48436.67	49342.33	49995.00
Central Asia	9572.10	34868.29	69929.08	125057.66	127130.67	120611.33	117890.67	125143.33
North West Pacific and East Asia	3644.67	14912.33	78756.33	138574.33	184745.00	204376.33	219310.67	224785.00
South Asia	2973.67	26429.33	29729.33	93639.33	151723.00	172348.00	177639.00	195473.67
South East Asia	5016.00	2845.33	11484.00	86166.67	117641.33	148302.00	156368.67	121095.33
South Pacific	0.00	0.00	0.00	146.67	157.67	161.33	161.33	161.33
EUROPE	128173.56	507167.71	1153571.58	1713335.67	1708685.00	1825875.33	1781131.00	1794320.00
Central Europe	26128.81	75659.21	147954.18	156110.99	145335.67	161934.67	155008.33	152584.67
Eastern Europe	79135.42	288266.50	578124.41	1033888.68	910323.33	927289.00	909447.00	891491.33
Western Europe	22909.33	143242.00	427493.00	523336.00	653026.00	736651.67	716675.67	750244.00
LATIN AMERICA AND CARIBBEAN	20757.00	59242.33	116053.67	169480.67	208820.33	221679.33	240137.33	253590.33
Caribbean	1470.33	3589.67	7267.33	10054.00	13046.00	14740.00	15686.00	16397.33
Meso-America	6383.67	20625.00	45617.00	48044.33	32281.33	47732.67	53636.00	62788.00
South America	12903.00	35027.67	63169.33	111382.33	163493.00	159206.67	170815.33	174405.00
NORTH AMERICA	683100.00	1206927.33	1098687.33	1077798.33	1344992.00	1353960.67	1368950.00	1299162.33
North America	683100.00	1206927.33	1098687.33	1077798.33	1344992.00	1353960.67	1368950.00	1299162.33
POLAR	0.00	0.00	0.00	0.00	0.00	0.00	0.00	0.00
Antarctic	----	----	----	----	----	----	----	----
Arctic	0.00	0.00	0.00	0.00	0.00	0.00	0.00	0.00
WEST ASIA	2966.33	8462.67	42933.00	140411.33	196779.00	209458.33	231792.00	246972.00
Arabian Peninsula	1807.67	6955.67	40377.33	135798.67	185760.67	197728.67	218529.67	230945.00
Mashriq	1158.67	1507.00	2555.67	4612.67	11018.33	11729.67	13262.33	16027.00
GLOBAL TOTALS	856251.00	1864635.67	2655308.33	3663788.33	4189500.33	4414655.67	4455102.67	4421281.33
REPORTED GLOBAL TOTALS FROM ORIGINAL DATA SOURCES	861666.67	1892000.00	2662000.00	3696000.00	4205666.67	4425666.67	4436666.67	4473333.33

Comments:

Emissions of CO_2 - from gas fuel consumption (CDIAC): annual estimations of CO_2 emissions from gas gas fuel consumption, primarily derived from U.N. data. Thereafter supplemented with data from the U.S. Department of Energy's Energy Information Administration, R. M. Rotty (1974), and with a few national estimates provided by G. Marland. For information about the data collection methodology used by R. M. Rotty (1974) and G. Marland see http://cdiac.esd.ornl.gov .

The sum of emissions estimates for all countries is not equal to the estimate of global total emissions because:

1. The global total includes emissions from bunker fuels (that is to say fuels used by ships and aircraft during international trade) whereas these are not included in any national totals.

2. The global total includes estimates for the oxidation of non-fuel hydrocarbon products (for example asphalt) whereas national totals do not.

3. National totals include annual changes in fuel stocks whereas the global total does not.

4. Due to statistical differences in the international statistics, the sum of exports from all exporters is not identical to the sum of imports by all importers.

Emissions of CO₂ - from Liquid Fuel Consumption (CDIAC)

Units: thousand metric tonnes of CO_2

Data Source: Energy Statistics Published by the United Nations and the U.S. Bureau of Mines and Energy Information Administratic

Data Provider: Carbon Dioxide Information Analysis Centre (CDIAC)

Years: 1960-1998

	1960	1970	1980	1990	1995	1996	1997	1998
AFRICA	47270.67	94614.67	192815.33	264828.67	310185.33	295482.00	294231.67	309646.
Central Africa	1994.67	4158.00	9669.00	9665.33	7062.00	6647.67	5954.67	6849.
Eastern Africa	2962.67	5390.00	9243.67	9298.67	9984.33	9680.00	9408.67	12202.
Northern Africa	24867.33	38613.67	94277.33	127482.67	145669.33	146010.33	159694.33	161142.
Southern Africa	10681.00	29230.67	40040.00	53903.67	89232.00	74910.00	67580.33	71390
Western Africa	6156.33	15583.33	36637.33	61240.67	53735.00	53423.33	46702.33	53020
Western Indian Ocean	608.67	1639.00	2948.00	3237.67	4502.67	4810.67	4891.33	5041
ASIA AND PACIFIC	272235.39	1001040.47	1641610.46	2033103.94	2399925.00	2535481.67	2618157.67	2545785
Australia and New Zealand	32464.67	71811.67	90387.00	93001.33	101603.33	107715.67	103007.67	102949
Central Asia	46478.72	95010.81	155110.79	154611.94	65600.33	58868.33	68812.33	57746
North West Pacific and East Asia	111701.33	616619.67	1011571.00	1180956.33	1380727.33	1450790.00	1490720.00	1427081
South Asia	49753.00	92796.00	183040.00	327055.67	429018.33	460515.00	461945.00	481664
South East Asia	31023.67	120358.33	194036.33	269518.33	412954.67	447571.67	483629.67	466389.
South Pacific	814.00	4444.00	7465.33	7960.33	10021.00	10021.00	10043.00	9955
EUROPE	903873.61	2390153.19	3065205.21	2749032.39	2226983.00	2242225.33	2200289.67	2212672
Central Europe	58861.56	188319.42	333143.55	302729.29	269782.33	277992.00	273089.67	265331
Eastern Europe	308204.72	630025.44	1028554.00	1025246.10	518191.67	467558.67	451491.33	448602
Western Europe	536807.33	1571808.33	1703507.67	1421057.00	1439009.00	1496674.67	1475708.67	1498739.
LATIN AMERICA AND CARIBBEAN	198289.67	340915.67	638685.67	682843.33	802556.33	821260.00	856610.33	896874
Caribbean	25047.00	53401.33	88502.33	80901.33	92084.67	92671.33	93258.00	103909
Meso-America	46783.00	73128.00	194997.00	250602.00	281119.67	273056.67	294323.33	307362
South America	126459.67	214386.33	355186.33	351340.00	429352.00	455532.00	469029.00	485602
NORTH AMERICA	1398397.00	2049446.67	2341218.00	2237634.67	2333815.00	2385287.67	2489017.67	2468290
North America	1398397.00	2049446.67	2341218.00	2237634.67	2333815.00	2385287.67	2489017.67	2468290
POLAR	102.67	304.33	553.67	564.67	520.67	535.33	531.67	535
Antarctic	----	----	----	----	----	----	----	
Arctic	102.67	304.33	553.67	553.67	502.33	517.00	520.67	528
WEST ASIA	25593.33	32703.00	127919.00	243184.33	358215.00	379302.00	386650.00	404070
Arabian Peninsula	13024.00	12045.00	80388.00	163489.33	229944.00	251265.67	255376.00	267795
Mashriq	12569.33	20658.00	47531.00	79695.00	128271.00	128036.33	131274.00	136275
GLOBAL TOTALS	2845762.33	5909178.00	8008007.33	8211192.00	8432200.33	8659574.00	8845488.67	8837873
REPORTED GLOBAL TOTALS FROM ORIGINAL DATA SOURCES	3116666.67	6739333.33	8825666.67	9159333.33	9507666.67	9691000.00	9984333.33	10167666

Comments:

Emissions of CO₂ - from liquid fuel consumption (CDIAC): annual estimations of CO_2 emissions from gas liquid fuel consumption, primarily derived from U.N. data. Thereafter supplemented with data from the U.S. Department of Energy's Energy Information Administration, R. M. Rotty (1974), and with a few national estimates provided by G. Marland. For information about the data collection methodology used by R. M. Rotty (1974) and G. Marland see http://cdiac.esd.ornl.gov.

The sum of emissions estimates for all countries is not equal to the estimate of global total emissions because:

1. The global total includes emissions from bunker fuels (that is to say fuels used by ships and aircraft during international trade) whereas these are not included in any national totals.

2. The global total includes estimates for the oxidation of non-fuel hydrocarbon products (for example asphalt) whereas national totals do not.

3. National totals include annual changes in fuel stocks whereas the global total does not.

4. Due to statistical differences in the international statistics, the sum of exports from all exporters is not identical to the sum of imports by all importer

Emissions of CO$_2$ - from Solid Fuel Consumption (CDIAC)

Units: thousand metric tonnes of CO$_2$

Data Source: Energy Statistics Published by the United Nations and the U.S. Bureau of Mines and Energy Information Administration

Data Provider: Carbon Dioxide Information Analysis Centre (CDIAC)

Years: 1960-1998

	1960	1970	1980	1990	1995	1996	1997	1998
AFRICA	94380.00	139645.00	197057.67	269859.33	289226.67	288673.00	303112.33	306812.00
Central Africa	1045.00	792.00	843.33	770.00	854.33	858.00	865.33	869.00
Eastern Africa	139.33	220.00	55.00	418.00	271.33	264.00	322.67	194.33
Northern Africa	2783.00	3776.67	5368.00	11352.00	11018.33	12932.33	12793.00	13746.33
Southern Africa	88715.00	134511.67	190186.33	256472.33	275993.67	273526.00	288031.33	291118.67
Western Africa	1642.67	289.67	546.33	608.67	869.00	869.00	872.67	656.33
Western Indian Ocean	55.00	55.00	58.67	238.33	220.00	223.67	227.33	227.33
ASIA AND PACIFIC	1270220.81	1428327.47	2177366.17	3510861.14	4314284.33	4491740.00	4418883.33	4233416.00
Australia and New Zealand	65571.00	79328.33	107195.00	150168.33	170841.00	184767.00	192056.33	205109.67
Central Asia	160104.48	194659.14	234729.50	203564.47	112973.67	102633.67	89408.00	86247.33
North West Pacific and East Asia	930625.67	1000820.33	1558326.00	2621428.33	3269929.67	3389749.00	3300333.67	3098073.00
South Asia	104522.00	143465.67	259842.00	491296.67	644332.33	692857.00	724375.67	729171.67
South East Asia	8697.33	8994.33	16899.67	43901.00	115716.33	121223.67	112200.00	114323.00
South Pacific	700.33	1059.67	374.00	502.33	491.33	509.67	509.67	491.33
EUROPE	2821361.19	2901543.86	3147627.50	2913542.20	2168070.67	2161217.67	2046894.67	1970569.33
Central Europe	440788.55	626262.38	784671.00	707926.37	594938.67	609854.67	602360.00	570636.00
Eastern Europe	790808.64	961485.48	1159406.17	1005471.83	628884.67	611398.33	562873.67	540173.33
Western Europe	1589764.00	1313796.00	1203550.33	1200144.00	944247.33	939964.67	881661.00	859760.00
LATIN AMERICA AND CARIBBEAN	24376.00	31925.67	50743.00	75870.67	91912.33	98739.67	103814.33	104005.00
Caribbean	179.67	319.00	256.67	1356.67	1712.33	1661.00	1789.33	1587.67
Meso-America	3494.33	6772.33	9518.67	9232.67	15829.00	18241.67	18868.67	19264.67
South America	20702.00	24834.33	40967.67	65281.33	74371.00	78837.00	83156.33	83152.67
NORTH AMERICA	968333.67	1227805.33	1562414.33	1880123.67	1925876.33	2012849.67	2032418.67	2087290.33
North America	968333.67	1227805.33	1562414.33	1880123.67	1925876.33	2012849.67	2032418.67	2087290.33
POLAR	121.00	77.00	3.67	0.00	0.00	0.00	0.00	0.00
Antarctic	----	----	----	----	----	----	----	----
Arctic	121.00	77.00	3.67	0.00	0.00	0.00	0.00	0.00
WEST ASIA	29.33	18.33	29.33	3.67	517.00	608.67	619.67	326.33
Arabian Peninsula	0.00	0.00	0.00	0.00	0.00	0.00	0.00	0.00
Mashriq	29.33	18.33	29.33	3.67	517.00	608.67	619.67	326.33
GLOBAL TOTALS	5178822.00	5729342.67	7135241.67	8650260.67	8789887.33	9053828.67	8905743.00	8702419.00
REPORTED GLOBAL TOTALS FROM ORIGINAL DATA SOURCES	5170000.00	5705333.33	7168333.33	8715666.67	8734000.00	8877000.00	8935666.67	8660666.67

Comments:

Emissions of CO$_2$ - from solid fuel consumption (CDIAC): annual estimations of CO$_2$ emissions from solid fuel consumption, primarily derived from U.N. data. Thereafter supplemented with data from the U.S. Department of Energy's Energy Information Administration, R. M. Rotty (1974), and with a few national estimates provided by G. Marland. For information about the data collection methodology used by R. M. Rotty (1974) and G. Marland see http://cdiac.esd.ornl.gov.

The sum of emissions estimates for all countries is not equal to the estimate of global total emissions because:

1. The global total includes emissions from bunker fuels (that is to say fuels used by ships and aircraft during international trade) whereas these are not included in any national totals.

2. The global total includes estimates for the oxidation of non-fuel hydrocarbon products (for example asphalt) whereas national totals do not.

3. National totals include annual changes in fuel stocks whereas the global total does not.

4. Due to statistical differences in the international statistics, the sum of exports from all exporters is not identical to the sum of imports by all importers.

AGGREGATIONS

Emissions of CO$_2$ - from Public Electricity and Heat Production (IEA)

Units: thousand metric tonnes of CO$_2$

Data Source: Energy Balances of OECD Countries (1960-1999) and Non-OECD Countries (1971-1999)

Data Provider: International Energy Agency (IEA)

Years: 1960-1999

	1960	1970	1980	1990	1995	1996	1997	1998	1999
AFRICA	----	----	120770.00	200920.00	233830.00	242490.00	254890.00	270290.00	267790.00
Central Africa	----	----	380.00	250.00	270.00	290.00	300.00	340.00	310.00
Eastern Africa	----	----	----	----	----	----	----	----	----
Northern Africa	----	----	23940.00	50410.00	61300.00	61270.00	64690.00	70840.00	76990.00
Southern Africa	----	----	89000.00	139080.00	160730.00	168640.00	177090.00	183820.00	173580.00
Western Africa	----	----		7000.00	6890.00	7460.00	8430.00	10040.00	11440.00
Western Indian Ocean	----	----	----	----	----	----	----	----	----
ASIA AND PACIFIC	----	----	978227.39	1673017.57	2272020.00	2422960.00	2610980.00	2652080.00	2724330.00
Australia and New Zealand	29280.00	53240.00	84860.00	124230.00	132400.00	144640.00	149590.00	160900.00	170970.00
Central Asia	----	----	149797.39	177657.57	121790.00	113090.00	101270.00	103790.00	101260.00
North West Pacific and East Asia	----	----	576040.00	1014810.00	1482690.00	1581400.00	1723300.00	1730140.00	1768910.00
South Asia	----	----	123360.00	259770.00	384750.00	416880.00	447510.00	471200.00	497970.00
South East Asia	----	----	44170.00	96550.00	150390.00	166950.00	189310.00	186050.00	185220.00
South Pacific	----	----	----	----	----	----	----	----	----
EUROPE	----	----	2194422.61	2435522.43	2021310.00	2064400.00	1977900.00	1939820.00	1922990.00
Central Europe	----	----	386504.04	469769.77	427950.00	449510.00	442390.00	437350.00	420330.00
Eastern Europe	----	----	896918.57	1063732.66	715020.00	717180.00	667790.00	633580.00	647190.00
Western Europe	411300.00	781130.00	911000.00	902020.00	878340.00	897710.00	867720.00	868890.00	855470.00
LATIN AMERICA AND CARIBBEAN	----	----	106720.00	136570.00	161880.00	173130.00	193030.00	211910.00	216940.00
Caribbean	----	----	14530.00	20300.00	20640.00	21330.00	23350.00	24290.00	24290.00
Meso-America	----	----	40670.00	66210.00	82830.00	86650.00	96680.00	107710.00	106450.00
South America	----	----	51520.00	50060.00	58410.00	65150.00	73000.00	79910.00	86200.00
NORTH AMERICA	581450.00	1107080.00	1606130.00	1849860.00	1936220.00	2012670.00	2092830.00	2196820.00	2124220.00
North America	581450.00	1107080.00	1606130.00	1849860.00	1936220.00	2012670.00	2092830.00	2196820.00	2124220.00
POLAR	----	----	----	----	----	----	----	----	----
Antarctic	----	----	----	----	----	----	----	----	----
Arctic	----	----	----	----	----	----	----	----	----
WEST ASIA	----	----	43330.00	87370.00	138500.00	150040.00	151000.00	162080.00	166920.00
Arabian Peninsula	----	----	29240.00	65550.00	105000.00	114740.00	115000.00	123490.00	128030.00
Mashriq	----	----	14090.00	21820.00	33500.00	35300.00	36000.00	38590.00	38890.00
GLOBAL TOTALS	----	----	5049600.00	6383260.00	6763760.00	7065690.00	7280630.00	7433000.00	7423190.00
REPORTED GLOBAL TOTALS FROM ORIGINAL DATA SOURCES	----	----	5050770.00	6384450.00	6764980.00	7066910.00	7281920.00	7434240.00	7424440.00

Comments:

Emissions of CO$_2$ from public electricity and heat production contain the sum of emissions from public electricity generation, public combined heat and power generation, and public heat plants. Public utilities are defined as those undertakings whose primary activity is to supply electricity and heat to the public. They may be publicly or privately owned. Emissions from own on-site use of fuel should be included. This corresponds to IPCC source/sink category 1.A.1(a).

Data source: the estimates of CO$_2$ emissions from fuel combustion are calculated using the IEA energy data and the default methods and emission factors from the Revised 1996 IPCC Guidelines for National Greenhouse Gas Inventories, IPCC/OECD/IEA Paris, 1997 (Guidelines). There are many reasons why the IEA estimates may not be the same as the numbers that a country submits to the UNFCCC, even if a country has accounted for all of its energy use and correctly applied the IPCC guidelines.

1 thousand metric tonne of CO$_2$ (1 gigagram of CO$_2$), is equal to 272.72 tonnes of C.

Emissions of CO$_2$ - from Manufacturing Industries and Construction (IEA)

Units: thousand metric tonnes of CO$_2$

Data Source: Energy Balances of OECD Countries (1960-1999) and Non-OECD Countries (1971-1999)

Data Provider: International Energy Agency (IEA)

Years: 1960-1999

	1960	1970	1980	1990	1995	1996	1997	1998	1999
AFRICA	----	----	123940.00	131960.00	122220.00	126440.00	133430.00	143000.00	133870.00
Central Africa	----	----	1730.00	1370.00	1420.00	1300.00	1310.00	1330.00	1270.00
Eastern Africa	----	----	----	----	----	----	----	----	----
Northern Africa	----	----	32560.00	43490.00	43980.00	44410.00	47500.00	52820.00	51250.00
Southern Africa	----	----	80280.00	76630.00	63250.00	66470.00	68940.00	73610.00	66030.00
Western Africa	----	----	----	6750.00	10330.00	10790.00	11880.00	11510.00	11520.00
Western Indian Ocean	----	----	----	----	----	----	----	----	----
ASIA AND PACIFIC	----	----	1425613.93	2011768.94	2350080.00	2432030.00	2248730.00	2229310.00	2075560.00
Australia and New Zealand	34310.00	49810.00	52290.00	51600.00	56090.00	57020.00	57910.00	58370.00	58320.00
Central Asia	----	----	133143.93	127858.94	67250.00	58220.00	53300.00	56800.00	56800.00
North West Pacific and East Asia	----	----	1058350.00	1491890.00	1756490.00	1811580.00	1659060.00	1665990.00	1528140.00
South Asia	----	----	127220.00	260740.00	354030.00	371250.00	351350.00	329870.00	301280.00
South East Asia	----	----	53660.00	78600.00	115610.00	133350.00	126140.00	117330.00	130070.00
South Pacific	----	----	----	----	----	----	----	----	----
EUROPE	----	----	1904816.07	1646241.06	1170210.00	1135430.00	1121640.00	1091080.00	1066650.00
Central Europe	----	----	351695.42	282668.04	222630.00	230390.00	229310.00	212060.00	186690.00
Eastern Europe	----	----	761770.64	731533.02	372640.00	328840.00	306310.00	298940.00	312500.00
Western Europe	590450.00	919500.00	791350.00	632040.00	574940.00	576200.00	586020.00	580080.00	567460.00
LATIN AMERICA AND CARIBBEAN	----	----	215210.00	225500.00	259710.00	270670.00	285090.00	285130.00	275080.00
Caribbean	----	----	17500.00	17260.00	17850.00	19200.00	20270.00	21410.00	23420.00
Meso-America	----	----	62960.00	77950.00	78360.00	74980.00	74550.00	71150.00	68440.00
South America	----	----	134750.00	130290.00	163500.00	176490.00	190270.00	192570.00	183220.00
NORTH AMERICA	648110.00	835730.00	898750.00	702600.00	673510.00	687560.00	693250.00	652940.00	645570.00
North America	648110.00	835730.00	898750.00	702600.00	673510.00	687560.00	693250.00	652940.00	645570.00
POLAR	----	----	----	----	----	----	----	----	----
Antarctic	----	----	----	----	----	----	----	----	----
Arctic	----	----	----	----	----	----	----	----	----
WEST ASIA	----	----	73320.00	84250.00	112220.00	122810.00	123780.00	135650.00	139850.00
Arabian Peninsula	----	----	60910.00	65940.00	86140.00	96860.00	94950.00	104470.00	108080.00
Mashriq	----	----	12410.00	18310.00	26080.00	25950.00	28830.00	31180.00	31770.00
GLOBAL TOTALS	----	----	4641650.00	4802320.00	4687950.00	4774940.00	4605920.00	4537110.00	4336580.00
REPORTED GLOBAL TOTALS FROM ORIGINAL DATA SOURCES	----	----	4641610.00	4802360.00	4687960.00	4774930.00	4605960.00	4537140.00	4336570.00

Comments:

Emissions of CO$_2$ from manufacturing industries and construction contain the emissions from combustion of fuels (coal, oil and gas) in industry. The IPCC Source/Sink Category 1.A.2 includes these emissions. However, in the guidelines, the IPCC category also includes emissions from industry autoproducers that generate electricity and/or heat. The IEA data are not collected in a way that allows the energy consumption to be split by specific end-use and therefore, autoproducers are shown as a separate item (unallocated autoproducers). Manufacturing industries and construction also includes emissions from coke inputs into blast furnaces, which may be reported either in the transformation sector, the industry sector or the separate IPCC source/sink sategory 2, industrial processes.

Data source: the estimates of CO$_2$ emissions from fuel combustion are calculated using the IEA energy data and the default methods and emission factors from the Revised 1996 IPCC Guidelines for national greenhouse gas inventories, IPCC/OECD/IEA Paris, 1997 (Guidelines). There are many reasons why the IEA estimates may not be the same as the numbers that a country submits to the UNFCCC, even if a country has accounted for all of its energy use and correctly applied the IPCC guidelines.

1 thousand metric tonne of CO$_2$ (1 gigagram of CO$_2$), is equal to 272.72 tonnes of C

Emissions of CO$_2$ - from Transport (IEA)

Units: thousand metric tonnes of CO$_2$

Data Source: Energy Balances of OECD Countries (1960-1999) and Non-OECD Countries (1971-1999)

Data Provider: International Energy Agency (IEA)

Years: 1960-1999

	1960	1970	1980	1990	1995	1996	1997	1998	1999
AFRICA	----	----	83350.00	102710.00	111650.00	116040.00	120710.00	120870.00	123810.0
Central Africa	----	----	3620.00	3220.00	2940.00	3000.00	3050.00	3230.00	2930.0
Eastern Africa	----	----	----	----	----	----	----	----	--
Northern Africa	----	----	25360.00	40550.00	38860.00	39800.00	41470.00	41740.00	45120.0
Southern Africa	----	----	29660.00	33740.00	40620.00	40380.00	41400.00	41950.00	42780.0
Western Africa	----	----	----	15480.00	18810.00	22050.00	23240.00	22490.00¯	21560.0
Western Indian Ocean	----	----	----	----	----	----	----	----	--
ASIA AND PACIFIC	----	----	485685.58	725959.77	938090.00	994870.00	1045630.00	1056000.00	1108020.0
Australia and New Zealand	24790.00	37070.00	55970.00	70720.00	78310.00	80980.00	82750.00	82920.00	84400.0
Central Asia	----	----	27635.58	32229.77	19640.00	21780.00	20840.00	21740.00	21460.0
North West Pacific and East Asia	----	----	267380.00	396470.00	509850.00	531250.00	556630.00	568520.00	595980.0
South Asia	----	----	87790.00	132640.00	182470.00	197610.00	205670.00	213070.00	226270.0
South East Asia	----	----	46910.00	93900.00	147820.00	163250.00	179740.00	169750.00	179910.0
South Pacific	----	----	----	----	----	----	----	----	--
EUROPE	----	----	901174.42	1126980.23	1107460.00	1127530.00	1123410.00	1176030.00	1194820.0
Central Europe	----	----	96235.56	109812.32	109030.00	122970.00	125300.00	124790.00	127110.0
Eastern Europe	----	----	245058.86	285797.91	208510.00	197400.00	180170.00	208090.00	208040.0
Western Europe	231100.00	385690.00	559880.00	731370.00	789920.00	807160.00	817940.00	843150.00	859670.0
LATIN AMERICA AND CARIBBEAN	----	----	249080.00	286950.00	347930.00	360360.00	373350.00	388720.00	385770.0
Caribbean	----	----	12780.00	12710.00	12410.00	13270.00	14570.00	14800.00	15420.0
Meso-America	----	----	73410.00	93680.00	105890.00	103510.00	105950.00	110240.00	112370.0
South America	----	----	162890.00	180560.00	229630.00	243580.00	252830.00	263680.00	257980.0
NORTH AMERICA	715500.00	1116660.00	1365930.00	1548210.00	1674210.00	1717310.00	1747120.00	1788530.00	1844250.0
North America	715500.00	1116660.00	1365930.00	1548210.00	1674210.00	1717310.00	1747120.00	1788530.00	1844250.0
POLAR	----	----	----	----	----	----	----	----	--
Antarctic	----	----	----	----	----	----	----	----	--
Arctic	----	----	----	----	----	----	----	----	--
WEST ASIA	----	----	52520.00	67880.00	85160.00	88550.00	90610.00	91560.00	88880.0
Arabian Peninsula	----	----	30790.00	36090.00	48700.00	52180.00	53890.00	53930.00	51750.0
Mashriq	----	----	21730.00	31790.00	36460.00	36370.00	36720.00	37630.00	37130.0
GLOBAL TOTALS (excluding International Marine and Aviation Bunkers)	----	----	3137740.00	3858690.00	4264500.00	4404660.00	4500830.00	4621710.00	4745550.0
GLOBAL TOTALS (including International Marine and Aviation Bunkers)	----	----	3681020.00	4486410.00	4595790.00	4669060.00	4702480.00	4796740.00	4936360.0
REPORTED GLOBAL TOTALS FROM ORIGINAL DATA SOURCES	----	----	3681020.00	4486410.00	4936360.00	5095070.00	5214370.00	5345360.00	5503720.0

Comments:

Emissions of CO$_2$ from transport contain emissions from the combustion of fuel (coal, oil and gas) for all transport activity, regardless of the sector, except for international marine and aviation bunkers. This corresponds to IPCC Source/Sink Category 1.A.3. In addition, the IEA data are not collected in a way that allows the autoproducer consumption to be split by specific end-use.

Data source: the estimates of CO$_2$ emissions from fuel combustion are calculated using the IEA energy data and the default methods and emission factors from the Revised 1996 IPCC Guidelines for National Greenhouse Gas Inventories, IPCC/OECD/IEA Paris, 1997 (Guidelines). There are many reasons why the IEA estimates may not be the same as the numbers that a country submits to the UNFCCC, even if a country has accounted for all of its energy use and correctly applied the IPCC guidelines.

1 thousand metric tonne of CO$_2$ (1 gigagram of CO$_2$), is equal to 272.72 tonnes of C.

AGGREGATIONS

Units: thousand metric tonnes of CH₄

Data Source: Emission Database for Global Atmospheric Research (EDGAR 3.2)

Data Provider: Netherlands Organisation for Applied Scientific Research (TNO)
and National Institute of Public Health and the Environment (RIVM)

Years: 1970, 1975, ..., 1995

	1970	1975	1980	1985	1990	1995
AFRICA	11821.29	12155.78	12955.13	13210.91	14024.91	14515.88
Central Africa	1220.78	1211.23	1319.01	1336.82	1432.07	1473.97
Eastern Africa	2680.36	2765.35	2854.39	3114.41	3243.14	3276.84
Northern Africa	1870.95	2004.68	2127.55	2192.03	2290.54	2439.38
Southern Africa	3174.74	3332.35	3454.34	3391.61	3557.86	3513.91
Western Africa	2292.76	2256.03	2531.07	2530.87	2856.38	3154.41
Western Indian Ocean	581.71	586.14	668.77	645.18	644.93	657.37
ASIA AND PACIFIC	75997.76	76598.16	73373.47	71721.19	70438.17	72322.72
Australia and New Zealand	4489.17	4857.47	4393.80	4297.96	4410.03	4155.99
Central Asia	1816.95	2015.24	2078.82	2144.63	2079.41	1959.73
North West Pacific and East Asia	27916.59	28674.17	26598.70	24209.58	24014.19	24253.21
South Asia	29524.38	29363.25	28665.35	28922.59	28484.34	29160.83
South East Asia	12189.23	11618.80	11563.27	12067.12	11367.93	12702.87
South Pacific	61.44	69.22	73.53	79.31	82.27	90.10
EUROPE	19237.31	20932.70	21518.79	21504.90	20487.25	16464.28
Central Europe	4232.99	4725.41	4896.16	4679.19	4397.37	3151.30
Eastern Europe	6322.51	7168.35	7524.10	7797.70	7607.80	5624.72
Western Europe	8681.81	9038.95	9098.54	9028.02	8482.08	7688.26
LATIN AMERICA AND CARIBBEAN	15623.18	17511.83	19270.36	19975.25	20943.37	21822.32
Caribbean	690.73	680.39	667.85	671.35	643.34	583.94
Meso-America	2189.14	2360.46	2546.50	2800.56	2773.40	2637.87
South America	12743.31	14470.98	16056.00	16503.33	17526.63	18600.51
NORTH AMERICA	9203.85	10178.12	9232.13	8788.39	7993.34	8489.42
North America	9203.85	10178.12	9232.13	8788.39	7993.34	8489.42
POLAR	1.43	1.30	1.22	1.14	1.07	1.07
Antarctic	----	----	----	----	0.00	0.00
Arctic	1.43	1.30	1.22	1.14	1.07	1.07
WEST ASIA	405.61	385.22	437.76	455.61	492.41	494.79
Arabian Peninsula	116.96	146.24	165.65	196.84	211.36	227.76
Mashriq	288.65	238.99	272.11	258.78	281.05	267.03
GLOBAL TOTALS	132290.43	137763.11	136788.85	135657.39	134380.52	134110.48
REPORTED GLOBAL TOTALS FROM ORIGINAL DATA SOURCES	134445.94	137764.50	136790.33	135658.63	134368.53	134098.07

Comments:

Emissions of CH₄ (methane) from the agricultural sector include emissions from: rice cultivation (IPCC 4.C); animal breeding: enteric fermentation and animal waste management (IPCC 4.A and 4.B,); savannah burning (IPCC 4.E); agricultural waste burning (IPCC 4.F).

The emissions from deforestation (IPCC 5.A.1) and vegetation fires (IPCC 5.A.2,3) are not included.

Data source: The EDGAR set of inventories were compiled from the perspective of providing good quality reference estimates of anthropogenic emission sources per source category, based on scientifically sound input data. This was done by using (a) international statistics as activity data, since these are comparable between countries in definition and units, (b) emission factors from the scientific literature, also common across countries when judged comparable, and (c) grid maps for allocating sectoral emissions of a country to a grid.

EDGAR 3.2 by RIVM/TNO (Olivier, J.G.J. and Berdowski, J.J.M., 2001, Global Emission Sources and Sinks. In: J. Berdowski, R. Guicherit and B.J. Heij (eds.), The Climate System: 33-77. Lisse: Swets and Zeitlinger Publishers). For more detailed information about the EDGAR 3.2 database and data source see: http://www.rivm.nl/env/int/coredata/edgar/index.html.

1 thousand metric tonne corresponds to 1 gigagram.

Emissions of CH₄ - from Energy (RIVM)

Units: thousand metric tonnes of CH_4

Data Source: Emission Database for Global Atmospheric Research (EDGAR 3.2)

Data Provider: Netherlands Organisation for Applied Scientific Research (TNO)
and National Institute of Public Health and the Environment (RIVM)

Years: 1970, 1975, ..., 1995

	1970	1975	1980	1985	1990	1995
AFRICA	3798.73	4074.23	5025.52	5663.43	6423.98	7091.18
Central Africa	157.57	222.09	263.17	352.35	404.54	450.26
Eastern Africa	310.40	353.51	410.66	470.54	537.13	587.28
Northern Africa	1292.64	1076.49	1405.88	1473.49	1719.29	1888.63
Southern Africa	1015.20	1130.25	1461.97	1889.76	2043.73	2211.30
Western Africa	988.05	1252.89	1439.22	1425.98	1659.70	1885.09
Western Indian Ocean	34.88	38.99	44.62	51.31	59.59	68.62
ASIA AND PACIFIC	16590.02	19343.78	22097.33	28031.34	31212.75	35045.17
Australia and New Zealand	549.12	665.13	743.45	985.16	1184.59	1287.26
Central Asia	2622.23	3140.55	3783.09	5077.50	4188.35	3142.43
North West Pacific and East Asia	8001.40	8878.91	10523.88	13147.13	14994.04	17226.21
South Asia	4016.86	4910.28	4471.94	5350.72	6401.97	7583.80
South East Asia	1383.04	1730.76	2555.89	3449.85	4404.47	5756.98
South Pacific	17.38	18.14	19.08	20.99	39.33	48.47
EUROPE	26104.63	29104.51	32852.66	38437.09	39584.82	30673.96
Central Europe	5485.07	6595.36	7221.76	7382.97	6032.32	5063.45
Eastern Europe	12166.41	15223.10	18532.85	24971.57	28131.08	21144.49
Western Europe	8453.15	7286.05	7098.05	6082.55	5421.42	4466.01
LATIN AMERICA AND CARIBBEAN	2387.08	2308.30	2832.30	3217.72	3414.20	3901.70
Caribbean	78.06	117.91	135.76	156.08	145.55	163.09
Meso.America	426.57	583.07	968.97	1190.48	1173.88	1178.13
South America	1882.46	1607.32	1727.57	1871.16	2094.77	2560.47
NORTH AMERICA	22162.27	20339.56	21934.30	20915.42	23933.87	23887.08
North America	22162.27	20339.56	21934.30	20915.42	23933.87	23887.08
POLAR	0.00	0.00	0.00	0.00	0.00	0.00
Antarctic	0.00	0.00	0.00	0.00	0.00	0.00
Arctic	0.00	0.00	0.00	0.00	0.00	0.00
WEST ASIA	1275.33	1854.37	2628.79	2266.29	3464.59	4305.61
Arabian Peninsula	1074.34	1585.86	2311.82	2042.95	3148.90	4083.17
Mashriq	200.99	268.51	316.97	223.33	315.69	222.44
GLOBAL TOTALS	72318.07	77024.75	87370.91	98531.28	108034.21	104904.70
REPORTED GLOBAL TOTALS FROM ORIGINAL DATA SOURCES	72331.47	77028.37	87374.62	98535.25	108054.17	104927.74

Comments:

"Energy" comprises production, handling, transmission and combustion of fossil fuels and biofuels (IPCC category 1.A and 1.B) as follows: industrial sector (IPCC 1.A.2); power generation (IPCC 1.A.1(a)); other transformation sector (all) (IPCC 1.A.1(b, c)); residentials, commercials and other sector (IPCC 1.A.4); transport road (IPCC 1.A.3(b)); transport land non-road (IPCC 1.A.3(c),(d-ii),(e)); transport air (all) (IPCC 1.A.3(a)); transport international shipping (IPCC 1.A.3(d-I)); coal production (including CH4 recovery) (IPCC 1.B.1(a)); oil production, handling and flaring (IPCC 1.B.2(a,c)); gas production (IPCC 1.B.2 (b-i)); gas transmission (IPCC 1.B.2(b-ii)); industry v3(IPCC 1.A.2); power generation (IPCC 1.A.1(a)); charcoal production (IPCC 1.B.1(b)); residential (IPCC 1.A.4); transport road (IPCC 1.A.3(b)).

Data source: the EDGAR set of inventories were compiled from the perspective of providing good quality reference estimates of anthropogenic emission sources per source category, based on scientifically sound input data. This was done by using (a) international statistics as activity data, since these are comparable between countries in definition and units, (b) emission factors from the scientific literature, also common across countries when judged comparable, and (c) grid maps for allocating sectoral emissions of a country to a grid.

EDGAR 3.2 by RIVM/TNO (Olivier, J.G.J. and Berdowski, J.J.M., 2001, Global Emission Sources and Sinks. In: J. Berdowski, R. Guicherit and B.J. Heij (eds.), The Climate System: 33-77. Lisse: Swets and Zeitlinger Publishers). For more detailed information about the EDGAR 3.2 database and data source see: http://www.rivm.nl/env/int/coredata/edgar/index.html.

1 thousand metric tonne corresponds to 1 gigagram.

AGGREGATIONS

Emissions of CH$_4$ - Total (RIVM)

Units: thousand metric tonnes of CH$_4$

Data Source: Emission Database for Global Atmospheric Research (EDGAR 3.2)

Data Provider: Netherlands Organisation for Applied Scientific Research (TNO)
and National Institute of Public Health and the Environment (RIVM)

Years: 1970, 1975, ..., 1995

	1970	1975	1980	1985	1990	1995
AFRICA	18661.16	19566.27	21687.21	22967.82	25039.45	26737.70
Central Africa	2157.44	2241.43	2425.80	2578.39	2777.37	2932.61
Eastern Africa	3283.11	3453.01	3655.78	4042.47	4319.53	4488.93
Northern Africa	3722.34	3711.73	4253.91	4477.32	4919.13	5341.44
Southern Africa	4874.92	5208.30	5740.54	6164.64	6593.89	6809.52
Western Africa	3930.46	4244.44	4807.37	4908.89	5613.99	6314.68
Western Indian Ocean	692.90	707.36	803.82	796.09	815.54	850.52
ASIA AND PACIFIC	107573.40	112662.46	113874.56	119804.47	123825.06	131586.73
Australia and New Zealand	5459.74	6000.11	5653.56	5838.95	6258.76	6082.78
Central Asia	4627.51	5375.22	6111.13	7504.02	6595.79	5452.69
North West Pacific and East Asia	41897.33	44274.04	44502.87	45310.27	47774.97	51026.26
South Asia	38448.02	39830.64	39386.29	41296.20	42777.89	45535.73
South East Asia	16989.87	17020.59	18050.94	19670.88	20203.28	23249.63
South Pacific	150.93	161.86	169.78	184.15	214.37	239.64
EUROPE	53574.70	59003.17	64006.78	69935.26	70645.58	57413.73
Central Europe	10853.16	12564.32	13484.57	13551.14	11988.76	9801.26
Eastern Europe	20676.32	24850.48	28780.46	35765.72	39300.61	30192.74
Western Europe	22045.22	21588.37	21741.76	20618.39	19356.21	17419.73
LATIN AMERICA AND CARIBBEAN	24717.82	27100.21	30128.50	31393.52	32924.74	35058.99
Caribbean	1039.89	1101.84	1139.93	1196.75	1189.99	1183.68
Meso-America	3723.39	4216.60	4972.18	5619.87	5766.51	5894.46
South America	19954.53	21781.78	24016.38	24576.90	25968.24	27980.86
NORTH AMERICA	40117.42	40496.40	42562.78	41650.06	44797.91	45624.21
North America	40117.42	40496.40	42562.78	41650.06	44797.91	45624.21
POLAR	2.03	2.03	1.89	1.85	1.81	1.84
Antarctic	0.38	0.48	0.38	0.39	0.39	0.40
Arctic	1.66	1.56	1.51	1.46	1.42	1.44
WEST ASIA	1917.11	2541.56	3437.79	3182.37	4516.27	5455.97
Arabian Peninsula	1286.50	1854.94	2638.83	2448.58	3623.79	4628.06
Mashriq	630.61	686.62	798.95	733.79	892.48	827.91
GLOBAL TOTALS	246563.64	261372.11	275699.50	288935.36	301750.81	301879.18
REPORTED GLOBAL TOTALS FROM ORIGINAL DATA SOURCES	248772.97	261417.54	275751.24	288992.98	301811.67	301948.95

Comments:

Emissions of CH$_4$ - total (RIVM) include "energy", "agriculture", "waste" and "others" EDGAR subdivisions. Energy comprises production, handling, transmission and combustion of fossil fuels and biofuels (IPCC category 1.A and 1.B); agriculture comprises animals, animal waste, rice production, agricultural waste burning (non-energy, on-site) and savannah burning (IPCC category 4); waste comprises landfills, wastewater treatment, human wastewater disposal and waste incineration (non-energy) (IPCC category 6); others include industrial process emissions and tropical and temperate forest fires (IPCC categories 2 and 5).

Data source: the EDGAR set of inventories were compiled from the perspective of providing good quality reference estimates of anthropogenic emission sources per source category, based on scientifically sound input data. This was done by using (a) international statistics as activity data, since these are comparable between countries in definition and units, (b) emission factors from the scientific literature, also common across countries when judged comparable, and (c) grid maps for allocating sectoral emissions of a country to a grid.

EDGAR 3.2 by RIVM/TNO (Olivier, J.G.J. and Berdowski, J.J.M., 2001, Global Emission Sources and Sinks. In: J. Berdowski, R. Guicherit and B.J. Heij (eds.), The Climate System: 33-77. Lisse: Swets and Zeitlinger Publishers). For more detailed information about the EDGAR 3.2 database and data source see: http://www.rivm.nl/env/int/coredata/edgar/index.html.

1 thousand metric tonne corresponds to 1 gigagram.

Emissions of CH₄ - Waste (RIVM)

Units: thousand metric tonnes of CH₄

Data Source: Emission Database for Global Atmospheric Research (EDGAR 3.2)

Data Provider: Netherlands Organisation for Applied Scientific Research (TNO)
and National Institute of Public Health and the Environment (RIVM)

Years: 1970, 1975, ..., 1995

	1970	1975	1980	1985	1990	1995
AFRICA	1876.21	2169.77	2537.54	2980.35	3532.71	4118.79
Central Africa	177.52	182.99	206.53	242.04	288.20	340.23
Eastern Africa	272.54	280.52	314.33	370.92	433.92	511.75
Northern Africa	452.02	465.66	523.44	612.97	720.61	836.40
Southern Africa	445.03	456.35	504.57	580.95	668.31	806.65
Western Africa	486.60	502.55	572.47	674.03	802.59	958.66
Western Indian Ocean	42.51	43.68	48.43	56.63	66.71	79.03
ASIA AND PACIFIC	12801.13	14485.18	16135.46	17992.44	20149.71	22266.17
Australia and New Zealand	412.03	422.94	465.65	504.79	548.43	593.52
Central Asia	178.59	184.49	207.65	235.92	268.53	310.89
North West Pacific and East Asia	5627.83	5776.45	6325.27	6953.34	7593.50	8403.53
South Asia	4691.72	4816.10	5340.04	6031.59	6865.45	7764.23
South East Asia	1874.06	1923.40	2127.29	2387.88	2691.22	3048.46
South Pacific	16.90	17.13	19.28	21.95	25.31	29.07
EUROPE	7870.58	8576.26	9213.35	9606.15	9804.98	9826.31
Central Europe	1093.47	1111.20	1191.14	1300.24	1425.86	1494.13
Eastern Europe	2069.58	2119.89	2318.09	2565.54	2835.40	3152.18
Western Europe	4707.53	4795.33	5067.02	5347.56	5344.89	5158.67
LATIN AMERICA AND CARIBBEAN	3231.81	3795.72	4530.70	5200.68	6070.17	7248.61
Caribbean	238.64	244.14	271.08	303.85	340.79	376.02
Meso-America	827.73	859.64	992.33	1173.73	1358.51	1564.19
South America	2165.44	2235.18	2532.30	3053.11	3501.38	4129.96
NORTH AMERICA	8432.34	9669.68	11094.72	11816.38	12143.25	11492.76
North America	8432.34	8682.74	9669.68	11094.72	11816.38	12143.25
POLAR	0.60	0.74	0.67	0.71	0.74	0.77
Antarctic	0.38	0.38	0.48	0.38	0.39	0.39
Arctic	0.22	0.23	0.26	0.29	0.32	0.35
WEST ASIA	234.31	300.10	369.35	458.38	556.12	652.18
Arabian Peninsula	94.86	99.78	122.50	161.00	208.51	262.89
Mashriq	139.45	146.96	177.60	208.36	249.87	293.23
GLOBAL TOTALS	34446.99	38997.43	43881.79	48055.08	52257.68	55605.58
REPORTED GLOBAL TOTALS FROM ORIGINAL DATA SOURCES	34482.28	39037.84	43928.34	48107.50	52311.25	55665.42

Comments:

Emissions of CH₄ - from waste (RIVM) include emissions from: landfills (including CH4 recovery) (IPCC 6.A.1,2); wastewater treatment (including CH4 recovery) (IPCC 6.B.1,2); human wastewater disposal (IPCC 6.B.2); waste incineration (non-energy) (IPCC 6.C).

Data source: the EDGAR set of inventories were compiled from the perspective of providing good quality reference estimates of anthropogenic emission sources per source category, based on scientifically sound input data. This was done by using (a) international statistics as activity data, since these are comparable between countries in definition and units, (b) emission factors from the scientific literature, also common across countries when judged comparable, and (c) grid maps for allocating sectoral emissions of a country to a grid.

EDGAR 3.2 by RIVM/TNO (Olivier, J.G.J. and Berdowski, J.J.M., 2001, Global Emission Sources and Sinks. In: J. Berdowski, R. Guicherit and B.J. Heij (eds.), The Climate System: 33-77. Lisse: Swets and Zeitlinger Publishers). For more detailed information about the EDGAR 3.2 database and data source see: http://www.rivm.nl/env/int/coredata/edgar/index.html

1 thousand metric tonne corresponds to 1 gigagram

Units: gigagrams of CH$_4$

Data Source: Greenhouse Gas (GHG) Inventory Submission 1998, 1999, and 2000

Data Provider: United Nations Framework Convention on Climate Change (UNFCCC) Secretariat

Years: 1990-1998

	1990	1991	1992	1993	1994	1995	1996	1997	1998
AFRICA	----	----	----	----	----	----	----	----	----
Central Africa	----	----	----	----	----	----	----	----	----
Eastern Africa	----	----	----	----	----	----	----	----	----
Northern Africa	----	----	----	----	----	----	----	----	----
Southern Africa	----	----	----	----	----	----	----	----	----
Western Africa	----	----	----	----	----	----	----	----	----
Western Indian Ocean	----	----	----	----	----	----	----	----	----
ASIA AND PACIFIC	----	----	----	----	----	----	----	----	----
Australia and New Zealand	7254.01	7122.38	7087.68	7020.11	6950.42	7028.03	7068.77	7076.56	7187.69
Central Asia	----	----	----	----	----	----	----	----	----
North West Pacific and East Asia	----	----	----	----	----	----	----	----	----
South Asia	----	----	----	----	----	----	----	----	----
South East Asia	----	----	----	----	----	----	----	----	----
South Pacific	----	----	----	----	----	----	----	----	----
EUROPE	66849.27	63041.23	59727.91	56400.19	53391.00	52311.42	50719.14	49530.00	47884.52
Central Europe	9484.02	8638.70	8026.62	7622.00	----	----	----	----	----
Eastern Europe	35902.30	----	----	----	27180.30	26358.90	25603.40	----	----
Western Europe	21462.95	20971.23	20298.59	19513.49	19015.67	18889.51	18394.84	18159.49	18041.47
LATIN AMERICA AND CARIBBEAN	----	----	----	----	----	----	----	----	----
Caribbean	----	----	----	----	----	----	----	----	----
Meso-America	----	----	----	----	----	----	----	----	----
South America	----	----	----	----	----	----	----	----	----
NORTH AMERICA	34595.99	34688.91	35117.82	35119.22	35729.37	36309.95	36266.18	36401.01	35853.75
North America	34595.99	34688.91	35117.82	35119.22	35729.37	36309.95	36266.18	36401.01	35853.75
POLAR	----	----	----	----	----	----	----	----	----
Antarctic	----	----	----	----	----	----	----	----	----
Arctic	----	----	----	----	----	----	----	----	----
WEST ASIA	----	----	----	----	----	----	----	----	----
Arabian Peninsula	----	----	----	----	----	----	----	----	----
Mashriq	----	----	----	----	----	----	----	----	----
GLOBAL TOTALS	----	----	----	----	----	----	----	----	----
REPORTED GLOBAL TOTALS FROM ORIGINAL DATA SOURCES	****	****	****	****	****	****	****	****	****
SUM OF ANNEX I PARTIES (INCLUDES ESTIMATIONS)	110242.15	106375.51	103446.42	100044.05	97560.16	97126.07	95501.37	94396.13	90925.96

Comments:

Emissions of CH$_4$ - total anthropogenic (UNFCCC): total anthropogenic emissions of CH$_4$ (methane) includes emissions from the energy (fuel combustion and fugitive fuel) and agricultural (livestock and other) sectors, waste and other. The data is based on the national greenhouse gas inventories given by the parties of the Climate Change Convention Secretariat. This data is annually submitted to the Climate Change Secretariat by national communications under the Convention by the Annex I parties.

Annex I parties - the industrialized countries listed in this annex to the Climate Change Convention tried to return their greenhouse gas emissions to 1990 levels by the year 2000. They have also accepted emission targets for the period 2008-12. They include the following members Australia, Austria, Belarus, Belgium, Bulgaria, Canada, Czech Republic, Denmark, European Economic Community, Estonia, Finland, France, Germany, Greece, Hungary, Iceland, Ireland, Italy, Japan, Latvia, Lithuania, Luxembourg, Netherlands, New Zealand, Norway, Poland, Portugal, Romania, Russian Federation, Slovakia, Slovenia, Spain, Sweden, Switzerland, Turkey, Ukraine, United Kingdom of Great Britain and Northern Ireland and the United States of America.

Greenhouse gases (GHGs) - the major anthropogenic GHGs responsible for causing climate change are carbon dioxide (CO$_2$), methane (CH$_4$), and nitrous oxide (N$_2$O). The Kyoto Protocol also addresses hydrofluorocarbons (HFCs), perfluorocarbons (PFCs), and sulphur hexafluoride (SF$_6$). For more detailed information about the GHG inventory database see: http://ghg. unfccc.int .

1 gigagram corresponds to 1 thousand metric tonne.

AGGREGATIONS

Emissions of CO_2 - Anthropogenic Emissions and Removals from Land-Use Change and Forestry (UNFCCC)

Units: gigagrams of CO_2

Data Source: Greenhouse Gas (GHG) Inventory Submission 1998, 1999, and 2000

Data Provider: United Nations Framework Convention on Climate Change (UNFCCC) Secretariat

Years: 1990-1998

	1990	1991	1992	1993	1994	1995	1996	1997	1998
AFRICA	----	----	----	----	----	----	----	----	----
Central Africa	----	----	----	----	----	----	----	----	----
Eastern Africa	----	----	----	----	----	----	----	----	----
Northern Africa	----	----	----	----	----	----	----	----	----
Southern Africa	----	----	----	----	----	----	----	----	----
Western Africa	----	----	----	----	----	----	----	----	----
Western Indian Ocean	----	----	----	----	----	----	----	----	----
ASIA AND PACIFIC	----	----	----	----	----	----	----	----	----
Australia and New Zealand	48562.15	22924.00	21658.81	21228.08	20286.40	19213.76	21415.67	18379.18	14277.09
Central Asia	----	----	----	----	----	----	----	----	----
North West Pacific and East Asia	----	----	----	----	----	----	----	----	----
South Asia	----	----	----	----	----	----	----	----	----
South East Asia	----	----	----	----	----	----	----	----	----
South Pacific	----	----	----	----	----	----	----	----	----
EUROPE	-749657.27	-829343.43	-861358.23	-906792.31	-943125.06	-1208772.02	-1232993.99	-1270243.05	-1254320.51
Central Europe	-83415.99	-99893.21	-96975.72	-93817.97	-92375.62	-91942.90	-89954.76	-68593.40	-59074.83
Eastern Europe	-444107.10	----	----	----	-620720.90	-892939.60	-906150.60	----	----
Western Europe	-222134.17	-240486.72	-231724.51	-236056.94	-230028.53	-223889.52	-236888.63	-229986.11	-224967.47
LATIN AMERICA AND CARIBBEAN	----	----	----	----	----	----	----	----	----
Caribbean	----	----	----	----	----	----	----	----	----
Meso-America	----	----	----	----	----	----	----	----	----
South America	----	----	----	----	----	----	----	----	----
NORTH AMERICA	-1199134.68	-1216915.77	-1204650.65	-814513.30	-808013.91	-797787.23	-803611.25	-797707.94	-794851.94
North America	-1199134.68	-1216915.77	-1204650.65	-814513.30	-808013.91	-797787.23	-803611.25	-797707.94	-794851.94
POLAR	----	----	----	----	----	----	----	----	----
Antarctic	----	----	----	----	----	----	----	----	----
Arctic	----	----	----	----	----	----	----	----	----
WEST ASIA	----	----	----	----	----	----	----	----	----
Arabian Peninsula	----	----	----	----	----	----	----	----	----
Mashriq	----	----	----	----	----	----	----	----	----
GLOBAL TOTALS	----	----	----	----	----	----	----	----	----
REPORTED GLOBAL TOTALS FROM ORIGINAL DATA SOURCES	****	****	****	****	****	****	****	****	****

Comments:

Emissions of CO_2 - total anthropogenic and removals from land-use change and forestry (UNFCCC): are all the emissions of CO_2 associated with human activities including land-use change and forestry. These include emissions from fuel combustion, industrial processes, agriculture, waste, and other. The data is based on the national greenhouse gas inventories given by the parties of the Climate Change Convention Secretariat. This data is annually submitted to the Climate Change Secretariat by national communications under the Convention by the Annex I parties.

Annex I parties - the industrialized countries listed in this annex to the Climate Change Convention tried to return their greenhouse gas emissions to 1990 levels by the year 2000. They have also accepted emission targets for the period 2008-12. They include the following members Australia, Austria, Belarus, Belgium, Bulgaria, Canada, Czech Republic, Denmark, European Economic Community, Estonia, Finland, France, Germany, Greece, Hungary, Iceland, Ireland, Italy, Japan, Latvia, Lithuania, Luxembourg, Netherlands, New Zealand, Norway, Poland, Portugal, Romania, Russian Federation, Slovakia, Slovenia, Spain, Sweden, Switzerland, Turkey, Ukraine, United Kingdom of Great Britain and Northern Ireland and the United States of America.

Greenhouse gases (GHGs) - the major anthropogenic GHGs responsible for causing climate change are carbon dioxide (CO_2), methane (CH_4), and nitrous oxide (N_2O). The Kyoto Protocol also addresses hydrofluorocarbons (HFCs), perfluorocarbons (PFCs), and sulphur hexafluoride (SF6). For more detailed information about the GHG inventory database see: http://ghg.unfccc.int.

1 gigagram corresponds to 1 thousand metric tonne.

AGGREGATIONS

Emissions of CO$_2$ - from Fuel Combustion (UNFCCC)

Units: gigagrams of CO$_2$

Data Source: Greenhouse Gas (GHG) Inventory Submission 1998, 1999, and 2000

Data Provider: United Nations Framework Convention on Climate Change (UNFCCC) Secretariat

Years: 1990-1998

	1990	1991	1992	1993	1994	1995	1996	1997	1998
AFRICA	----	----	----	----	----	----	----	----	----
Central Africa	----	----	----	----	----	----	----	----	----
Eastern Africa	----	----	----	----	----	----	----	----	----
Northern Africa	----	----	----	----	----	----	----	----	----
Southern Africa	----	----	----	----	----	----	----	----	----
Western Africa	----	----	----	----	----	----	----	----	----
Western Indian Ocean	----	----	----	----	----	----	----	----	----
ASIA AND PACIFIC	----	----	----	----	----	----	----	----	----
Australia and New Zealand	287685.80	290148.39	294520.73	296948.22	300617.90	312609.38	323443.29	332860.12	349734.36
Central Asia	----	----	----	----	----	----	----	----	----
North West Pacific and East Asia	----	----	----	----	----	----	----	----	----
South Asia	----	----	----	----	----	----	----	----	----
South East Asia	----	----	----	----	----	----	----	----	----
South Pacific	----	----	----	----	----	----	----	----	----
EUROPE	7322621.98	6808567.76	6506392.09	6216096.34	5910282.95	5850766.55	5840446.21	5680862.23	5668242.21
Central Europe	1152940.20	893547.44	842499.70	818483.60	808018.12	784991.62	813507.67	800120.74	750006.50
Eastern Europe	2970974.80	2680747.40	2497438.70	2289956.20	1991218.00	1915332.90	1794975.90	----	----
Western Europe	3198706.98	3234272.93	3166453.69	3107656.54	3111046.82	3150442.03	3231962.65	3173717.26	3219813.37
LATIN AMERICA AND CARIBBEAN	----	----	----	----	----	----	----	----	----
Caribbean	----	----	----	----	----	----	----	----	----
Meso-America	----	----	----	----	----	----	----	----	----
South America	----	----	----	----	----	----	----	----	----
NORTH AMERICA	5256172.78	5193764.29	5296617.18	5409016.56	5497170.73	5544802.93	5737846.05	5820719.11	5859928.05
North America	5256172.78	5193764.29	5296617.18	5409016.56	5497170.73	5544802.93	5737846.05	5820719.11	5859928.05
POLAR	----	----	----	----	----	----	----	----	----
Antarctic	----	----	----	----	----	----	----	----	----
Arctic	----	----	----	----	----	----	----	----	----
WEST ASIA	----	----	----	----	----	----	----	----	----
Arabian Peninsula	----	----	----	----	----	----	----	----	----
Mashriq	----	----	----	----	----	----	----	----	----
GLOBAL TOTALS	----	----	----	----	----	----	----	----	----
REPORTED GLOBAL TOTALS FROM ORIGINAL DATA SOURCES	****	****	****	****	****	****	****	****	****
SUM OF ANNEX I PARTIES (INCLUDES ESTIMATIONS)	13919444.63	13365242.15	13182740.79	12986502.05	12841362.73	12846530.64	13055277.86	12985116.65	13037751.12

Comments:

Emissions of CO$_2$ - fuel combustion (UNFCCC): emissions of CO$_2$ from fuel combustion include emissions from energy industries, manufacturing industries and construction, transport and other fuel combustion sectors. The data is based on the national greenhouse gas inventories given by the parties of the Climate Change Convention Secretariat. This data is annually submitted to the Climate Change Secretariat by national communications under the Convention by the Annex I parties.

Annex I parties - the industrialized countries listed in this annex to the Climate Change Convention tried to return their greenhouse gas emissions to 1990 levels by the year 2000. They have also accepted emission targets for the period 2008-12. They include the following members Australia, Austria, Belarus, Belgium, Bulgaria, Canada, Czech Republic, Denmark, European Economic Community, Estonia, Finland, France, Germany, Greece, Hungary, Iceland, Ireland, Italy, Japan, Latvia, Lithuania, Luxembourg, Netherlands, New Zealand, Norway, Poland, Portugal, Romania, Russian Federation, Slovakia, Slovenia, Spain, Sweden, Switzerland, Turkey, Ukraine, United Kingdom of Great Britain and Northern Ireland and the United States of America.

Greenhouse gases (GHGs) - the major anthropogenic GHGs responsible for causing climate change are carbon dioxide (CO$_2$), methane (CH$_4$), and nitrous oxide (N$_2$O). The Kyoto Protocol also addresses hydrofluorocarbons (HFCs), perfluorocarbons (PFCs), and sulphur hexafluoride (SF$_6$). For more detailed information about the GHG inventory database see: http://ghg.unfccc.int.

1 gigagram corresponds to 1 thousand metric tonne.

AGGREGATIONS

Emissions of CO$_2$ - from Industrial Processes (UNFCCC)

Units: gigagrams of CO$_2$

Data Source: Greenhouse Gas (GHG) Inventory Submission 1998, 1999, and 2000

Data Provider: United Nations Framework Convention on Climate Change (UNFCCC) Secretariat

Years: 1990-1998

	1990	1991	1992	1993	1994	1995	1996	1997	1998
AFRICA	----	----	----	----	----	----	----	----	----
Central Africa	----	----	----	----	----	----	----	----	----
Eastern Africa	----	----	----	----	----	----	----	----	----
Northern Africa	----	----	----	----	----	----	----	----	----
Southern Africa	----	----	----	----	----	----	----	----	----
Western Africa	----	----	----	----	----	----	----	----	----
Western Indian Ocean	----	----	----	----	----	----	----	----	----
ASIA AND PACIFIC	----	----	----	----	----	----	----	----	----
Australia and New Zealand	9040.20	8840.27	8857.90	9290.43	9948.29	9755.02	9851.80	9927.52	10562.87
Central Asia	----	----	----	----	----	----	----	----	----
North West Pacific and East Asia	58794.68	60381.39	60998.36	60332.69	61302.69	61236.14	61078.72	59500.61	60763.87
South Asia	----	----	----	----	----	----	----	----	----
South East Asia	----	----	----	----	----	----	----	----	----
South Pacific	----	----	----	----	----	----	----	----	----
EUROPE	288760.95	256453.30	243357.41	220536.06	222944.55	226122.84	214672.75	219634.00	221056.45
Central Europe	49747.57	32550.03	31406.07	28758.29	29627.85	31002.12	27677.57	29661.63	30811.26
Eastern Europe	78016.80	72333.50	64178.80	49067.70	40720.00	38675.20	33711.70	----	----
Western Europe	160996.58	151569.77	147772.54	142710.06	152596.71	156445.52	153283.49	155236.76	158142.52
LATIN AMERICA AND CARIBBEAN	----	----	----	----	----	----	----	----	----
Caribbean	----	----	----	----	----	----	----	----	----
Meso-America	----	----	----	----	----	----	----	----	----
South America	----	----	----	----	----	----	----	----	----
NORTH AMERICA	87150.79	86704.98	86633.44	90022.96	94217.62	98199.54	101236.09	104420.11	105512.53
North America	87150.79	86704.98	86633.44	90022.96	94217.62	98199.54	101236.09	104420.11	105512.53
POLAR	----	----	----	----	----	----	----	----	----
Antarctic	----	----	----	----	----	----	----	----	----
Arctic	----	----	----	----	----	----	----	----	----
WEST ASIA	----	----	----	----	----	----	----	----	----
Arabian Peninsula	----	----	----	----	----	----	----	----	----
Mashriq	----	----	----	----	----	----	----	----	----
GLOBAL TOTALS	----	----	----	----	----	----	----	----	----
REPORTED GLOBAL TOTALS FROM ORIGINAL DATA SOURCES	****	****	****	****	****	****	****	****	****
SUM OF ANNEX I PARTIES (INCLUDES ESTIMATIONS)	443746.62	412379.94	399847.11	380182.13	388413.15	395313.55	386839.37	393482.23	397895.72

Comments:

Emissions of CO$_2$ - industrial processes (UNFCCC): emissions of CO$_2$ from industrial processes of mineral products, chemical industry, metal production, other production and other. The data is based on the national greenhouse gas inventories given by the parties of the Climate Change Convention Secretariat. This data is annually submitted to the Climate Change Secretariat by national communications under the Convention by the Annex I parties.

Annex I parties - the industrialized countries listed in this annex to the Climate Change Convention tried to return their greenhouse gas emissions to 1990 levels by the year 2000. They have also accepted emission targets for the period 2008-12. They include the following members Australia, Austria, Belarus, Belgium, Bulgaria, Canada, Czech Republic, Denmark, European Economic Community, Estonia, Finland, France, Germany, Greece, Hungary, Iceland, Ireland, Italy, Japan, Latvia, Lithuania, Luxembourg, Netherlands, New Zealand, Norway, Poland, Portugal, Romania, Russian Federation, Slovakia, Slovenia, Spain, Sweden, Switzerland, Turkey, Ukraine, United Kingdom of Great Britain and Northern Ireland and the United States of America.

Greenhouse gases (GHGs) - the major anthropogenic GHGs responsible for causing climate change are carbon dioxide (CO$_2$), methane (CH$_4$), and nitrous oxide (N$_2$O). The Kyoto Protocol also addresses hydrofluorocarbons (HFCs), perfluorocarbons (PFCs), and sulphur hexafluoride (SF$_6$). For more detailed information about the GHG inventory database see: http://ghg.unfccc.int.

1 gigagram corresponds to 1 thousand metric tonne.

AGGREGATIONS

Emissions of CO_2 - Total Anthropogenic (UNFCCC) Excluding Land-Use Change and Forestry

Units: gigagrams of CO_2

Data Source: Greenhouse Gas (GHG) Inventory Submission 1998, 1999, and 2000

Data Provider: United Nations Framework Convention on Climate Change (UNFCCC) Secretariat

Years: 1990-1998

	1990	1991	1992	1993	1994	1995	1996	1997	1998
AFRICA	----	----	----	----	----	----	----	----	----
Central Africa	----	----	----	----	----	----	----	----	----
Eastern Africa	----	----	----	----	----	----	----	----	----
Northern Africa	----	----	----	----	----	----	----	----	----
Southern Africa	----	----	----	----	----	----	----	----	----
Western Africa	----	----	----	----	----	----	----	----	----
Western Indian Ocean	----	----	----	----	----	----	----	----	----
ASIA AND PACIFIC	----	----	----	----	----	----	----	----	----
Australia and New Zealand	304066.83	306087.03	310531.27	313431.60	317564.02	329483.61	340562.73	349903.68	366913.73
Central Asia	----	----	----	----	----	----	----	----	----
North West Pacific and East Asia	----	----	----	----	----	----	----	----	----
South Asia	----	----	----	----	----	----	----	----	----
South East Asia	----	----	----	----	----	----	----	----	----
South Pacific	----	----	----	----	----	----	----	----	----
EUROPE	7685006.76	6896824.39	6659333.49	6423123.66	6218560.00	6148034.71	6118122.60	5852301.81	5820719.01
Central Europe	1202747.85	933984.50	885259.76	856406.49	842555.82	825325.55	843600.54	834506.75	783559.99
Eastern Europe	3076091.65	----	----	----	2066838.04	1971348.14	1842687.64	----	----
Western Europe	3406167.25	3429791.95	3358083.20	3293999.82	3309166.14	3351361.01	3431834.41	3374430.14	3421729.30
LATIN AMERICA AND CARIBBEAN	----	----	----	----	----	----	----	----	----
Caribbean	----	----	----	----	----	----	----	----	----
Meso-America	----	----	----	----	----	----	----	----	----
South America	----	----	----	----	----	----	----	----	----
NORTH AMERICA	5380106.55	5318718.45	5421099.34	5540309.25	5632846.01	5687682.03	5882635.54	5968350.21	6007482.24
North America	5380106.55	5318718.45	5421099.34	5540309.25	5632846.01	5687682.03	5882635.54	5968350.21	6007482.24
POLAR	----	----	----	----	----	----	----	----	----
Antarctic	----	----	----	----	----	----	----	----	----
Arctic	----	----	----	----	----	----	----	----	----
WEST ASIA	----	----	----	----	----	----	----	----	----
Arabian Peninsula	----	----	----	----	----	----	----	----	----
Mashriq	----	----	----	----	----	----	----	----	----
GLOBAL TOTALS	----	----	----	----	----	----	----	----	----
REPORTED GLOBAL TOTALS FROM ORIGINAL DATA SOURCES	****	****	****	****	****	****	****	****	****

Comments:

Emissions of CO_2 - total anthropogenic (UNFCCC) excluding land-use change and forestry: all the emissions of CO_2 associated with human activities excluding land-use change and forestry. Emissions included are emitted from fuel combustion, industrial processes, agriculture, waste, and other. The data is based on the national greenhouse gas inventories given by the parties of the Climate Change Convention. This data is annually submitted to the Climate Change secretariat by national communications under the Convention by the Annex I parties.

Annex I parties - the industrialized countries listed in this annex to the Climate Change Convention tried to return their greenhouse gas emissions to 1990 levels by the year 2000. They have also accepted emission targets for the period 2008-12. They include the following members Australia, Austria, Belarus, Belgium, Bulgaria, Canada, Czech Republic, Denmark, European Economic Community, Estonia, Finland, France, Germany, Greece, Hungary, Iceland, Ireland, Italy, Japan, Latvia, Lithuania, Luxembourg, Netherlands, New Zealand, Norway, Poland, Portugal, Romania, Russian Federation, Slovakia, Slovenia, Spain, Sweden, Switzerland, Turkey, Ukraine, United Kingdom of Great Britain and Northern Ireland and the United States of America.

Greenhouse gases (GHGs) - the major anthropogenic GHGs responsible for causing climate change are carbon dioxide (CO_2), methane (CH_4), and nitrous oxide (N_2O). The Kyoto Protocol also addresses hydrofluorocarbons (HFCs), perfluorocarbons (PFCs), and sulphur hexafluoride (SF_6). For more detailed information about the GHG inventory database see: http://ghg.unfccc.int.

1 gigagram corresponds to 1 thousand metric tonne.

AGGREGATIONS

Emissions of CO$_2$ - from Transport (UNFCCC)

Units: gigagrams of CO$_2$

Data Source: Greenhouse Gas (GHG) Inventory Submission 1998, 1999, and 2000

Data Provider: United Nations Framework Convention on Climate Change (UNFCCC) Secretariat

Years: 1990-1998

	1990	1991	1992	1993	1994	1995	1996	1997	1998
AFRICA	----	----	----	----	----	----	----	----	----
Central Africa	----	----	----	----	----	----	----	----	----
Eastern Africa	----	----	----	----	----	----	----	----	----
Northern Africa	----	----	----	----	----	----	----	----	----
Southern Africa	----	----	----	----	----	----	----	----	----
Western Africa	----	----	----	----	----	----	----	----	----
Western Indian Ocean	----	----	----	----	----	----	----	----	----
ASIA AND PACIFIC	----	----	----	----	----	----	----	----	----
Australia and New Zealand	67948.06	67238.44	69141.91	70290.10	72538.03	75738.77	77904.03	79729.21	79868.91
Central Asia	----	----	----	----	----	----	----	----	----
North West Pacific and East Asia	----	----	----	----	----	----	----	----	----
South Asia	----	----	----	----	----	----	----	----	----
South East Asia	----	----	----	----	----	----	----	----	----
South Pacific	----	----	----	----	----	----	----	----	----
EUROPE	804471.05	800990.62	826393.70	831154.06	839519.68	844287.87	865668.98	878585.32	901567.73
Central Europe	85103.76	69138.72	70620.23	69332.83	70188.90	65914.72	68734.71	69767.27	71801.95
Eastern Europe	----	----	----	----	----	----	----	----	----
Western Europe	719367.30	731851.90	755773.47	761821.24	769330.78	778373.15	796934.27	808818.05	829765.78
LATIN AMERICA AND CARIBBEAN	----	----	----	----	----	----	----	----	----
Caribbean	----	----	----	----	----	----	----	----	----
Meso-America	----	----	----	----	----	----	----	----	----
South America	----	----	----	----	----	----	----	----	----
NORTH AMERICA	1559196.16	1522096.75	1565630.72	1599083.90	1649829.85	1683237.79	1734147.63	1752909.20	1781833.27
North America	1559196.16	1522096.75	1565630.72	1599083.90	1649829.85	1683237.79	1734147.63	1752909.20	1781833.27
POLAR	----	----	----	----	----	----	----	----	----
Antarctic	----	----	----	----	----	----	----	----	----
Arctic	----	----	----	----	----	----	----	----	----
WEST ASIA	----	----	----	----	----	----	----	----	----
Arabian Peninsula	----	----	----	----	----	----	----	----	----
Mashriq	----	----	----	----	----	----	----	----	----
GLOBAL TOTALS	----	----	----	----	----	----	----	----	----
REPORTED GLOBAL TOTALS FROM ORIGINAL DATA SOURCES	****	****	****	****	****	****	****	****	****
SUM OF ANNEX I PARTIES (INCLUDES ESTIMATIONS)	2637247.88	2605638.51	2681639.29	2723002.11	2795312.89	2843556.88	2924594.19	2962599.63	3014645.81

Comments:

Emissions of CO$_2$ - Transport (UNFCCC): emissions of CO$_2$ (carbon dioxide) from transport exclude emissions from international bunkers (aviation and marine). The data is based on the national greenhouse gas inventories given by the parties of the Climate Change Convention Secretariat. This data is annually submitted to the Climate Change Secretariat by national communications under the Convention by the Annex I parties.

Annex I parties - the industrialized countries listed in this annex to the Climate Change Convention tried to return their greenhouse gas emissions to 1990 levels by the year 2000. They have also accepted emission targets for the period 2008-12. They include the following members Australia, Austria, Belarus, Belgium, Bulgaria, Canada, Czech Republic, Denmark, European Economic Community, Estonia, Finland, France, Germany, Greece, Hungary, Iceland, Ireland, Italy, Japan, Latvia, Lithuania, Luxembourg, Netherlands, New Zealand, Norway, Poland, Portugal, Romania, Russian Federation, Slovakia, Slovenia, Spain, Sweden, Switzerland, Turkey, Ukraine, United Kingdom of Great Britain and Northern Ireland and the United States of America.

Greenhouse gases (GHGs) - the major anthropogenic GHGs responsible for causing climate change are carbon dioxide (CO$_2$), methane (CH$_4$), and nitrous oxide (N$_2$O). The Kyoto Protocol also addresses hydrofluorocarbons (HFCs), perfluorocarbons (PFCs), and sulphur hexafluoride (SF$_6$). For more detailed information about the GHG inventory database see: http://ghg.unfccc.int.

1 gigagram corresponds to 1 thousand metric tonne.

AGGREGATIONS

Emissions of CO_2, CH_4, N_2O, HFCs, PFCs and SF_6 - Aggregated (UNFCCC) Excluding CO_2 from Land-Use Change and Forestry

Units: gigagrams equivalent of CO_2

Data Source: Greenhouse Gas (GHG) Inventory Submission 1998, 1999, and 2000

Data Provider: United Nations Framework Convention on Climate Change (UNFCCC) Secretariat

Years: 1990-1998

	1990	1991	1992	1993	1994	1995	1996	1997	1998
AFRICA	----	----	----	----	----	----	----	----	----
Central Africa	----	----	----	----	----	----	----	----	----
Eastern Africa	----	----	----	----	----	----	----	----	----
Northern Africa	----	----	----	----	----	----	----	----	----
Southern Africa	----	----	----	----	----	----	----	----	----
Western Africa	----	----	----	----	----	----	----	----	----
Western Indian Ocean	----	----	----	----	----	----	----	----	----
ASIA AND PACIFIC	----	----	----	----	----	----	----	----	----
Australia and New Zealand	496305.09	495600.86	498337.75	499831.32	502131.60	515567.58	527793.41	538393.55	559585.09
Central Asia	----	----	----	----	----	----	----	----	----
North West Pacific and East Asia	----	----	----	----	----	----	----	----	----
South Asia	----	----	----	----	----	----	----	----	----
South East Asia	----	----	----	----	----	----	----	----	----
South Pacific	----	----	----	----	----	----	----	----	----
EUROPE	9776986.93	8874585.97	8369437.98	8051615.89	7914174.70	7828985.68	7796659.57	7681010.56	7646411.44
Central Europe	1501177.15	1172220.54	1107464.93	1067735.84	1046607.09	1028259.91	1053132.55	1026756.07	985177.21
Eastern Europe	3959281.95	----	----	----	2722014.34	2604544.34	2462074.94	----	----
Western Europe	4316527.84	4318254.69	4216882.81	4125579.52	4145553.27	4196181.44	4281452.08	4225342.96	4243859.60
LATIN AMERICA AND CARIBBEAN	----	----	----	----	----	----	----	----	----
Caribbean	----	----	----	----	----	----	----	----	----
Meso-America	----	----	----	----	----	----	----	----	----
South America	----	----	----	----	----	----	----	----	----
NORTH AMERICA	6660556.21	6605138.55	6731554.42	6855919.87	6998058.16	7069807.65	7292616.29	7389494.89	7419226.56
North America	6660556.21	6605138.55	6731554.42	6855919.87	6998058.16	7069807.65	7292616.29	7389494.89	7419226.56
POLAR	----	----	----	----	----	----	----	----	----
Antarctic	----	----	----	----	----	----	----	----	----
Arctic	----	----	----	----	----	----	----	----	----
WEST ASIA	----	----	----	----	----	----	----	----	----
Arabian Peninsula	----	----	----	'	----	----	----	----	----
Mashriq	----	----	----	----	----	----	----	----	----
GLOBAL TOTALS	----	----	----	----	----	----	----	----	----
REPORTED GLOBAL TOTALS FROM ORIGINAL DATA SOURCES	****	****	****	****	****	****	****	****	****
SUM OF ANNEX I PARTIES (INCLUDES ESTIMATIONS)	18147110.45	14615740.57	14578931.34	14465828.47	16723873.82	16736654.04	16943096.65	14966790.30	14977419.21

Comments:

The Global Warming Potential (GWP) is an index used to translate the level of emissions of various gases into a common measure in order to compare the relative radiative forcing of different gases without directly calculating the changes in atmospheric concentrations. GWPs are calculated as the ratio of the radiative forcing that would result from the emissions of one kilogram of a greenhouse gas to that from the emission of one kilogram of carbon dioxide over a period of time (usually 100 years). Gases involved in complex atmospheric chemical processes have not been assigned GWPs.

The data is based on the national greenhouse gas inventories given by the parties of the Climate Change Convention Secretariat. This data is annually submitted to the Climate Change Secretariat by national communications under the Convention by the Annex I parties: the industrialized countries listed in this annex to the Climate Change Convention tried to return their greenhouse gas emissions to 1990 levels by the year 2000. They have also accepted emission targets for the period 2008-12. They include the following members Australia, Austria, Belarus, Belgium, Bulgaria, Canada, Czech Republic, Denmark, European Economic Community, Estonia, Finland, France, Germany, Greece, Hungary, Iceland, Ireland, Italy, Japan, Latvia, Lithuania, Luxembourg, Netherlands, New Zealand, Norway, Poland, Portugal, Romania, Russian Federation, Slovakia, Slovenia, Spain, Sweden, Switzerland, Turkey, Ukraine, United Kingdom of Great Britain and Northern Ireland and the United States of America. Greenhouse gases (GHGs) - the major anthropogenic GHGs responsible for causing climate change are water vapor, carbon dioxide (CO_2), methane (CH_4), and nitrous oxide (N_2O). The Kyoto Protocol also addresses hydrofluorocarbons (HFCs), perfluorocarbons (PFCs), and sulphur hexafluoride (SF_6). For more detailed information about the GHG inventory database see: http://ghg.unfccc.int.

1 gigagram corresponds to 1 thousand metric tonne. CO_2 equivalent is a metric measure used to compare the emissions of the different greenhouse gases based upon their global warming potential (GWP).

AGGREGATIONS

Emissions of CO_2, CH_4, N_2O, HFCs, PFCs and SF_6 - Aggregated (UNFCCC) Including CO_2 from Land-Use Change and Forestry

Units: gigagrams equivalent of CO_2

Data Source: Greenhouse Gas (GHG) Inventory Submission 1998, 1999, and 2000

Data Provider: United Nations Framework Convention on Climate Change (UNFCCC) Secretariat

Years: 1990-1998

	1990	1991	1992	1993	1994	1995	1996	1997	1998
AFRICA	----	----	----	----	----	----	----	----	----
Central Africa	----	----	----	----	----	----	----	----	----
Eastern Africa	----	----	----	----	----	----	----	----	----
Northern Africa	----	----	----	----	----	----	----	----	----
Southern Africa	----	----	----	----	----	----	----	----	----
Western Africa	----	----	----	----	----	----	----	----	----
Western Indian Ocean	----	----	----	----	----	----	----	----	----
ASIA AND PACIFIC	----	----	----	----	----	----	----	----	----
Australia and New Zealand	544866.78	518525.02	519996.33	521058.68	522417.59	534781.42	549208.46	556772.19	573862.22
Central Asia	----	----	----	----	----	----	----	----	----
North West Pacific and East Asia	----	----	----	----	----	----	----	----	----
South Asia	----	----	----	----	----	----	----	----	----
South East Asia	----	----	----	----	----	----	----	----	----
South Pacific	----	----	----	----	----	----	----	----	----
EUROPE	9027330.95	7996103.50	7672604.80	7184100.21	6974078.51	6619794.37	6565114.34	6467973.91	6448369.48
Central Europe	1417761.19	1087114.58	1024898.94	988950.18	957260.63	935897.93	964666.72	952583.60	920913.20
Eastern Europe	3515174.85	----	----	----	2101293.44	1711604.44	----	----	----
Western Europe	4094394.91	4077841.67	3985273.61	3889766.89	3915524.44	3972292.00	4044523.18	3995285.06	4018789.86
LATIN AMERICA AND CARIBBEAN	----	----	----	----	----	----	----	----	----
Caribbean	----	----	----	----	----	----	----	----	----
Meso-America	----	----	----	----	----	----	----	----	----
South America	----	----	----	----	----	----	----	----	----
NORTH AMERICA	5461420.78	5388222.88	5526904.18	6041406.20	6190043.96	6272021.16	6489004.94	6591786.92	6624374.41
North America	5461420.78	5388222.88	5526904.18	6041406.20	6190043.96	6272021.16	6489004.94	6591786.92	6624374.41
POLAR	----	----	----	----	----	----	----	----	----
Antarctic	----	----	----	----	----	----	----	----	----
Arctic	----	----	----	----	----	----	----	----	----
WEST ASIA	----	----	----	----	----	----	----	----	----
Arabian Peninsula	----	----	----	----	----	----	----	----	----
Mashriq	----	----	----	----	----	----	----	----	----
GLOBAL TOTALS	----	----	----	----	----	----	----	----	----
REPORTED GLOBAL TOTALS FROM ORIGINAL DATA SOURCES	****	****	****	****	****	****	****	****	****
SUM OF ANNEX I PARTIES (INCLUDES ESTIMATIONS)	16162977.65	15059401.43	14893538.74	14894942.12	14902504.59	14652185.03	14838539.37	14861368.18	14901064.84

Comments:

The Global Warming Potential (GWP) is an index used to translate the level of emissions of various gases into a common measure in order to compare the relative radiative forcing of different gases without directly calculating the changes in atmospheric concentrations. GWPs are calculated as the ratio of the radiative forcing that would result from the emissions of one kilogram of a greenhouse gas to that from the emission of one kilogram of carbon dioxide over a period of time (usually 100 years). Gases involved in complex atmospheric chemical processes have not been assigned GWPs.

The data is based on the national greenhouse gas inventories given by the parties of the Climate Change Convention Secretariat. This data is annually submitted to the Climate Change Secretariat by national communications under the Convention by the Annex I parties: the industrialized countries listed in this annex to the Climate Change Convention tried to return their greenhouse gas emissions to 1990 levels by the year 2000. They have also accepted emission targets for the period 2008-12. They include the following members Australia, Austria, Belarus, Belgium, Bulgaria, Canada, Czech Republic, Denmark, European Economic Community, Estonia, Finland, France, Germany, Greece, Hungary, Iceland, Ireland, Italy, Japan, Latvia, Lithuania, Luxembourg, Netherlands, New Zealand, Norway, Poland, Portugal, Romania, Russian Federation, Slovakia, Slovenia, Spain, Sweden, Switzerland, Turkey, Ukraine, United Kingdom of Great Britain and Northern Ireland and the United States of America. Greenhouse gases (GHGs) - the major anthropogenic GHGs responsible for causing climate change are water vapor, carbon dioxide (CO_2), methane (CH_4), and nitrous oxide (N_2O). The Kyoto Protocol also addresses hydrofluorocarbons (HFCs), perfluorocarbons (PFCs), and sulphur hexafluoride (SF_6). For more detailed information about the GHG inventory database see: http://ghg.unfccc.int.

1 gigagram corresponds to 1 thousand metric tonne. CO_2 equivalent is a metric measure used to compare the emissions of the different greenhouse gases based upon their global warming potential (GWP).

AGGREGATIONS

Emissions of HFCs, PFCs and SF$_6$ - Aggregated (UNFCCC)

Units: gigagrams equivalent of CO$_2$

Data Source: Greenhouse Gas (GHG) Inventory Submission 1998, 1999, and 2000

Data Provider: United Nations Framework Convention on Climate Change (UNFCCC) Secretariat

Years: 1990-1998

	1990	1991	1992	1993	1994	1995	1996	1997	1998
AFRICA	----	----	----	----	----	----	----	----	----
Central Africa	----	----	----	----	----	----	----	----	----
Eastern Africa	----	----	----	----	----	----	----	----	----
Northern Africa	----	----	----	----	----	----	----	----	----
Southern Africa	----	----	----	----	----	----	----	----	----
Western Africa	----	----	----	----	----	----	----	----	----
Western Indian Ocean	----	----	----	----	----	----	----	----	----
ASIA AND PACIFIC	----	----	----	----	----	----	----	----	----
Australia and New Zealand	5431.46	5478.85	4220.23	3388.72	2400.41	1849.92	1799.62	1607.54	1884.96
Central Asia	----	----	----	----	----	----	----	----	----
North West Pacific and East Asia	----	----	----	----	----	----	----	----	----
South Asia	----	----	----	----	----	----	----	----	----
South East Asia	----	----	----	----	----	----	----	----	----
South Pacific	----	----	----	----	----	----	----	----	----
EUROPE	----	----	----	----	----	----	----	----	----
Central Europe	----	----	----	----	----	----	----	----	----
Eastern Europe	----	----	----	----	----	----	----	----	----
Western Europe	51137.75	48686.07	45836.46	47943.47	53092.07	61786.92	65371.98	72250.41	75636.81
LATIN AMERICA AND CARIBBEAN	----	----	----	----	----	----	----	----	----
Caribbean	----	----	----	----	----	----	----	----	----
Meso-America	----	----	----	----	----	----	----	----	----
South America	----	----	----	----	----	----	----	----	----
NORTH AMERICA	94287.75	90192.90	95060.58	96795.37	100984.28	114759.26	131050.10	137788.04	156199.21
North America	94287.75	90192.90	95060.58	96795.37	100984.28	114759.26	131050.10	137788.04	156199.21
POLAR	----	----	----	----	----	----	----	----	----
Antarctic	----	----	----	----	----	----	----	----	----
Arctic	----	----	----	----	----	----	----	----	----
WEST ASIA	----	----	----	----	----	----	----	----	----
Arabian Peninsula	----	----	----	----	----	----	----	----	----
Mashriq	----	----	----	----	----	----	----	----	----
GLOBAL TOTALS	----	----	----	----	----	----	----	----	----
REPORTED GLOBAL TOTALS FROM ORIGINAL DATA SOURCES	****	****	****	****	****	****	****	****	****
SUM OF ANNEX I PARTIES (INCLUDES ESTIMATIONS)	230663.93	223894.98	230746.34	231693.45	240634.20	263005.63	284679.36	298126.98	321766.08

Comments:

The Global Warming Potential (GWP) is an index used to translate the level of emissions of various gases into a common measure in order to compare the relative radiative forcing of different gases without directly calculating the changes in atmospheric concentrations. GWPs are calculated as the ratio of the radiative forcing that would result from the emissions of one kilogram of a greenhouse gas to that from the emission of one kilogram of carbon dioxide over a period of time (usually 100 years). Gases involved in complex atmospheric chemical processes have not been assigned GWPs.

The data is based on the national greenhouse gas inventories given by the parties of the Climate Change Convention Secretariat. This data is annually submitted to the Climate Change Secretariat by national communications under the Convention by the Annex I parties: the industrialized countries listed in this annex to the Climate Change Convention tried to return their greenhouse gas emissions to 1990 levels by the year 2000. They have also accepted emission targets for the period 2008-12. They include the following members Australia, Austria, Belarus, Belgium, Bulgaria, Canada, Czech Republic, Denmark, European Economic Community, Estonia, Finland, France, Germany, Greece, Hungary, Iceland, Ireland, Italy, Japan, Latvia, Lithuania, Luxembourg, Netherlands, New Zealand, Norway, Poland, Portugal, Romania, Russian Federation, Slovakia, Slovenia, Spain, Sweden, Switzerland, Turkey, Ukraine, United Kingdom of Great Britain and Northern Ireland and the United States of America. Greenhouse gases (GHGs) - the major anthropogenic GHGs responsible for causing climate change are water vapor, carbon dioxide (CO$_2$), methane (CH$_4$), and nitrous oxide (N$_2$O). The Kyoto Protocol also addresses hydrofluorocarbons (HFCs), perfluorocarbons (PFCs), and sulphur hexafluoride (SF$_6$). For more detailed information about the GHG inventory database see: http://ghg.unfccc.int.

1 gigagram corresponds to 1 thousand metric tonne. CO$_2$ equivalent is a metric measure used to compare the emissions of the different greenhouse gases based upon their global warming potential (GWP).

AGGREGATIONS

Units: thousand metric tonnes of N$_2$O

Data Source: Emission Database for Global Atmospheric Research (EDGAR 3.2)

Data Provider: Netherlands Organisation for Applied Scientific Research (TNO)
                and National Institute of Public Health and the Environment (RIVM)

Years: 1970, 1975, ..., 1995

	1970	1975	1980	1985	1990	1995
AFRICA	892.92	939.71	1047.85	1096.98	1204.27	1243.17
Central Africa	57.30	57.32	68.91	70.48	84.47	90.12
Eastern Africa	280.54	291.00	307.97	336.03	348.04	338.76
Northern Africa	182.30	205.66	229.98	245.54	260.96	279.25
Southern Africa	193.58	208.04	226.31	222.38	241.67	231.55
Western Africa	150.36	148.51	181.23	189.05	235.80	270.03
Western Indian Ocean	28.85	29.18	33.45	33.49	33.33	33.47
ASIA AND PACIFIC	1994.91	2253.38	2654.69	2980.41	3574.45	4027.27
Australia and New Zealand	438.56	441.72	413.23	424.01	444.86	398.26
Central Asia	120.60	141.59	149.09	164.65	154.39	118.71
North West Pacific and East Asia	671.54	800.33	1089.59	1148.26	1501.51	1785.64
South Asia	583.09	673.64	764.53	934.97	1109.57	1285.04
South East Asia	173.26	186.47	224.49	290.64	337.99	402.40
South Pacific	7.86	9.63	13.77	17.89	26.13	37.21
EUROPE	1562.11	1794.58	1933.65	2088.03	2007.63	1571.49
Central Europe	391.15	462.56	492.46	492.91	457.40	345.86
Eastern Europe	445.97	534.91	579.07	667.64	613.62	328.07
Western Europe	724.98	797.11	862.12	927.49	936.61	897.55
LATIN AMERICA AND CARIBBEAN	973.09	1101.91	1251.64	1351.99	1443.26	1512.73
Caribbean	65.74	60.92	67.69	72.24	70.81	56.72
Meso-America	186.46	208.25	229.71	260.42	263.19	248.33
South America	720.89	832.74	954.23	1019.34	1109.26	1207.69
NORTH AMERICA	962.41	1173.51	1153.23	1149.00	1106.46	1193.46
North America	962.41	1173.51	1153.23	1149.00	1106.46	1193.46
POLAR	3.84	5.16	7.42	10.59	13.70	20.26
Antarctic	3.84	5.16	7.42	10.59	13.70	20.26
Arctic	----	----	----	----	0.00	0.00
WEST ASIA	57.08	60.99	74.35	87.31	98.66	100.16
Arabian Peninsula	18.68	24.09	30.30	40.85	48.18	49.02
Mashriq	38.40	36.90	44.05	46.46	50.48	51.15
GLOBAL TOTALS	6446.36	7329.25	8122.82	8764.31	9448.42	9668.54
REPORTED GLOBAL TOTALS FROM ORIGINAL DATA SOURCES	6447.04	7329.37	8123.00	8764.47	9430.24	9649.86

Comments:

Emissions of N$_2$O - from agriculture (RIVM). Emissions of CH4 (methane) from the agricultural sector include emissions from: arable land (fertiliser use) (IPCC 4.D); animal waste management (IPCC 4.B); savannah burning (IPCC 4.E); agricultural waste burning (IPCC 4.F); crop production (IPCC 4.D); animal waste (deposited on soil - N$_2$O) (IPCC 4.B); atmospheric deposition (IPCC 4.D); leaching and run-off (IPCC 4.D). The emissions from deforestation (IPCC 5.A.1), vegetation fires (IPCC 5.A.2,3) and deforestation post burn effects (IPCC 5B1) are not included.

The EDGAR set of inventories were compiled from the perspective of providing good quality reference estimates of anthropogenic emission sources per source category, based on scientifically sound input data. This was done by using (a) international statistics as activity data, since these are comparable between countries in definition and units, (b) emission factors from the scientific literature, also common across countries when judged comparable, and (c) grid maps for allocating sectoral emissions of a country to a grid. EDGAR 3.2 by RIVM/TNO (Olivier, J.G.J. and Berdowski, J.J.M., 2001, Global emission sources and sinks. In: J. Berdowski, R. Guicherit and B.J. Heij (eds.), The Climate System: 33-77. Lisse: Swets and Zeitlinger Publishers).

For more detailed information about the EDGAR 3.2 database and data source see:
http://www.rivm.nl/env/int/coredata/edgar/index.html.

1 thousand metric tonne corresponds to 1 gigagram.

Units: thousand metric tonnes of N$_2$O

Data Source: Emission Database for Global Atmospheric Research (EDGAR 3.2)

Data Provider: Netherlands Organisation for Applied Scientific Research (TNO)
and National Institute of Public Health and the Environment (RIVM)

Years: 1970, 1975, ..., 1995

	1970	1975	1980	1985	1990	1995
AFRICA	22.91	26.04	30.09	34.14	38.44	43.15
Central Africa	1.42	1.59	1.83	2.10	2.40	2.81
Eastern Africa	3.18	3.61	4.18	4.77	5.47	5.99
Northern Africa	3.44	3.97	4.84	5.70	6.32	6.96
Southern Africa	7.08	8.06	9.19	10.28	11.49	12.96
Western Africa	7.39	8.35	9.52	10.68	12.06	13.63
Western Indian Ocean	0.41	0.46	0.53	0.60	0.70	0.80
ASIA AND PACIFIC	99.05	112.78	129.35	148.87	175.06	200.92
Australia and New Zealand	1.93	2.35	2.53	2.73	4.34	5.78
Central Asia	1.83	2.86	2.93	2.81	3.37	3.25
North West Pacific and East Asia	43.79	49.55	59.45	70.22	84.16	98.06
South Asia	34.30	38.78	42.98	49.42	56.55	64.12
South East Asia	16.94	18.95	21.16	23.35	26.27	29.30
South Pacific	0.26	0.28	0.31	0.34	0.37	0.40
EUROPE	72.64	77.32	84.18	81.92	83.76	73.94
Central Europe	12.00	13.93	15.76	15.80	14.26	12.86
Eastern Europe	26.35	31.03	33.40	32.98	34.07	22.02
Western Europe	34.29	32.35	35.02	33.15	35.44	39.06
LATIN AMERICA AND CARIBBEAN	13.83	14.94	16.77	17.86	18.97	21.04
Caribbean	1.94	1.78	1.92	1.89	2.14	2.02
Meso-America	3.14	3.82	4.38	4.85	5.46	5.70
South America	8.75	9.34	10.48	11.12	11.37	13.31
NORTH AMERICA	40.96	43.74	48.16	75.58	116.78	122.81
North America	40.96	43.74	48.16	75.58	116.78	122.81
POLAR	0.00	0.00	0.00	0.00	0.00	0.00
Antarctic	0.00	0.00	0.00	0.00	0.00	0.00
Arctic	0.00	0.00	0.00	0.00	0.00	0.00
WEST ASIA	1.58	1.85	2.78	3.41	3.76	4.27
Arabian Peninsula	1.02	1.15	1.86	2.21	2.28	2.55
Mashriq	0.56	0.70	0.92	1.19	1.48	1.71
GLOBAL TOTALS	250.96	276.68	311.34	361.78	436.77	466.13
REPORTED GLOBAL TOTALS FROM ORIGINAL DATA SOURCES	220.51	276.88	311.59	362.20	439.58	469.38

Comments:

Emissions of N$_2$O - energy (RIVM). "energy" comprises combustion of fossil fuels and biofuels (IPCC category 1.A and 1.B) as follows: industrial sector (IPCC 1.A.2); power generation (IPCC 1.A.1(a)); other transformation sector (all) (IPCC 1.A.1(b,c)); residentials, commercials and other sector (IPCC 1.A.4); transport road (IPCC 1.A.3(b)); transport land non-road (IPCC 1.A.3(c, d-ii,e)); transport air (all) (IPCC 1.A.3(a)); transport international shipping (IPCC 1.A.3(d-I)); oil production, handling and flaring (IPCC 1.B.2(a,c)); industry v3 (IPCC 1.A.2); power generation (IPCC 1.A.1(a)); charcoal production (IPCC 1.B.1(b)); residential (IPCC 1.A.4); transport road (IPCC 1.A.3(b)).

The EDGAR set of inventories were compiled from the perspective of providing good quality reference estimates of anthropogenic emission sources per source category, based on scientifically sound input data. This was done by using (a) international statistics as activity data, since these are comparable between countries in definition and units, (b) emission factors from the scientific literature, also common across countries when judged comparable, and (c) grid maps for allocating sectoral emissions of a country to a grid. EDGAR 3.2 by RIVM/TNO (Olivier, J.G.J. and Berdowski, J.J.M., 2001, Global Emission Sources and Sinks. In: J. Berdowski, R. Guicherit and B.J. Heij (eds.), The Climate System: 33-77. Lisse: Swets and Zeitlinger Publishers).

For more detailed information about the EDGAR 3.2 database and data source see: http://www.rivm.nl/env/int/coredata/edgar/index.html.

1 thousand metric tonne corresponds to 1 gigagram.

AGGREGATIONS

Emissions of N$_2$O - Total (RIVM)

Units: thousand metric tonnes of N$_2$O

Data Source: Emission Database for Global Atmospheric Research (EDGAR 3.2)

Data Provider: Netherlands Organisation for Applied Scientific Research (TNO)
and National Institute of Public Health and the Environment (RIVM)

Years: 1970, 1975, ..., 1995

	1970	1975	1980	1985	1990	1995
AFRICA	1015.51	1067.26	1179.86	1230.28	1342.55	1385.61
Central Africa	107.45	107.64	119.47	121.28	135.62	141.63
Eastern Africa	285.33	296.23	313.76	342.72	355.76	347.27
Northern Africa	196.13	219.84	244.66	262.23	278.08	294.70
Southern Africa	223.28	240.01	260.22	254.29	272.75	263.18
Western Africa	171.34	170.82	204.42	212.53	263.24	301.42
Western Indian Ocean	32.00	32.72	37.33	37.23	37.09	37.41
ASIA AND PACIFIC	2312.84	2607.88	3031.14	3361.93	3967.81	4436.71
Australia and New Zealand	444.38	449.27	420.10	430.31	452.38	406.99
Central Asia	142.81	165.61	172.09	182.57	170.21	125.24
North West Pacific and East Asia	758.99	902.69	1208.01	1279.82	1647.21	1947.29
South Asia	635.28	742.26	838.98	1011.36	1186.75	1367.93
South East Asia	318.61	333.51	373.24	434.90	479.57	546.11
South Pacific	12.78	14.54	18.73	22.98	31.69	43.15
EUROPE	2325.48	2707.30	2810.56	2857.47	2615.42	2118.25
Central Europe	530.87	638.88	653.12	633.61	556.23	439.92
Eastern Europe	525.75	622.48	666.08	745.22	689.28	373.27
Western Europe	1268.86	1445.94	1491.36	1478.63	1369.91	1305.06
LATIN AMERICA AND CARIBBEAN	1292.38	1437.09	1593.71	1646.79	1690.91	1738.34
Caribbean	73.50	68.64	78.40	81.71	78.54	63.88
Meso-America	216.82	243.04	264.67	293.00	291.61	275.92
South America	1002.06	1125.40	1250.64	1272.08	1320.76	1398.54
NORTH AMERICA	1383.68	1604.12	1611.57	1555.27	1546.10	1729.02
North America	1383.68	1604.12	1611.57	1555.27	1546.10	1729.02
POLAR	3.84	5.16	7.42	10.59	13.70	20.26
Antarctic	3.84	5.16	7.42	10.59	13.70	20.26
Arctic	0.00	0.00	0.00	0.00	0.00	0.00
WEST ASIA	58.81	64.30	81.20	94.90	105.82	108.24
Arabian Peninsula	19.73	25.27	35.54	46.03	53.14	54.59
Mashriq	39.08	39.03	45.66	48.87	52.68	53.64
GLOBAL TOTALS	8392.53	9493.11	10315.46	10757.24	11282.31	11536.43
REPORTED GLOBAL TOTALS FROM ORIGINAL DATA SOURCES	8362.76	9493.43	10315.89	10757.82	11264.24	11518.30

Comments:

Emissions of N$_2$O - total (RIVM) include "energy", "agriculture", "waste" and "others" EDGAR subdivisions. Energy comprises combustion of fossil fuels and biofuels (IPCC category 1.A and 1.B); agriculture comprises fertiliser use (synthetic and animal manure), animal waste management, agricultural waste burning (non-energy, on-site) and savannah burning (IPCC category 4); waste comprises human sewage discharge and waste incineration (non-energy) (IPCC category 6); others include industrial process emissions, N$_2$O usage and tropical and temperate forest fires (IPCC categories 2, 3 and 5).

The EDGAR set of inventories were compiled from the perspective of providing good quality reference estimates of anthropogenic emission sources per source category, based on scientifically sound input data. This was done by using (a) international statistics as activity data, since these are comparable between countries in definition and units, (b) emission factors from the scientific literature, also common across countries when judged comparable, and (c) grid maps for allocating sectoral emissions of a country to a grid. EDGAR 3.2 by RIVM/TNO (Olivier, J.G.J. and Berdowski, J.J.M., 2001, Global Emission Sources and Sinks. In: J. Berdowski, R. Guicherit and B.J. Heij (eds.), The Climate System: 33-77. Lisse: Swets and Zeitlinger Publishers).

For more detailed information about the EDGAR 3.2 database and data source see:
http://www.rivm.nl/env/int/coredata/edgar/index.html.

1 thousand metric tonne corresponds to 1 gigagram.

AGGREGATIONS

Units: thousand metric tonnes of CO

Data Source: Emission Database for Global Atmospheric Research (EDGAR 3.2)

Data Provider: Netherlands Organisation for Applied Scientific Research (TNO)
and National Institute of Public Health and the Environment (RIVM)

Years: 1990, 1995

	1990	1995
AFRICA	177468.21	184603.43
Central Africa	36714.27	37419.44
Eastern Africa	19137.86	20223.88
Northern Africa	21067.59	21718.50
Southern Africa	55989.87	57883.29
Western Africa	41183.16	43833.86
Western Indian Ocean	3375.46	3524.45
ASIA AND PACIFIC	274307.85	288107.51
Australia and New Zealand	16022.50	16179.23
Central Asia	6994.38	3876.09
North West Pacific and East Asia	106349.58	111533.03
South Asia	76204.25	83512.85
South East Asia	66509.03	70655.21
South Pacific	2228.11	2351.09
EUROPE	145379.89	101944.91
Central Europe	21327.64	18071.53
Eastern Europe	70170.26	36817.67
Western Europe	53881.99	47055.71
LATIN AMERICA AND CARIBBEAN	128398.25	124563.16
Caribbean	3309.77	3182.36
Meso-America	20260.73	20927.23
South America	104827.75	100453.58
NORTH AMERICA	106910.35	142733.53
North America	106910.35	142733.53
POLAR	0.79	0.81
Antarctic	----	----
Arctic	0.63	0.65
WEST ASIA	8616.72	10461.30
Arabian Peninsula	5175.38	6603.13
Mashriq	3441.33	3858.17
GLOBAL TOTALS	841082.05	852414.64
REPORTED GLOBAL TOTALS FROM ORIGINAL DATA SOURCES	841082.16	852414.69

Comments:

Emissions of CO (carbon monoxide) - total (RIVM) include the following EDGAR subdivisions: "fuel combustion", "biofuel combustion", "fugitive", "industry", "solvent use", "agriculture", "waste" and "others". Fuel combustion refers to fossil fuel combustion and evaporation of NMVOC in road transport (part of IPCC category 1.A); biofuel combustion refers to traditional biofuels as well as to wood waste, paper, ethanol, etc. (part of IPCC category 1.A); fugitive comprises flaring and venting of associated gas in oil and gas production, handling/transmission losses of oil and charcoal production (IPCC category 1.B); industry refers to non-combustion industrial processes, excluding solvent use (IPCC category 2); solvent use refers to solvent use in industry and non-industry sectors (IPCC category 3); agriculture comprises agricultural waste burning (non-energy, on-site) and savannah burning (IPCC category 4); waste comprises waste incineration (non-energy) (uncontrolled residential burning and controlled non-residential burning) and hazardous waste handling (IPCC category 6); others comprises tropical forest fires and temperate forest fires (IPCC category 5.A).

The EDGAR set of inventories were compiled from the perspective of providing good quality reference estimates of anthropogenic emission sources per source category, based on scientifically sound input data. This was done by using (a) international statistics as activity data, since these are comparable between countries in definition and units, (b) emission factors from the scientific literature, also common across countries when judged comparable, and (c) grid maps for allocating sectoral emissions of a country to a grid. EDGAR 3.2 by RIVM/TNO (Olivier, J.G.J. and Berdowski, J.J.M., 2001, Global Emission Sources and Sinks. In: J. Berdowski, R. Guicherit and B.J. Heij (eds.), The Climate System: 33-77. Lisse: Swets and Zeitlinger Publishers).

For more detailed information about the EDGAR 3.2 database and data source see:
http://www.rivm.nl/env/int/coredata/edgar/index.html.

1 thousand metric tonne corresponds to 1 gigagram.

AGGREGATIONS

Units: thousand metric tonnes of NMVOC

Data Source: Emission Database for Global Atmospheric Research (EDGAR 3.2)

Data Provider: Netherlands Organisation for Applied Scientific Research (TNO)
and National Institute of Public Health and the Environment (RIVM)

Years: 1990, 1995

	1990	1995
AFRICA	19058.02	20142.40
Central Africa	3003.52	3096.99
Eastern Africa	1867.72	1993.57
Northern Africa	3966.55	4119.15
Southern Africa	4943.08	5214.22
Western Africa	4978.38	5398.63
Western Indian Ocean	298.78	319.85
ASIA AND PACIFIC	43955.51	48075.03
Australia and New Zealand	2536.40	2563.94
Central Asia	1857.01	1311.08
North West Pacific and East Asia	17780.46	19512.72
South Asia	12217.14	13676.18
South East Asia	9349.32	10768.79
South Pacific	215.19	242.31
EUROPE	43404.23	36782.04
Central Europe	4072.57	3775.40
Eastern Europe	22917.49	16054.36
Western Europe	16414.17	16952.28
LATIN AMERICA AND CARIBBEAN	15691.05	16823.42
Caribbean	675.72	646.91
Meso-America	3857.39	4033.43
South America	11157.95	12143.08
NORTH AMERICA	23228.31	28686.80
North America	23228.31	28686.80
POLAR	1.29	1.33
Antarctic	0.04	0.04
Arctic	1.25	1.28
WEST ASIA	7906.03	9123.39
Arabian Peninsula	6303.80	8026.61
Mashriq	1602.23	1096.79
GLOBAL TOTALS	153244.45	159634.41
REPORTED GLOBAL TOTALS FROM ORIGINAL DATA SOURCES	153244.49	159634.43

Comments:

Emissions of NMVOC (Non-Methane Volatile Organic Compounds) - total (RIVM) include the following EDGAR subdivisions: "fuel combustion", "biofuel combustion", "fugitive", "industry", "solvent use", "agriculture", "waste" and "others". Fuel combustion refers to fossil fuel combustion and evaporation of NMVOC in road transport (part of IPCC category 1.A); biofuel combustion refers to traditional biofuels as well as to wood waste, paper, ethanol, etc. (part of IPCC category 1.A); fugitive comprises flaring and venting of associated gas in oil and gas production, handling/transmission losses of oil and charcoal production (IPCC category 1.B); industry refers to non-combustion industrial processes, excluding solvent use (IPCC category 2); solvent use refers to solvent use in industry and non-industry sectors (IPCC category 3); agriculture comprises agricultural waste burning (non-energy, on-site) and savannah burning (IPCC category 4); waste comprises waste incineration (non-energy) (uncontrolled residential burning and controlled non-residential burning) and hazardous waste handling (IPCC category 6); others comprises tropical forest fires and temperate forest fires (IPCC category 5.A).

The EDGAR set of inventories were compiled from the perspective of providing good quality reference estimates of anthropogenic emission sources per source category, based on scientifically sound input data. This was done by using (a) international statistics as activity data, since these are comparable between countries in definition and units, (b) emission factors from the scientific literature, also common across countries when judged comparable, and (c) grid maps for allocating sectoral emissions of a country to a grid. EDGAR 3.2 by RIVM/TNO (Olivier, J.G.J. and Berdowski, J.J.M., 2001, Global Emission Sources and Sinks. In: J. Berdowski, R. Guicherit and B.J. Heij (eds.), The Climate System: 33-77. Lisse: Swets and Zeitlinger Publishers).

For more detailed information about the EDGAR 3.2 database and data source see:
http://www.rivm.nl/env/int/coredata/edgar/index.html.

1 thousand metric tonne corresponds to 1 gigagram.

Emissions of NO$_x$ - Total (RIVM)

Units: thousand metric tonnes of NO$_x$

Data Source: Emission Database for Global Atmospheric Research (EDGAR 3.2)

Data Provider: Netherlands Organisation for Applied Scientific Research (TNO)
and National Institute of Public Health and the Environment (RIVM)

Years: 1990, 1995

	1990	1995
AFRICA	9981.26	10416.71
Central Africa	1616.65	1637.07
Eastern Africa	890.22	923.99
Northern Africa	1547.47	1699.34
Southern Africa	3877.70	4023.63
Western Africa	1892.86	1972.93
Western Indian Ocean	156.36	159.76
ASIA AND PACIFIC	25431.70	32646.08
Australia and New Zealand	1967.58	2076.68
Central Asia	1158.15	1071.08
North West Pacific and East Asia	13199.12	17789.17
South Asia	5799.27	7519.21
South East Asia	3218.34	4095.72
South Pacific	89.24	94.22
EUROPE	32218.14	23126.50
Central Europe	4571.51	3727.64
Eastern Europe	13371.82	7984.11
Western Europe	14274.82	11414.75
LATIN AMERICA AND CARIBBEAN	8529.24	9311.25
Caribbean	539.42	531.24
Meso-America	1861.11	2057.79
South America	6128.70	6722.22
NORTH AMERICA	21513.29	21799.77
North America	21513.29	21799.77
POLAR	2.91	2.99
Antarctic	0.53	0.53
Arctic	2.38	2.46
WEST ASIA	1605.71	1967.36
Arabian Peninsula	986.63	1210.21
Mashriq	619.09	757.14
GLOBAL TOTALS	99282.25	99270.66
REPORTED GLOBAL TOTALS FROM ORIGINAL DATA SOURCES	99282.45	99270.83

Comments:

Emissions of NO$_x$ (nitrogen oxides) - total (RIVM) include the following EDGAR subdivisions: "fuel combustion", "biofuel combustion", "fugitive", "industry", "solvent use", "agriculture", "waste" and "others". Fuel combustion refers to fossil fuel combustion and evaporation of NMVOC in road transport (part of IPCC category 1.A); biofuel combustion refers to traditional biofuels as well as to wood waste, paper, ethanol, etc. (part of IPCC category 1.A); fugitive comprises flaring and venting of associated gas in oil and gas production, handling/transmission losses of oil and charcoal production (IPCC category 1.B); industry refers to non-combustion industrial processes, excluding solvent use (IPCC category 2); Solvent use refers to solvent use in industry and non-industry sectors (IPCC category 3); agriculture comprises agricultural waste burning (non-energy, on-site) and savannah burning (IPCC category 4); waste comprises waste incineration (non-energy) (uncontrolled residential burning and controlled non-residential burning) and hazardous waste handling (IPCC category 6); others comprises tropical forest fires and temperate forest fires (IPCC category 5.A).

The EDGAR set of inventories were compiled from the perspective of providing good quality reference estimates of anthropogenic emission sources per source category, based on scientifically sound input data. This was done by using (a) international statistics as activity data, since these are comparable between countries in definition and units, (b) emission factors from the scientific literature, also common across countries when judged comparable, and (c) grid maps for allocating sectoral emissions of a country to a grid. EDGAR 3.2 by RIVM/TNO (Olivier, J.G.J. and Berdowski, J.J.M., 2001, Global emission sources and sinks. In: J. Berdowski, R. Guicherit and B.J. Heij (eds.), The Climate System: 33-77. Lisse: Swets and Zeitlinger Publishers).

For more detailed information about the EDGAR 3.2 database and data source see:
http://www.rivm.nl/env/int/coredata/edgar/index.html.

1 thousand metric tonne corresponds to 1 gigagram.

AGGREGATIONS

Emissions of SO$_2$ - Total (RIVM)

Units: thousand metric tonnes of SO$_2$

Data Source: Emission Database for Global Atmospheric Research (EDGAR 3.2)

Data Provider: Netherlands Organisation for Applied Scientific Research (TNO)
and National Institute of Public Health and the Environment (RIVM)

Years: 1990, 1995

	1990	1995
AFRICA	7682.18	6876.34
Central Africa	701.91	474.54
Eastern Africa	284.36	303.20
Northern Africa	1599.29	1663.79
Southern Africa	4049.52	3280.59
Western Africa	997.21	1097.88
Western Indian Ocean	49.90	56.34
ASIA AND PACIFIC	47375.05	59169.43
Australia and New Zealand	1561.05	1658.55
Central Asia	3329.71	2624.51
North West Pacific and East Asia	32407.40	42316.80
South Asia	6990.55	8727.12
South East Asia	3049.66	3802.54
South Pacific	36.67	39.91
EUROPE	63614.57	42591.14
Central Europe	15566.82	12684.38
Eastern Europe	24248.16	13065.50
Western Europe	23799.58	16841.26
LATIN AMERICA AND CARIBBEAN	8804.55	10083.71
Caribbean	630.29	586.42
Meso-America	2240.82	2435.39
South America	5933.44	7061.89
NORTH AMERICA	24982.24	20902.83
North America	24982.24	20902.83
POLAR	0.19	0.18
Antarctic	0.02	0.02
Arctic	0.17	0.16
WEST ASIA	1821.31	2251.59
Arabian Peninsula	1067.02	1309.33
Mashriq	754.30	942.26
GLOBAL TOTALS	154280.08	141875.21
REPORTED GLOBAL TOTALS FROM ORIGINAL DATA SOURCES	154280.31	141875.42

Comments:

Emissions of SO$_2$ (sulphur dioxide) - total (RIVM) include the following EDGAR subdivisions: "fuel combustion", "biofuel combustion", "fugitive", "industry", "solvent use", "agriculture", "waste" and "others". Fuel combustion refers to fossil fuel combustion and evaporation of NMVOC in road transport (part of IPCC category 1.A); biofuel combustion refers to traditional biofuels as well as to wood waste, paper, ethanol, etc. (part of IPCC category 1.A); fugitive comprises flaring and venting of associated gas in oil and gas production, handling / transmission losses of oil and charcoal production (IPCC category 1.B); industry refers to non-combustion industrial processes, excluding solvent use (IPCC category 2); solvent use refers to solvent use in industry and non-industry sectors (IPCC category 3); agriculture comprises agricultural waste burning (non-energy, on-site) and savannah burning (IPCC category 4); waste comprises waste incineration (non-energy) (uncontrolled residential burning and controlled non-residential burning) and hazardous waste handling (IPCC category 6); others comprises tropical forest fires and temperate forest fires (IPCC category 5.A).

The EDGAR set of inventories were compiled from the perspective of providing good quality reference estimates of anthropogenic emission sources per source category, based on scientifically sound input data. This was done by using (a) international statistics as activity data, since these are comparable between countries in definition and units, (b) emission factors from the scientific literature, also common across countries when judged comparable, and (c) grid maps for allocating sectoral emissions of a country to a grid. EDGAR 3.2 by RIVM/TNO (Olivier, J.G.J. and Berdowski, J.J.M., 2001, Global emission sources and sinks. In: J. Berdowski, R. Guicherit and B.J. Heij (eds.), The Climate System: 33-77. Lisse: Swets and Zeitlinger Publishers).

For more detailed information about the EDGAR 3.2 database and data source see: http://www.rivm.nl/env/int/coredata/edgar/index.html.

1 thousand metric tonne corresponds to 1 gigagram.

Emissions of N$_2$O - from Agricultural Soils (UNFCCC)

Units: gigagrams of N$_2$O

Data Source: Greenhouse Gas (GHG) Inventory Submission 1998, 1999, and 2000

Data Provider: United Nations Framework Convention on Climate Change (UNFCCC) Secretariat

Years: 1990-1998

	1990	1991	1992	1993	1994	1995	1996	1997	1998
AFRICA	----	----	----	----	----	----	----	----	----
Central Africa	----	----	----	----	----	----	----	----	----
Eastern Africa	----	----	----	----	----	----	----	----	----
Northern Africa	----	----	----	----	----	----	----	----	----
Southern Africa	----	----	----	----	----	----	----	----	----
Western Africa	----	----	----	----	----	----	----	----	----
Western Indian Ocean	----	----	----	----	----	----	----	----	----
ASIA AND PACIFIC	----	----	----	----	----	----	----	----	----
Australia and New Zealand	83.72	83.70	83.30	84.55	85.25	84.52	83.83	86.89	89.05
Central Asia	----	----	----	----	----	----	----	----	----
North West Pacific and East Asia	----	----	----	----	----	----	----	----	----
South Asia	----	----	----	----	----	----	----	----	----
South East Asia	----	----	----	----	----	----	----	----	----
South Pacific	----	----	----	----	----	----	----	----	----
EUROPE	1015.16	897.01	842.55	805.14	811.13	811.41	821.18	799.72	856.96
Central Europe	179.19	131.82	121.32	110.97	107.83	108.70	99.72	74.29	133.54
Eastern Europe	199.00	----	----	----	111.00	111.00	105.00	----	----
Western Europe	636.96	613.19	593.23	580.17	592.29	591.71	616.46	620.42	618.42
LATIN AMERICA AND CARIBBEAN	----	----	----	----	----	----	----	----	----
Caribbean	----	----	----	----	----	----	----	----	----
Meso-America	----	----	----	----	----	----	----	----	----
South America	----	----	----	----	----	----	----	----	----
NORTH AMERICA	1006.95	1018.98	1039.17	1032.98	1112.71	1077.11	1106.83	1126.96	1122.66
North America	1006.95	1018.98	1039.17	1032.98	1112.71	1077.11	1106.83	1126.96	1122.66
POLAR	----	----	----	----	----	----	----	----	----
Antarctic	----	----	----	----	----	----	----	----	----
Arctic	----	----	----	----	----	----	----	----	----
WEST ASIA	----	----	----	----	----	----	----	----	----
Arabian Peninsula	----	----	----	----	----	----	----	----	----
Mashriq	----	----	----	----	----	----	----	----	----
GLOBAL TOTALS	----	----	----	----	----	----	----	----	----
REPORTED GLOBAL TOTALS FROM ORIGINAL DATA SOURCES	****	****	****	****	****	****	****	****	****
SUM OF ANNEX I PARTIES (INCLUDES ESTIMATIONS)	2109.61	2003.26	1968.59	1926.22	2012.58	1976.31	2013.94	2014.62	2069.56

Comments:

The 39 countries included in the database allow aggregations for five GEO sub-regions (Central, Eastern and Western Europe, North America and Australia + New Zealand) and two regions (Europe and North America) only. Data for the sub-region of Central Europe excludes data for: Albania, Bosnia and Herzegovina, Cyprus, the former Yugoslav Republic of Macedonia, Turkey and Yugoslavia. Data for the sub-region of Eastern Europe excludes data for: Armenia, Azerbaijan, Belarus, Georgia and the Republic of Moldova. Emissions of N$_2$O (Nitrous Oxide) from agriculture include emissions from manure management, agricultural soils, prescribed burning of savannas, field burning of agricultural residues and other.

Annex I Parties - the industrialized countries listed in this annex to the Climate Change Convention tried to return their greenhouse gas emissions to 1990 levels by the year 2000. They have also accepted emission targets for the period 2008-12. They include the following members Australia, Austria, Belarus, Belgium, Bulgaria, Canada, Czech Republic, Denmark, European Economic Community, Estonia, Finland, France, Germany, Greece, Hungary, Iceland, Ireland, Italy, Japan, Latvia, Lithuania, Luxembourg, Netherlands, New Zealand, Norway, Poland, Portugal, Romania, Russian Federation, Slovakia, Slovenia, Spain, Sweden, Switzerland, Turkey, Ukraine, United Kingdom of Great Britain and Northern Ireland and the United States of America.

Greenhouse gases (GHGs) - the major anthropogenic GHGs responsible for causing climate change are carbon dioxide (CO$_2$), methane (CH$_4$), and nitrous oxide (N$_2$O). The Kyoto Protocol also addresses hydrofluorocarbons (HFCs), perfluorocarbons (PFCs), and sulphur hexafluoride (SF$_6$).

For more detailed information about the GHG inventory database see: http://ghg.unfccc.int.

1 gigagram corresponds to 1 thousand metric tonne.

AGGREGATIONS

Emissions of N$_2$O - from Agriculture (UNFCCC)

Units: gigagrams of N$_2$O

Data Source: Greenhouse Gas (GHG) Inventory Submission 1998, 1999, and 2000

Data Provider: United Nations Framework Convention on Climate Change (UNFCCC) Secretariat

Years: 1990-1998

	1990	1991	1992	1993	1994	1995	1996	1997	1998
AFRICA	----	----	----	----	----	----	----	----	----
Central Africa	----	----	----	----	----	----	----	----	----
Eastern Africa	----	----	----	----	----	----	----	----	----
Northern Africa	----	----	----	----	----	----	----	----	----
Southern Africa	----	----	----	----	----	----	----	----	----
Western Africa	----	----	----	----	----	----	----	----	----
Western Indian Ocean	----	----	----	----	----	----	----	----	----
ASIA AND PACIFIC	----	----	----	----	----	----	----	----	----
Australia and New Zealand	97.88	97.92	97.31	99.28	100.53	100.86	100.91	104.19	108.34
Central Asia	----	----	----	----	----	----	----	----	----
North West Pacific and East Asia	----	----	----	----	----	----	----	----	----
South Asia	----	----	----	----	----	----	----	----	----
South East Asia	----	----	----	----	----	----	----	----	----
South Pacific	----	----	----	----	----	----	----	----	----
EUROPE	1151.81	1035.73	966.47	925.92	918.15	917.68	938.55	911.03	969.25
Central Europe	187.36	139.27	127.73	116.39	112.50	113.42	111.05	80.04	140.29
Eastern Europe	228.00	----	----	----	119.60	120.40	110.90	----	----
Western Europe	736.46	707.06	687.44	675.83	686.04	683.87	716.59	719.49	717.56
LATIN AMERICA AND CARIBBEAN	----	----	----	----	----	----	----	----	----
Caribbean	----	----	----	----	----	----	----	----	----
Meso-America	----	----	----	----	----	----	----	----	----
South America	----	----	----	----	----	----	----	----	----
NORTH AMERICA	1061.51	1076.11	1096.27	1091.86	1173.95	1138.40	1169.59	1191.11	1187.63
North America	1061.51	1076.11	1096.27	1091.86	1173.95	1138.40	1169.59	1191.11	1187.63
POLAR	----	----	----	----	----	----	----	----	----
Antarctic	----	----	----	----	----	----	----	----	----
Arctic	----	----	----	----	----	----	----	----	----
WEST ASIA	----	----	----	----	----	----	----	----	----
Arabian Peninsula	----	----	----	----	----	----	----	----	----
Mashriq	----	----	----	----	----	----	----	----	----
GLOBAL TOTALS	----	----	----	----	----	----	----	----	----
REPORTED GLOBAL TOTALS FROM ORIGINAL DATA SOURCES	****	****	****	****	****	****	****	****	****
SUM OF ANNEX I PARTIES (INCLUDES ESTIMATIONS)	2320.46	2218.81	2169.08	2126.07	2201.45	2165.46	2216.68	2212.83	2271.73

Comments:

Emissions of N$_2$O (nitrous oxide) from agriculture include emissions from manure management, agricultural soils, prescribed burning of savannas, field burning of agricultural residues and other. The data is based on the national greenhouse gas inventories given by the parties of the Climate Change Convention Secretariat. This data is submitted to the Climate Change Secretariat by national communications under the Convention by the Annex I parties.

Annex I Parties - the industrialized countries listed in this annex to the Climate Change Convention tried to return their greenhouse gas emissions to 1990 levels by the year 2000. They have also accepted emission targets for the period 2008-12. They include the following members Australia, Austria, Belarus, Belgium, Bulgaria, Canada, Czech Republic, Denmark, European Economic Community, Estonia, Finland, France, Germany, Greece, Hungary, Iceland, Ireland, Italy, Japan, Latvia, Lithuania, Luxembourg, Netherlands, New Zealand, Norway, Poland, Portugal, Romania, Russian Federation, Slovakia, Slovenia, Spain, Sweden, Switzerland, Turkey, Ukraine, United Kingdom of Great Britain and Northern Ireland and the United States of America.

Greenhouse gases (GHGs) - the major anthropogenic GHGs responsible for causing climate change are carbon dioxide (CO$_2$), methane (CH$_4$), and nitrous oxide (N$_2$O). The Kyoto Protocol also addresses hydrofluorocarbons (HFCs), perfluorocarbons (PFCs), and sulphur hexafluoride (SF$_6$).

For more detailed information about the GHG inventory database see: http://ghg.unfccc.int.

1 gigagram corresponds to 1 thousand metric tonne.

AGGREGATIONS

Emissions of N₂O - from Fuel Combustion (UNFCCC)

Units: gigagrams of N_2O

Data Source: Greenhouse Gas (GHG) Inventory Submission 1998, 1999, and 2000

Data Provider: United Nations Framework Convention on Climate Change (UNFCCC) Secretariat

Years: 1990-1998

	1990	1991	1992	1993	1994	1995	1996	1997	1998
AFRICA	----	----	----	----	----	----	----	----	----
Central Africa	----	----	----	----	----	----	----	----	----
Eastern Africa	----	----	----	----	----	----	----	----	----
Northern Africa	----	----	----	----	----	----	----	----	----
Southern Africa	----	----	----	----	----	----	----	----	----
Western Africa	----	----	----	----	----	----	----	----	----
Western Indian Ocean	----	----	----	----	----	----	----	----	----
ASIA AND PACIFIC	----	----	----	----	----	----	----	----	----
Australia and New Zealand	8.48	9.21	10.45	11.44	12.45	13.48	14.28	15.14	15.98
Central Asia	----	----	----	----	----	----	----	----	----
North West Pacific and East Asia	----	----	----	----	----	----	----	----	----
South Asia	----	----	----	----	----	----	----	----	----
South East Asia	----	----	----	----	----	----	----	----	----
South Pacific	----	----	----	----	----	----	----	----	----
EUROPE	264.65	244.46	242.43	240.76	241.36	251.98	244.53	247.06	227.11
Central Europe	71.16	52.53	52.20	51.56	50.53	50.74	38.86	37.86	34.42
Eastern Europe	24.10	16.90	15.00	14.30	11.80	11.80	10.26	----	----
Western Europe	169.39	175.03	175.24	174.91	179.03	189.44	195.41	199.04	182.53
LATIN AMERICA AND CARIBBEAN	----	----	----	----	----	----	----	----	----
Caribbean	----	----	----	----	----	----	----	----	----
Meso-America	----	----	----	----	----	----	----	----	----
South America	----	----	----	----	----	----	----	----	----
NORTH AMERICA	235.70	245.36	261.46	273.95	284.59	289.72	292.51	291.23	289.69
North America	235.70	245.36	261.46	273.95	284.59	289.72	292.51	291.23	289.69
POLAR	----	----	----	----	----	----	----	----	----
Antarctic	----	----	----	----	----	----	----	----	----
Arctic	----	----	----	----	----	----	----	----	----
WEST ASIA	----	----	----	----	----	----	----	----	----
Arabian Peninsula	----	----	----	----	----	----	----	----	----
Mashriq	----	----	----	----	----	----	----	----	----
GLOBAL TOTALS	----	----	----	----	----	----	----	----	----
REPORTED GLOBAL TOTALS FROM ORIGINAL DATA SOURCES	****	****	****	****	****	****	****	****	****
SUM OF ANNEX I PARTIES (INCLUDES ESTIMATIONS)	528.17	519.02	534.87	546.84	560.12	578.54	574.90	577.25	556.85

Comments:

Emissions of N_2O (nitrous oxide) - from fuel combustion derives from energy industries, manufacturing industries and construction, transport, other sectors. The data is based on the national greenhouse gas inventories given by the parties of the Climate Change Convention Secretariat. This data is submitted to the Climate Change Secretariat by national communications under the Convention by the Annex I parties.

Annex I Parties - the industrialized countries listed in this annex to the Climate Change Convention tried to return their greenhouse gas emissions to 1990 levels by the year 2000. They have also accepted emission targets for the period 2008-12. They include the following members Australia, Austria, Belarus, Belgium, Bulgaria, Canada, Czech Republic, Denmark, European Economic Community, Estonia, Finland, France, Germany, Greece, Hungary, Iceland, Ireland, Italy, Japan, Latvia, Lithuania, Luxembourg, Netherlands, New Zealand, Norway, Poland, Portugal, Romania, Russian Federation, Slovakia, Slovenia, Spain, Sweden, Switzerland, Turkey, Ukraine, United Kingdom of Great Britain and Northern Ireland and the United States of America.

Greenhouse gases (GHGs) - the major anthropogenic GHGs responsible for causing climate change are carbon dioxide (CO_2), methane (CH_4), and nitrous oxide (N_2O). The Kyoto Protocol also addresses hydrofluorocarbons (HFCs), perfluorocarbons (PFCs), and sulphur hexafluoride (SF_6).

For more detailed information about the GHG inventory database see: http://ghg.unfccc.int.

1 gigagram corresponds to 1 thousand metric tonne.

AGGREGATIONS

Emissions of N₂O - from Industrial Processes (UNFCCC)

Units: gigagrams of N_2O

Data Source: Greenhouse Gas (GHG) Inventory Submission 1998, 1999, and 2000

Data Provider: United Nations Framework Convention on Climate Change (UNFCCC) Secretariat

Years: 1990-1998

	1990	1991	1992	1993	1994	1995	1996	1997	1998
AFRICA	----	----	----	----	----	----	----	----	----
Central Africa	----	----	----	----	----	----	----	----	----
Eastern Africa	----	----	----	----	----	----	----	----	----
Northern Africa	----	----	----	----	----	----	----	----	----
Southern Africa	----	----	----	----	----	----	----	----	----
Western Africa	----	----	----	----	----	----	----	----	----
Western Indian Ocean	----	----	----	----	----	----	----	----	----
ASIA AND PACIFIC	----	----	----	----	----	----	----	----	----
Australia and New Zealand	1.62	1.51	1.83	1.60	1.39	1.40	1.56	1.56	1.65
Central Asia	----	----	----	----	----	----	----	----	----
North West Pacific and East Asia	----	----	----	----	----	----	----	----	----
South Asia	----	----	----	----	----	----	----	----	----
South East Asia	----	----	----	----	----	----	----	----	----
South Pacific	----	----	----	----	----	----	----	----	----
EUROPE	449.26	414.72	384.96	344.78	363.94	363.46	374.74	----	----
Central Europe	58.83	31.16	31.37	29.58	32.23	37.01	36.39	37.92	----
Eastern Europe	25.90	25.55	17.31	8.89	10.32	5.73	8.41	9.47	9.57
Western Europe	364.53	358.02	336.28	306.31	321.39	320.72	329.94	318.93	235.86
LATIN AMERICA AND CARIBBEAN	----	----	----	----	----	----	----	----	----
Caribbean	----	----	----	----	----	----	----	----	----
Meso-America	----	----	----	----	----	----	----	----	----
South America	----	----	----	----	----	----	----	----	----
NORTH AMERICA	153.68	154.19	150.52	153.16	166.56	166.86	173.43	158.09	110.26
North America	153.68	154.19	150.52	153.16	166.56	166.86	173.43	158.09	110.26
POLAR	----	----	----	----	----	----	----	----	----
Antarctic	----	----	----	----	----	----	----	----	----
Arctic	----	----	----	----	----	----	----	----	----
WEST ASIA	----	----	----	----	----	----	----	----	----
Arabian Peninsula	----	----	----	----	----	----	----	----	----
Mashriq	----	----	----	----	----	----	----	----	----
GLOBAL TOTALS	----	----	----	----	----	----	----	----	----
REPORTED GLOBAL TOTALS FROM ORIGINAL DATA SOURCES	****	****	****	****	****	****	****	****	****
SUM OF ANNEX I PARTIES (INCLUDES ESTIMATIONS)	628.48	592.26	558.90	520.70	555.91	555.49	576.36	554.09	418.16

Comments:

Emissions of N₂O (nitrous oxide) from industrial processes include emissions from mineral products, chemical industry, metal production, other production and other sector. The data is based on the national greenhouse gas inventories given by the parties of the Climate Change Convention Secretariat. This data is submitted to the Climate Change Secretariat by national communications under the Convention by the Annex I parties.

Annex I Parties - the industrialized countries listed in this annex to the Climate Change Convention tried to return their greenhouse gas emissions to 1990 levels by the year 2000. They have also accepted emission targets for the period 2008-12. They include the following members Australia, Austria, Belarus, Belgium, Bulgaria, Canada, Czech Republic, Denmark, European Economic Community, Estonia, Finland, France, Germany, Greece, Hungary, Iceland, Ireland, Italy, Japan, Latvia, Lithuania, Luxembourg, Netherlands, New Zealand, Norway, Poland, Portugal, Romania, Russian Federation, Slovakia, Slovenia, Spain, Sweden, Switzerland, Turkey, Ukraine, United Kingdom of Great Britain and Northern Ireland and the United States of America.

Greenhouse gases (GHGs) - the major anthropogenic GHGs responsible for causing climate change are carbon dioxide (CO₂), methane (CH₄), and nitrous oxide (N₂O). The Kyoto Protocol also addresses hydrofluorocarbons (HFCs), perfluorocarbons (PFCs), and sulphur hexafluoride (SF₆).

For more detailed information about the GHG inventory database see: http://ghg.unfccc.int.

1 gigagram corresponds to 1 thousand metric tonne.

AGGREGATIONS

Emissions of N$_2$O - Total Anthropogenic (UNFCCC)

Units: gigagrams of N$_2$O

Data Source: Greenhouse Gas (GHG) Inventory Submission 1998, 1999, and 2000

Data Provider: United Nations Framework Convention on Climate Change (UNFCCC) Secretariat

Years: 1990-1998

	1990	1991	1992	1993	1994	1995	1996	1997	1998
AFRICA	----	----	----	----	----	----	----	----	----
Central Africa	----	----	----	----	----	----	----	----	----
Eastern Africa	----	----	----	----	----	----	----	----	----
Northern Africa	----	----	----	----	----	----	----	----	----
Southern Africa	----	----	----	----	----	----	----	----	----
Western Africa	----	----	----	----	----	----	----	----	----
Western Indian Ocean	----	----	----	----	----	----	----	----	----
ASIA AND PACIFIC	----	----	----	----	----	----	----	----	----
Australia and New Zealand	111.20	111.18	112.08	114.80	116.80	118.21	119.31	123.47	128.53
Central Asia	----	----	----	----	----	----	----	----	----
North West Pacific and East Asia	----	----	----	----	----	----	----	----	----
South Asia	----	----	----	----	----	----	----	----	----
South East Asia	----	----	----	----	----	----	----	----	----
South Pacific	----	----	----	----	----	----	----	----	----
EUROPE	1920.78	1733.38	1635.08	1549.28	1571.05	1590.77	1619.87	1580.70	1543.10
Central Europe	319.33	211.62	199.18	186.87	184.73	190.26	177.73	145.88	209.60
Eastern Europe	283.70	----	----	----	147.70	154.20	146.90	----	----
Western Europe	1317.74	1288.45	1247.60	1206.21	1238.62	1246.30	1295.24	1287.22	1185.70
LATIN AMERICA AND CARIBBEAN	----	----	----	----	----	----	----	----	----
Caribbean	----	----	----	----	----	----	----	----	----
Meso-America	----	----	----	----	----	----	----	----	----
South America	----	----	----	----	----	----	----	----	----
NORTH AMERICA	1482.73	1508.90	1541.68	1552.62	1657.78	1628.57	1668.84	1673.99	1621.34
North America	1482.73	1508.90	1541.68	1552.62	1657.78	1628.57	1668.84	1673.99	1621.34
POLAR	----	----	----	----	----	----	----	----	----
Antarctic	----	----	----	----	----	----	----	----	----
Arctic	----	----	----	----	----	----	----	----	----
WEST ASIA	----	----	----	----	----	----	----	----	----
Arabian Peninsula	----	----	----	----	----	----	----	----	----
Mashriq	----	----	----	----	----	----	----	----	----
GLOBAL TOTALS	----	----	----	----	----	----	----	----	----
REPORTED GLOBAL TOTALS FROM ORIGINAL DATA SOURCES	****	****	****	****	****	****	****	****	****
SUM OF ANNEX I PARTIES (INCLUDES ESTIMATIONS)	3573.06	3410.13	3346.00	3273.52	3406.54	3399.76	3472.78	3443.87	3359.66

Comments:

Emissions of N$_2$O (nitrous oxide) - total anthropogenic are all the emissions of N$_2$O associated with human activities. These include emissions from fuel combustion, industrial processes, agriculture, waste, and other. The data is based on the national greenhouse gas inventories given by the parties of the Climate Change Convention Secretariat. This data is submitted to the Climate Change Secretariat by national communications under the Convention by the Annex I parties.

Annex I Parties - The industrialized countries listed in this annex to the Climate Change Convention tried to return their greenhouse gas emissions to 1990 levels by the year 2000. They have also accepted emission targets for the period 2008-12. They include the following members Australia, Austria, Belarus, Belgium, Bulgaria, Canada, Czech Republic, Denmark, European Economic Community, Estonia, Finland, France, Germany, Greece, Hungary, Iceland, Ireland, Italy, Japan, Latvia, Lithuania, Luxembourg, Netherlands, New Zealand, Norway, Poland, Portugal, Romania, Russian Federation, Slovakia, Slovenia, Spain, Sweden, Switzerland, Turkey, Ukraine, United Kingdom of Great Britain and Northern Ireland and the United States of America.

Greenhouse gases (GHGs) - the major anthropogenic GHGs responsible for causing climate change are carbon dioxide (CO$_2$), methane (CH$_4$), and nitrous oxide (N$_2$O). The Kyoto Protocol also addresses hydrofluorocarbons (HFCs), perfluorocarbons (PFCs), and sulphur hexafluoride (SF$_6$).

For more detailed information about the GHG inventory database see: http://ghg.unfccc.int.

1 gigagram corresponds to 1 thousand metric tonne.

AGGREGATIONS

Production of ODS: Carbon Tetrachloride

Units: ozone depleting potential (ODP) tonnes

Data Source: Production and Consumption of Ozone Depleting Substances 1986-2000

Data Provider: Secretariat for the Vienna Convention and the Montreal Protocol (the Ozone Secretariat)

Years: 1989-2000

	1989	1990	1991	1992	1993	1994	1995	1996	1997	1998	1999	2000
AFRICA	----	----	----	----	----	----	4931.30	0.00	0.00	0.00	----	----
Central Africa	----	----	----	----	----	----	----	----	----	----	----	----
Eastern Africa	0.00	----	----	----	----	----	0.00	0.00	0.00	0.00	----	----
Northern Africa	0.00	----	0.00	----	0.00	0.00	0.00	0.00	0.00	0.00	0.00	----
Southern Africa	----	----	----	----	----	----	----	----	----	----	----	----
Western Africa	----	----	----	----	0.00	0.00	0.00	0.00	0.00	0.00	0.00	----
Western Indian Ocean	----	----	----	----	----	----	0.00	0.00	0.00	0.00	0.00	----
ASIA AND PACIFIC	54169.50	----	----	-15577.10	-11515.82	-31055.31	-63132.00	-54727.20	7916.71	41738.49	15935.40	----
Australia and New Zealand	0.00	----	----	-11485.10	-11464.12	-5639.81	-3377.80	0.00	0.00	0.00	0.00	----
Central Asia	0.00	----	----	----	0.00	----	----	0.00	0.00	0.00	----	----
North West Pacific and East Asia	49412.00	----	-1727.00	-6050.00	984.50	-33848.10	-37966.50	-34940.40	40.70	35124.10	38.50	----
South Asia	4757.50	----	----	1958.00	-1036.20	8432.60	-21787.70	-19786.80	7876.01	6614.39	15896.90	----
South East Asia	0.00	----	----	0.00	----	0.00	0.00	0.00	0.00	0.00	0.00	----
South Pacific	0.00	----	----	----	0.00	0.00	0.00	0.00	0.00	0.00	----	----
EUROPE	213596.52	----	----	35074.60	7908.01	21901.44	-6312.46	-17849.59	-1549.16	5895.65	634.21	----
Central Europe	21124.02	----	----	----	12535.71	11199.54	----	2838.11	-192.53	1993.62	-672.70	----
Eastern Europe	103290.00	1980.00	-26950.00	2200.00	2200.00	3372.60	2749.93	746.90	-2178.55	2853.73	346.61	----
Western Europe	89182.50	----	----	32874.60	-6827.70	7329.30	-13728.00	-21434.60	821.92	1048.30	----	----
LATIN AMERICA AND CARIBBEAN	56830.40	21436.36	----	32757.12	30638.52	2281.40	-5242.08	-2907.30	-1245.20	16525.30	11350.90	----
Caribbean	----	----	----	----	----	----	----	----	----	----	----	----
Meso-America	4668.40	-5470.30	-6512.00	-6546.10	-10472.00	-9212.50	-9887.90	-8890.20	0.00	0.00	0.00	----
South America	52162.00	26906.66	----	39303.22	41110.52	11493.90	4645.82	5982.90	-1245.20	16525.30	11350.90	----
NORTH AMERICA	85345.70	----	----	18973.90	23203.40	18868.30	11485.10	-689.70	14.30	2.20	18.70	----
North America	85345.70	----	----	18973.90	23203.40	18868.30	11485.10	-689.70	14.30	2.20	18.70	----
POLAR	----	----	----	----	----	----	----	----	----	----	----	----
Antarctic	----	----	----	----	----	----	----	----	----	----	----	----
Arctic	----	----	----	----	----	----	----	----	----	----	----	----
WEST ASIA	----	----	----	----	----	----	----	----	----	----	----	----
Arabian Peninsula	----	----	----	----	----	----	0.00	0.00	0.00	0.00	----	----
Mashriq	----	----	----	----	----	----	----	----	----	----	----	----
GLOBAL TOTALS	----	----	----	----	----	----	----	----	----	----	----	----
REPORTED GLOBAL TOTALS FROM ORIGINAL DATA SOURCES	****	****	****	****	****	****	****	****	****	****	****	****

Comments:

Production means the amount of controlled substances produced, minus the amount destroyed by technologies approved by the parties (to the Montreal Protocol on Substances that Deplete the Ozone Layer) and minus the amount entirely used as feedstock in the manufacture of other chemicals. The amount recycled and reused is not to be considered as production. The data forms prescribe reporting of feedstock use and of quantities destroyed separately, and reporting of total production without deduction. The Ozone Secretariat would make the necessary deduction.

The data is reported to the Ozone Secretariat by the parties to the Montreal Protocol on Substances that Deplete the Ozone Layer. Annex B, group II: carbon tetrachloride comprises CCl_4.

Some of the figures may be negative since the figures are for each calendar year, it is quite possible that in some years the feedstock figure may exceed the production figure of that year, if the feedstock use is from a carry-over stock. The production could be negative in such cases. For the same reason, the consumption could also be negative.

Units: ozone depleting potential (ODP) tonnes

Data Source: Production and Consumption of Ozone Depleting Substances 1986-2000

Data Provider: Secretariat for the Vienna Convention and the Montreal Protocol (the Ozone Secretariat)

Years: 1989-2000

	1989	1990	1991	1992	1993	1994	1995	1996	1997	1998	1999	2000
AFRICA	----	----	----	----	----	----	56.76	0.00	0.00	0.00	----	----
Central Africa	----	----	----	----	----	----	----	----	----	----	----	----
Eastern Africa	0.00	----	----	----	----	----	0.00	0.00	0.00	0.00	----	----
Northern Africa	0.00	----	0.00	----	0.00	0.00	0.00	0.00	0.00	0.00	0.00	----
Southern Africa	----	----	----	----	----	----	----	----	----	----	----	----
Western Africa	----	----	0.00	0.00	0.00	0.00	0.00	0.00	0.00	0.00	0.00	----
Western Indian Ocean	----	----	----	----	----	----	0.00	0.00	0.00	0.00	0.00	----
ASIA AND PACIFIC	2511.80	----	----	3225.93	5689.17	5077.90	7382.33	5605.31	5750.74	4966.49	8702.37	----
Australia and New Zealand	132.36	----	----	123.86	170.01	95.39	84.48	0.00	0.00	0.00	0.00	----
Central Asia	0.00	----	----	----	0.00	----	----	0.00	0.00	0.00	----	----
North West Pacific and East Asia	2260.51	----	2697.27	2896.31	5254.12	4716.53	6983.66	5324.86	5750.74	4966.49	8702.37	----
South Asia	118.93	----	----	205.76	265.04	265.98	314.19	280.45	0.00	0.00	0.00	----
South East Asia	0.00	----	----	0.00	----	0.00	0.00	0.00	0.00	0.00	0.00	----
South Pacific	0.00	----	----	----	----	----	0.00	0.00	0.00	0.00	----	----
EUROPE	3905.05	----	----	4575.50	5463.95	7316.48	9048.27	8951.72	9204.80	10335.94	9650.42	----
Central Europe	0.00	----	----	----	0.00	0.00	----	-46.26	0.00	0.00	0.00	----
Eastern Europe	1194.00	436.50	425.50	267.16	172.00	198.12	184.20	74.41	72.26	67.10	146.27	----
Western Europe	2711.05	----	----	4308.34	5291.95	7118.36	8864.19	8923.57	9132.54	10268.84	----	----
LATIN AMERICA AND CARIBBEAN	499.52	391.45	----	435.79	515.10	453.04	415.37	504.96	523.36	438.54	371.96	----
Caribbean	----	----	----	----	----	----	----	----	----	----	----	----
Meso-America	208.45	138.27	160.49	102.96	158.13	126.39	117.76	296.67	303.32	238.54	328.96	----
South America	291.07	253.18	----	332.83	356.97	326.65	297.61	208.29	220.04	200.00	43.00	----
NORTH AMERICA	6810.67	----	----	5657.37	7227.81	12419.00	14951.75	12607.54	12602.61	15043.50	14728.61	----
North America	6810.67	----	----	5657.37	7227.81	12419.00	14951.75	12607.54	12602.61	15043.50	14728.61	----
POLAR	----	----	----	----	----	----	----	----	----	----	----	----
Antarctic	----	----	----	----	----	----	----	----	----	----	----	----
Arctic	----	----	----	----	----	----	----	----	----	----	----	----
WEST ASIA	----	----	----	----	----	----	----	----	----	----	----	----
Arabian Peninsula	----	----	----	----	----	----	0.00	0.00	0.00	0.00	----	----
Mashriq	----	----	----	----	----	----	----	----	----	----	----	----
GLOBAL TOTALS	----	----	----	----	----	----	----	----	----	----	----	----
REPORTED GLOBAL TOTALS FROM ORIGINAL DATA SOURCES	****	****	****	****	****	****	****	****	****	****	****	****

Comments:

Production means the amount of controlled substances produced, minus the amount destroyed by technologies approved by the parties (to the Montreal Protocol on Substances that Deplete the Ozone Layer) and minus the amount entirely used as feedstock in the manufacture of other chemicals. The amount recycled and reused is not to be considered as production. The data forms prescribe reporting of feedstock use and of quantities destroyed separately, and reporting of total production without deduction. The Ozone Secretariat would make the necessary deduction.

The data is reported to the Ozone Secretariat by the parties to the Montreal Protocol on Substances that Deplete the Ozone Layer. The Annex A, group I: chlorofluorocarbons (CFCs) comprise $CFCl_3$, CF_2Cl_2, $C_2F_3Cl_3$, $C_2F_4Cl_2$, C_2F_5Cl.

Some of the figures may be negative since the figures are for each calendar year, it is quite possible that in some years the feedstock figure may exceed the production figure of that year, if the feedstock use is from a carry-over stock. The production could be negative in such cases. For the same reason, the consumption could also be negative.

AGGREGATIONS

Production of ODS: Hydrochlorofluorocarbons

Units: ozone depleting potential (ODP) tonnes

Data Source: Production and Consumption of Ozone Depleting Substances 1986-2000

Data Provider: Secretariat for the Vienna Convention and the Montreal Protocol (the Ozone Secretariat)

Years: 1989-2000

	1989	1990	1991	1992	1993	1994	1995	1996	1997	1998	1999	2000
AFRICA	----	----	----	----	----	----	56.76	0.00	0.00	0.00	----	----
Central Africa	----	----	----	----	----	----	----	----	----	----	----	----
Eastern Africa	0.00	----	----	----	----	----	0.00	0.00	0.00	0.00	----	----
Northern Africa	0.00	----	0.00	----	0.00	0.00	0.00	0.00	0.00	0.00	0.00	----
Southern Africa	----	----	----	----	----	----	----	----	----	----	----	----
Western Africa	----	----	0.00	0.00	0.00	0.00	0.00	0.00	0.00	0.00	0.00	----
Western Indian Ocean	----	----	----	----	----	----	0.00	0.00	0.00	0.00	0.00	----
ASIA AND PACIFIC	2511.80	----	----	3225.93	5689.17	5077.90	7382.33	5605.31	5750.74	4966.49	8702.37	----
Australia and New Zealand	132.36	----	----	123.86	170.01	95.39	84.48	0.00	0.00	0.00	0.00	----
Central Asia	0.00	----	----	----	0.00	----	----	0.00	0.00	0.00	----	----
North West Pacific and East Asia	2260.51	----	2697.27	2896.31	5254.12	4716.53	6983.66	5324.86	5750.74	4966.49	8702.37	----
South Asia	118.93	----	----	205.76	265.04	265.98	314.19	280.45	0.00	0.00	0.00	----
South East Asia	0.00	----	----	0.00	----	0.00	0.00	0.00	0.00	0.00	0.00	----
South Pacific	0.00	----	----	----	----	----	0.00	0.00	0.00	0.00	----	----
EUROPE	3905.05	----	----	4575.50	5463.95	7316.48	9048.27	8951.72	9204.80	10335.94	9650.42	----
Central Europe	0.00	----	----	----	0.00	0.00	----	-46.26	0.00	0.00	0.00	----
Eastern Europe	1194.00	436.50	425.50	267.16	172.00	198.12	184.20	74.41	72.26	67.10	146.27	----
Western Europe	2711.05	----	----	4308.34	5291.95	7118.36	8864.19	8923.57	9132.54	10268.84	----	----
LATIN AMERICA AND CARIBBEAN	499.52	391.45	----	435.79	515.10	453.04	415.37	504.96	523.36	438.54	371.96	----
Caribbean	----	----	----	----	----	----	----	----	----	----	----	----
Meso- America	208.45	138.27	160.49	102.96	158.13	126.39	117.76	296.67	303.32	238.54	328.96	----
South America	291.07	253.18	----	332.83	356.97	326.65	297.61	208.29	220.04	200.00	43.00	----
NORTH AMERICA	6810.67	----	----	5657.37	7227.81	12419.00	14951.75	12607.54	12602.61	15043.50	14728.61	----
North America	6810.67	----	----	5657.37	7227.81	12419.00	14951.75	12607.54	12602.61	15043.50	14728.61	----
POLAR	----	----	----	----	----	----	----	----	----	----	----	----
Antarctic	----	----	----	----	----	----	----	----	----	----	----	----
Arctic	----	----	----	----	----	----	----	----	----	----	----	----
WEST ASIA	----	----	----	----	----	----	----	----	----	----	----	----
Arabian Peninsula	----	----	----	----	----	----	0.00	0.00	0.00	0.00	----	----
Mashriq	----	----	----	----	----	----	----	----	----	----	----	----
GLOBAL TOTALS	----	----	----	----	----	----	----	----	----	----	----	----
REPORTED GLOBAL TOTALS FROM ORIGINAL DATA SOURCES	****	****	****	****	****	****	****	****	****	****	****	****

Comments:

Production means the amount of controlled substances produced, minus the amount destroyed by technologies approved by the parties (to the Montreal Protocol on Substances that Deplete the Ozone Layer) and minus the amount entirely used as feedstock in the manufacture of other chemicals. The amount recycled and reused is not to be considered as production. The data forms prescribe reporting of feedstock use and of quantities destroyed separately, and reporting of total production without deduction. The Ozone Secretariat would make the necessary deduction.

The data is reported to the Ozone Secretariat by the parties to the Montreal Protocol on Substances that Deplete the Ozone Layer. The Annex C, group I: hydrochlorofluorocarbons (HCFCs) comprise $CHFCl_2$, CHF_2Cl, CH_2FCl, C_2HFCl_4, $C_2HF_2Cl_3$, $C_2HF_3Cl_2$, $CHCl_2CF_3$, C_2HF_4Cl, $CHFClCF_3$, $C_2H_2FCl_3$, $C_2H_2F_2Cl_2$, $C_2H_2F_3Cl$, $C_2H_3FCl_2$, CH_3CFCl_2, $C_2H_3F_2Cl$, CH_3CF_2Cl, C_2H_4FCl, C_3HFCl_6, $C_3HF_2Cl_5$, $C_3HF_3Cl_4$, $C_3HF_4Cl_3$, $C_3HF_5Cl_2$, $CF_3CF_2CHCl_2$, CF_2ClCF_2CHClF, C_3HF_6Cl, $C_3H_2FCl_5$, $C_3H_2F_2Cl_4$, $C_3H_2F_3Cl_3$, $C_3H_2F_4Cl_2$, $C_3H_2F_5Cl$, $C_3H_3FCl_4$, $C_3H_3F_2Cl_3$, $C_3H_3F_3Cl_2$, $C_3H_3F_4Cl$, $C_3H_4FCl_3$, $C_3H_4F_2Cl_2$, $C_3H_4F_3Cl$, $C_3H_5FCl_2$, $C_3H_5F_2Cl$, C_3H_6FCl.

Some of the figures may be negative since the figures are for each calendar year, it is quite possible that in some years the feedstock figure may exceed the production figure of that year, if the feedstock use is from a carry-over stock. The production could be negative in such cases. For the same reason, the consumption could also be negative.

AGGREGATIONS

Production of ODS: Methyl Chloroform

Units: ozone depleting potential (ODP) tonnes

Data Source: Production and Consumption of Ozone Depleting Substances 1986-2000

Data Provider: Secretariat for the Vienna Convention and the Montreal Protocol (the Ozone Secretariat)

Years: 1989-2000

	1989	1990	1991	1992	1993	1994	1995	1996	1997	1998	1999	2000
AFRICA	----	----	----	----	----	----	0.00	0.00	0.00	0.00	----	----
Central Africa	----	----	----	----	----	----	----	----	----	----	----	----
Eastern Africa	0.00	----	----	----	----	----	0.00	0.00	0.00	0.00	----	----
Northern Africa	0.00	----	0.00	----	0.00	0.00	0.00	0.00	0.00	0.00	0.00	----
Southern Africa	----	----	----	----	----	----	----	----	----	----	----	----
Western Africa	----	----	----	----	0.00	0.00	0.00	0.00	0.00	0.00	0.00	----
Western Indian Ocean	----	----	----	----	----	----	0.00	0.00	0.00	0.00	0.00	----
ASIA AND PACIFIC	15683.10	----	----	15773.70	7221.35	388.90	5149.10	967.00	1183.10	1033.70	1170.30	----
Australia and New Zealand	0.00	----	0.00	0.00	0.00	0.00	0.00	0.00	0.00	0.00	0.00	----
Central Asia	0.00	----	----	----	0.00	----	----	0.00	0.00	0.00	----	----
North West Pacific and East Asia	15636.40	44.00	17119.10	15725.40	7165.45	388.90	5149.10	967.00	1183.10	1033.70	1170.30	----
South Asia	46.70	----	----	48.30	55.90	0.00	0.00	0.00	0.00	0.00	0.00	----
South East Asia	0.00	----	----	----	----	0.00	0.00	0.00	0.00	0.00	0.00	----
South Pacific	0.00	----	----	----	0.00	0.00	0.00	0.00	0.00	0.00	----	----
EUROPE	13421.90	----	----	11361.40	5534.70	3044.47	1641.95	79.86	225.91	184.30	259.40	----
Central Europe	27.20	----	----	----	9.90	7.80	----	12.90	2.91	0.00	0.00	----
Eastern Europe	330.00	310.00	310.00	400.00	100.00	196.57	196.90	0.00	0.00	0.00	0.00	----
Western Europe	13064.70	----	----	10961.40	5424.80	2840.10	1439.00	66.96	223.00	184.30	----	----
LATIN AMERICA AND CARIBBEAN	1130.00	660.00	----	750.00	764.00	39.20	0.00	0.00	0.00	0.00	97.20	----
Caribbean	----	----	----	----	----	----	----	----	----	----	----	----
Meso-America	0.00	0.00	0.00	0.00	0.00	0.00	0.00	0.00	0.00	0.00	0.00	----
South America	1130.00	660.00	----	750.00	764.00	39.20	0.00	0.00	0.00	0.00	97.20	----
NORTH AMERICA	32649.10	----	----	25722.60	20637.10	5794.60	4598.60	447.50	437.30	262.30	245.80	----
North America	32649.10	----	----	25722.60	20637.10	5794.60	4598.60	447.50	437.30	262.30	245.80	----
POLAR	----	----	----	----	----	----	----	----	----	----	----	----
Antarctic	----	----	----	----	----	----	----	----	----	----	----	----
Arctic	----	----	----	----	----	----	----	----	----	----	----	----
WEST ASIA	----	----	----	----	----	----	----	----	----	----	----	----
Arabian Peninsula	----	----	----	----	----	----	0.00	0.00	0.00	0.00	----	----
Mashriq	----	----	----	----	----	----	----	----	----	----	----	----
GLOBAL TOTALS	----	----	----	----	----	----	----	----	----	----	----	----
REPORTED GLOBAL TOTALS FROM ORIGINAL DATA SOURCES	****	****	****	****	****	****	****	****	****	****	****	****

Comments:

Production means the amount of controlled substances produced, minus the amount destroyed by technologies approved by the parties (to the Montreal Protocol on Substances that Deplete the Ozone Layer) and minus the amount entirely used as feedstock in the manufacture of other chemicals. The amount recycled and reused is not to be considered as production. The data forms prescribe reporting of feedstock use and of quantities destroyed separately, and reporting of total production without deduction. The Ozone Secretariat would make the necessary deduction.

The data is reported to the Ozone Secretariat by the Parties to the Montreal Protocol on Substances that Deplete the Ozone Layer. The Annex B, group III: methyl chloroform comprises $C_2H_3Cl_3$ (this formula does not refer to 1,1,2-trichloroethane).

Some of the figures may be negative since the figures are for each calendar year, it is quite possible that in some years the feedstock figure may exceed the production figure of that year, if the feedstock use is from a carry-over stock. The production could be negative in such cases. For the same reason, the consumption could also be negative.

Units: ozone depleting potential (ODP) tonnes

Data Source: Production and Consumption of Ozone Depleting Substances 1986-2000

Data Provider: Secretariat for the Vienna Convention and the Montreal Protocol (the Ozone Secretariat)

Years: 1989-2000

	1989	1990	1991	1992	1993	1994	1995	1996	1997	1998	1999	2000
AFRICA	----	----	----	----	----	----	0.00	0.00	0.00	0.00	----	----
Central Africa	----	----	----	----	----	----	----	----	----	----	----	----
Eastern Africa	0.00	----	----	----	----	----	0.00	0.00	0.00	0.00	----	----
Northern Africa	0.00	----	0.00	----	0.00	0.00	0.00	0.00	0.00	0.00	0.00	----
Southern Africa	----	----	----	----	----	----	----	----	----	----	----	----
Western Africa	----	----	----	----	0.00	0.00	0.00	0.00	0.00	0.00	0.00	----
Western Indian Ocean	----	----	----	----	----	----	0.00	0.00	0.00	0.00	0.00	----
ASIA AND PACIFIC	2342.00	----	----	1600.00	808.00	136.00	35.00	17.00	27.00	26.00	27.00	----
Australia and New Zealand	0.00	----	----	0.00	0.00	0.00	0.00	0.00	0.00	0.00	0.00	----
Central Asia	0.00	----	----	----	0.00	----	----	0.00	0.00	0.00	----	----
North West Pacific and East Asia	2342.00	----	1585.00	1600.00	808.00	136.00	35.00	17.00	27.00	26.00	27.00	----
South Asia	0.00	----	----	0.00	0.00	0.00	0.00	0.00	0.00	0.00	0.00	----
South East Asia	0.00	----	----	----	----	0.00	0.00	0.00	0.00	0.00	0.00	----
South Pacific	0.00	----	----	----	0.00	0.00	0.00	0.00	0.00	0.00	----	----
EUROPE	384.00	----	----	----	51.60	-145.50	-257.77	-126.00	75.00	13.05	16.50	----
Central Europe	0.00	----	----	----	0.00	----	----	0.00	0.00	0.00	0.00	----
Eastern Europe	300.00	300.00	250.00	17.00	0.60	25.00	24.78	20.00	75.00	13.05	16.50	
Western Europe	84.00	----	----	20.00	51.00	-170.50	-282.00	-146.00	0.00	0.00	----	----
LATIN AMERICA AND CARIBBEAN	0.00	0.00	----	10.80	0.00	0.00	0.00	0.00	0.00	0.00	0.00	----
Caribbean	----	----	----	----	----	----	----	----	----	----	----	----
Meso-America	0.00	0.00	0.00	0.00	0.00	0.00	0.00	0.00	0.00	0.00	0.00	----
South America	0.00	0.00	----	10.80	0.00	0.00	0.00	0.00	0.00	0.00	0.00	----
NORTH AMERICA	577.00	----	----	75.00	106.00	101.00	38.00	0.00	0.00	-1.00	0.00	----
North America	577.00	----	----	75.00	106.00	101.00	38.00	0.00	0.00	-1.00	0.00	----
POLAR	----	----	----	----	----	----	----	----	----	----	----	----
Antarctic	----	----	----	----	----	----	----	----	----	----	----	----
Arctic	----	----	----	----	----	----	----	----	----	----	----	----
WEST ASIA	----	----	----	----	----	----	----	----	----	----	----	----
Arabian Peninsula	----	----	----	----	----	----	0.00	0.00	0.00	0.00	----	----
Mashriq	----	----	----	----	----	----	----	----	----	----	----	----
GLOBAL TOTALS	----	----	----	----	----	----	----	----	----	----	----	----
REPORTED GLOBAL TOTALS FROM ORIGINAL DATA SOURCES	****	****	****	****	****	****	****	****	****	****	****	****

Comments:

The data is reported to the Ozone Secretariat by the parties to the Montreal Protocol on Substances that Deplete the Ozone Layer. Annex B, group I: other fully halogenated CFCs comprise CF_3Cl, C_2FCl_5, $C_2F_2Cl_4$, C_3FCl_7, $C_3F_2Cl_6$, $C_3F_3Cl_5$, $C_3F_4Cl_4$, $C_3F_5Cl_3$, $C_3F_6Cl_2$, C_3F_7Cl.

1.8 Urban Areas

1.8.1 Health

Urban Population Water Comfort

The "Global Water Supply and Sanitation Assessment 2000 Report" did not provide standardized definitions of urban and rural, as none could be found that would be consistent with the range of definitions adopted locally. For more information see: http://www.who.int/water_sanitation_health/Globassessment/GlobalTOC.htm

Units: percent

Data Source: Global Water Supply and Sanitation Assessment 2000 Report

Data Provider: World Health Organization and United Nations Children's Fund Joint Monitoring
Programme for Water Supply and Sanitation (JMP)

Years: 1990, 2000

Urban Population with Access to Improved Sanitation

	1990	2000
AFRICA	86.40	84.36
Central Africa	----	62.30
Eastern Africa	77.61	77.72
Northern Africa	94.78	95.05
Southern Africa	----	92.49
Western Africa	75.58	78.59
Western Indian Ocean	75.01	73.85
ASIA AND PACIFIC	64.02	76.08
Australia and New Zealand	100.00	100.00
Central Asia	----	100.00
North West Pacific and East Asia	57.00	69.48
South Asia	65.07	77.26
South East Asia	80.73	87.22
South Pacific	----	88.21
EUROPE	----	----
Central Europe	----	----
Eastern Europe	----	----
Western Europe	----	----
LATIN AMERICA AND CARIBBEAN	84.48	86.92
Caribbean	----	82.41
Meso-America	85.80	88.53
South America	86.20	86.78
NORTH AMERICA	100.00	100.00
North America	100.00	100.00
POLAR	----	----
Antarctic	----	----
Arctic	----	----
WEST ASIA	----	96.48
Arabian Peninsula	----	97.38
Mashriq	----	95.76
GLOBAL TOTALS	81.58	85.94
REPORTED GLOBAL TOTALS FROM ORIGINAL DATA SOURCES	82.00	86.00

Comments:

The access to sanitation is defined in terms of the types of technology and levels of service afforded. This includes connection to a public sewer, connection to septic system, pour-flush latrine, simple pit latrine, ventilated improved pit latrine allowance was also made for other locally-defined technologies. Access to sanitation, does not imply that the level of service is "adequate" or "safe"; these terms were replaced with "improved".

Urban Population with Access to Improved Water Supply

	1990	2000
AFRICA	85.38	85.48
Central Africa	----	81.02
Eastern Africa	82.58	80.37
Northern Africa	92.90	94.53
Southern Africa	----	85.13
Western Africa	78.48	79.82
Western Indian Ocean	87.42	86.88
ASIA AND PACIFIC	95.66	93.21
Australia and New Zealand	100.00	100.00
Central Asia	----	97.08
North West Pacific and East Asia	99.00	94.36
South Asia	93.18	92.88
South East Asia	89.51	89.57
South Pacific	----	78.51
EUROPE	----	----
Central Europe	----	----
Eastern Europe	----	100.00
Western Europe	----	----
LATIN AMERICA AND CARIBBEAN	91.25	92.66
Caribbean	----	85.41
Meso-America	91.77	94.06
South America	92.63	92.82
NORTH AMERICA	100.00	100.00
North America	100.00	100.00
POLAR	----	----
Antarctic	----	----
Arctic	----	----
WEST ASIA	----	94.35
Arabian Peninsula	----	91.96
Mashriq	----	96.29
GLOBAL TOTALS	94.47	93.79
REPORTED GLOBAL TOTALS FROM ORIGINAL DATA SOURCES	95.00	94.00

Comments:

The access to water supply is defined in terms of the types of technology and levels of service afforded. This included house connections, public standpipes, boreholes with handpumps, protected dug wells, protected springs and rainwater collection; allowance was also made for other locally-defined technologies. "Reasonable access" was broadly defined as the availability of at least 20 litres per person per day from a source within one kilometre of the user's dwelling. Access to water, does not imply that the level of service or quality of water is "adequate" or "safe"; these terms were replaced with "improved"

AGGREGATIONS

1.8.2 Urbanisation

Urban Population - Average Annual Growth Rate (Projection)

Units: percent per year

Data Source: World Urbanization Prospects: The 1999 Revision

Data Provider: United Nations Population Division/Department of Economic and Social Affairs

Years: 1950-1955, 1955-1960, ..., 2025-2030

	1950-55	1970-75	1990-95	1995-00	2000-05	2005-10	2010-15	2015-20	2020-25	2025-30
AFRICA	4.48	4.34	4.25	3.98	3.72	3.47	3.26	3.07	2.84	2.58
Central Africa	4.07	3.77	4.50	3.88	4.18	4.25	4.08	3.90	3.59	3.22
Eastern Africa	5.58	5.16	5.18	5.13	4.75	4.54	4.34	4.10	3.83	3.53
Northern Africa	4.28	3.62	3.16	3.09	2.88	2.60	2.31	2.12	1.94	1.75
Southern Africa	3.69	4.20	4.66	4.04	3.43	3.04	2.97	2.94	2.74	2.51
Western Africa	5.59	5.56	4.76	4.47	4.15	3.85	3.52	3.18	2.88	2.58
Western Indian Ocean	5.36	4.38	4.87	4.56	4.30	4.07	3.87	3.58	3.25	2.87
ASIA AND PACIFIC	3.52	3.23	2.69	2.59	2.49	2.40	2.31	2.19	1.98	1.76
Australia and New Zealand	2.93	2.23	1.25	1.04	0.95	0.95	0.97	0.97	0.87	0.76
Central Asia	4.04	2.98	0.18	0.38	1.00	1.39	1.74	2.05	2.18	2.01
North West Pacific and East Asia	3.92	2.55	2.20	2.02	1.91	1.85	1.79	1.64	1.45	1.24
South Asia	2.84	4.02	3.13	3.12	3.05	3.00	2.89	2.80	2.53	2.27
South East Asia	3.72	4.19	3.84	3.57	3.17	2.82	2.50	2.24	1.97	1.78
South Pacific	5.51	4.51	3.40	3.44	3.47	3.42	3.35	3.23	2.94	2.68
EUROPE	2.13	1.55	0.79	0.59	0.51	0.44	0.37	0.28	0.20	0.12
Central Europe	4.12	2.96	1.60	1.44	1.26	1.08	0.93	0.76	0.65	0.54
Eastern Europe	3.55	1.96	0.41	0.20	0.21	0.24	0.20	0.08	0.00	-0.05
Western Europe	1.25	0.98	0.68	0.46	0.35	0.26	0.21	0.17	0.08	-0.02
LATIN AMERICA AND CARIBBEAN	4.46	3.74	2.36	2.11	1.89	1.69	1.51	1.34	1.20	1.07
Caribbean	3.09	3.22	1.89	1.66	1.56	1.46	1.38	1.27	1.17	1.03
Meso-America	4.39	4.22	2.32	2.16	2.01	1.85	1.73	1.61	1.47	1.31
South America	4.64	3.65	2.41	2.13	1.88	1.65	1.45	1.25	1.11	0.99
NORTH AMERICA	2.65	1.02	1.24	1.11	1.02	0.99	0.99	0.93	0.82	0.70
North America	2.65	1.02	1.24	1.11	1.02	0.99	0.99	0.93	0.82	0.70
POLAR	3.02	1.59	0.45	0.41	0.44	0.46	0.48	0.50	0.52	0.48
Antarctic	----	----	----	----	----	----	----	----	----	----
Arctic	3.02	1.59	0.45	0.41	0.44	0.46	0.48	0.50	0.52	0.48
WEST ASIA	5.43	5.87	3.52	3.67	3.36	3.07	2.79	2.52	2.23	1.94
Arabian Peninsula	7.87	8.33	3.48	4.02	3.48	3.19	2.95	2.75	2.46	2.14
Mashriq	4.74	4.69	3.55	3.38	3.25	2.97	2.65	2.32	2.02	1.75
GLOBAL TOTALS	3.01	2.59	2.21	2.11	2.03	1.95	1.90	1.80	1.66	1.50
REPORTED GLOBAL TOTALS FROM ORIGINAL DATA SOURCES	3.01	2.59	2.21	2.11	2.03	1.96	1.89	1.80	1.66	1.50

Comments:

Figures for 1950-1995 are estimates, and those for years beyond 1995 are projections. Estimates and projections are presented only for years that are multiples of five. The urban population data presented are consistent with the medium variant of the *1998 Revision of World Population Prospects*, the official United Nations population estimates and projections at the national level.

The urban population estimates are based on the varying national definitions of urban areas.

AGGREGATIONS

Units: percent

Data Source: World Urbanization Prospects: The 1999 Revision

Data Provider: United Nations Population Division/Department of Economic and Social Affairs

Years: 1960-2000

	1960	1970	1980	1990	1991	1992	1993	1994	1995	1996	1997	1998	1999	2000
AFRICA	18.48	23.11	27.40	32.10	32.67	33.25	33.83	34.41	34.98	35.57	36.16	36.74	37.31	37.89
Central Africa	19.20	26.47	29.34	31.69	32.03	32.37	32.70	33.04	33.38	33.82	34.27	34.72	35.17	35.60
Eastern Africa	6.66	9.03	11.93	15.53	16.02	16.51	17.02	17.51	17.98	18.47	18.94	19.39	19.84	20.29
Northern Africa	30.12	36.28	40.31	45.19	45.72	46.24	46.77	47.29	47.82	48.36	48.90	49.45	49.99	50.53
Southern Africa	20.32	23.32	27.76	32.24	32.88	33.53	34.19	34.86	35.54	36.24	36.95	37.65	38.35	39.05
Western Africa	14.46	19.57	25.68	32.43	33.16	33.89	34.62	35.36	36.09	36.85	37.62	38.38	39.14	39.90
Western Indian Ocean	14.03	18.63	22.40	26.79	27.32	27.85	28.38	28.91	29.44	29.99	30.55	31.10	31.66	32.21
ASIA AND PACIFIC	20.55	22.86	26.01	31.09	31.47	31.85	32.22	32.60	32.97	33.39	33.81	34.23	34.65	35.07
Australia and New Zealand	79.74	84.45	85.36	85.03	84.99	84.94	84.89	84.85	84.80	84.82	84.83	84.85	84.87	84.88
Central Asia	39.10	42.89	45.16	44.94	44.56	44.18	43.79	43.40	43.01	42.77	42.52	42.27	42.03	41.79
North West Pacific and East Asia	22.28	24.43	27.03	33.98	34.38	34.79	35.20	35.61	36.03	36.45	36.88	37.31	37.74	38.18
South Asia	17.36	19.60	23.42	26.62	26.94	27.25	27.56	27.87	28.18	28.56	28.93	29.31	29.68	30.05
South East Asia	17.63	20.42	24.36	30.22	30.88	31.54	32.21	32.87	33.54	34.26	34.98	35.70	36.41	37.13
South Pacific	11.96	18.34	21.47	23.52	23.78	24.04	24.30	24.56	24.83	25.14	25.46	25.78	26.10	26.42
EUROPE	56.67	63.07	67.73	71.32	71.66	72.01	72.36	72.72	73.07	73.39	73.70	74.01	74.33	74.64
Central Europe	37.77	44.95	52.38	59.77	60.44	61.13	61.84	62.55	63.26	63.87	64.47	65.07	65.68	66.29
Eastern Europe	50.26	58.58	65.82	70.42	70.72	71.02	71.33	71.63	71.93	72.23	72.53	72.84	73.14	73.44
Western Europe	67.97	73.34	75.87	77.52	77.73	77.94	78.14	78.35	78.55	78.74	78.92	79.11	79.29	79.48
LATIN AMERICA AND CARIBBEAN	49.28	57.39	64.89	71.00	71.47	71.94	72.40	72.86	73.33	73.73	74.12	74.52	74.92	75.32
Caribbean	40.55	47.43	54.14	59.25	59.63	60.02	60.41	60.80	61.20	61.57	61.94	62.31	62.68	63.05
Meso-America	46.67	53.83	60.12	65.30	65.49	65.67	65.86	66.04	66.22	66.42	66.63	66.84	67.05	67.26
South America	51.36	59.93	67.96	74.51	75.09	75.68	76.26	76.84	77.42	77.90	78.38	78.86	79.34	79.83
NORTH AMERICA	69.90	73.80	73.89	75.34	75.50	75.67	75.83	76.00	76.16	76.37	76.57	76.78	76.98	77.19
North America	69.90	73.80	73.89	75.34	75.50	75.67	75.83	76.00	76.16	76.37	76.57	76.78	76.98	77.19
POLAR	74.90	73.40	76.00	79.40	79.70	80.00	80.30	80.60	80.90	81.12	81.34	81.56	81.78	82.00
Antarctic	----	----	----	----	----	----	----	----	----	----	----	----	----	----
Arctic	74.90	73.40	76.00	79.40	79.70	80.00	80.30	80.60	80.90	81.12	81.34	81.56	81.78	82.00
WEST ASIA	33.42	45.40	55.83	63.48	63.88	64.22	64.53	64.87	65.25	65.57	65.94	66.34	66.73	67.11
Arabian Peninsula	20.15	33.27	48.84	60.10	60.39	60.50	60.54	60.63	60.84	61.03	61.34	61.71	62.09	62.44
Mashriq	42.50	53.62	61.30	66.52	67.02	67.54	68.08	68.62	69.15	69.62	70.08	70.53	70.99	71.44
GLOBAL TOTALS	33.62	36.70	39.57	43.52	43.84	44.17	44.50	44.83	45.17	45.52	45.88	46.23	46.59	46.95
REPORTED GLOBAL TOTALS FROM ORIGINAL DATA SOURCES	33.60	36.70	39.60	43.50	43.84	44.18	44.52	44.86	45.20	45.56	45.92	46.28	46.64	47.00

Comments:

Figures for 1960-1995 are estimates and those for years beyond 1995 are projections. The urban population data presented are consistent with the medium variant of the *1998 Revision* of *World Population Prospects*, the official United Nations population estimates and projections at the national level.

The urban population estimates are based on the varying national definitions of urban areas.

Urban Population - Percent of Total Population (Population)

Units: percent

Data Source: World Urbanization Prospects: The 1999 Revision

Data Provider: United Nations Population Division/Department of Economic and Social Affairs

Years: 1950, 1955, ..., 2030

	1950	1960	1970	1980	1990	1995	2000	2005	2010	2015	2020	2025	2030
AFRICA	14.66	18.48	23.11	27.40	32.10	34.98	37.89	40.81	43.66	46.38	49.00	51.56	54.06
Central Africa	15.44	19.20	26.47	29.34	31.69	33.38	35.60	38.07	41.01	43.91	46.81	49.72	52.53
Eastern Africa	4.80	6.66	9.03	11.93	15.53	17.98	20.29	22.79	25.37	27.99	30.72	33.52	36.43
Northern Africa	24.66	30.12	36.28	40.31	45.19	47.82	50.53	53.24	56.00	58.61	61.15	63.59	65.97
Southern Africa	17.51	20.32	23.32	27.76	32.24	35.54	39.05	42.46	45.60	48.54	51.52	54.34	57.10
Western Africa	10.11	14.46	19.57	25.68	32.43	36.09	39.90	43.53	46.90	50.05	52.82	55.50	58.06
Western Indian Ocean	10.53	14.03	18.63	22.40	26.79	29.44	32.21	35.20	38.33	41.51	44.61	47.73	50.77
ASIA AND PACIFIC	17.40	20.56	22.87	26.02	31.10	32.98	35.07	37.43	40.04	42.85	45.81	48.75	51.68
Australia and New Zealand	74.61	79.74	84.45	85.36	85.03	84.80	84.88	85.06	85.57	86.27	87.13	87.92	88.70
Central Asia	35.04	39.10	42.89	45.16	44.94	43.01	41.79	41.80	42.27	43.70	46.02	48.88	51.71
North West Pacific and East Asia	17.81	22.31	24.46	27.06	34.01	36.05	38.20	40.54	43.12	45.86	48.70	51.56	54.44
South Asia	15.99	17.36	19.60	23.42	26.62	28.18	30.05	32.35	35.00	38.00	41.25	44.48	47.68
South East Asia	14.78	17.63	20.42	24.36	30.22	33.54	37.13	40.69	44.13	47.32	50.29	53.03	55.76
South Pacific	8.53	11.95	18.33	21.46	23.51	24.82	26.42	28.24	30.29	32.58	35.20	37.88	40.65
EUROPE	51.16	56.67	63.07	67.73	71.32	73.07	74.64	76.19	77.69	79.14	80.53	81.87	83.13
Central Europe	29.95	37.77	44.95	52.38	59.77	63.26	66.29	69.04	71.48	73.64	75.56	77.38	79.07
Eastern Europe	41.78	50.26	58.58	65.82	70.42	71.93	73.44	74.89	76.32	77.74	79.08	80.44	81.70
Western Europe	63.94	67.97	73.34	75.87	77.52	78.55	79.48	80.49	81.58	82.71	83.87	84.97	86.00
LATIN AMERICA AND CARIBBEAN	41.43	49.28	57.39	64.89	71.00	73.33	75.32	77.05	78.59	79.93	81.11	82.24	83.33
Caribbean	35.55	40.55	47.43	54.14	59.25	61.20	63.05	64.88	66.74	68.50	70.21	71.90	73.47
Meso-America	39.79	46.67	53.83	60.12	65.30	66.22	67.26	68.45	69.79	71.20	72.84	74.41	75.96
South America	42.85	51.36	59.93	67.96	74.51	77.42	79.83	81.77	83.36	84.64	85.58	86.47	87.32
NORTH AMERICA	63.93	69.90	73.80	73.89	75.34	76.16	77.19	78.34	79.60	80.89	82.10	83.30	84.41
North America	63.93	69.90	73.80	73.89	75.34	76.16	77.19	78.34	79.60	80.89	82.10	83.30	84.41
POLAR	76.30	74.90	73.40	76.00	79.40	80.90	82.00	83.20	84.20	85.20	86.20	87.10	87.90
Antarctic	----	----	----	----	----	----	----	----	----	----	----	----	----
Arctic	76.30	74.90	73.40	76.00	79.40	80.90	82.00	83.20	84.20	85.20	86.20	87.10	87.90
WEST ASIA	24.34	33.42	45.40	55.83	63.48	65.25	67.11	68.54	69.76	70.78	71.73	72.60	73.44
Arabian Peninsula	11.60	20.15	33.27	48.84	60.10	60.84	62.44	63.42	64.22	64.89	65.58	66.28	67.05
Mashriq	33.52	42.50	53.62	61.30	66.52	69.15	71.44	73.45	75.28	76.90	78.46	79.93	81.34
GLOBAL TOTALS	29.74	33.62	36.70	39.57	43.52	45.17	46.95	48.89	50.98	53.19	55.47	57.74	60.00
REPORTED GLOBAL TOTALS FROM ORIGINAL DATA SOURCES	29.70	33.60	36.70	39.60	43.50	45.20	47.00	49.00	51.10	53.40	55.70	58.00	60.30

Comments:

Figures for 1950-1995 are estimates and those for years beyond 1995 are projections. Estimates and projections are presented only for years that are multiples of five. The urban population data presented are consistent with the medium variant of the *1998 Revision of World Population Prospects*, the official United Nations population estimates and projections at the national level.

The urban population estimates are based on the varying national definitions of urban areas.

AGGREGATIONS

Urban Population - Total

Units: thousand people

Data Source: World Urbanization Prospects: The 1999 Revision

Data Provider: United Nations Population Division/Department of Economic and Social Affairs

Years: 1960-2000

	1960	1970	1980	1990	1995	1996	1997	1998	1999	2000
AFRICA	51144.02	82367.27	127933.36	198838.38	246087.76	256475.34	267095.71	277979.14	289171.69	300710.24
Central Africa	5186.29	9179.97	13236.61	19335.36	24165.47	25171.09	26155.96	27146.15	28181.32	29291.92
Eastern Africa	3185.45	5653.67	9844.38	17320.20	22468.37	23701.06	24998.12	26353.51	27756.70	29200.19
Northern Africa	20258.01	31185.11	44676.55	64694.24	75892.86	78235.67	80605.88	83014.54	85477.43	88006.07
Southern Africa	10671.30	15861.83	24920.13	38177.67	47961.17	50086.25	52234.05	54398.65	56576.03	58763.98
Western Africa	10910.43	18893.13	32797.68	55503.14	70815.62	74271.53	77857.07	81577.47	85438.22	89444.20
Western Indian Ocean	932.54	1593.57	2458.00	3807.78	4784.27	5009.74	5244.64	5488.83	5741.99	6003.89
ASIA AND PACIFIC	338527.79	473576.24	659392.79	943872.88	1080673.39	1109984.05	1139620.76	1169542.65	1199682.96	1229994.09
Australia and New Zealand	10085.56	12966.80	15093.45	17217.89	18381.18	18609.28	18828.84	19041.03	19247.92	19451.11
Central Asia	9541.88	14228.66	18657.27	22748.46	23185.38	23310.94	23430.79	23543.53	23646.12	23737.05
North West Pacific and East Asia	176036.37	240429.40	317377.97	457116.34	509802.95	520493.58	531216.17	541940.28	552614.67	563202.91
South Asia	103308.10	147006.89	220184.04	312482.55	366015.92	377981.45	390168.79	402572.98	415184.69	427997.53
South East Asia	39192.45	58239.66	87038.30	132877.87	161593.84	167834.36	174160.29	180566.20	187046.62	193596.44
South Pacific	363.44	704.83	1041.75	1429.77	1694.11	1754.43	1815.89	1878.63	1942.95	2009.05
EUROPE	365642.15	446380.99	512367.52	570076.54	593934.50	597696.21	601192.99	604451.81	607510.49	610401.12
Central Europe	50476.38	67733.40	88721.55	111648.83	121011.30	122743.68	124513.28	126307.89	128105.71	129889.91
Eastern Europe	92377.94	118757.49	142627.00	162653.69	166257.58	166521.52	166666.07	166697.46	166619.36	166438.95
Western Europe	222787.82	259890.10	281018.97	295774.01	306665.62	308431.01	310013.64	311446.47	312785.43	314072.26
LATIN AMERICA AND CARIBBEAN	107536.00	163433.45	234474.21	312669.19	351794.42	359497.08	367248.47	375046.34	382888.03	390771.06
Caribbean	8279.46	11784.76	15832.67	20085.48	21997.97	22374.10	22752.34	23132.51	23514.18	23897.04
Meso-America	23046.25	36339.23	54004.14	72752.10	81594.27	83439.80	85293.97	87154.45	89019.29	90886.70
South America	76210.29	115309.46	164637.40	219831.61	248202.18	253683.18	259202.16	264759.38	270354.56	275987.32
NORTH AMERICA	142654.34	171086.15	188373.64	212816.40	227034.93	230110.44	233209.72	236306.90	239369.21	242373.47
North America	142654.34	171086.15	188373.64	212816.40	227034.93	230110.44	233209.72	236306.90	239369.21	242373.47
POLAR	24.72	34.08	38.15	44.13	45.12	45.28	45.47	45.68	45.89	46.10
Antarctic	----	----	----	----	----	----	----	----	----	----
Arctic	24.72	34.08	38.15	44.13	45.12	45.28	45.47	45.68	45.89	46.10
WEST ASIA	8592.35	15843.87	27886.41	45489.11	54539.02	56542.51	58669.87	60902.55	63209.23	65566.88
Arabian Peninsula	2105.35	4690.47	10704.83	20359.07	23891.18	24788.32	25820.26	26953.78	28134.11	29320.78
Mashriq	6487.00	11153.41	17181.58	25130.04	30647.84	31754.19	32849.62	33948.77	35075.12	36246.10
GLOBAL TOTALS	1014121.35	1352722.05	1750466.09	2283806.63	2554109.14	2610350.91	2667083.00	2724275.08	2781877.52	2839862.96
REPORTED GLOBAL TOTALS FROM ORIGINAL DATA SOURCES	1013541.00	1352856.09	1751903.08	2282981.35	2555910.39	2612575.04	2669654.81	2727120.23	2784918.36	2843017.59

Comments:

Figures for 1960-1995 are estimates and those for years beyond 1995 are projections. The urban population data presented are consistent with the medium variant of the *1998 Revision of World Population Prospects*, the official United Nations population estimates and projections at the national level.

The urban population estimates are based on the varying national definitions of urban areas.

Units: thousand people

Data Source: World Urbanization Prospects: The 1999 Revision

Data Provider: United Nations Population Division/Department of Economic and Social Affairs

Years: 1950, 1955, ..., 2030

	1950	1960	1970	1980	1990	1995	2000	2010	2020	2030
AFRICA	32437	51152	82319	127248	197091	243627	297141	425671	583974	765713
Central Africa	3423	5185	9141	13245	19369	24257	29455	44898	66911	94067
Eastern Africa	1805	3183	5673	9892	17436	22590	29187	46443	70836	102308
Northern Africa	13145	20234	31097	44566	64384	75379	87949	115701	144373	173658
Southern Africa	7326	10694	15606	24159	36545	46076	56368	77941	104717	136108
Western Africa	6193	10924	19220	32958	55624	70567	88207	131601	183946	241659
Western Indian Ocean	545	932	1582	2428	3733	4758	5975	9087	13191	17913
ASIA AND PACIFIC	238435	341326	477992	666875	953759	1090636	1241192	1584212	1982095	2387551
Australia and New Zealand	7557	10086	12965	15088	17217	18331	19308	21241	23401	25375
Central Asia	6135	9542	14225	18658	22752	22953	23389	26362	31862	39292
North West Pacific and East Asia	120839	178341	243815	322666	463095	516799	571674	689195	816628	932937
South Asia	76783	103405	147686	221767	316191	369671	432102	584514	776745	987291
South East Asia	26915	39589	58586	87631	133060	161172	192687	260033	329477	397381
South Pacific	206	363	715	1065	1444	1710	2032	2867	3982	5275
EUROPE	295502	365494	446150	512355	570317	593223	610990	640797	662217	672713
Central Europe	34319	50445	67704	88625	111677	120923	129937	146156	159050	168833
Eastern Europe	65565	92414	118740	142554	162624	166011	167668	171476	173839	173421
Western Europe	195618	222635	259706	281176	296016	306289	313385	323165	329328	330459
LATIN AMERICA AND CARIBBEAN	69146	107568	163488	234591	312664	351821	390869	467377	539042	603871
Caribbean	6052	8280	11784	15834	20104	22103	24017	27941	31899	35611
Meso-America	14695	23034	36359	54042	72723	81650	90925	110264	130295	149682
South America	48399	76254	115345	164715	219837	248068	275927	329172	376848	418578
NORTH AMERICA	109604	142648	170785	188511	212448	226056	238938	264095	290538	313533
North America	109604	142648	170785	188511	212448	226056	238938	264095	290538	313533
POLAR	18	25	34	38	44	45	46	48	51	53
Antarctic	----	----	----	----	----	----	----	----	----	----
Arctic	18	25	34	38	44	45	46	48	51	53
WEST ASIA	4727	8388	15566	27414	45927	54743	65773	90697	118289	145678
Arabian Peninsula	967	2108	4689	10700	20830	24775	30281	42256	56158	70655
Mashriq	3760	6280	10877	16714	25097	29968	35492	48441	62131	75023
GLOBAL TOTALS	749869	1016601	1356334	1757032	2292250	2560151	2844949	3472897	4176206	4889112
REPORTED GLOBAL TOTALS FROM ORIGINAL DATA SOURCES	749934	1016665	1356396	1757102	2292321	2560235	2845049	3473058	4176428	4889393

Comments:

Figures for 1950-1995 are estimates and those for years beyond 1995 are projections. Estimates and projections are presented only for years that are multiples of five. The urban population data presented are consistent with the medium variant of the *1998 Revision of World Population Prospects,* the official United Nations population estimates and projections at the national level.

The urban population estimates are based on the varying national definitions of urban areas.

1.9 Disasters

Disaster: situation or event, which overwhelms local capacity, necessitating a request to national or international level for external assistance; an unforeseen and often sudden event that causes great damage, destruction and human suffering. Though often caused by nature, disasters can have human origins. Wars and civil disturbances that destroy homelands and displace people are included among the causes of disasters. Other causes can be: building collapse, blizzard, drought, epidemic, earthquake, explosion, fire, flood, hazardous material or transportation incident (such as a chemical spill), hurricane, nuclear incident, tornado, or volcano.

1.9.1 Environmental Hazards

For all variables, if not specified otherwise:

Units: number of people

Data Source: EM-DAT: The OFDA/CRED International Disaster Database (data as of December 2001)

Data Provider: The OFDA/CRED International Disaster Database - www.cred.be/emdat - Université Catholique de Louvain - Brussels - Belgium

Years: 1975-2000

Droughts – Affected People

	1975	1980	1990	1995	1996	1997	1998	1999	2000
AFRICA	325000	18135000	13655000	8312904	3260000	3216100	3306700	2822145	19238890
Central Africa	0	0	0	0	0	0	0	0	0
Eastern Africa	0	685000	6500000	1200000	100000	2586200	126000	2522145	17007600
Northern Africa	0	0	600000	0	160000	0	0	275000	700000
Southern Africa	0	6000000	3705000	7112904	3000000	600000	3160000	25000	1300000
Western Africa	325000	1450000	2600000	0	0	29900	20700	0	0
Western Indian Ocean	0	0	250000	0	0	0	0	0	231290
ASIA AND PACIFIC	0	3500000	0	12810000	0	4290000	5268635	26504000	155730000
Australia and New Zealand	0	0	0	1750000	0	0	0	0	0
Central Asia	0	0	0	0	0	0	0	0	3500000
North West Pacific and East Asia	0	0	0	11060000	0	0	0	19000000	20450000
South Asia	0	3500000	0	0	0	0	0	0	131780000
South East Asia	0	0	0	0	0	3090000	4970000	7420000	0
South Pacific	0	0	0	0	0	1200000	298635	84000	0
EUROPE	0	0	0	6000000	0	0	0	0	993000
Central Europe	0	0	0	0	0	0	0	0	0
Eastern Europe	0	0	0	0	0	0	0	0	993000
Western Europe	0	0	0	6000000	0	0	0	0	0
LATIN AMERICA AND CARIBBEAN	0	0	2483160	0	0	358000	11459000	667000	36125
Caribbean	0	0	0	0	0	0	600000	0	0
Meso-America	0	0	0	0	0	290000	0	65000	1125
South America	0	0	2483160	0	0	68000	10859000	602000	35000
NORTH AMERICA	0	0	0	0	0	0	0	0	0
North America	0	0	0	0	0	0	0	0	0
POLAR	0	0	0	0	0	0	0	0	0
Antarctic	0	0	0	0	0	0	0	0	0
Arctic	0	0	0	0	0	0	0	0	0
WEST ASIA	0	0	0	0	0	0	0	509000	479000
Arabian Peninsula	0	0	0	0	0	0	0	0	0
Mashriq	0	0	0	0	0	0	0	509000	479000
GLOBAL TOTALS	325000	21635000	16138160	27122904	3260000	7864100	20034335	30502145	176477015
REPORTED GLOBAL TOTALS FROM ORIGINAL DATA SOURCES	****	****	****	****	****	****	****	****	****

Comments:

Drought: period of deficiency of moisture in the soil such that there is inadequate water required for plants, animals and human beings.

Affected people: people requiring immediate assistance during a period of emergency, for example requiring basic survival needs such as food, water, shelter, sanitation and immediate medical assistance. Appearance of a significant number of cases of an infectious disease introduced in a region or a population that is usually free from that disease.

AGGREGATIONS

Droughts – Killed People

	1975	1980	1990	1991	1992	1993	1994	1995	1996	1997	1998	1999	2000
AFRICA	0	0	0	0	0	0	0	0	0	0	12	0	227
Central Africa	0	0	0	0	0	0	0	0	0	0	0	0	0
Eastern Africa	0	0	0	0	0	0	0	0	0	0	0	0	227
Northern Africa	0	0	0	0	0	0	0	0	0	0	0	0	0
Southern Africa	0	0	0	0	0	0	0	0	0	0	0	0	0
Western Africa	0	0	0	0	0	0	0	0	0	0	12	0	0
Western Indian Ocean	0	0	0	0	0	0	0	0	0	0	0	0	0
ASIA AND PACIFIC	0	0	0	2000	0	0	0	0	0	530	248	0	143
Australia and New Zealand	0	0	0	0	0	0	0	0	0	0	0	0	0
Central Asia	0	0	0	0	0	0	0	0	0	0	0	0	0
North West Pacific and East Asia	0	0	0	2000	0	0	0	0	0	0	0	0	0
South Asia	0	0	0	0	0	0	0	0	0	0	0	0	143
South East Asia	0	0	0	0	0	0	0	0	0	460	220	0	0
South Pacific	0	0	0	0	0	0	0	0	0	70	28	0	0
EUROPE	0	0	0	0	0	0	0	0	0	0	0	0	0
Central Europe	0	0	0	0	0	0	0	0	0	0	0	0	0
Eastern Europe	0	0	0	0	0	0	0	0	0	0	0	0	0
Western Europe	0	0	0	0	0	0	0	0	0	0	0	0	0
LATIN AMERICA AND CARIBBEAN	0	0	0	0	0	0	0	0	0	0	0	0	0
Caribbean	0	0	0	0	0	0	0	0	0	0	0	0	0
Meso-America	0	0	0	0	0	0	0	0	0	0	0	0	0
South America	0	0	0	0	0	0	0	0	0	0	0	0	0
NORTH AMERICA	0	0	0	0	0	0	0	0	0	0	0	0	0
North America	0	0	0	0	0	0	0	0	0	0	0	0	0
POLAR	0	0	0	0	0	0	0	0	0	0	0	0	0
Antarctic	0	0	0	0	0	0	0	0	0	0	0	0	0
Arctic	0	0	0	0	0	0	0	0	0	0	0	0	0
WEST ASIA	0	0	0	0	0	0	0	0	0	0	0	0	0
Arabian Peninsula	0	0	0	0	0	0	0	0	0	0	0	0	0
Mashriq	0	0	0	0	0	0	0	0	0	0	0	0	0
GLOBAL TOTALS	0	0	0	2000	0	0	0	0	0	530	260	0	370
REPORTED GLOBAL TOTALS FROM ORIGINAL DATA SOURCES	****	****	****	****	****	****	****	****	****	****	****	****	****

Comments:

Drought: period of deficiency of moisture in the soil such that there is inadequate water required for plants, animals and human beings.

Killed: persons confirmed as dead and persons missing and presumed dead.

Droughts – Number

	1975	1980	1990	1991	1992	1993	1994	1995	1996	1997	1998	1999	2000
AFRICA	10	13	7	3	17	5	2	8	3	5	7	10	13
Central Africa	1	0	0	0	0	1	0	0	0	0	0	0	0
Eastern Africa	1	4	1	1	2	0	0	1	1	2	2	5	8
Northern Africa	0	1	1	1	1	0	0	0	1	0	0	1	2
Southern Africa	0	2	3	0	12	3	2	7	1	1	3	3	2
Western Africa	8	6	1	1	1	1	0	0	0	2	2	0	0
Western Indian Ocean	0	0	1	0	1	0	0	0	0	0	0	1	1
ASIA AND PACIFIC	1	3	0	5	4	5	3	3	1	4	8	7	12
Australia and New Zealand	1	0	0	1	1	1	1	1	0	0	0	0	0
Central Asia	0	0	0	0	0	0	0	0	0	0	0	0	2
North West Pacific and East Asia	0	1	0	1	1	1	2	2	0	0	0	2	5
South Asia	0	2	0	0	0	1	0	0	1	0	0	1	5
South East Asia	0	0	0	3	2	2	0	0	0	2	4	3	0
South Pacific	0	0	0	0	0	0	0	0	0	2	4	1	0
EUROPE	0	1	1	2	3	1	0	1	0	1	1	2	7
Central Europe	0	0	0	1	2	0	0	0	0	0	0	0	3
Eastern Europe	0	0	0	0	0	1	0	0	0	0	1	0	4
Western Europe	0	1	1	1	1	0	0	1	0	1	0	2	0
LATIN AMERICA AND CARIBBEAN	0	0	2	1	1	0	3	1	1	3	5	4	9
Caribbean	0	0	0	0	0	0	0	0	0	0	1	0	2
Meso-America	0	0	0	0	0	0	1	1	1	1	1	1	4
South America	0	0	2	1	1	0	2	0	0	2	3	3	3
NORTH AMERICA	0	1	0	2	0	0	0	0	0	0	0	1	3
North America	0	1	0	2	0	0	0	0	0	0	0	1	3
POLAR	0	0	0	0	0	0	0	0	0	0	0	0	0
Antarctic	0	0	0	0	0	0	0	0	0	0	0	0	0
Arctic	0	0	0	0	0	0	0	0	0	0	0	0	0
WEST ASIA	1	0	0	0	0	0	0	0	0	0	0	2	3
Arabian Peninsula	1	0	0	0	0	0	0	0	0	0	0	0	0
Mashriq	0	0	0	0	0	0	0	0	0	0	0	2	3
GLOBAL TOTALS	12	18	10	13	25	11	8	13	5	13	21	26	47
REPORTED GLOBAL TOTALS FROM ORIGINAL DATA SOURCES	****	****	****	****	****	****	****	****	****	****	****	****	****

Comments:

Drought: period of deficiency of moisture in the soil such that there is inadequate water required for plants, animals and human beings

Earthquakes – Affected People

	1975	1980	1990	1991	1992	1993	1994	1995	1996	1997	1998	1999	2000
AFRICA	0	478948	10000	0	2250	0	62500	0	0	0	0	25000	75
Central Africa	0	0	0	0	0	0	0	0	0	0	0	0	
Eastern Africa	0	0	0	0	0	0	50000	0	0	0	0	0	
Northern Africa	0	478948	10000	0	2250	0	12500	0	0	0	0	25000	
Southern Africa	0	0	0	0	0	0	0	0	0	0	0	0	75
Western Africa	0	0	0	0	0	0	0	0	0	0	0	0	
Western Indian Ocean	0	0	0	0	0	0	0	0	0	0	0	0	
ASIA AND PACIFIC	0	260000	2213254	609570	260000	207800	569819	1930194	1449198	270805	965655	545685	221495
Australia and New Zealand	0	0	0	0	0	0	5000	0	0	0	0	0	
Central Asia	0	0	0	3000	230000	0	0	0	0	0	0	0	1700
North West Pacific and East Asia	0	0	54574	282205	16000	43800	0	1711512	1443369	0	854300	50600	205309
South Asia	0	240000	521000	324205	6500	154000	116510	0	2500	268000	106355	467925	350
South East Asia	0	20000	1637680	160	7500	0	448309	218682	3329	2805	5000	15160	14136
South Pacific	0	0	0	0	0	10000	0	0	0	0	0	12000	
EUROPE	50000	426300	0	105000	232100	1500	0	75350	1500	130000	745335	986250	2769
Central Europe	50000	5000	0	0	230000	0	0	52850	0	0	45335	865000	2440
Eastern Europe	0	0	0	105000	0	0	0	0	0	0	700000	21250	269
Western Europe	0	421300	0	0	2100	1500	0	22500	1500	130000	0	100000	60
LATIN AMERICA AND CARIBBEAN	0	9000	70300	85520	48500	3331	12461	68257	88013	72000	19250	792205	565
Caribbean	0	0	0	0	7000	0	0	0	0	0	0	0	
Meso-America	0	6000	300	38450	40500	200	0	54735	0	17000	0	40205	565
South America	0	3000	70000	47070	1000	3131	12461	13522	88013	55000	19250	752000	
NORTH AMERICA	0	0	0	0	0	0	288	0	0	0	0	0	
North America	0	0	0	0	0	0	288	0	0	0	0	0	
POLAR	0	0	0	0	0	0	0	0	0	0	0	0	
Antarctic	0	0	0	0	0	0	0	0	0	0	0	0	
Arctic	0	0	0	0	0	0	0	0	0	0	0	0	
WEST ASIA	0	0	0	217000	0	0	0	0	0	0	0	0	
Arabian Peninsula	0	0	0	217000	0	0	0	0	0	0	0	0	
Mashriq	0	0	0	0	0	0	0	0	0	0	0	0	
GLOBAL TOTALS	50000	1174248	2293554	1017090	542850	212631	645068	2073801	1538711	472805	1730240	2349140	224904
REPORTED GLOBAL TOTALS FROM ORIGINAL DATA SOURCES	****	****	****	****	****	****	****	****	****	****	****	****	***

Comments:

Earthquake: sudden break within the upper layers of the earth, sometimes breaking the surface, resulting in the vibration of the ground, which where strong enough will cause the collapse of buildings and destruction of life and property.

Affected people: people requiring immediate assistance during a period of emergency, for example requiring basic survival needs such as food, water, shelter, sanitation and immediate medical assistance. Appearance of a significant number of cases of an infectious disease introduced in a region or a population that is usually free from that disease.

AGGREGATIONS

Earthquakes – Homeless People

	1975	1980	1990	1991	1992	1993	1994	1995	1996	1997	1998	1999	2000
AFRICA	0	443000	0	0	25000	0	10000	0	0	0	0	7500	35
Central Africa	0	0	0	0	0	0	0	0	0	0	0	0	0
Eastern Africa	0	0	0	0	0	0	0	0	0	0	0	0	0
Northern Africa	0	443000	0	0	25000	0	10000	0	0	0	0	7500	0
Southern Africa	0	0	0	0	0	0	0	0	0	0	0	0	35
Western Africa	0	0	0	0	0	0	0	0	0	0	0	0	0
Western Indian Ocean	0	0	0	0	0	0	0	0	0	0	0	0	0
ASIA AND PACIFIC	0	30000	163987	35915	110800	40000	10350	890000	409920	57000	119850	281045	177791
Australia and New Zealand	0	0	0	0	0	0	20	0	0	0	0	0	0
Central Asia	0	0	21500	0	16800	0	0	0	0	0	0	0	0
North West Pacific and East Asia	0	0	7696	600	0	0	0	825000	396170	3000	70600	121000	110791
South Asia	0	30000	0	29915	4000	30000	0	0	750	54000	49250	155345	500
South East Asia	0	0	134791	5400	90000	0	10330	65000	13000	0	0	2700	61500
South Pacific	0	0	0	0	0	10000	0	0	0	0	0	2000	5000
EUROPE	0	0	3000	164000	90000	0	2100	6550	2000	38000	7265	730240	19133
Central Europe	0	0	0	4000	90000	0	0	250	2000	0	6265	655000	0
Eastern Europe	0	0	0	160000	0	0	2100	0	0	0	0	5240	19100
Western Europe	0	0	3000	0	0	0	0	6300	0	38000	1000	70000	33
LATIN AMERICA AND CARIBBEAN	0	0	44000	17295	2500	0	24797	16755	27000	13000	750	453873	1785
Caribbean	0	0	0	0	0	0	0	0	0	0	0	0	0
Meso-America	0	0	14000	17000	0	0	0	12275	0	0	0	453	1785
South America	0	0	30000	295	2500	0	24797	4480	27000	13000	750	453420	0
NORTH AMERICA	0	0	0	0	247	0	25000	0	0	0	0	0	0
North America	0	0	0	0	247	0	25000	0	0	0	0	0	0
POLAR	0	0	0	0	0	0	0	0	0	0	0	0	0
Antarctic	0	0	0	0	0	0	0	0	0	0	0	0	0
Arctic	0	0	0	0	0	0	0	0	0	0	0	0	0
WEST ASIA	0	0	0	150500	0	0	0	0	0	0	0	0	0
Arabian Peninsula	0	0	0	150000	0	0	0	0	0	0	0	0	0
Mashriq	0	0	0	500	0	0	0	0	0	0	0	0	0
GLOBAL TOTALS	0	473000	210987	367710	228547	40000	72247	913305	438920	108000	127865	1472658	198744
REPORTED GLOBAL TOTALS FROM ORIGINAL DATA SOURCES	****	****	****	****	****	****	****	****	****	****	****	****	****

Comments:

Earthquake: sudden break within the upper layers of the earth, sometimes breaking the surface, resulting in the vibration of the ground, which where strong enough will cause the collapse of buildings and destruction of life and property.

Homeless: people needing immediate assistance in the form of shelter.

Earthquakes – Injured People

	1980	1990	1991	1992	1993	1994	1995	1996	1997	1998	1999	2000
AFRICA	8459	0	0	9929	15	289	69	0	0	0	175	6
Central Africa	0	0	0	0	0	0	0	0	0	0	0	0
Eastern Africa	0	0	0	0	0	0	0	0	0	0	0	0
Northern Africa	8459	0	0	9929	15	289	69	0	0	0	175	0
Southern Africa	0	0	0	0	0	0	0	0	0	0	0	6
Western Africa	0	0	0	0	0	0	0	0	0	0	0	0
Western Indian Ocean	0	. 0	0	0	0	0	0	0	0	0	0	0
ASIA AND PACIFIC	5811	111029	2748	2385	16927	3796	39769	17922	11813	16628	10602	5984
Australia and New Zealand	0	0	0	0	0	5	0	0	0	0	0	0
Central Asia	0	16	6	100	0	0	0	0	0	0	0	0
North West Pacific and East Asia	67	2065	553	32	1021	1178	37391	17490	106	13627	8994	3116
South Asia	5744	105346	2008	150	15615	461	0	0	11407	2913	1248	15
South East Asia	0	3600	181	2103	0	2152	2378	432	300	88	260	2853
South Pacific	0	2	0	0	291	0	0	0	0	0	100	0
EUROPE	8417	1097	1705	2055	16	242	1233	0	100	2633	51847	689
Central Europe	100	820	5	2000	0	0	363	0	0	2533	49792	81
Eastern Europe	0	0	1700	10	0	242	750	0	0	0	35	608
Western Europe	8317	277	0	45	16	0	120	0	100	100	2020	0
LATIN AMERICA AND CARIBBEAN	400	1119	1636	599	125	158	717	765	781	190	9043	42
Caribbean	0	0	0	50	0	0	0	0	0	0	0	0
Meso-America	360	309	849	489	40	0	300	0	0	0	400	42
South America	40	810	787	60	85	158	417	765	781	190	8643	0
NORTH AMERICA	0	30	104	400	0	8500	0	0	0	0	0	41
North America	0	30	104	400	0	8500	0	0	0	0	0	41
POLAR	0	0	0	0	0	0	0	0	0	0	0	0
Antarctic	0	0	0	0	0	0	0	0	0	0	0	0
Arctic	0	0	0	0	0	0	0	0	0	0	0	0
WEST ASIA	0	0	40	0	0	0	0	0	0	0	0	0
Arabian Peninsula	0	0	40	0	0	0	0	0	0	0	0	0
Mashriq	0	0	0	0	0	0	0	0	0	0	0	0
GLOBAL TOTALS	23087	113275	6233	15368	17083	12985	41788	18687	12694	19451	71667	6762
REPORTED GLOBAL TOTALS FROM ORIGINAL DATA SOURCES	****	****	****	****	****	****	****	****	****	****	****	****

Comments:

Earthquake: sudden break within the upper layers of the earth, sometimes breaking the surface, resulting in the vibration of the ground, which where strong enough will cause the collapse of buildings and destruction of life and property.

Injured: people suffering from physical injuries, trauma or an illness requiring medical treatment as a direct result of a disaster.

AGGREGATIONS

Earthquakes – Killed People

	1975	1980	1990	1991	1992	1993	1994	1995	1996	1997	1998	1999	2000
AFRICA	0	2635	31	0	561	3	178	10	0	0	0	22	1
Central Africa	0	0	0	0	0	0	0	0	0	0	0	0	0
Eastern Africa	0	0	0	0	0	0	7	0	0	0	0	0	0
Northern Africa	0	2635	31	0	561	3	171	10	0	0	0	22	0
Southern Africa	0	0	0	0	0	0	0	0	0	0	0	0	1
Western Africa	0	0	0	0	0	0	0	0	0	0	0	0	0
Western Indian Ocean	0	0	0	0	0	0	0	0	0	0	0	0	0
ASIA AND PACIFIC	347	215	42597	2397	2626	10098	722	5681	541	2973	7149	2491	147
Australia and New Zealand	0	0	0	0	0	0	0	0	0	0	0	0	0
Central Asia	0	0	1	4	67	0	0	0	0	0	0	0	0
North West Pacific and East Asia	300	2	127	1	0	234	13	5571	353	58	54	2264	10
South Asia	47	213	40046	2369	59	9806	174	0	0	2895	7045	204	1
South East Asia	0	0	2423	23	2500	0	535	110	188	20	50	11	135
South Pacific	0	0	0	0	0	58	0	0	0	0	0	12	1
EUROPE	2385	4759	45	278	549	1	11	2144	0	14	155	18122	34
Central Europe	2385	0	18	0	547	0	0	103	0	0	145	17980	3
Eastern Europe	0	0	4	278	1	0	11	1989	0	0	0	2	31
Western Europe	0	4759	23	0	1	1	0	52	0	14	10	140	0
LATIN AMERICA AND CARIBBEAN	0	80	140	155	190	11	271	131	41	89	108	1235	7
Caribbean	0	0	0	0	0	0	0	0	0	0	0	0	0
Meso-America	0	67	1	91	179	3	0	69	0	1	0	48	7
South America	0	13	139	64	11	8	271	62	41	88	108	1187	0
NORTH AMERICA	0	0	0	2	1	0	60	0	0	0	0	0	0
North America	0	0	0	2	1	0	60	0	0	0	0	0	0
POLAR	0	0	0	0	0	0	0	0	0	0	0	0	0
Antarctic	0	0	0	0	0	0	0	0	0	0	0	0	0
Arctic	0	0	0	0	0	0	0	0	0	0	0	0	0
WEST ASIA	0	0	0	31	0	0	0	0	0	0	0	0	0
Arabian Peninsula	0	0	0	11	0	0	0	0	0	0	0	0	0
Mashriq	0	0	0	20	0	0	0	0	0	0	0	0	0
GLOBAL TOTALS	2732	7689	42813	2863	3927	10113	1242	7966	582	3076	7412	21870	189
REPORTED GLOBAL TOTALS FROM ORIGINAL DATA SOURCES	****	****	****	****	****	****	****	****	****	****	****	****	****

Comments:

Earthquake: sudden break within the upper layers of the earth, sometimes breaking the surface, resulting in the vibration of the ground, which where strong enough will cause the collapse of buildings and destruction of life and property.

Killed: persons confirmed as dead and persons missing and presumed dead.

	1975	1980	1990	1991	1992	1993	1994	1995	1996	1997	1998	1999	2000
AFRICA	0	2	3	0	2	1	2	1	0	0	0	1	1
Central Africa	0	0	0	0	0	0	0	0	0	0	0	0	0
Eastern Africa	0	0	2	0	0	0	1	0	0	0	0	0	0
Northern Africa	0	2	1	0	2	1	1	1	0	0	0	1	0
Southern Africa	0	0	0	0	0	0	0	0	0	0	0	0	1
Western Africa	0	0	0	0	0	0	0	0	0	0	0	0	0
Western Indian Ocean	0	0	0	0	0	0	0	0	0	0	0	0	0
ASIA AND PACIFIC	2	15	21	12	11	8	15	10	7	9	9	14	17
Australia and New Zealand	0	0	0	0	0	0	1	0	0	0	0	0	0
Central Asia	0	0	2	2	3	0	0	0	0	0	0	0	2
North West Pacific and East Asia	1	4	2	2	2	3	4	5	4	2	3	5	7
South Asia	1	9	5	5	4	2	4	0	1	6	5	6	3
South East Asia	0	2	5	3	2	0	5	5	2	1	1	2	4
South Pacific	0	0	7	0	0	3	1	0	0	0	0	1	1
EUROPE	1	5	9	4	5	1	1	5	2	1	5	11	6
Central Europe	1	2	3	1	1	0	0	2	1	0	3	6	2
Eastern Europe	0	0	1	3	1	0	1	1	0	0	1	3	2
Western Europe	0	3	5	0	3	1	0	2	1	1	1	2	2
LATIN AMERICA AND CARIBBEAN	0	4	5	7	3	4	2	9	2	4	2	7	1
Caribbean	0	0	0	0	1	0	0	0	0	0	0	0	0
Meso-America	0	2	3	4	1	2	0	3	0	2	0	4	1
South America	0	2	2	3	1	2	2	6	2	2	2	3	0
NORTH AMERICA	0	0	1	1	2	0	2	0	0	0	0	0	1
North America	0	0	1	1	2	0	2	0	0	0	0	0	1
POLAR	0	0	0	0	0	0	0	0	0	0	0	0	0
Antarctic	0	0	0	0	0	0	0	0	0	0	0	0	0
Arctic	0	0	0	0	0	0	0	0	0	0	0	0	0
WEST ASIA	0	0	0	2	0	0	0	0	0	0	0	0	0
Arabian Peninsula	0	0	0	1	0	0	0	0	0	0	0	0	0
Mashriq	0	0	0	1	0	0	0	0	0	0	0	0	0
GLOBAL TOTALS	3	26	39	26	23	14	22	25	11	14	16	33	26
REPORTED GLOBAL TOTALS FROM ORIGINAL DATA SOURCES	****	****	****	****	****	****	****	****	****	****	****	****	****

Comments:

Earthquake: sudden break within the upper layers of the earth, sometimes breaking the surface, resulting in the vibration of the ground, which where strong enough will cause the collapse of buildings and destruction of life and property.

Earthquakes – Total Affected

	1975	1980	1990	1991	1992	1993	1994	1995	1996	1997	1998	1999	2000
AFRICA	0	930407	10000	0	37179	15	72789	69	0	0	0	32675	791
Central Africa	0	0	0	0	0	0	0	0	0	0	0	0	0
Eastern Africa	0	0	0	0	0	0	50000	0	0	0	0	0	0
Northern Africa	0	930407	10000	0	37179	15	22789	69	0	0	0	32675	0
Southern Africa	0	0	0	0	0	0	0	0	0	0	0	0	791
Western Africa	0	0	0	0	0	0	0	0	0	0	0	0	0
Western Indian Ocean	0	0	0	0	0	0	0	0	0	0	0	0	0
ASIA AND PACIFIC	0	295811	2488270	648073	373185	264727	585125	2859963	1877040	339618	1102133	837332	2398725
Australia and New Zealand	0	0	0	0	0	0	5025	0	0	0	0	0	0
Central Asia	0	0	21516	3006	246900	0	0	0	0	0	0	0	17000
North West Pacific and East Asia	0	67	64335	283358	16032	44821	2338	2573903	1857029	3106	938527	180594	2166997
South Asia	0	275744	626346	356128	10650	199615	116971	0	3250	333407	158518	624518	4015
South East Asia	0	20000	1776071	5581	99603	0	460791	286060	16761	3105	5088	18120	205713
South Pacific	0	0	2	0	0	20291	0	0	0	0	0	14100	5000
EUROPE	53372	434717	4097	270705	324145	1516	2342	83133	3500	168100	755233	1768337	47516
Central Europe	53372	5100	820	4005	322000	0	0	53463	2000	0	54133	1569792	24481
Eastern Europe	0	0	0	266700	0	0	2342	750	0	0	700000	26525	22402
Western Europe	0	429617	3277	0	2145	1516	0	28920	1500	168100	1100	172020	633
LATIN AMERICA AND CARIBBEAN	0	9400	115419	104451	51599	3456	37416	85729	115778	85781	20190	1255121	7477
Caribbean	0	0	0	0	7050	0	0	0	0	0	0	0	0
Meso-America	0	6360	14609	56299	40989	240	0	67310	0	17000	0	41058	7477
South America	0	3040	100810	48152	3560	3216	37416	18419	115778	68781	20190	1214063	0
NORTH AMERICA	0	0	30	104	647	0	33788	0	0	0	0	0	41
North America	0	0	30	104	647	0	33788	0	0	0	0	0	41
POLAR	0	0	0	0	0	0	0	0	0	0	0	0	0
Antarctic	0	0	0	0	0	0	0	0	0	0	0	0	0
Arctic	0	0	0	0	0	0	0	0	0	0	0	0	0
WEST ASIA	0	0	0	367540	0	0	0	0	0	0	0	0	0
Arabian Peninsula	0	0	0	367040	0	0	0	0	0	0	0	0	0
Mashriq	0	0	0	500	0	0	0	0	0	0	0	0	0
GLOBAL TOTALS	53372	1670335	2617816	1390873	786755	269714	731460	3028894	1996318	593499	1877556	3893465	2454550
REPORTED GLOBAL TOTALS FROM ORIGINAL DATA SOURCES	****	****	****	****	****	****	****	****	****	****	****	****	****

Comments:

Earthquake: sudden break within the upper layers of the earth, sometimes breaking the surface, resulting in the vibration of the ground, which where strong enough will cause the collapse of buildings and destruction of life and property.

Total affected: people that have been injured affected and left homeless after a disaster are included in this category.

Extreme Temperatures - Affected People

	1975	1980	1990	1991	1992	1993	1994	1995	1996	1997	1998	1999	2000
AFRICA	0	0	1000000	0	0	0	0	0	0	0	0	0	0
Central Africa	0	0	0	0	0	0	0	0	0	0	0	0	0
Eastern Africa	0	0	0	0	0	0	0	0	0	0	0	0	0
Northern Africa	0	0	0	0	0	0	0	0	0	0	0	0	0
Southern Africa	0	0	0	0	0	0	0	0	0	0	0	0	0
Western Africa	0	0	1000000	0	0	0	0	0	0	0	0	0	0
Western Indian Ocean	0	0	0	0	0	0	0	0	0	0	0	0	0
ASIA AND PACIFIC	0	0	0	450	0	3000000	1100000	500000	0	603180	34000	0	0
Australia and New Zealand	0	0	0	0	0	3000000	1100000	500000	0	0	0	0	0
Central Asia	0	0	0	0	0	0	0	0	0	600000	0	0	0
North West Pacific and East Asia	0	0	0	0	0	0	0	0	0	3180	0	0	0
South Asia	0	0	0	450	0	0	0	0	0	0	34000	0	0
South East Asia	0	0	0	0	0	0	0	0	0	0	0	0	0
South Pacific	0	0	0	0	0	0	0	0	0	0	0	0	0
EUROPE	0	0	0	0	240	200	8000	0	0	10000	2000	725000	0
Central Europe	0	0	0	0	0	0	8000	0	0	0	2000	0	0
Eastern Europe	0	0	0	0	0	200	0	0	0	0	0	725000	0
Western Europe	0	0	0	0	240	0	0	0	0	10000	0	0	0
LATIN AMERICA AND CARIBBEAN	0	0	0	0	0	0	0	35000	0	1400	0	0	25977
Caribbean	0	0	0	0	0	0	0	0	0	0	0	0	0
Meso-America	0	0	0	0	0	0	0	0	0	1400	0	0	0
South America	0	0	0	0	0	0	0	35000	0	0	0	0	25977
NORTH AMERICA	0	0	0	0	0	0	0	0	0	0	0	0	0
North America	0	0	0	0	0	0	0	0	0	0	0	0	0
POLAR	0	0	0	0	0	0	0	0	0	0	0	0	0
Antarctic	0	0	0	0	0	0	0	0	0	0	0	0	0
Arctic	0	0	0	0	0	0	0	0	0	0	0	0	0
WEST ASIA	0	0	0	0	0	0	0	0	0	0	0	0	0
Arabian Peninsula	0	0	0	0	0	0	0	0	0	0	0	0	0
Mashriq	0	0	0	0	0	0	0	0	0	0	0	0	0
GLOBAL TOTALS	0	0	1000000	450	240	3000200	1108000	535000	0	614580	36000	725000	25977
REPORTED GLOBAL TOTALS FROM ORIGINAL DATA SOURCES	****	****	****	****	****	****	****	****	****	****	****	****	****

Comments:

Extreme temperature: disaster type term comprising the two disaster subsets "heat wave" and "cold wave" (long lasting period with extremely high or low surface temperature).

Affected people: people requiring immediate assistance during a period of emergency, for example requiring basic survival needs such as food, water, shelter, sanitation and immediate medical assistance. Appearance of a significant number of cases of an infectious disease introduced in a region or a population that is usually free from that disease.

Extreme Temperatures - Homeless People

	1975	1980	1990	1991	1992	1993	1994	1995	1996	1997	1998	1999	2000
AFRICA	0	0	0	0	0	0	0	0	0	0	0	0	0
Central Africa	0	0	0	0	0	0	0	0	0	0	0	0	0
Eastern Africa	0	0	0	0	0	0	0	0	0	0	0	0	0
Northern Africa	0	0	0	0	0	0	0	0	0	0	0	0	0
Southern Africa	0	0	0	0	0	0	0	0	0	0	0	0	0
Western Africa	0	0	0	0	0	0	0	0	0	0	0	0	0
Western Indian Ocean	0	0	0	0	0	0	0	0	0	0	0	0	0
ASIA AND PACIFIC	0	0	0	0	0	0	0	0	0	0	0	0	0
Australia and New Zealand	0	0	0	0	0	0	0	0	0	0	0	0	0
Central Asia	0	0	0	0	0	0	0	0	0	0	0	0	0
North West Pacific and East Asia	0	0	0	0	0	0	0	0	0	0	0	0	0
South Asia	0	0	0	0	0	0	0	0	0	0	0	0	0
South East Asia	0	0	0	0	0	0	0	0	0	0	0	0	0
South Pacific	0	0	0	0	0	0	0	0	0	0	0	0	0
EUROPE	0	0	0	0	0	0	0	0	0	0	0	0	0
Central Europe	0	0	0	0	0	0	0	0	0	0	0	0	0
Eastern Europe	0	0	0	0	0	0	0	0	0	0	0	0	0
Western Europe	0	0	0	0	0	0	0	0	0	0	0	0	0
LATIN AMERICA AND CARIBBEAN	0	0	0	0	16000	0	0	0	0	0	0	0	0
Caribbean	0	0	0	0	0	0	0	0	0	0	0	0	0
Meso-America	0	0	0	0	16000	0	0	0	0	0	0	0	0
South America	0	0	0	0	0	0	0	0	0	0	0	0	0
NORTH AMERICA	0	0	0	0	0	0	0	0	0	0	0	0	0
North America	0	0	0	0	0	0	0	0	0	0	0	0	0
POLAR	0	0	0	0	0	0	0	0	0	0	0	0	0
Antarctic	0	0	0	0	0	0	0	0	0	0	0	0	0
Arctic	0	0	0	0	0	0	0	0	0	0	0	0	0
WEST ASIA	0	0	0	0	0	0	0	0	0	0	0	0	0
Arabian Peninsula	0	0	0	0	0	0	0	0	0	0	0	0	0
Mashriq	0	0	0	0	0	0	0	0	0	0	0	0	0
GLOBAL TOTALS	0	0	0	0	16000	0	0	0	0	0	0	0	0
REPORTED GLOBAL TOTALS FROM ORIGINAL DATA SOURCES	****	****	****	****	****	****	****	****	****	****	****	****	****

Comments:

Extreme temperature: disaster type term comprising the two disaster subsets "heat wave" and "cold wave" (long lasting period with extremely high or low surface temperature).

Homeless: people needing immediate assistance in the form of shelter.

AGGREGATIONS

Extreme Temperatures - Injured People

	1975	1980	1990	1991	1992	1993	1994	1995	1996	1997	1998	1999	2000
AFRICA	0	0	0	0	0	0	0	0	0	0	0	0	105
Central Africa	0	0	0	0	0	0	0	0	0	0	0	0	0
Eastern Africa	0	0	0	0	0	0	0	0	0	0	0	0	0
Northern Africa	0	0	0	0	0	0	0	0	0	0	0	0	105
Southern Africa	0	0	0	0	0	0	0	0	0	0	0	0	0
Western Africa	0	0	0	0	0	0	0	0	0	0	0	0	0
Western Indian Ocean	0	0	0	0	0	0	0	0	0	0	0	0	0
ASIA AND PACIFIC	0	0	0	0	0	500	184	100	0	0	0	0	0
Australia and New Zealand	0	0	0	0	0	500	184	100	0	0	0	0	0
Central Asia	0	0	0	0	0	0	0	0	0	0	0	0	0
North West Pacific and East Asia	0	0	0	0	0	0	0	0	0	0	0	0	0
South Asia	0	0	0	0	0	0	0	0	0	0	0	0	0
South East Asia	0	0	0	0	0	0	0	0	0	0	0	0	0
South Pacific	0	0	0	0	0	0	0	0	0	0	0	0	0
EUROPE	0	0	0	0	0	60	0	178	200	0	386	246	1428
Central Europe	0	0	0	0	0	0	0	0	200	0	123	0	1070
Eastern Europe	0	0	0	0	0	60	0	108	0	0	263	246	346
Western Europe	0	0	0	0	0	0	0	70	0	0	0	0	12
LATIN AMERICA AND CARIBBEAN	600	0	0	0	0	0	0	0	0	0	0	0	0
Caribbean	0	0	0	0	0	0	0	0	0	0	0	0	0
Meso-America	0	0	0	0	0	0	0	0	0	0	0	0	0
South America	600	0	0	0	0	0	0	0	0	0	0	0	0
NORTH AMERICA	0	0	0	0	0	0	0	0	0	0	0	0	0
North America	0	0	0	0	0	0	0	0	0	0	0	0	0
POLAR	0	0	0	0	0	0	0	0	0	0	0	0	0
Antarctic	0	0	0	0	0	0	0	0	0	0	0	0	0
Arctic	0	0	0	0	0	0	0	0	0	0	0	0	0
WEST ASIA	0	0	0	0	0	0	0	0	0	0	0	0	12
Arabian Peninsula	0	0	0	0	0	0	0	0	0	0	0	0	0
Mashriq	0	0	0	0	0	0	0	0	0	0	0	0	12
GLOBAL TOTALS	600	0	0	0	0	560	184	278	200	0	386	246	1545
REPORTED GLOBAL TOTALS FROM ORIGINAL DATA SOURCES	****	****	****	****	****	****	****	****	****	****	****	****	****

Comments:

Extreme temperature: disaster type term comprising the two disaster subsets "heat wave" and "cold wave" (long lasting period with extremely high or low surface temperature).

Injured: people suffering from physical injuries, trauma or an illness requiring medical treatment as a direct result of a disaster.

AGGREGATIONS

Extreme Temperatures - Killed People

	1975	1980	1990	1991	1992	1993	1994	1995	1996	1997	1998	1999	2000
AFRICA	0	0	0	0	18	0	0	32	52	0	0	0	3
Central Africa	0	0	0	0	0	0	0	0	0	0	0	0	0
Eastern Africa	0	0	0	0	0	0	0	0	0	0	0	0	0
Northern Africa	0	0	0	0	0	0	0	32	22	0	0	0	3
Southern Africa	0	0	0	0	0	0	0	0	30	0	0	0	0
Western Africa	0	0	0	0	18	0	0	0	0	0	0	0	0
Western Indian Ocean	0	0	0	0	0	0	0	0	0	0	0	0	0
ASIA AND PACIFIC	14	129	596	785	275	17	299	679	232	188	2761	151	366
Australia and New Zealand	0	0	0	0	0	17	5	1	0	0	0	0	0
Central Asia	0	0	0	0	0	0	0	0	0	0	0	0	11
North West Pacific and East Asia	0	0	103	0	0	0	104	0	0	55	0	0	0
South Asia	14	129	493	785	275	0	190	678	232	133	2761	151	355
South East Asia	0	0	0	0	0	0	0	0	0	0	0	0	0
South Pacific	0	0	0	0	0	0	0	0	0	0	0	0	0
EUROPE	0	0	5	25	0	41	42	258	16	139	220	348	337
Central Europe	0	0	0	0	0	0	0	0	16	50	199	186	161
Eastern Europe	0	0	0	0	0	41	42	228	0	22	11	162	173
Western Europe	0	0	5	25	0	0	0	30	0	67	10	0	3
LATIN AMERICA AND CARIBBEAN	126	0	380	25	80	0	0	51	0	292	127	15	109
Caribbean	0	0	0	0	0	0	0	0	0	0	0	0	0
Meso-America	0	0	380	0	80	0	0	49	0	292	127	15	66
South America	126	0	0	25	0	0	0	2	0	0	0	0	43
NORTH AMERICA	0	362	11	0	0	48	75	710	0	0	117	257	35
North America	0	362	11	0	0	48	75	710	0	0	117	257	35
POLAR	0	0	0	0	0	0	0	0	0	0	0	0	0
Antarctic	0	0	0	0	0	0	0	0	0	0	0	0	0
Arctic	0	0	0	0	0	0	0	0	0	0	0	0	0
WEST ASIA	0	0	0	0	15	0	0	0	0	0	0	0	0
Arabian Peninsula	0	0	0	0	0	0	0	0	0	0	0	0	0
Mashriq	0	0	0	0	15	0	0	0	0	0	0	0	0
GLOBAL TOTALS	140	491	992	835	388	106	416	1730	300	619	3225	771	850
REPORTED GLOBAL TOTALS FROM ORIGINAL DATA SOURCES	****	****	****	****	****	****	****	****	****	****	****	****	****

Comments:

Extreme temperature: disaster type term comprising the two disaster subsets "heat wave" and "cold wave" (long lasting period with extremely high or low surface temperature).

Killed: persons confirmed as dead and persons missing and presumed dead.

AGGREGATIONS

197

Extreme Temperatures - Number

	1975	1980	1990	1991	1992	1993	1994	1995	1996	1997	1998	1999	2000
AFRICA	0	0	1	0	1	0	0	1	2	0	0	0	2
Central Africa	0	0	0	0	0	0	0	0	0	0	0	0	0
Eastern Africa	0	0	0	0	0	0	0	0	0	0	0	0	0
Northern Africa	0	0	0	0	0	0	0	1	1	0	0	0	2
Southern Africa	0	0	0	0	0	0	0	0	1	0	0	0	0
Western Africa	0	0	1	0	1	0	0	0	0	0	0	0	0
Western Indian Ocean	0	0	0	0	0	0	0	0	0	0	0	0	0
ASIA AND PACIFIC	2	2	8	3	1	1	5	3	2	6	3	2	5
Australia and New Zealand	0	0	0	0	0	1	2	1	0	0	0	0	0
Central Asia	0	0	0	0	0	0	0	0	0	1	0	0	1
North West Pacific and East Asia	1	0	3	0	0	0	1	0	0	3	0	0	0
South Asia	1	2	5	3	1	0	2	2	2	2	3	2	4
South East Asia	0	0	0	0	0	0	0	0	0	0	0	0	0
South Pacific	0	0	0	0	0	0	0	0	0	0	0	0	0
EUROPE	0	0	1	4	1	2	3	4	1	5	7	4	15
Central Europe	0	0	0	0	0	1	2	0	1	1	5	2	8
Eastern Europe	0	0	0	0	0	1	1	2	0	1	1	2	5
Western Europe	0	0	1	4	1	0	0	2	0	3	1	0	2
LATIN AMERICA AND CARIBBEAN	2	0	2	1	2	0	1	3	0	2	2	1	7
Caribbean	0	0	0	0	0	0	0	0	0	0	0	0	0
Meso-America	0	0	2	0	2	0	0	1	0	2	2	1	1
South America	2	0	0	1	0	0	1	2	0	0	0	0	6
NORTH AMERICA	0	3	2	0	1	1	1	2	0	0	1	1	1
North America	0	3	2	0	1	1	1	2	0	0	1	1	1
POLAR	0	0	0	0	0	0	0	0	0	0	0	0	0
Antarctic	0	0	0	0	0	0	0	0	0	0	0	0	0
Arctic	0	0	0	0	0	0	0	0	0	0	0	0	0
WEST ASIA	0	0	0	0	1	0	0	0	0	0	0	0	1
Arabian Peninsula	0	0	0	0	0	0	0	0	0	0	0	0	0
Mashriq	0	0	0	0	1	0	0	0	0	0	0	0	1
GLOBAL TOTALS	4	5	14	8	7	4	10	13	5	13	13	8	31
REPORTED GLOBAL TOTALS FROM ORIGINAL DATA SOURCES	****	****	****	****	****	****	****	****	****	****	****	****	****

Comments:

Extreme temperature: disaster type term comprising the two disaster subsets "heat wave" and "cold wave" (long lasting period with extremely high or low surface temperature).

Extreme Temperatures - Total Affected

	1975	1980	1990	1991	1992	1993	1994	1995	1996	1997	1998	1999	2000
AFRICA	0	0	1000000	0	0	0	0	0	0	0	0	0	105
Central Africa	0	0	0	0	0	0	0	0	0	0	0	0	0
Eastern Africa	0	0	0	0	0	0	0	0	0	0	0	0	0
Northern Africa	0	0	0	0	0	0	0	0	0	0	0	0	105
Southern Africa	0	0	0	0	0	0	0	0	0	0	0	0	0
Western Africa	0	0	1000000	0	0	0	0	0	0	0	0	0	0
Western Indian Ocean	0	0	0	0	0	0	0	0	0	0	0	0	0
ASIA AND PACIFIC	0	0	0	450	0	3000500	1100184	500100	0	603180	34000	0	0
Australia and New Zealand	0	0	0	0	0	3000500	1100184	500100	0	0	0	0	0
Central Asia	0	0	0	0	0	0	0	0	0	600000	0	0	0
North West Pacific and East Asia	0	0	0	0	0	0	0	0	0	3180	0	0	0
South Asia	0	0	0	450	0	0	0	0	0	0	34000	0	0
South East Asia	0	0	0	0	0	0	0	0	0	0	0	0	0
South Pacific	0	0	0	0	0	0	0	0	0	0	0	0	0
EUROPE	0	0	0	0	240	260	8000	178	200	10000	2386	725246	1428
Central Europe	0	0	0	0	0	0	8000	0	200	0	2123	0	1070
Eastern Europe	0	0	0	0	0	260	0	108	0	0	263	725246	346
Western Europe	0	0	0	0	240	0	0	70	0	10000	0	0	12
LATIN AMERICA AND CARIBBEAN	600	0	0	0	16000	0	0	35000	0	1400	0	0	25977
Caribbean	0	0	0	0	0	0	0	0	0	0	0	0	0
Meso-America	0	0	0	0	16000	0	0	0	0	1400	0	0	0
South America	600	0	0	0	0	0	0	35000	0	0	0	0	25977
NORTH AMERICA	0	0	0	0	0	0	0	0	0	0	0	0	0
North America	0	0	0	0	0	0	0	0	0	0	0	0	0
POLAR	0	0	0	0	0	0	0	0	0	0	0	0	0
Antarctic	0	0	0	0	0	0	0	0	0	0	0	0	0
Arctic	0	0	0	0	0	0	0	0	0	0	0	0	0
WEST ASIA	0	0	0	0	0	0	0	0	0	0	0	0	12
Arabian Peninsula	0	0	0	0	0	0	0	0	0	0	0	0	0
Mashriq	0	0	0	0	0	0	0	0	0	0	0	0	12
GLOBAL TOTALS	600	0	1000000	450	16240	3000760	1108184	535278	200	614580	36386	725246	27522
REPORTED GLOBAL TOTALS FROM ORIGINAL DATA SC	****	****	****	****	****	****	****	****	****	****	****	****	****

Comments:

Extreme temperature: disaster type term comprising the two disaster subsets "heat wave" and "cold wave" (Long lasting period with extremely high or low surface temperature).

Total affected: people that have been injured affected and left homeless after a disaster are included in this category.

Floods - Affected People

	1975	1980	1990	1995	1996	1997	1998	1999	2000
AFRICA	248000	0	555000	856583	770901	2503453	3592689	2798126	2296582
Central Africa	0	0	15000	42300	16000	26000	20000	253506	0
Eastern Africa	16000	0	350000	102375	90000	1517000	600500	23100	153500
Northern Africa	157000	0	48000	38000	106000	100000	900000	1920741	650
Southern Africa	75000	0	142000	8500	207000	808328	1315000	72000	2137776
Western Africa	0	0	0	665408	351901	51000	737189	528779	4656
Western Indian Ocean	0	0	0	0	0	1125	20000	0	0
ASIA AND PACIFIC	30003309	44338943	45391233	190695553	173199133	39070311	268514772	141932281	58036610
Australia and New Zealand	0	2900	6000	0	12700	600	13075	350	200
Central Asia	0	0	0	0	180000	0	62235	6366	2500
North West Pacific and East Asia	3210	80000	42189233	130713400	153301865	7000000	223292156	105010544	1435000
South Asia	27000000	42967300	2452500	54582331	9164150	31915328	45013776	28239485	44115173
South East Asia	3000099	1288743	743500	5399822	10540418	154383	133530	8675536	12483737
South Pacific	0	0	0	0	0	0	0	0	0
EUROPE	1000000	64800	4500	3634100	6630	357773	95490	539895	187191
Central Europe	1000000	64500	4500	26200	0	258225	25250	235705	71500
Eastern Europe	0	0	0	3203100	0	84148	66100	2960	58501
Western Europe	0	300	0	404800	6630	15400	4140	301230	57190
LATIN AMERICA AND CARIBBEAN	748000	503350	211336	365227	169066	235207	1391300	1904352	794541
Caribbean	0	0	5000	33180	11066	35000	0	0	675
Meso-America	0	141350	166211	54547	95000	0	405000	1145805	10600
South America	748000	362000	40125	277500	63000	200207	986300	758547	783266
NORTH AMERICA	0	111000	0	9690	164000	338890	26500	0	1846
North America	0	111000	0	9690	164000	338890	26500	0	1846
POLAR	0	0	0	0	0	0	0	0	0
Antarctic	0	0	0	0	0	0	0	0	0
Arctic	0	0	0	0	0	0	0	0	0
WEST ASIA	50000	0	0	0	124000	200	3000	12250	0
Arabian Peninsula	50000	0	0	0	124000	200	3000	12250	0
Mashriq	0	0	0	0	0	0	0	0	0
GLOBAL TOTALS	32049309	45018093	46162069	195561153	174433730	42505834	273623751	147186904	61316770
REPORTED GLOBAL TOTALS FROM ORIGINAL DATA SOURCES	****	****	****	****	****	****	****	****	****

Comments:

Flood: significant rise of water level in a stream, lake, reservoir or coastal region.

Affected people: people requiring immediate assistance during a period of emergency, for example requiring basic survival needs such as food, water, shelter, sanitation and immediate medical assistance. Appearance of a significant number of cases of an infectious disease introduced in a region or a population that is usually free from that disease.

AGGREGATIONS

Floods - Homeless People

	1975	1980	1990	1995	1996	1997	1998	1999	2000
AFRICA	0	0	80868	351792	55526	248220	268850	408037	372415
Central Africa	0	0	12000	42300	1500	9506	10200	86919	300
Eastern Africa	0	0	0	0	0	233500	2400	125000	250500
Northern Africa	0	0	48000	3000	47700	3000	100000	75700	815
Southern Africa	0	0	20868	23300	5500	2104	4600	1200	115800
Western Africa	0	0	0	283192	826	0	151650	119218	5000
Western Indian Ocean	0	0	0	0	0	110	0	0	0
ASIA AND PACIFIC	7000150	150000	552507	1237285	5152050	512633	17534714	1989834	274861
Australia and New Zealand	0	0	0	0	500	0	0	0	0
Central Asia	0	0	0	0	0	0	12244	3003	0
North West Pacific and East Asia	150	0	550507	502150	4468500	363115	17471000	1253816	20586
South Asia	7000000	150000	2000	643750	577050	115750	50000	471000	254000
South East Asia	0	0	0	91385	106000	33768	1470	224015	275
South Pacific	0	0	0	0	0	0	0	38000	0
EUROPE	0	0	0	156000	5200	196789	65096	16107	431
Central Europe	0	0	0	0	5200	196293	1756	13954	431
Eastern Europe	0	0	0	156000	0	296	60340	1953	0
Western Europe	0	0	0	0	0	200	3000	200	0
LATIN AMERICA AND CARIBBEAN	80000	0	14500	124880	94700	99009	117244	528649	143478
Caribbean	0	0	6000	0	1000	0	0	50	1200
Meso-America	0	0	6000	5080	0	0	106744	104861	1250
South America	80000	0	2500	119800	93700	99009	10500	423738	141028
NORTH AMERICA	0	6000	0	1700	5000	400	5000	0	2000
North America	0	6000	0	1700	5000	400	5000	0	2000
POLAR	0	0	0	0	0	0	0	0	0
Antarctic	0	0	0	0	0	0	0	0	0
Arctic	0	0	0	0	0	0	0	0	0
WEST ASIA	0	0	0	0	114210	0	0	7500	0
Arabian Peninsula	0	0	0	0	114210	0	0	7500	0
Mashriq	0	0	0	0	0	0	0	0	0
GLOBAL TOTALS	7080150	156000	647875	1871657	5426686	1057051	17990904	2950127	793185
REPORTED GLOBAL TOTALS FROM ORIGINAL DATA SOURCES	****	****	****	****	****	****	****	****	****

Comments:

Flood: significant rise of water level in a stream, lake, reservoir or coastal region.

Homeless: people needing immediate assistance in the form of shelter.

AGGREGATIONS

Floods - Injured People

	1975	1980	1990	1991	1992	1993	1994	1995	1996	1997	1998	1999	2000
AFRICA	0	0	500	22	8	46	14	0	18631	31	0	31	36
Central Africa	0	0	500	14	0	0	0	0	0	0	0	0	0
Eastern Africa	0	0	0	0	0	0	14	0	0	22	0	0	0
Northern Africa	0	0	0	8	8	16	0	0	18601	0	0	2	5
Southern Africa	0	0	0	0	0	30	0	0	30	7	0	0	28
Western Africa	0	0	0	0	0	0	0	0	0	0	0	29	3
Western Indian Ocean	0	0	0	0	0	0	0	0	0	2	0	0	0
ASIA AND PACIFIC	93	27123	7713	32403	251792	992	22201	70578	234196	26709	123623	25137	1194
Australia and New Zealand	0	0	0	0	0	30	0	0	26	0	0	0	0
Central Asia	0	0	0	0	0	0	0	0	0	0	39	23	0
North West Pacific and East Asia	0	27100	5913	32344	2041	413	22082	70571	234000	26229	123205	24122	432
South Asia	0	23	1600	45	200	291	116	0	142	86	374	305	120
South East Asia	93	0	200	14	249551	258	3	7	28	394	5	687	642
South Pacific	0	0	0	0	0	0	0	0	0	0	0	0	0
EUROPE	0	13	0	0	0	453	0	91	0	2409	291	30	34
Central Europe	0	0	0	0	0	0	0	91	0	2409	61	25	0
Eastern Europe	0	0	0	0	0	453	0	0	0	0	230	0	0
Western Europe	0	13	0	0	0	0	0	0	0	0	0	5	34
LATIN AMERICA AND CARIBBEAN	0	50	100	82	320	92	198	55	0	321	750	4469	112
Caribbean	0	0	0	0	0	0	0	0	0	0	0	0	0
Meso-America	0	0	100	82	0	1	0	55	0	0	0	161	36
South America	0	50	0	0	320	91	198	0	0	321	750	4308	76
NORTH AMERICA	0	0	0	200	0	0	70	0	0	24	0	0	0
North America	0	0	0	200	0	0	70	0	0	24	0	0	0
POLAR	0	0	0	0	0	0	0	0	0	0	0	0	0
Antarctic	0	0	0	0	0	0	0	0	0	0	0	0	0
Arctic	0	0	0	0	0	0	0	0	0	0	0	0	0
WEST ASIA	0	0	0	0	0	0	0	0	0	0	0	0	0
Arabian Peninsula	0	0	0	0	0	0	0	0	0	0	0	0	0
Mashriq	0	0	0	0	0	0	0	0	0	0	0	0	0
GLOBAL TOTALS	93	27186	8313	32707	252120	1583	22483	70724	252827	29494	124664	29667	1376
REPORTED GLOBAL TOTALS FROM ORIGINAL DATA SOURCES	****	****	****	****	****	****	****	****	****	****	****	****	****

Comments:

Flood: significant rise of water level in a stream, lake, reservoir or coastal region.

Injured: people suffering from physical injuries, trauma or an illness requiring medical treatment as a direct result of a disaster.

AGGREGATIONS

Floods - Killed People

	1975	1980	1990	1991	1992	1993	1994	1995	1996	1997	1998	1999	2000
AFRICA	26	0	315	522	21	111	1080	1246	340	2907	353	387	1196
Central Africa	0	0	23	41	0	0	45	2	7	0	0	28	3
Eastern Africa	0	0	44	0	0	2	249	47	40	2762	126	48	69
Northern Africa	25	0	37	1	21	43	645	798	81	22	103	50	50
Southern Africa	1	0	211	472	0	66	31	231	161	118	96	42	1053
Western Africa	0	0	0	8	0	0	110	168	51	0	28	219	21
Western Indian Ocean	0	0	0	0	0	0	0	0	0	5	0	0	0
ASIA AND PACIFIC	593	10002	1579	5114	5133	4703	4818	5770	7219	3077	8590	3019	4554
Australia and New Zealand	0	0	6	6	0	1	0	0	5	0	4	0	0
Central Asia	0	0	0	0	1346	15	0	0	0	0	154	27	0
North West Pacific and East Asia	4	6323	649	1915	430	1303	1315	1611	4296	651	4617	1079	385
South Asia	350	3397	497	2682	3150	3088	2508	2829	2424	2277	3766	895	2910
South East Asia	239	282	427	511	207	296	995	1330	494	149	49	1016	1255
South Pacific	0	0	0	0	0	0	0	0	0	0	0	2	4
EUROPE	60	81	77	189	17	150	177	183	42	156	369	95	92
Central Europe	60	75	63	138	12	0	0	152	0	104	182	55	12
Eastern Europe	0	0	14	41	0	142	73	6	0	24	31	2	3
Western Europe	0	6	0	10	5	8	104	25	42	28	156	38	77
LATIN AMERICA AND CARIBBEAN	118	323	189	55	196	842	392	298	46	384	1786	30865	457
Caribbean	0	0	4	10	21	103	12	0	6	0	0	13	18
Meso-America	0	105	54	36	2	487	163	91	13	0	1414	728	178
South America	118	218	131	9	173	252	217	207	27	384	372	30124	261
NORTH AMERICA	0	36	43	48	0	93	37	28	55	76	21	0	8
North America	0	36	43	48	0	93	37	28	55	76	21	0	8
POLAR	0	0	0	0	0	0	0	0	0	0	0	0	0
Antarctic	0	0	0	0	0	0	0	0	0	0	0	0	0
Arctic	0	0	0	0	0	0	0	0	0	0	0	0	0
WEST ASIA	52	0	0	8	0	31	0	0	338	2	67	0	0
Arabian Peninsula	52	0	0	0	0	31	0	0	338	2	67	0	0
Mashriq	0	0	0	8	0	0	0	0	0	0	0	0	0
GLOBAL TOTALS	849	10442	2203	5936	5367	5930	6504	7525	8040	6602	11186	34366	6307
REPORTED GLOBAL TOTALS FROM ORIGINAL DATA SOURCES	****	****	****	****	****	****	****	****	****	****	****	****	****

Comments:

Flood: significant rise of water level in a stream, lake, reservoir or coastal region.

   Killed: persons confirmed as dead and persons missing and presumed dead.

Floods - Number

	1975	1980	1990	1991	1992	1993	1994	1995	1996	1997	1998	1999	2000
AFRICA	6	0	7	6	1	10	18	18	12	17	22	34	37
Central Africa	0	0	1	2	0	0	3	2	1	1	3	6	1
Eastern Africa	1	0	2	0	0	3	3	2	1	6	3	6	6
Northern Africa	3	0	1	1	1	2	4	5	3	3	1	5	6
Southern Africa	2	0	3	1	0	3	2	5	3	5	5	4	15
Western Africa	0	0	0	2	0	2	6	4	4	1	9	13	9
Western Indian Ocean	0	0	0	0	0	0	0	0	0	1	1	0	0
ASIA AND PACIFIC	10	21	20	41	33	36	33	34	32	22	34	41	43
Australia and New Zealand	1	2	1	4	0	1	1	0	2	2	7	1	2
Central Asia	0	0	0	0	1	3	1	0	1	0	5	1	1
North West Pacific and East Asia	6	5	5	4	8	6	5	5	6	6	7	7	11
South Asia	1	10	9	21	16	15	14	14	11	10	12	14	15
South East Asia	2	4	5	12	7	10	12	15	12	4	3	17	13
South Pacific	0	0	0	0	1	1	0	0	0	0	0	1	1
EUROPE	1	4	5	7	6	10	14	17	9	15	17	15	34
Central Europe	1	2	3	2	2	0	0	6	2	4	7	7	8
Eastern Europe	0	0	2	3	1	5	5	5	1	4	4	3	8
Western Europe	0	2	0	2	3	5	9	6	6	7	6	5	18
LATIN AMERICA AND CARIBBEAN	2	11	17	8	15	18	13	16	12	11	9	21	31
Caribbean	1	0	2	2	3	7	1	1	5	1	0	1	3
Meso-America	0	4	4	4	1	5	4	6	2	0	2	11	8
South America	1	7	11	2	11	6	8	9	5	10	7	9	20
NORTH AMERICA	0	3	3	18	2	7	2	3	3	10	6	0	8
North America	0	3	3	18	2	7	2	3	3	10	6	0	8
POLAR	0	0	0	0	0	0	0	0	0	0	0	0	0
Antarctic	0	0	0	0	0	0	0	0	0	0	0	0	0
Arctic	0	0	0	0	0	0	0	0	0	0	0	0	0
WEST ASIA	1	0	0	2	0	1	0	0	1	1	2	1	0
Arabian Peninsula	1	0	0	1	0	1	0	0	1	1	2	1	0
Mashriq	0	0	0	1	0	0	0	0	0	0	0	0	0
GLOBAL TOTALS	20	39	52	82	57	82	80	88	69	76	90	112	153
REPORTED GLOBAL TOTALS FROM ORIGINAL DATA SOURCES	****	****	****	****	****	****	****	****	****	****	****	****	****

Comments:

Flood: significant rise of water level in a stream, lake, reservoir or coastal region.

AGGREGATIONS

Floods - Total Affected

	1975	1980	1990	1995	1996	1997	1998	1999	2000
AFRICA	248000	0	636368	1208375	845058	2751704	3861539	3206194	2669033
Central Africa	0	0	27500	84600	17500	35506	30200	340425	300
Eastern Africa	16000	0	350000	102375	90000	1750522	602900	148100	404000
Northern Africa	157000	0	96000	41000	172301	103000	1000000	1996443	1470
Southern Africa	75000	0	162868	31800	212530	810439	1319600	73200	2253604
Western Africa	0	0	0	948600	352727	51000	888839	648026	9659
Western Indian Ocean	0	0	0	0	0	1237	20000	0	0
ASIA AND PACIFIC	37003552	44516066	45951453	180927416	178585379	39609653	286158809	143947252	58312665
Australia and New Zealand	0	2900	6000	0	13226	600	13075	350	200
Central Asia	0	0	0	0	180000	0	74518	9392	2500
North West Pacific and East Asia	3360	107100	42745653	120216121	158004365	7389344	240872061	106288482	1456018
South Asia	34000000	43117323	2456100	55220081	9741342	32031164	45064150	28710790	44369293
South East Asia	3000192	1288743	743700	5491214	10646446	188545	135005	8900238	12484654
South Pacific	0	0	0	0	0	0	0	38000	0
EUROPE	1000000	64813	4500	2090191	11830	556971	160877	556032	187656
Central Europe	1000000	64500	4500	26291	5200	456927	27067	249684	71931
Eastern Europe	0	0	0	1659100	0	84444	126670	4913	58501
Western Europe	0	313	0	404800	6630	15600	7140	301435	57224
LATIN AMERICA AND CARIBBEAN	828000	503400	225936	490162	263766	334537	1509294	2437470	938131
Caribbean	0	0	11000	33180	12066	35000	0	50	1875
Meso-America	0	141350	172311	59682	95000	0	511744	1250827	11886
South America	828000	362050	42625	397300	156700	299537	997550	1186593	924370
NORTH AMERICA	0	117000	0	11390	169000	339314	31500	0	3846
North America	0	117000	0	11390	169000	339314	31500	0	3846
POLAR	0	0	0	0	0	0	0	0	0
Antarctic	0	0	0	0	0	0	0	0	0
Arctic	0	0	0	0	0	0	0	0	0
WEST ASIA	50000	0	0	0	238210	200	3000	19750	0
Arabian Peninsula	50000	0	0	0	238210	200	3000	19750	0
Mashriq	0	0	0	0	0	0	0	0	0
GLOBAL TOTALS	39129552	45201279	46818257	184727534	180113243	43592379	291725019	150166698	62111331
REPORTED GLOBAL TOTALS FROM ORIGINAL DATA SOURCES	****	****	****	****	****	****	****	****	****

Comments:

Flood: significant rise of water level in a stream, lake, reservoir or coastal region.

Total affected: people that have been injured affected and left homeless after a disaster are included in this category.

Natural Disasters - Affected People

	1975	1980	1990	1995	1996	1997	1998	1999	2000
AFRICA	1419199	8620948	15220000	9171647	4131131	6239553	6900589	5757434	226437
Central Africa	0	0	15000	42385	16000	26000	20000	256006	
Eastern Africa	16000	685000	6850000	1303150	190000	4103200	726500	2576055	171611
Northern Africa	157000	478948	658000	38000	266000	100000	900000	2220741	7006
Southern Africa	75000	6000000	3847000	7121404	3207230	1408328	4475000	110000	34398
Western Africa	335000	1450000	3600000	666708	351901	80900	759089	594632	46
Western Indian Ocean	836199	7000	250000	0	100000	521125	20000	0	13374
ASIA AND PACIFIC	30106683	64529318	67248557	249605748	199152882	54786491	284646434	184868662	2264824
Australia and New Zealand	0	2900	6000	2681570	636300	740	13075	12350	15
Central Asia	0	0	0	0	180000	608000	62235	6366	35195
North West Pacific and East Asia	4066	115309	47946862	149738912	169958134	13641907	224197586	126074944	331163
South Asia	27075000	46708300	10987370	56837846	16584250	34976066	50246571	42531106	1763761
South East Asia	3004617	17674809	8132225	40347420	11792398	4346178	9812662	16147896	134688
South Pacific	23000	28000	176100	0	1800	1213600	314305	96000	
EUROPE	1050000	491100	204500	9713038	21290	520478	1197415	5763445	38781
Central Europe	1050000	69500	4500	79050	500	258225	72585	1101705	963
Eastern Europe	0	0	0	3204425	500	105193	866100	749210	37020
Western Europe	0	421600	200000	6429563	20290	157060	258730	3912530	798
LATIN AMERICA AND CARIBBEAN	748000	962350	2772296	720544	1299580	1459312	17159618	3633459	9993
Caribbean	0	450000	8500	246440	318100	39000	1624000	235357	6
Meso-America	0	147350	166511	142082	823467	1062955	3652468	1259855	1542
South America	748000	365000	2597285	332022	158013	357357	11883150	2138247	8443
NORTH AMERICA	200	116000	0	94190	682500	346160	741771	9057251	615
North America	200	116000	0	94190	682500	346160	741771	9057251	615
POLAR	0	0	0	0	0	0	0	0	
Antarctic	0	0	0	0	0	0	0	0	
Arctic	0	0	0	0	0	0	0	0	
WEST ASIA	50000	0	0	0	124000	200	3000	521250	4790
Arabian Peninsula	50000	0	0	0	124000	200	3000	12250	
Mashriq	0	0	0	0	0	0	0	509000	4790
GLOBAL TOTALS	33374082	74719716	85445353	269305167	205411383	63352194	310648827	209601501	2545443
REPORTED GLOBAL TOTALS FROM ORIGINAL DATA SOURCES	****	****	****	****	****	****	****	****	**

Comments:

Natural disasters include: droughts, earthquakes, extreme temperatures, floods, slides, volcanic eruptions, wave/surges, wild fires and wind storms.

Affected: people requiring immediate assistance during a period of emergency, for example requiring basic survival needs such as food, water, shelter, sanitation and immediate medical assistance (included in the field "total affected"); Appearance of a significant number of cases of an infectious disease introduced in a region or a population that is usually free from that disease. (100 or more people affected).

AGGREGATIONS

Natural Disasters - Homeless People

	1975	1980	1990	1995	1996	1997	1998	1999	2000
AFRICA	0	443000	95868	358792	55546	328220	271650	454507	385815
Central Africa	0	0	12000	42300	1500	9506	10200	87429	300
Eastern Africa	0	0	0	0	0	233500	2400	125000	260665
Northern Africa	0	443000	48000	3000	47700	3000	100000	83200	815
Southern Africa	0	0	35868	23300	5520	2104	7100	9500	117135
Western Africa	0	0	0	290192	826	0	151950	149378	6300
Western Indian Ocean	0	0	0	0	0	80110	0	0	600
ASIA AND PACIFIC	7004142	1397479	2071711	4808830	5576714	2241472	28755235	3284795	1652109
Australia and New Zealand	0	0	0	200	1730	0	0	702	0
Central Asia	0	0	21500	0	0	0	12244	4503	0
North West Pacific and East Asia	1444	2100	768398	1352600	4874934	637409	28543976	1566215	350593
South Asia	7002000	180000	6000	1880645	577800	1171750	102250	1439942	344500
South East Asia	698	1208379	1246811	1575385	122250	416813	96765	231661	952016
South Pacific	0	7000	29002	0	0	15500	0	41772	5000
EUROPE	0	0	3000	163750	7200	235139	73344	747683	20726
Central Europe	0	0	0	250	7200	196293	8021	670224	1243
Eastern Europe	0	0	0	157200	0	596	61023	7193	19100
Western Europe	0	0	3000	6300	0	38250	4300	70266	383
LATIN AMERICA AND CARIBBEAN	80000	15500	58500	205672	208156	164709	241814	1028099	288973
Caribbean	0	15500	6000	16037	68596	200	122000	31127	1200
Meso-America	0	0	20000	63355	18750	52000	108564	107314	3035
South America	80000	0	32500	126280	120810	112509	11250	889658	284738
NORTH AMERICA	300	8900	0	3700	5000	400	6125	10009	6864
North America	300	8900	0	3700	5000	400	6125	10009	6864
POLAR	0	0	0	0	0	0	0	0	0
Antarctic	0	0	0	0	0	0	0	0	0
Arctic	0	0	0	0	0	0	0	0	0
WEST ASIA	0	0	0	0	114210	0	0	7500	0
Arabian Peninsula	0	0	0	0	114210	0	0	7500	0
Mashriq	0	0	0	0	0	0	0	0	0
GLOBAL TOTALS	7084442	1864879	2229079	5540744	5966826	2969940	29348168	5532593	2354487
REPORTED GLOBAL TOTALS FROM ORIGINAL DATA SOURCES	****	****	****	****	****	****	****	****	****

Comments:

Natural disasters include: droughts, earthquakes, extreme temperatures, floods, slides, volcanic eruptions, wave/surges, wild fires and wind storms.

Homeless: people needing immediate assistance in the form of shelter (included in the field "total affected"). They are always part of the affected population. Reporting from the field should give the number of individuals that are homeless; if only the number of families or houses is reported, the figure is multiplied by the average family size for the affected area (times 5 for the developing countries, times 3 for the industrialised countries, according to UNDP country list). (100 or more people homeless).

AGGREGATIONS

	1975	1980	1990	1991	1992	1993	1994	1995	1996	1997	1998	1999	2000
AFRICA	109	8459	500	322	9937	107	2364	75	18631	91	163	1711	172
Central Africa	0	0	500	14	0	0	0	0	0	0	0	0	0
Eastern Africa	0	0	0	0	0	0	24	0	0	22	0	0	10
Northern Africa	0	8459	0	8	9937	77	289	69	18601	60	0	177	120
Southern Africa	0	0	0	0	0	30	2000	0	30	7	163	485	38
Western Africa	0	0	0	300	0	0	0	6	0	0	0	49	3
Western Indian Ocean	109	0	0	0	0	0	51	0	0	2	0	1000	1
ASIA AND PACIFIC	199	33855	121851	186628	275258	29379	34986	118805	294806	55341	147761	45681	9731
Australia and New Zealand	0	40	0	100	0	555	444	169	103	0	0	24	0
Central Asia	0	0	16	6	100	0	0	0	0	0	39	23	0
North West Pacific and East Asia	98	27276	8607	40293	21999	5503	27260	110960	257241	29862	139178	37979	4734
South Asia	0	5840	107646	141202	750	21965	4706	1903	36337	23653	7680	6596	974
South East Asia	101	449	5577	5027	251857	1017	2574	5773	1088	1606	185	959	4023
South Pacific	0	250	5	0	552	339	2	0	37	220	679	100	0
EUROPE	3372	8430	1097	1708	2139	549	448	1586	329	2668	3491	52159	2931
Central Europe	3372	100	820	8	2084	0	6	454	200	2469	2717	49817	1171
Eastern Europe	0	0	0	1700	10	513	322	868	0	45	653	281	1672
Western Europe	0	8330	277	0	45	36	120	264	129	154	121	2061	88
LATIN AMERICA AND CARIBBEAN	600	514	1219	1858	1204	833	356	993	977	1455	13855	13588	779
Caribbean	0	16	0	0	50	105	0	199	3	0	649	67	0
Meso-America	0	360	409	931	614	41	0	377	80	200	12236	561	648
South America	600	138	810	927	540	687	356	417	894	1255	970	12960	131
NORTH AMERICA	40	170	30	554	455	458	9026	167	181	246	572	951	495
North America	40	170	30	554	455	458	9026	167	181	246	572	951	495
POLAR	0	0	0	0	0	0	0	0	0	0	0	0	0
Antarctic	0	0	0	0	0	0	0	0	0	0	0	0	0
Arctic	0	0	0	0	0	0	0	0	0	0	0	0	0
WEST ASIA	0	0	0	40	75	0	0	0	0	0	0	0	212
Arabian Peninsula	0	0	0	40	0	0	0	0	0	0	0	0	0
Mashriq	0	0	0	0	75	0	0	0	0	0	0	0	212
GLOBAL TOTALS	4320	51428	124697	191110	289068	31326	47180	121626	314924	59801	165842	114090	14320
REPORTED GLOBAL TOTALS FROM ORIGINAL DATA SOURCES	****	****	****	****	****	****	****	****	****	****	****	****	****

Comments:

Natural disasters include: droughts, earthquakes, extreme temperatures, floods, slides, volcanic eruptions, wave/surges, wild fires and wind storms.

Injured: people suffering from physical injuries, trauma or an illness requiring medical treatment as a direct result of a disaster (included in the field "total affected"). The number of injured is entered when the term "injured" is written in the source. Injured people are always part of the affected population. Any related word like "hospitalized" is considered as injured. If there is no precise number like "hundreds of injured", 200 injured will be entered (although it is probably underestimated). (100 or more people injured).

Natural Disasters - Killed People

	1975	1980	1990	1991	1992	1993	1994	1995	1996	1997	1998	1999	2000
AFRICA	54	2660	346	575	600	184	1792	1288	452	3119	545	740	1633
Central Africa	0	0	23	41	0	0	45	2	7	0	0	28	3
Eastern Africa	0	0	44	13	0	2	308	47	40	2782	126	48	300
Northern Africa	25	2635	68	1	582	95	838	840	117	40	150	72	57
Southern Africa	12	0	211	472	0	73	277	231	225	118	226	72	1082
Western Africa	1	0	0	12	18	0	110	168	51	0	43	520	57
Western Indian Ocean	16	25	0	36	0	14	214	0	12	179	0	0	134
ASIA AND PACIFIC	2134	11728	49388	158339	9937	17858	8510	16715	11960	12226	26464	17029	6789
Australia and New Zealand	0	0	6	7	2	18	36	9	22	24	4	8	1
Central Asia	0	0	1	54	1669	15	162	112	0	40	165	51	11
North West Pacific and East Asia	405	6687	1713	4636	1357	2095	3010	7970	5803	1272	5014	3908	860
South Asia	1051	4071	43645	145001	3763	14422	3332	5631	4349	6367	17878	11829	3744
South East Asia	478	952	4003	8413	3142	1206	1965	2993	1744	4377	1180	1175	2168
South Pacific	200	18	20	228	4	102	5	0	42	146	2223	58	5
EUROPE	2485	4880	223	540	991	658	316	2784	166	414	795	18819	626
Central Europe	2445	115	81	139	885	144	30	255	16	166	526	18261	210
Eastern Europe	0	0	61	319	32	482	141	2388	19	65	53	178	255
Western Europe	40	4765	81	82	74	32	145	141	131	183	216	380	161
LATIN AMERICA AND CARIBBEAN	321	740	788	541	566	1680	1819	865	498	1426	21711	32256	762
Caribbean	26	322	12	10	25	110	1144	26	93	32	550	26	18
Meso-America	31	172	473	163	278	587	163	397	135	571	20472	800	306
South America	264	246	303	368	263	983	512	442	270	823	689	31430	438
NORTH AMERICA	195	500	209	127	97	406	269	869	367	256	338	477	173
North America	195	500	209	127	97	406	269	869	367	256	338	477	173
POLAR	0	0	0	0	0	0	0	0	0	0	0	0	0
Antarctic	0	0	0	0	0	0	0	0	0	0	0	0	0
Arctic	0	0	0	0	0	0	0	0	0	0	0	0	0
WEST ASIA	52	0	0	41	40	31	0	0	338	2	67	0	9
Arabian Peninsula	52	0	0	11	0	31	0	0	338	2	67	0	0
Mashriq	0	0	0	30	40	0	0	0	0	0	0	0	9
GLOBAL TOTALS	5241	20508	50954	160163	12231	20817	12706	22521	13781	17443	49920	69321	9992
REPORTED GLOBAL TOTALS FROM ORIGINAL DATA SOURCES	****	****	****	****	****	****	****	****	****	****	****	****	****

Comments:

Natural disasters include : droughts, earthquakes, extreme temperatures, floods, slides, volcanic eruptions, wave/surges, wild fires and wind storms.

Killed: persons confirmed as dead and persons missing and presumed dead.

Natural Disasters – Number

	1975	1980	1990	1991	1992	1993	1994	1995	1996	1997	1998	1999	2000
AFRICA	20	16	19	14	21	21	32	32	23	27	34	56	70
Central Africa	1	0	1	2	0	1	3	3	1	1	3	7	1
Eastern Africa	2	4	5	2	2	3	6	4	2	9	5	12	18
Northern Africa	3	3	3	2	4	5	6	7	6	4	2	7	11
Southern Africa	3	2	7	1	12	7	7	12	8	7	11	10	23
Western Africa	9	6	2	4	2	4	6	6	4	3	12	18	13
Western Indian Ocean	2	1	1	3	1	1	4	0	2	3	1	2	4
ASIA AND PACIFIC	36	77	103	110	98	123	106	87	92	85	106	109	148
Australia and New Zealand	5	8	4	7	5	5	11	7	10	6	7	5	7
Central Asia	0	0	2	3	7	3	3	1	1	3	6	3	6
North West Pacific and East Asia	13	21	29	23	26	41	23	21	26	23	25	28	48
South Asia	10	29	31	36	29	37	28	27	27	31	39	36	41
South East Asia	4	18	21	37	25	28	39	31	25	14	18	31	44
South Pacific	4	1	16	4	6	9	2	0	3	8	11	6	2
EUROPE	5	12	72	24	25	26	28	38	20	32	39	53	99
Central Europe	2	5	6	5	9	5	4	8	5	7	15	19	31
Eastern Europe	0	0	4	6	5	9	9	13	3	8	9	11	31
Western Europe	3	7	62	13	11	12	15	17	12	17	15	23	37
LATIN AMERICA AND CARIBBEAN	9	27	39	29	32	42	30	53	37	36	46	63	67
Caribbean	2	10	12	2	5	10	5	13	13	3	6	15	5
Meso-America	2	7	10	12	9	13	7	18	11	11	21	22	24
South America	5	10	17	15	18	19	18	22	13	22	19	26	38
NORTH AMERICA	6	13	21	37	35	29	18	14	15	36	29	24	33
North America	6	13	21	37	35	29	18	14	15	36	29	24	33
POLAR	0	0	0	0	0	0	0	0	0	0	0	0	0
Antarctic	0	0	0	0	0	0	0	0	0	0	0	0	0
Arctic	0	0	0	0	0	0	0	0	0	0	0	0	0
WEST ASIA	2	0	0	5	2	1	0	0	1	1	2	3	5
Arabian Peninsula	2	0	0	2	0	1	0	0	1	1	2	1	0
Mashriq	0	0	0	3	2	0	0	0	0	0	0	2	5
GLOBAL TOTALS	78	145	254	219	213	242	214	224	188	217	256	308	422
REPORTED GLOBAL TOTALS FROM ORIGINAL DATA SOURCES	****	****	****	****	****	****	****	****	****	****	****	****	****

Comments:

Natural disasters include: droughts, earthquakes, extreme temperatures, floods, slides, volcanic eruptions, wave/surges, wild fires and wind storms.

AGGREGATIONS

	1975	1980	1990	1995	1996	1997	1998	1999	2000
AFRICA	1419308	9072407	15316368	9530514	4205308	6567864	7172402	6213652	23029767
Central Africa	0	0	27500	84685	17500	35506	30200	343435	300
Eastern Africa	16000	685000	6850000	1303150	190000	4336722	728900	2701055	17421775
Northern Africa	157000	930407	706000	41069	332301	103060	1000000	2304118	701585
Southern Africa	75000	6000000	3882868	7144704	3212780	1410439	4482263	119985	3557049
Western Africa	335000	1450000	3600000	956906	352727	80900	911039	744059	10959
Western Indian Ocean	836308	7000	250000	0	100000	601237	20000	1000	1338099
ASIA AND PACIFIC	37111024	65960652	69442119	243457383	205024317	57083304	313535130	188199138	228144330
Australia and New Zealand	0	2940	6000	2681939	638133	740	13075	13076	1550
Central Asia	0	0	21516	0	180000	608000	74518	10892	3519500
North West Pacific and East Asia	5608	144685	48723867	140132472	175090309	14309178	252866440	127679138	33471717
South Asia	34077000	46894140	11101016	58714394	17198302	36171469	50356501	43977644	176721647
South East Asia	3005416	18883637	9384613	41928578	11915736	4764597	9909612	16380516	14424916
South Pacific	23000	35250	205107	0	1837	1229320	314984	137872	5000
EUROPE	1053372	499530	208597	8178374	28819	758285	1274250	6563287	3901852
Central Europe	1053372	69600	5320	79754	7900	456987	83323	1821746	98714
Eastern Europe	0	0	0	1662493	500	105834	927776	756684	3722867
Western Europe	0	429930	203277	6436127	20419	195464	263151	3984857	80271
LATIN AMERICA AND CARIBBEAN	828600	978364	2832015	927209	1508713	1625476	17415287	4675146	1289105
Caribbean	0	465516	14500	262676	386699	39200	1746649	266551	1875
Meso-America	0	147710	186920	205814	842297	1115155	3773268	1367730	157968
South America	828600	365138	2630595	458719	279717	471121	11895370	3040865	1129262
NORTH AMERICA	540	125070	30	98057	687681	346806	748468	9068211	68873
North America	540	125070	30	98057	687681	346806	748468	9068211	68873
POLAR	0	0	0	0	0	0	0	0	0
Antarctic	0	0	0	0	0	0	0	0	0
Arctic	0	0	0	0	0	0	0	0	0
WEST ASIA	50000	0	0	0	238210	200	3000	528750	479212
Arabian Peninsula	50000	0	0	0	238210	200	3000	19750	0
Mashriq	0	0	0	0	0	0	0	509000	479212
GLOBAL TOTALS	40462844	76636023	87799129	262191537	211693048	66381935	340148537	215248184	256913139
REPORTED GLOBAL TOTALS FROM ORIGINAL DATA SOURCES	****	****	****	****	****	****	****	****	****

Comments:

Natural disasters include: droughts, earthquakes, extreme temperatures, floods, slides, volcanic eruptions, wave/surges, wild fires and wind storms.

Total affected: people that have been injured affected and left homeless after a disaster are included in this category.

AGGREGATIONS

Slides - Affected People

	1975	1980	1990	1991	1992	1993	1994	1995	1996	1997	1998	1999	2000
AFRICA	0	0	0	0	0	300	0	0	0	0	0	0	0
Central Africa	0	0	0	0	0	0	0	0	0	0	0	0	0
Eastern Africa	0	0	0	0	0	0	0	0	0	0	0	0	0
Northern Africa	0	0	0	0	0	300	0	0	0	0	0	0	0
Southern Africa	0	0	0	0	0	0	0	0	0	0	0	0	0
Western Africa	0	0	0	0	0	0	0	0	0	0	0	0	0
Western Indian Ocean	0	0	0	0	0	0	0	0	0	0	0	0	0
ASIA AND PACIFIC	75000	0	5000	0	37000	390	95862	5515	600	1527	205630	0	56885
Australia and New Zealand	0	0	0	0	0	0	0	0	100	0	0	0	0
Central Asia	0	0	0	0	0	0	45000	0	0	0	0	0	0
North West Pacific and East Asia	0	0	5000	0	0	60	0	0	500	1527	130	0	0
South Asia	75000	0	0	0	0	330	0	5515	0	0	205000	0	0
South East Asia	0	0	0	0	37000	0	50862	0	0	0	0	0	56885
South Pacific	0	0	0	0	0	0	0	0	0	0	500	0	0
EUROPE	0	0	0	0	2750	0	0	63	400	0	0	10200	2400
Central Europe	0	0	0	0	1000	0	0	0	0	0	0	0	400
Eastern Europe	0	0	0	0	1750	0	0	0	300	0	0	0	500
Western Europe	0	0	0	0	0	0	0	63	100	0	0	10200	1500
LATIN AMERICA AND CARIBBEAN	0	0	0	65221	0	78328	165000	0	7175	30000	2600	200	350
Caribbean	0	0	0	0	0	0	0	0	175	0	0	0	0
Meso-America	0	0	0	0	0	0	0	0	0	0	0	0	200
South America	0	0	0	65221	0	78328	165000	0	7000	30000	2600	200	150
NORTH AMERICA	0	0	0	0	0	0	0	0	0	0	0	0	0
North America	0	0	0	0	0	0	0	0	0	0	0	0	0
POLAR	0	0	0	0	0	0	0	0	0	0	0	0	0
Antarctic	0	0	0	0	0	0	0	0	0	0	0	0	0
Arctic	0	0	0	0	0	0	0	0	0	0	0	0	0
WEST ASIA	0	0	0	0	0	0	0	0	0	0	0	0	0
Arabian Peninsula	0	0	0	0	0	0	0	0	0	0	0	0	0
Mashriq	0	0	0	0	0	0	0	0	0	0	0	0	0
GLOBAL TOTALS	75000	0	5000	65221	39750	79018	260862	5578	8175	31527	208230	10400	59635
REPORTED GLOBAL TOTALS FROM ORIGINAL DATA SOURCES	****	****	****	****	****	****	****	****	****	****	****	****	****

Comments:

Slide: disaster type comprising the two disaster subsets "avalanche" (rapid and sudden sliding and flowage of masses of usually unsorted mixtures of snow/ice/rock material) and "landslide" (in general, all varieties of slope movement, under the influence of gravity. More strictly refers to down-slope movement of rock and/or earth masses along one or several slide surfaces).

Affected people: people requiring immediate assistance during a period of emergency, for example requiring basic survival needs such as food, water, shelter, sanitation and immediate medical assistance. Appearance of a significant number of cases of an infectious disease introduced in a region or a population that is usually free from that disease.

AGGREGATIONS

	1975	1980	1990	1991	1992	1993	1994	1995	1996	1997	1998	1999	2000
AFRICA	0	0	0	0	0	650	19	0	0	0	2500	0	465
Central Africa	0	0	0	0	0	0	0	0	0	0	0	0	0
Eastern Africa	0	0	0	0	0	0	19	0	0	0	0	0	165
Northern Africa	0	0	0	0	0	650	0	0	0	0	0	0	0
Southern Africa	0	0	0	0	0	0	0	0	0	0	2500	0	0
Western Africa	0	0	0	0	0	0	0	0	0	0	0	0	300
Western Indian Ocean	0	0	0	0	0	0	0	0	0	0	0	0	0
ASIA AND PACIFIC	0	3000	0	5500	38400	300	37477	1114395	250	2294	3000	70	4645
Australia and New Zealand	0	0	0	0	0	0	0	0	0	0	0	0	0
Central Asia	0	0	0	0	400	0	13500	0	0	0	0	0	0
North West Pacific and East Asia	0	0	0	500	0	0	0	0	0	294	0	35	2000
South Asia	0	0	0	0	0	300	0	1114395	0	2000	3000	0	0
South East Asia	0	3000	0	0	38000	0	23977	0	250	0	0	35	2645
South Pacific	0	0	0	5000	0	0	0	0	0	0	0	0	0
EUROPE	0	0	0	0	0	0	0	0	0	50	0	390	0
Central Europe	0	0	0	0	0	0	0	0	0	0	0	330	0
Eastern Europe	0	0	0	0	0	0	0	0	0	0	0	0	0
Western Europe	0	0	0	0	0	0	0	0	0	50	0	60	0
LATIN AMERICA AND CARIBBEAN	0	0	0	18271	500	125	0	2000	110	0	120	4400	143180
Caribbean	0	0	0	0	0	0	0	0	0	0	0	0	0
Meso-America	0	0	0	0	0	0	0	0	0	0	120	0	0
South America	0	0	0	18271	500	125	0	2000	110	0	0	4400	143180
NORTH AMERICA	0	0	0	0	0	0	0	0	0	0	0	0	0
North America	0	0	0	0	0	0	0	0	0	0	0	0	0
POLAR	0	0	0	0	0	0	0	0	0	0	0	0	0
Antarctic	0	0	0	0	0	0	0	0	0	0	0	0	0
Arctic	0	0	0	0	0	0	0	0	0	0	0	0	0
WEST ASIA	0	0	0	0	0	0	0	0	0	0	0	0	0
Arabian Peninsula	0	0	0	0	0	0	0	0	0	0	0	0	0
Mashriq	0	0	0	0	0	0	0	0	0	0	0	0	0
GLOBAL TOTALS	0	3000	0	23771	38900	1075	37496	1116395	360	2344	5620	4860	148290
REPORTED GLOBAL TOTALS FROM ORIGINAL DATA SOURCES	****	****	****	****	****	****	****	****	****	****	****	****	****

Comments:

Slide: disaster type comprising the two disaster subsets "avalanche" (rapid and sudden sliding and flowage of masses of usually unsorted mixtures of snow/ice/rock material) and "landslide" (in general, all varieties of slope movement, under the influence of gravity. More strictly refers to down-slope movement of rock and/or earth masses along one or several slide surfaces).

Homeless: people needing immediate assistance in the form of shelter.

AGGREGATIONS

	1975	1980	1990	1991	1992	1993	1994	1995	1996	1997	1998	1999	2000
AFRICA	0	0	0	0	0	46	10	0	0	0	0	0	0
Central Africa	0	0	0	0	0	0	0	0	0	0	0	0	0
Eastern Africa	0	0	0	0	0	0	10	0	0	0	0	0	0
Northern Africa	0	0	0	0	0	46	0	0	0	0	0	0	0
Southern Africa	0	0	0	0	0	0	0	0	0	0	0	0	0
Western Africa	0	0	0	0	0	0	0	0	0	0	0	0	0
Western Indian Ocean	0	0	0	0	0	0	0	0	0	0	0	0	0
ASIA AND PACIFIC	0	10	115	0	0	114	38	356	167	75	222	22	185
Australia and New Zealand	0	0	0	0	0	0	0	0	1	0	0	0	0
Central Asia	0	0	0	0	0	0	0	0	0	0	0	0	0
North West Pacific and East Asia	0	0	115	0	0	75	0	0	134	45	15	10	27
South Asia	0	0	0	0	0	39	0	333	0	30	196	0	100
South East Asia	0	10	0	0	0	0	38	23	24	0	0	12	58
South Pacific	0	0	0	0	0	0	0	0	8	0	11	0	0
EUROPE	0	0	0	0	84	0	0	20	129	5	21	0	11
Central Europe	0	0	0	0	84	0	0	0	0	0	0	0	3
Eastern Europe	0	0	0	0	0	0	0	0	0	0	0	0	8
Western Europe	0	0	0	0	0	0	0	20	129	5	21	0	0
LATIN AMERICA AND CARIBBEAN	0	0	0	140	160	105	0	0	105	0	38	9	55
Caribbean	0	0	0	0	0	0	0	0	0	0	0	0	0
Meso-America	0	0	0	0	0	0	0	0	0	0	8	0	0
South America	0	0	0	140	160	105	0	0	105	0	30	9	55
NORTH AMERICA	0	0	0	0	0	0	0	0	0	0	0	0	0
North America	0	0	0	0	0	0	0	0	0	0	0	0	0
POLAR	0	0	0	0	0	0	0	0	0	0	0	0	0
Antarctic	0	0	0	0	0	0	0	0	0	0	0	0	0
Arctic	0	0	0	0	0	0	0	0	0	0	0	0	0
WEST ASIA	0	0	0	0	0	0	0	0	0	0	0	0	0
Arabian Peninsula	0	0	0	0	0	0	0	0	0	0	0	0	0
Mashriq	0	0	0	0	0	0	0	0	0	0	0	0	0
GLOBAL TOTALS	0	10	115	140	244	265	48	376	401	80	281	31	251
REPORTED GLOBAL TOTALS FROM ORIGINAL DATA SOURCES	****	****	****	****	****	****	****	****	****	****	****	****	****

Comments:

Slide: disaster type comprising the two disaster subsets "avalanche" (rapid and sudden sliding and flowage of masses of usually unsorted mixtures of snow/ice/rock material) and "landslide" (in general, all varieties of slope movement, under the influence of gravity. More strictly refers to down-slope movement of rock and/or earth masses along one or several slide surfaces).

Injured: people suffering from physical injuries, trauma or an illness requiring medical treatment as a direct result of a disaster.

AGGREGATIONS

Slides - Killed People

	1975	1980	1990	1991	1992	1993	1994	1995	1996	1997	1998	1999	2000
AFRICA	0	0	0	13	0	49	22	0	34	20	87	0	49
Central Africa	0	0	0	0	0	0	0	0	0	0	0	0	0
Eastern Africa	0	0	0	13	0	0	22	0	0	20	0	0	4
Northern Africa	0	0	0	0	0	49	0	0	0	0	0	0	0
Southern Africa	0	0	0	0	0	0	0	0	34	0	87	0	13
Western Africa	0	0	0	0	0	0	0	0	0	0	0	0	32
Western Indian Ocean	0	0	0	0	0	0	0	0	0	0	0	0	0
ASIA AND PACIFIC	155	300	124	475	700	448	258	1281	821	440	626	178	845
Australia and New Zealand	0	0	0	0	0	0	0	0	9	19	0	0	0
Central Asia	0	0	0	50	256	0	162	0	0	40	11	24	0
North West Pacific and East Asia	30	0	73	200	49	84	48	37	564	210	100	23	85
South Asia	125	150	51	25	120	271	0	1224	92	171	502	0	227
South East Asia	0	150	0	0	275	93	48	20	118	0	0	131	533
South Pacific	0	0	0	200	0	0	0	0	38	0	13	0	0
EUROPE	40	40	43	0	291	434	0	51	84	29	9	61	73
Central Europe	0	40	0	0	291	135	0	0	0	0	0	0	6
Eastern Europe	0	0	43	0	0	299	0	17	0	0	0	0	38
Western Europe	40	0	0	0	0	0	0	34	84	29	9	61	29
LATIN AMERICA AND CARIBBEAN	0	0	33	293	79	617	0	165	190	312	272	112	132
Caribbean	0	0	0	0	0	0	0	0	0	0	0	0	0
Meso-America	0	0	0	23	0	22	0	0	0	12	63	0	7
South America	0	0	33	270	79	595	0	165	190	300	209	112	125
NORTH AMERICA	0	0	0	0	0	0	0	0	0	0	0	0	0
North America	0	0	0	0	0	0	0	0	0	0	0	0	0
POLAR	0	0	0	0	0	0	0	0	0	0	0	0	0
Antarctic	0	0	0	0	0	0	0	0	0	0	0	0	0
Arctic	0	0	0	0	0	0	0	0	0	0	0	0	0
WEST ASIA	0	0	0	0	0	0	0	0	0	0	0	0	0
Arabian Peninsula	0	0	0	0	0	0	0	0	0	0	0	0	0
Mashriq	0	0	0	0	0	0	0	0	0	0	0	0	0
GLOBAL TOTALS	195	340	200	781	1070	1548	280	1497	1129	801	994	351	1099
REPORTED GLOBAL TOTALS FROM ORIGINAL DATA SOURCES	****	****	****	****	****	****	****	****	****	****	****	****	****

Comments:

Slide: disaster type comprising the two disaster subsets "avalanche" (rapid and sudden sliding and flowage of masses of usually unsorted mixtures of snow/ice/rock material) and "landslide" (in general, all varieties of slope movement, under the influence of gravity. More strictly refers to down-slope movement of rock and/or earth masses along one or several slide surfaces).

Killed: persons confirmed as dead and persons missing and presumed dead.

AGGREGATIONS

Slides - Number

	1975	1980	1990	1991	1992	1993	1994	1995	1996	1997	1998	1999	2000
AFRICA	0	0	0	1	0	2	1	0	1	1	1	0	4
Central Africa	0	0	0	0	0	0	0	0	0	0	0	0	0
Eastern Africa	0	0	0	1	0	0	1	0	0	1	0	0	1
Northern Africa	0	0	0	0	0	2	0	0	0	0	0	0	0
Southern Africa	0	0	0	0	0	0	0	0	1	0	1	0	0
Western Africa	0	0	0	0	0	0	0	0	0	0	0	0	2
Western Indian Ocean	0	0	0	0	0	0	0	0	0	0	0	0	0
ASIA AND PACIFIC	2	3	4	4	8	11	6	8	14	9	14	7	13
Australia and New Zealand	0	0	0	0	0	0	0	0	1	1	0	0	0
Central Asia	0	0	0	1	3	0	2	0	0	1	1	1	0
North West Pacific and East Asia	1	0	2	1	1	3	1	1	6	3	2	1	2
South Asia	1	1	2	1	2	6	0	6	2	4	9	0	3
South East Asia	0	2	0	0	2	2	3	1	4	0	0	5	8
South Pacific	0	0	0	1	0	0	0	0	1	0	2	0	0
EUROPE	3	1	1	0	4	3	0	3	3	1	1	4	6
Central Europe	0	1	0	0	3	1	0	0	0	0	0	1	1
Eastern Europe	0	0	1	0	1	2	0	1	1	0	0	0	3
Western Europe	3	0	0	0	0	0	0	2	2	1	1	3	2
LATIN AMERICA AND CARIBBEAN	0	0	1	6	2	6	1	4	6	2	6	4	6
Caribbean	0	0	0	0	0	0	0	0	1	0	0	0	0
Meso-America	0	0	0	1	0	1	0	0	0	1	2	0	1
South America	0	0	1	5	2	5	1	4	5	1	4	4	5
NORTH AMERICA	0	0	0	0	0	0	0	0	0	0	0	0	0
North America	0	0	0	0	0	0	0	0	0	0	0	0	0
POLAR	0	0	0	0	0	0	0	0	0	0	0	0	0
Antarctic	0	0	0	0	0	0	0	0	0	0	0	0	0
Arctic	0	0	0	0	0	0	0	0	0	0	0	0	0
WEST ASIA	0	0	0	0	0	0	0	0	0	0	0	0	0
Arabian Peninsula	0	0	0	0	0	0	0	0	0	0	0	0	0
Mashriq	0	0	0	0	0	0	0	0	0	0	0	0	0
GLOBAL TOTALS	5	4	6	11	14	22	8	15	24	13	22	15	29
REPORTED GLOBAL TOTALS FROM ORIGINAL DATA SOURCES	****	****	****	****	****	****	****	****	****	****	****	****	****

Comments:

Slide: disaster type comprising the two disaster subsets "avalanche" (rapid and sudden sliding and flowage of masses of usually unsorted mixtures of snow/ice/rock material) and "landslide" (in general, all varieties of slope movement, under the influence of gravity. More strictly refers to down-slope movement of rock and/or earth masses along one or several slide surfaces).

AGGREGATIONS

Slides - Total Affected

	1975	1980	1990	1991	1992	1993	1994	1995	1996	1997	1998	1999	2000
AFRICA	0	0	0	0	0	996	29	0	0	0	2500	0	465
Central Africa	0	0	0	0	0	0	0	0	0	0	0	0	0
Eastern Africa	0	0	0	0	0	0	29	0	0	0	0	0	165
Northern Africa	0	0	0	0	0	996	0	0	0	0	0	0	0
Southern Africa	0	0	0	0	0	0	0	0	0	0	2500	0	0
Western Africa	0	0	0	0	0	0	0	0	0	0	0	0	300
Western Indian Ocean	0	0	0	0	0	0	0	0	0	0	0	0	0
ASIA AND PACIFIC	75000	3010	5115	5500	75400	804	133377	1120266	1017	3896	208852	92	61715
Australia and New Zealand	0	0	0	0	0	0	0	0	101	0	0	0	0
Central Asia	0	0	0	0	400	0	58500	0	0	0	0	0	0
North West Pacific and East Asia	0	0	5115	500	0	135	0	0	634	1866	145	45	2027
South Asia	75000	0	0	0	0	669	0	1120243	0	2030	208196	0	100
South East Asia	0	3010	0	0	75000	0	74877	23	274	0	0	47	59588
South Pacific	0	0	0	5000	0	0	0	0	8	0	511	0	0
EUROPE	0	0	0	0	2834	0	0	83	529	55	21	10590	2411
Central Europe	0	0	0	0	1084	0	0	0	0	0	0	330	403
Eastern Europe	0	0	0	0	1750	0	0	0	300	0	0	0	508
Western Europe	0	0	0	0	0	0	0	83	229	55	21	10260	1500
LATIN AMERICA AND CARIBBEAN	0	0	0	83632	660	78558	165000	2000	7390	30000	2758	4609	143585
Caribbean	0	0	0	0	0	0	0	0	175	0	0	0	0
Meso-America	0	0	0	0	0	0	0	0	0	0	128	0	200
South America	0	0	0	83632	660	78558	165000	2000	7215	30000	2630	4609	143385
NORTH AMERICA	0	0	0	0	0	0	0	0	0	0	0	0	0
North America	0	0	0	0	0	0	0	0	0	0	0	0	0
POLAR	0	0	0	0	0	0	0	0	0	0	0	0	0
Antarctic	0	0	0	0	0	0	0	0	0	0	0	0	0
Arctic	0	0	0	0	0	0	0	0	0	0	0	0	0
WEST ASIA	0	0	0	0	0	0	0	0	0	0	0	0	0
Arabian Peninsula	0	0	0	0	0	0	0	0	0	0	0	0	0
Mashriq	0	0	0	0	0	0	0	0	0	0	0	0	0
GLOBAL TOTALS	75000	3010	5115	89132	78894	80358	298406	1122349	8936	33951	214131	15291	208176
REPORTED GLOBAL TOTALS FROM ORIGINAL DATA SOURCES	****	****	****	****	****	****	****	****	****	****	****	****	****

Comments:

Slide: disaster type comprising the two disaster subsets "avalanche" (rapid and sudden sliding and flowage of masses of usually unsorted mixtures of snow/ice/rock material) and "landslide" (in general, all varieties of slope movement, under the influence of gravity. More strictly refers to down-slope movement of rock and/or earth masses along one or several slide surfaces).

Total affected: people that have been injured affected and left homeless after a disaster are included in this category.

AGGREGATIONS

Volcanic Eruptions - Affected People

	1975	1980	1990	1991	1992	1993	1994	1995	1996	1997	1998	1999	2000
AFRICA	0	0	0	0	0	0	0	1300	0	0	0	2500	0
Central Africa	0	0	0	0	0	0	0	0	0	0	0	2500	0
Eastern Africa	0	0	0	0	0	0	0	0	0	0	0	0	0
Northern Africa	0	0	0	0	0	0	0	0	0	0	0	0	0
Southern Africa	0	0	0	0	0	0	0	0	0	0	0	0	0
Western Africa	0	0	0	0	0	0	0	1300	0	0	0	0	0
Western Indian Ocean	0	0	0	0	0	0	0	0	0	0	0	0	0
ASIA AND PACIFIC	0	52235	42770	1053549	41578	116382	114026	70	1800	3000	6000	0	77196
Australia and New Zealand	0	0	0	0	0	0	0	70	0	0	0	0	0
Central Asia	0	0	0	0	0	0	0	0	0	0	0	0	0
North West Pacific and East Asia	0	0	0	10000	40000	4930	0	0	0	0	0	0	16400
South Asia	0	0	0	0	0	0	0	0	0	0	0	0	0
South East Asia	0	52235	42770	1043549	1578	111452	8026	0	0	3000	6000	0	60796
South Pacific	0	0	0	0	0	0	106000	0	1800	0	0	0	0
EUROPE	0	0	0	7000	0	0	0	0	0	0	0	0	0
Central Europe	0	0	0	0	0	0	0	0	0	0	0	0	0
Eastern Europe	0	0	0	0	0	0	0	0	0	0	0	0	0
Western Europe	0	0	0	7000	0	0	0	0	0	0	0	0	0
LATIN AMERICA AND CARIBBEAN	0	0	4000	124000	305000	350	75700	17000	4743	4000	1808	28845	41800
Caribbean	0	0	0	0	0	0	0	5000	4000	4000	0	0	0
Meso-America	0	0	0	0	305000	0	75700	12000	743	0	1808	6845	41800
South America	0	0	4000	124000	0	350	0	0	0	0	0	22000	0
NORTH AMERICA	0	0	0	0	0	0	0	0	0	0	0	0	0
North America	0	0	0	0	0	0	0	0	0	0	0	0	0
POLAR	0	0	0	0	0	0	0	0	0	0	0	0	0
Antarctic	0	0	0	0	0	0	0	0	0	0	0	0	0
Arctic	0	0	0	0	0	0	0	0	0	0	0	0	0
WEST ASIA	0	0	0	0	0	0	0	0	0	0	0	0	0
Arabian Peninsula	0	0	0	0	0	0	0	0	0	0	0	0	0
Mashriq	0	0	0	0	0	0	0	0	0	0	0	0	0
GLOBAL TOTALS	0	52235	46770	1184549	346578	116732	189726	18370	6543	7000	7808	31345	118996
REPORTED GLOBAL TOTALS FROM ORIGINAL DATA SOURCES	****	****	****	****	****	****	****	****	****	****	****	****	****

Comments:

Volcanic eruption: discharge (aerially explosive) of fragmentary ejecta, lava and gases from a volcanic vent.

Affected people: people requiring immediate assistance during a period of emergency, for example requiring basic survival needs such as food, water, shelter, sanitation and immediate medical assistance. Appearance of a significant number of cases of an infectious disease introduced in a region or a population that is usually free from that disease.

AGGREGATIONS

Volcanic Eruptions - Homeless People

	1975	1980	1990	1991	1992	1993	1994	1995	1996	1997	1998	1999	2000
AFRICA	0	0	0	200	0	0	0	5000	0	0	0	510	0
Central Africa	0	0	0	0	0	0	0	0	0	0	0	510	0
Eastern Africa	0	0	0	0	0	0	0	0	0	0	0	0	0
Northern Africa	0	0	0	0	0	0	0	0	0	0	0	0	0
Southern Africa	0	0	0	0	0	0	0	0	0	0	0	0	0
Western Africa	0	0	0	0	0	0	0	5000	0	0	0	0	0
Western Indian Ocean	0	0	0	200	0	0	0	0	0	0	0	0	0
ASIA AND PACIFIC	0	0	0	5800	0	57000	46000	0	0	0	0	0	0
Australia and New Zealand	0	0	0	0	0	0	0	0	0	0	0	0	0
Central Asia	0	0	0	0	0	0	0	0	0	0	0	0	0
North West Pacific and East Asia	0	0	0	0	0	0	0	0	0	0	0	0	0
South Asia	0	0	0	0	0	0	0	0	0	0	0	0	0
South East Asia	0	0	0	5800	0	57000	0	0	0	0	0	0	0
South Pacific	0	0	0	0	0	0	46000	0	0	0	0	0	0
EUROPE	0	0	0	0	0	0	0	0	0	0	0	0	0
Central Europe	0	0	0	0	0	0	0	0	0	0	0	0	0
Eastern Europe	0	0	0	0	0	0	0	0	0	0	0	0	0
Western Europe	0	0	0	0	0	0	0	0	0	0	0	0	0
LATIN AMERICA AND CARIBBEAN	0	0	0	2400	10000	0	0	0	0	200	0	2200	0
Caribbean	0	0	0	0	0	0	0	0	0	200	0	0	0
Meso-America	0	0	0	0	10000	0	0	0	0	0	0	0	0
South America	0	0	0	2400	0	0	0	0	0	0	0	2200	0
NORTH AMERICA	0	2500	0	0	0	0	0	0	0	0	0	0	0
North America	0	2500	0	0	0	0	0	0	0	0	0	0	0
POLAR	0	0	0	0	0	0	0	0	0	0	0	0	0
Antarctic	0	0	0	0	0	0	0	0	0	0	0	0	0
Arctic	0	0	0	0	0	0	0	0	0	0	0	0	0
WEST ASIA	0	0	0	0	0	0	0	0	0	0	0	0	0
Arabian Peninsula	0	0	0	0	0	0	0	0	0	0	0	0	0
Mashriq	0	0	0	0	0	0	0	0	0	0	0	0	0
GLOBAL TOTALS	0	2500	0	8400	10000	57000	46000	5000	0	200	0	2710	0
REPORTED GLOBAL TOTALS FROM ORIGINAL DATA SOURCES	****	****	****	****	****	****	****	****	****	****	****	****	****

Comments:

Volcanic eruption: discharge (aerially explosive) of fragmentary ejecta, lava and gases from a volcanic vent.

Homeless: people needing immediate assistance in the form of shelter.

Volcanic Eruptions - Injured People

	1975	1980	1990	1991	1992	1993	1994	1995	1996	1997	1998	1999	2000
AFRICA	0	0	0	0	0	0	0	6	0	0	0	0	0
Central Africa	0	0	0	0	0	0	0	0	0	0	0	0	0
Eastern Africa	0	0	0	0	0	0	0	0	0	0	0	0	0
Northern Africa	0	0	0	0	0	0	0	0	0	0	0	0	0
Southern Africa	0	0	0	0	0	0	0	0	0	0	0	0	0
Western Africa	0	0	0	0	0	0	0	6	0	0	0	0	0
Western Indian Ocean	0	0	0	0	0	0	0	0	0	0	0	0	0
ASIA AND PACIFIC	0	0	81	215	0	21	24	0	29	0	0	0	0
Australia and New Zealand	0	0	0	0	0	0	0	0	0	0	0	0	0
Central Asia	0	0	0	0	0	0	0	0	0	0	0	0	0
North West Pacific and East Asia	0	0	0	20	0	0	0	0	0	0	0	0	0
South Asia	0	0	0	0	0	0	0	0	0	0	0	0	0
South East Asia	0	0	81	195	0	21	22	0	0	0	0	0	0
South Pacific	0	0	0	0	0	0	2	0	29	0	0	0	0
EUROPE	0	0	0	0	0	0	0	0	0	0	0	0	0
Central Europe	0	0	0	0	0	0	0	0	0	0	0	0	0
Eastern Europe	0	0	0	0	0	0	0	0	0	0	0	0	0
Western Europe	0	0	0	0	0	0	0	0	0	0	0	0	0
LATIN AMERICA AND CARIBBEAN	0	0	0	0	75	6	0	0	0	0	0	0	0
Caribbean	0	0	0	0	0	0	0	0	0	0	0	0	0
Meso-America	0	0	0	0	75	0	0	0	0	0	0	0	0
South America	0	0	0	0	0	6	0	0	0	0	0	0	0
NORTH AMERICA	0	0	0	0	0	0	0	0	0	0	0	0	0
North America	0	0	0	0	0	0	0	0	0	0	0	0	0
POLAR	0	0	0	0	0	0	0	0	0	0	0	0	0
Antarctic	0	0	0	0	0	0	0	0	0	0	0	0	0
Arctic	0	0	0	0	0	0	0	0	0	0	0	0	0
WEST ASIA	0	0	0	0	0	0	0	0	0	0	0	0	0
Arabian Peninsula	0	0	0	0	0	0	0	0	0	0	0	0	0
Mashriq	0	0	0	0	0	0	0	0	0	0	0	0	0
GLOBAL TOTALS	0	0	81	215	75	27	24	6	29	0	0	0	0
REPORTED GLOBAL TOTALS FROM ORIGINAL DATA SOURCES	****	****	****	****	****	****	****	****	****	****	****	****	****

Comments:

Volcanic eruption: discharge (aerially explosive) of fragmentary ejecta, lava and gases from a volcanic vent.

Injured: people suffering from physical injuries, trauma or an illness requiring medical treatment as a direct result of a disaster.

AGGREGATIONS

Volcanic Eruptions - Killed People

	1975	1980	1990	1991	1992	1993	1994	1995	1996	1997	1998	1999	2000
AFRICA	0	0	0	0	0	0	0	0	0	0	0	0	0
Central Africa	0	0	0	0	0	0	0	0	0	0	0	0	0
Eastern Africa	0	0	0	0	0	0	0	0	0	0	0	0	0
Northern Africa	0	0	0	0	0	0	0	0	0	0	0	0	0
Southern Africa	0	0	0	0	0	0	0	0	0	0	0	0	0
Western Africa	0	0	0	0	0	0	0	0	0	0	0	0	0
Western Indian Ocean	0	0	0	0	0	0	0	0	0	0	0	0	0
ASIA AND PACIFIC	0	0	33	683	0	83	101	0	4	1	0	0	0
Australia and New Zealand	0	0	0	0	0	0	0	0	0	0	0	0	0
Central Asia	0	0	0	0	0	0	0	0	0	0	0	0	0
North West Pacific and East Asia	0	0	0	43	0	1	0	0	0	0	0	0	0
South Asia	0	0	0	0	0	0	0	0	0	0	0	0	0
South East Asia	0	0	33	640	0	82	96	0	0	1	0	0	0
South Pacific	0	0	0	0	0	0	5	0	4	0	0	0	0
EUROPE	0	0	0	0	0	0	0	0	0	0	0	0	0
Central Europe	0	0	0	0	0	0	0	0	0	0	0	0	0
Eastern Europe	0	0	0	0	0	0	0	0	0	0	0	0	0
Western Europe	0	0	0	0	0	0	0	0	0	0	0	0	0
LATIN AMERICA AND CARIBBEAN	2	0	0	0	2	16	0	0	0	52	0	0	0
Caribbean	0	0	0	0	0	0	0	0	0	32	0	0	0
Meso-America	2	0	0	0	2	0	0	0	0	20	0	0	0
South America	0	0	0	0	0	16	0	0	0	0	0	0	0
NORTH AMERICA	0	90	0	0	0	0	0	0	0	0	0	0	0
North America	0	90	0	0	0	0	0	0	0	0	0	0	0
POLAR	0	0	0	0	0	0	0	0	0	0	0	0	0
Antarctic	0	0	0	0	0	0	0	0	0	0	0	0	0
Arctic	0	0	0	0	0	0	0	0	0	0	0	0	0
WEST ASIA	0	0	0	0	0	0	0	0	0	0	0	0	0
Arabian Peninsula	0	0	0	0	0	0	0	0	0	0	0	0	0
Mashriq	0	0	0	0	0	0	0	0	0	0	0	0	0
GLOBAL TOTALS	2	90	33	683	2	99	101	0	4	53	0	0	0
REPORTED GLOBAL TOTALS FROM ORIGINAL DATA SOURCES	****	****	****	****	****	****	****	****	****	****	****	****	****

Comments:

Volcanic eruption: discharge (aerially explosive) of fragmentary ejecta, lava and gases from a volcanic vent.

Killed: persons confirmed as dead and persons missing and presumed dead.

Volcanic Eruptions - Number

	1975	1980	1990	1991	1992	1993	1994	1995	1996	1997	1998	1999	2000
AFRICA	0	0	0	1	0	0	0	1	0	0	0	1	0
Central Africa	0	0	0	0	0	0	0	0	0	0	0	1	0
Eastern Africa	0	0	0	0	0	0	0	0	0	0	0	0	0
Northern Africa	0	0	0	0	0	0	0	0	0	0	0	0	0
Southern Africa	0	0	0	0	0	0	0	0	0	0	0	0	0
Western Africa	0	0	0	0	0	0	0	1	0	0	0	0	0
Western Indian Ocean	0	0	0	1	0	0	0	0	0	0	0	0	0
ASIA AND PACIFIC	0	1	1	5	3	4	4	1	2	1	1	0	3
Australia and New Zealand	0	0	0	0	0	0	0	1	0	0	0	0	0
Central Asia	0	0	0	0	0	0	0	0	0	0	0	0	0
North West Pacific and East Asia	0	0	0	1	1	1	0	0	0	0	0	0	2
South Asia	0	0	0	0	0	0	0	0	0	0	0	0	0
South East Asia	0	1	1	4	1	3	3	0	0	1	1	0	1
South Pacific	0	0	0	0	1	0	1	0	2	0	0	0	0
EUROPE	0	0	0	1	0	0	0	0	1	0	0	0	0
Central Europe	0	0	0	0	0	0	0	0	0	0	0	0	0
Eastern Europe	0	0	0	0	0	0	0	0	0	0	0	0	0
Western Europe	0	0	0	1	0	0	0	0	1	0	0	0	0
LATIN AMERICA AND CARIBBEAN	2	0	1	3	2	2	2	2	2	3	3	4	2
Caribbean	0	0	0	0	0	0	0	1	1	2	0	0	0
Meso-America	1	0	0	1	2	0	2	1	1	1	3	3	2
South America	1	0	1	2	0	2	0	0	0	0	0	1	0
NORTH AMERICA	0	1	0	0	0	0	0	0	0	0	0	0	0
North America	0	1	0	0	0	0	0	0	0	0	0	0	0
POLAR	0	0	0	0	0	0	0	0	0	0	0	0	0
Antarctic	0	0	0	0	0	0	0	0	0	0	0	0	0
Arctic	0	0	0	0	0	0	0	0	0	0	0	0	0
WEST ASIA	0	0	0	0	0	0	0	0	0	0	0	0	0
Arabian Peninsula	0	0	0	0	0	0	0	0	0	0	0	0	0
Mashriq	0	0	0	0	0	0	0	0	0	0	0	0	0
GLOBAL TOTALS	2	2	2	10	5	6	6	4	5	4	4	5	5
REPORTED GLOBAL TOTALS FROM ORIGINAL DATA SOURCES	****	****	****	****	****	****	****	****	****	****	****	****	****

Comments:

Volcanic eruption: discharge (aerially explosive) of fragmentary ejecta, lava and gases from a volcanic vent.

AGGREGATIONS

Volcanic Eruptions - Total Affected

	1975	1980	1990	1991	1992	1993	1994	1995	1996	1997	1998	1999	2000
AFRICA	0	0	0	200	0	0	0	6306	0	0	0	3010	0
Central Africa	0	0	0	0	0	0	0	0	0	0	0	3010	0
Eastern Africa	0	0	0	0	0	0	0	0	0	0	0	0	0
Northern Africa	0	0	0	0	0	0	0	0	0	0	0	0	0
Southern Africa	0	0	0	0	0	0	0	0	0	0	0	0	0
Western Africa	0	0	0	0	0	0	0	6306	0	0	0	0	0
Western Indian Ocean	0	0	0	200	0	0	0	0	0	0	0	0	0
ASIA AND PACIFIC	0	52235	42851	1059564	41578	173403	160050	70	1829	3000	6000	0	77196
Australia and New Zealand	0	0	0	0	0	0	0	70	0	0	0	0	0
Central Asia	0	0	0	0	0	0	0	0	0	0	0	0	0
North West Pacific and East Asia	0	0	0	10020	40000	4930	0	0	0	0	0	0	16400
South Asia	0	0	0	0	0	0	0	0	0	0	0	0	0
South East Asia	0	52235	42851	1049544	1578	168473	8048	0	0	3000	6000	0	60796
South Pacific	0	0	0	0	0	0	152002	0	1829	0	0	0	0
EUROPE	0	0	0	7000	0	0	0	0	0	0	0	0	0
Central Europe	0	0	0	0	0	0	0	0	0	0	0	0	0
Eastern Europe	0	0	0	0	0	0	0	0	0	0	0	0	0
Western Europe	0	0	0	7000	0	0	0	0	0	0	0	0	0
LATIN AMERICA AND CARIBBEAN	0	0	4000	126400	315075	356	75700	17000	4743	4200	1808	31045	41800
Caribbean	0	0	0	0	0	0	0	5000	4000	4200	0	0	0
Meso-America	0	0	0	0	315075	0	75700	12000	743	0	1808	6845	41800
South America	0	0	4000	126400	0	356	0	0	0	0	0	24200	0
NORTH AMERICA	0	2500	0	0	0	0	0	0	0	0	0	0	0
North America	0	2500	0	0	0	0	0	0	0	0	0	0	0
POLAR	0	0	0	0	0	0	0	0	0	0	0	0	0
Antarctic	0	0	0	0	0	0	0	0	0	0	0	0	0
Arctic	0	0	0	0	0	0	0	0	0	0	0	0	0
WEST ASIA	0	0	0	0	0	0	0	0	0	0	0	0	0
Arabian Peninsula	0	0	0	0	0	0	0	0	0	0	0	0	0
Mashriq	0	0	0	0	0	0	0	0	0	0	0	0	0
GLOBAL TOTALS	0	54735	46851	1193164	356653	173759	235750	23376	6572	7200	7808	34055	118996
REPORTED GLOBAL TOTALS FROM ORIGINAL DATA SOURCES	****	****	****	****	****	****	****	****	****	****	****	****	****

Comments:

Volcanic eruption: discharge (aerially explosive) of fragmentary ejecta, lava and gases from a volcanic vent.

Total affected: people that have been injured affected and left homeless after a disaster are included in this category.

Waves/Surges - Affected People

	1975	1977	1979	1985	1987	1988	1989	1990	1996	1997	1998	1999	2000
AFRICA	0	0	0	0	0	0	0	0	0	0	0	0	0
Central Africa	0	0	0	0	0	0	0	0	0	0	0	0	0
Eastern Africa	0	0	0	0	0	0	0	0	0	0	0	0	0
Northern Africa	0	0	0	0	0	0	0	0	0	0	0	0	0
Southern Africa	0	0	0	0	0	0	0	0	0	0	0	0	0
Western Africa	0	0	0	0	0	0	0	0	0	0	0	0	0
Western Indian Ocean	0	0	0	0	0	0	0	0	0	0	0	0	0
ASIA AND PACIFIC	0	818	2000	9199	0	0	0	0	0	0	0	0	0
Australia and New Zealand	0	0	0	0	0	0	0	0	0	0	0	0	0
Central Asia	0	0	0	0	0	0	0	0	0	0	0	0	0
North West Pacific and East Asia	0	0	0	0	0	0	0	0	0	0	0	0	0
South Asia	0	0	0	0	0	0	0	0	0	0	0	0	0
South East Asia	0	818	2000	0	0	0	0	0	0	0	0	0	0
South Pacific	0	0	0	9199	0	0	0	0	0	0	0	0	0
EUROPE	0	0	0	0	0	0	0	0	0	0	0	0	0
Central Europe	0	0	0	0	0	0	0	0	0	0	0	0	0
Eastern Europe	0	0	0	0	0	0	0	0	0	0	0	0	0
Western Europe	0	0	0	0	0	0	0	0	0	0	0	0	0
LATIN AMERICA AND CARIBBEAN	0	0	0	0	0	0	0	0	0	0	0	0	0
Caribbean	0	0	0	0	0	0	0	0	0	0	0	0	0
Meso-America	0	0	0	0	0	0	0	0	0	0	0	0	0
South America	0	0	0	0	0	0	0	0	0	0	0	0	0
NORTH AMERICA	0	0	0	0	0	0	0	0	0	0	0	0	0
North America	0	0	0	0	0	0	0	0	0	0	0	0	0
POLAR	0	0	0	0	0	0	0	0	0	0	0	0	0
Antarctic	0	0	0	0	0	0	0	0	0	0	0	0	0
Arctic	0	0	0	0	0	0	0	0	0	0	0	0	0
WEST ASIA	0	0	0	0	0	0	0	0	0	0	0	0	0
Arabian Peninsula	0	0	0	0	0	0	0	0	0	0	0	0	0
Mashriq	0	0	0	0	0	0	0	0	0	0	0	0	0
GLOBAL TOTALS	0	818	2000	9199	0	0	0	0	0	0	0	0	0
REPORTED GLOBAL TOTALS FROM ORIGINAL DATA SOURCES	****	****	****	****	****	****	****	****	****	****	****	****	****

Comments:

Due to scattered data availability, the years shown are different from the Data Compendium's standard table for this theme.

Wave/surge: disaster type term comprising the two disaster subsets "tsunami" and "tidal wave".

Tsunami: series of large waves generated by sudden displacement of seawater (caused by earthquake, volcanic eruption or submarine landslide); capable of propagation over large distances and causing a destructive surge on reaching land.

Tidal wave: abrupt rise of tidal water (caused by atmospheric activities) moving rapidly inland from the mouth of an estuary or from the coast.

Affected people: people requiring immediate assistance during a period of emergency, for example requiring basic survival needs such as food, water, shelter, sanitation and immediate medical assistance. Appearance of a significant number of cases of an infectious disease introduced in a region or a population that is usually free from that disease.

	1975	1980	1984	1985	1986	1987	1990	1995	1996	1997	1998	1999	2000
AFRICA	0	0	0	0	0	0	0	0	0	0	0	0	0
Central Africa	0	0	0	0	0	0	0	0	0	0	0	0	0
Eastern Africa	0	0	0	0	0	0	0	0	0	0	0	0	0
Northern Africa	0	0	0	0	0	0	0	0	0	0	0	0	0
Southern Africa	0	0	0	0	0	0	0	0	0	0	0	0	0
Western Africa	0	0	0	0	0	0	0	0	0	0	0	0	0
Western Indian Ocean	0	0	0	0	0	0	0	0	0	0	0	0	0
ASIA AND PACIFIC	0	0	29000	0	200	17250	0	0	0	0	0	0	0
Australia and New Zealand	0	0	0	0	0	0	0	0	0	0	0	0	0
Central Asia	0	0	0	0	0	0	0	0	0	0	0	0	0
North West Pacific and East Asia	0	0	29000	0	0	0	0	0	0	0	0	0	0
South Asia	0	0	0	0	0	12000	0	0	0	0	0	0	0
South East Asia	0	0	0	0	200	5250	0	0	0	0	0	0	0
South Pacific	0	0	0	0	0	0	0	0	0	0	0	0	0
EUROPE	0	0	0	0	0	0	0	0	0	0	0	0	0
Central Europe	0	0	0	0	0	0	0	0	0	0	0	0	0
Eastern Europe	0	0	0	0	0	0	0	0	0	0	0	0	0
Western Europe	0	0	0	0	0	0	0	0	0	0	0	0	0
LATIN AMERICA AND CARIBBEAN	0	0	0	0	1100	0	0	0	0	0	0	0	0
Caribbean	0	0	0	0	0	0	0	0	0	0	0	0	0
Meso-America	0	0	0	0	0	0	0	0	0	0	0	0	0
South America	0	0	0	0	1100	0	0	0	0	0	0	0	0
NORTH AMERICA	0	0	0	0	0	0	0	0	0	0	0	0	0
North America	0	0	0	0	0	0	0	0	0	0	0	0	0
POLAR	0	0	0	0	0	0	0	0	0	0	0	0	0
Antarctic	0	0	0	0	0	0	0	0	0	0	0	0	0
Arctic	0	0	0	0	0	0	0	0	0	0	0	0	0
WEST ASIA	0	0	0	0	0	0	0	0	0	0	0	0	0
Arabian Peninsula	0	0	0	0	0	0	0	0	0	0	0	0	0
Mashriq	0	0	0	0	0	0	0	0	0	0	0	0	0
GLOBAL TOTALS	0	0	29000	0	1300	17250	0	0	0	0	0	0	0
REPORTED GLOBAL TOTALS FROM ORIGINAL DATA SOURCES	****	****	****	****	****	****	****	****	****	****	****	****	****

Comments:

Due to scattered data availability, the years shown are different from the Data Compendium's standard table for this theme.

Wave/surge: disaster type term comprising the two disaster subsets "tsunami" and "tidal wave".

Tsunami: series of large waves generated by sudden displacement of seawater (caused by earthquake, volcanic eruption or submarine landslide); capable of propagation over large distances and causing a destructive surge on reaching land.

Tidal wave: abrupt rise of tidal water (caused by atmospheric activities) moving rapidly inland from the mouth of an estuary or from the coast.

Homeless: people needing immediate assistance in the form of shelter.

AGGREGATIONS

	1975	1976	1980	1983	1985	1987	1990	1995	1996	1997	1998	1999	2000
AFRICA	0	0	0	0	0	0	0	0	0	0	0	0	0
Central Africa	0	0	0	0	0	0	0	0	0	0	0	0	0
Eastern Africa	0	0	0	0	0	0	0	0	0	0	0	0	0
Northern Africa	0	0	0	0	0	0	0	0	0	0	0	0	0
Southern Africa	0	0	0	0	0	0	0	0	0	0	0	0	0
Western Africa	0	0	0	0	0	0	0	0	0	0	0	0	0
Western Indian Ocean	0	0	0	0	0	0	0	0	0	0	0	0	0
ASIA AND PACIFIC	0	23	0	0	668	10	0	0	0	0	0	0	0
Australia and New Zealand	0	0	0	0	0	0	0	0	0	0	0	0	0
Central Asia	0	0	0	0	0	0	0	0	0	0	0	0	0
North West Pacific and East Asia	0	0	0	0	0	0	0	0	0	0	0	0	0
South Asia	0	0	0	0	0	10	0	0	0	0	0	0	0
South East Asia	0	23	0	0	0	0	0	0	0	0	0	0	0
South Pacific	0	0	0	0	668	0	0	0	0	0	0	0	0
EUROPE	0	2	0	0	0	0	0	0	0	0	0	0	0
Central Europe	0	0	0	0	0	0	0	0	0	0	0	0	0
Eastern Europe	0	0	0	0	0	0	0	0	0	0	0	0	0
Western Europe	0	2	0	0	0	0	0	0	0	0	0	0	0
LATIN AMERICA AND CARIBBEAN	0	0	0	24	0	0	0	0	0	0	0	0	0
Caribbean	0	0	0	0	0	0	0	0	0	0	0	0	0
Meso-America	0	0	0	0	0	0	0	0	0	0	0	0	0
South America	0	0	0	24	0	0	0	0	0	0	0	0	0
NORTH AMERICA	0	0	0	0	0	0	0	0	0	0	0	0	0
North America	0	0	0	0	0	0	0	0	0	0	0	0	0
POLAR	0	0	0	0	0	0	0	0	0	0	0	0	0
Antarctic	0	0	0	0	0	0	0	0	0	0	0	0	0
Arctic	0	0	0	0	0	0	0	0	0	0	0	0	0
WEST ASIA	0	0	0	0	0	0	0	0	0	0	0	0	0
Arabian Peninsula	0	0	0	0	0	0	0	0	0	0	0	0	0
Mashriq	0	0	0	0	0	0	0	0	0	0	0	0	0
GLOBAL TOTALS	0	25	0	24	668	10	0	0	0	0	0	0	0
REPORTED GLOBAL TOTALS FROM ORIGINAL DATA SOURCES	****	****	****	****	****	****	****	****	****	****	****	****	****

Comments:

Due to scattered data availability, the years shown are different from the Data Compendium's standard table for this theme.

Wave/surge: disaster type term comprising the two disaster subsets "tsunami" and "tidal wave".

Tsunami: series of large waves generated by sudden displacement of seawater (caused by earthquake, volcanic eruption or submarine landslide); capable of propagation over large distances and causing a destructive surge on reaching land.

Tidal wave: abrupt rise of tidal water (caused by atmospheric activities) moving rapidly inland from the mouth of an estuary or from the coast.

Injured: people suffering from physical injuries, trauma or an illness requiring medical treatment as a direct result of a disaster.

	1975	1980	1990	1991	1992	1993	1994	1995	1996	1997	1998	1999	2000
AFRICA	0	0	0	0	0	0	0	0	0	0	0	0	0
Central Africa	0	0	0	0	0	0	0	0	0	0	0	0	0
Eastern Africa	0	0	0	0	0	0	0	0	0	0	0	0	0
Northern Africa	0	0	0	0	0	0	0	0	0	0	0	0	0
Southern Africa	0	0	0	0	0	0	0	0	0	0	0	0	0
Western Africa	0	0	0	0	0	0	0	0	0	0	0	0	0
Western Indian Ocean	0	0	0	0	0	0	0	0	0	0	0	0	0
ASIA AND PACIFIC	200	10	0	0	0	0	0	0	0	0	0	0	0
Australia and New Zealand	0	0	0	0	0	0	0	0	0	0	0	0	0
Central Asia	0	0	0	0	0	0	0	0	0	0	0	0	0
North West Pacific and East Asia	0	0	0	0	0	0	0	0	0	0	0	0	0
South Asia	0	0	0	0	0	0	0	0	0	0	0	0	0
South East Asia	0	10	0	0	0	0	0	0	0	0	0	0	0
South Pacific	200	0	0	0	0	0	0	0	0	0	0	0	0
EUROPE	0	0	0	0	0	0	0	0	0	0	0	0	0
Central Europe	0	0	0	0	0	0	0	0	0	0	0	0	0
Eastern Europe	0	0	0	0	0	0	0	0	0	0	0	0	0
Western Europe	0	0	0	0	0	0	0	0	0	0	0	0	0
LATIN AMERICA AND CARIBBEAN	0	0	0	0	0	0	0	0	0	0	0	0	0
Caribbean	0	0	0	0	0	0	0	0	0	0	0	0	0
Meso-America	0	0	0	0	0	0	0	0	0	0	0	0	0
South America	0	0	0	0	0	0	0	0	0	0	0	0	0
NORTH AMERICA	0	0	0	0	0	0	0	0	0	0	0	0	0
North America	0	0	0	0	0	0	0	0	0	0	0	0	0
POLAR	0	0	0	0	0	0	0	0	0	0	0	0	0
Antarctic	0	0	0	0	0	0	0	0	0	0	0	0	0
Arctic	0	0	0	0	0	0	0	0	0	0	0	0	0
WEST ASIA	0	0	0	0	0	0	0	0	0	0	0	0	0
Arabian Peninsula	0	0	0	0	0	0	0	0	0	0	0	0	0
Mashriq	0	0	0	0	0	0	0	0	0	0	0	0	0
GLOBAL TOTALS	200	10	0	0	0	0	0	0	0	0	0	0	0
REPORTED GLOBAL TOTALS FROM ORIGINAL DATA SOURCES	****	****	****	****	****	****	****	****	****	****	****	****	****

Comments:

Wave/surge: disaster type term comprising the two disaster subsets "tsunami" and "tidal wave".

Tsunami: series of large waves generated by sudden displacement of seawater (caused by earthquake, volcanic eruption or submarine landslide); capable of propagation over large distances and causing a destructive surge on reaching land.

Tidal wave: abrupt rise of tidal water (caused by atmospheric activities) moving rapidly inland from the mouth of an estuary or from the coast.

Killed: persons confirmed as dead and persons missing and presumed dead.

Waves/Surges - Number

	1975	1980	1990	1991	1992	1993	1994	1995	1996	1997	1998	1999	2000
AFRICA	0	0	0	0	0	0	0	0	0	0	0	0	0
Central Africa	0	0	0	0	0	0	0	0	0	0	0	0	0
Eastern Africa	0	0	0	0	0	0	0	0	0	0	0	0	0
Northern Africa	0	0	0	0	0	0	0	0	0	0	0	0	0
Southern Africa	0	0	0	0	0	0	0	0	0	0	0	0	0
Western Africa	0	0	0	0	0	0	0	0	0	0	0	0	0
Western Indian Ocean	0	0	0	0	0	0	0	0	0	0	0	0	0
ASIA AND PACIFIC	1	0	0	1	0	1	1	0	1	2	1	1	2
Australia and New Zealand	0	0	0	0	0	0	0	0	0	0	0	0	0
Central Asia	0	0	0	0	0	0	0	0	0	0	0	0	0
North West Pacific and East Asia	0	0	0	0	0	1	0	0	1	1	0	0	0
South Asia	0	0	0	0	0	0	0	0	0	1	0	0	1
South East Asia	0	0	0	1	0	0	1	0	0	0	0	1	1
South Pacific	1	0	0	0	0	0	0	0	0	0	1	0	0
EUROPE	0	0	0	0	0	0	0	0	0	0	0	0	0
Central Europe	0	0	0	0	0	0	0	0	0	0	0	0	0
Eastern Europe	0	0	0	0	0	0	0	0	0	0	0	0	0
Western Europe	0	0	0	0	0	0	0	0	0	0	0	0	0
LATIN AMERICA AND CARIBBEAN	0	0	0	0	0	0	0	0	1	0	0	1	0
Caribbean	0	0	0	0	0	0	0	0	0	0	0	0	0
Meso-America	0	0	0	0	0	0	0	0	0	0	0	0	0
South America	0	0	0	0	0	0	0	0	1	0	0	1	0
NORTH AMERICA	0	0	0	0	0	0	0	0	0	0	0	0	0
North America	0	0	0	0	0	0	0	0	0	0	0	0	0
POLAR	0	0	0	0	0	0	0	0	0	0	0	0	0
Antarctic	0	0	0	0	0	0	0	0	0	0	0	0	0
Arctic	0	0	0	0	0	0	0	0	0	0	0	0	0
WEST ASIA	0	0	0	0	0	0	0	0	0	0	0	0	0
Arabian Peninsula	0	0	0	0	0	0	0	0	0	0	0	0	0
Mashriq	0	0	0	0	0	0	0	0	0	0	0	0	0
GLOBAL TOTALS	1	0	0	1	0	1	1	0	2	2	1	2	2
REPORTED GLOBAL TOTALS FROM ORIGINAL DATA SOURCES	****	****	****	****	****	****	****	****	****	****	****	****	****

Comments:

Wave/surge: disaster type term comprising the two disaster subsets "tsunami" and "tidal wave".

Tsunami: series of large waves generated by sudden displacement of seawater (caused by earthquake, volcanic eruption or submarine landslide); capable of propagation over large distances and causing a destructive surge on reaching land.

Tidal wave: abrupt rise of tidal water (caused by atmospheric activities) moving rapidly inland from the mouth of an estuary or from the coast.

	1975	1976	1977	1979	1980	1983	1984	1985	1986	1987	1990	1995	2000
AFRICA	0	0	0	0	0	0	0	0	0	0	0	0	0
Central Africa	0	0	0	0	0	0	0	0	0	0	0	0	0
Eastern Africa	0	0	0	0	0	0	0	0	0	0	0	0	0
Northern Africa	0	0	0	0	0	0	0	0	0	0	0	0	0
Southern Africa	0	0	0	0	0	0	0	0	0	0	0	0	0
Western Africa	0	0	0	0	0	0	0	0	0	0	0	0	0
Western Indian Ocean	0	0	0	0	0	0	0	0	0	0	0	0	0
ASIA AND PACIFIC	0	23	818	2000	0	0	29000	9867	200	17260	0	0	0
Australia and New Zealand	0	0	0	0	0	0	0	0	0	0	0	0	0
Central Asia	0	0	0	0	0	0	0	0	0	0	0	0	0
North West Pacific and East Asia	0	0	0	0	0	0	29000	0	0	0	0	0	0
South Asia	0	0	0	0	0	0	0	0	0	12010	0	0	0
South East Asia	0	23	818	2000	0	0	0	0	200	5250	0	0	0
South Pacific	0	0	0	0	0	0	0	9867	0	0	0	0	0
EUROPE	0	2	0	0	0	0	0	0	0	0	0	0	0
Central Europe	0	0	0	0	0	0	0	0	0	0	0	0	0
Eastern Europe	0	0	0	0	0	0	0	0	0	0	0	0	0
Western Europe	0	2	0	0	0	0	0	0	0	0	0	0	0
LATIN AMERICA AND CARIBBEAN	0	0	0	0	0	24	0	0	1100	0	0	0	0
Caribbean	0	0	0	0	0	0	0	0	0	0	0	0	0
Meso-America	0	0	0	0	0	0	0	0	0	0	0	0	0
South America	0	0	0	0	0	24	0	0	1100	0	0	0	0
NORTH AMERICA	0	0	0	0	0	0	0	0	0	0	0	0	0
North America	0	0	0	0	0	0	0	0	0	0	0	0	0
POLAR	0	0	0	0	0	0	0	0	0	0	0	0	0
Antarctic	0	0	0	0	0	0	0	0	0	0	0	0	0
Arctic	0	0	0	0	0	0	0	0	0	0	0	0	0
WEST ASIA	0	0	0	0	0	0	0	0	0	0	0	0	0
Arabian Peninsula	0	0	0	0	0	0	0	0	0	0	0	0	0
Mashriq	0	0	0	0	0	0	0	0	0	0	0	0	0
GLOBAL TOTALS	0	25	818	2000	0	24	29000	9867	1300	17260	0	0	0
REPORTED GLOBAL TOTALS FROM ORIGINAL DATA	****	****	****	****	****	****	****	****	****	****	****	****	****

Comments:

Wave/surge: disaster type term comprising the two disaster subsets "tsunami" and "tidal wave".

Tsunami: series of large waves generated by sudden displacement of seawater (caused by earthquake, volcanic eruption or submarine landslide); capable of propagation over large distances and causing a destructive surge on reaching land.

Tidal wave: abrupt rise of tidal water (caused by atmospheric activities) moving rapidly inland from the mouth of an estuary or from the coast.

Total affected: people that have been injured affected and left homeless after a disaster are included in this category.

	1975	1980	1990	1991	1992	1993	1994	1995	1996	1997	1998	1999	2000
AFRICA	0	0	0	0	0	0	0	85	0	0	1200	0	1000
Central Africa	0	0	0	0	0	0	0	85	0	0	0	0	0
Eastern Africa	0	0	0	0	0	0	0	0	0	0	0	0	0
Northern Africa	0	0	0	0	0	0	0	0	0	0	0	0	0
Southern Africa	0	0	0	0	0	0	0	0	0	0	0	0	1000
Western Africa	0	0	0	0	0	0	0	0	0	0	1200	0	0
Western Indian Ocean	0	0	0	0	0	0	0	0	0	0	0	0	0
ASIA AND PACIFIC	0	0	0	0	0	0	3045000	0	5000	48000	2300	2000	200
Australia and New Zealand	0	0	0	0	0	0	45000	0	0	0	0	2000	200
Central Asia	0	0	0	0	0	0	0	0	0	8000	0	0	0
North West Pacific and East Asia	0	0	0	0	0	0	0	0	5000	0	0	0	0
South Asia	0	0	0	0	0	0	0	0	0	0	0	0	0
South East Asia	0	0	0	0	0	0	3000000	0	0	32000	2300	0	0
South Pacific	0	0	0	0	0	0	0	0	0	8000	0	0	0
EUROPE	0	0	0	0	0	0	16800	2200	700	1250	100900	0	0
Central Europe	0	0	0	0	0	0	0	0	500	0	0	0	0
Eastern Europe	0	0	0	0	0	0	0	0	200	0	100000	0	0
Western Europe	0	0	0	0	0	0	16800	2200	0	1250	900	0	0
LATIN AMERICA AND CARIBBEAN	0	0	0	0	1200	0	0	0	0	1300	12000	3500	0
Caribbean	0	0	0	0	0	0	0	0	0	0	0	0	0
Meso-America	0	0	0	0	1200	0	0	0	0	0	0	0	0
South America	0	0	0	0	0	0	0	0	0	1300	12000	3500	0
NORTH AMERICA	0	5000	0	0	450	0	4400	6500	0	2600	49032	2300	35400
North America	0	5000	0	0	450	0	4400	6500	0	2600	49032	2300	35400
POLAR	0	0	0	0	0	0	0	0	0	0	0	0	0
Antarctic	0	0	0	0	0	0	0	0	0	0	0	0	0
Arctic	0	0	0	0	0	0	0	0	0	0	0	0	0
WEST ASIA	0	0	0	0	0	0	0	0	0	0	0	0	0
Arabian Peninsula	0	0	0	0	0	0	0	0	0	0	0	0	0
Mashriq	0	0	0	0	0	0	0	0	0	0	0	0	0
GLOBAL TOTALS	0	5000	0	0	1650	0	3066200	8785	5700	53150	165432	7800	36600
REPORTED GLOBAL TOTALS FROM ORIGINAL DATA SOURCES	****	****	****	****	****	****	****	****	****	****	****	****	****

Comments:

Wild fire: disaster type term comprising the two disaster subsets "forest fire" and "scrub fire".

Forest fire: fires in forest that cover extensive damage. They may start by natural causes such as volcanic eruptions or lightning, or they may be caused by arsonists or careless smokers, by those burning wood, or by clearing a forest area.

Scrub fire: fires in scrub or bush that cover extensive damage. They may start by natural causes such as volcanic eruptions or lightning, or they may be caused by arsonists or careless smokers, by those burning wood, or by clearing a forest area.

Affected people: people requiring immediate assistance during a period of emergency, for example requiring basic survival needs such as food, water, shelter, sanitation and immediate medical assistance. Appearance of a significant number of cases of an infectious disease introduced in a region or a population that is usually free from that disease.

AGGREGATIONS

Wild Fires - Homeless People

	1975	1980	1990	1991	1992	1993	1994	1995	1996	1997	1998	1999	2000
AFRICA	0	0	0	0	0	0	0	0	0	0	300	3000	250
Central Africa	0	0	0	0	0	0	0	0	0	0	0	0	0
Eastern Africa	0	0	0	0	0	0	0	0	0	0	0	0	0
Northern Africa	0	0	0	0	0	0	0	0	0	0	0	0	0
Southern Africa	0	0	0	0	0	0	0	0	0	0	0	3000	250
Western Africa	0	0	0	0	0	0	0	0	0	0	300	0	0
Western Indian Ocean	0	0	0	0	0	0	0	0	0	0	0	0	0
ASIA AND PACIFIC	0	0	0	0	50000	0	1000	3000	0	0	0	4000	840
Australia and New Zealand	0	0	0	0	0	0	1000	0	0	0	0	0	0
Central Asia	0	0	0	0	0	0	0	0	0	0	0	0	0
North West Pacific and East Asia	0	0	0	0	0	0	0	0	0	0	0	0	840
South Asia	0	0	0	0	50000	0	0	0	0	0	0	4000	0
South East Asia	0	0	0	0	0	0	0	3000	0	0	0	0	0
South Pacific	0	0	0	0	0	0	0	0	0	0	0	0	0
EUROPE	0	0	0	0	0	0	0	0	0	0	983	6	602
Central Europe	0	0	0	0	0	0	0	0	0	0	0	0	512
Eastern Europe	0	0	0	0	0	0	0	0	0	0	683	0	0
Western Europe	0	0	0	0	0	0	0	0	0	0	300	6	90
LATIN AMERICA AND CARIBBEAN	0	0	0	0	0	0	0	0	0	0	0	3800	0
Caribbean	0	0	0	0	0	0	0	0	0	0	0	0	0
Meso-America	0	0	0	0	0	0	0	0	0	0	0	0	0
South America	0	0	0	0	0	0	0	0	0	0	0	3800	0
NORTH AMERICA	0	0	0	7083	540	0	30	0	0	0	65	219	685
North America	0	0	0	7083	540	0	30	0	0	0	65	219	685
POLAR	0	0	0	0	0	0	0	0	0	0	0	0	0
Antarctic	0	0	0	0	0	0	0	0	0	0	0	0	0
Arctic	0	0	0	0	0	0	0	0	0	0	0	0	0
WEST ASIA	0	0	0	0	0	0	0	0	0	0	0	0	0
Arabian Peninsula	0	0	0	0	0	0	0	0	0	0	0	0	0
Mashriq	0	0	0	0	0	0	0	0	0	0	0	0	0
GLOBAL TOTALS	0	0	0	7083	50540	0	1030	3000	0	0	1348	11025	2377
REPORTED GLOBAL TOTALS FROM ORIGINAL DATA SOURCES	****	****	****	****	****	****	****	****	****	****	****	****	****

Comments:

Wild fire: disaster type term comprising the two disaster subsets "forest fire" and "scrub fire".

Forest fire: fires in forest that cover extensive damage. They may start by natural causes such as volcanic eruptions or lightning, or they may be caused by arsonists or careless smokers, by those burning wood, or by clearing a forest area.

Scrub fire: fires in scrub or bush that cover extensive damage. They may start by natural causes such as volcanic eruptions or lightning, or they may be caused by arsonists or careless smokers, by those burning wood, or by clearing a forest area.

Homeless: people needing immediate assistance in the form of shelter.

Wild Fires - Injured People

	1975	1980	1990	1991	1992	1993	1994	1995	1996	1997	1998	1999	2000
AFRICA	0	0	0	0	0	0	0	0	0	0	0	5	5
Central Africa	0	0	0	0	0	0	0	0	0	0	0	0	0
Eastern Africa	0	0	0	0	0	0	0	0	0	0	0	0	5
Northern Africa	0	0	0	0	0	0	0	0	0	0	0	0	0
Southern Africa	0	0	0	0	0	0	0	0	0	0	0	5	0
Western Africa	0	0	0	0	0	0	0	0	0	0	0	0	0
Western Indian Ocean	0	0	0	0	0	0	0	0	0	0	0	0	0
ASIA AND PACIFIC	0	40	0	8	0	0	161	0	61	0	0	0	15
Australia and New Zealand	0	40	0	0	0	0	161	0	0	0	0	0	0
Central Asia	0	0	0	0	0	0	0	0	0	0	0	0	0
North West Pacific and East Asia	0	0	0	0	0	0	0	0	61	0	0	0	15
South Asia	0	0	0	0	0	0	0	0	0	0	0	0	0
South East Asia	0	0	0	8	0	0	0	0	0	0	0	0	0
South Pacific	0	0	0	0	0	0	0	0	0	0	0	0	0
EUROPE	0	0	0	0	0	0	20	54	0	9	0	0	36
Central Europe	0	0	0	0	0	0	0	0	0	0	0	0	17
Eastern Europe	0	0	0	0	0	0	0	0	0	0	0	0	0
Western Europe	0	0	0	0	0	0	20	54	0	9	0	0	19
LATIN AMERICA AND CARIBBEAN	0	0	0	0	0	0	0	0	0	0	0	0	0
Caribbean	0	0	0	0	0	0	0	0	0	0	0	0	0
Meso-America	0	0	0	0	0	0	0	0	0	0	0	0	0
South America	0	0	0	0	0	0	0	0	0	0	0	0	0
NORTH AMERICA	0	0	0	150	0	130	2	0	50	0	124	0	2
North America	0	0	0	150	0	130	2	0	50	0	124	0	2
POLAR	0	0	0	0	0	0	0	0	0	0	0	0	0
Antarctic	0	0	0	0	0	0	0	0	0	0	0	0	0
Arctic	0	0	0	0	0	0	0	0	0	0	0	0	0
WEST ASIA	0	0	0	0	0	0	0	0	0	0	0	0	0
Arabian Peninsula	0	0	0	0	0	0	0	0	0	0	0	0	0
Mashriq	0	0	0	0	0	0	0	0	0	0	0	0	0
GLOBAL TOTALS	0	40	0	158	0	130	183	54	111	9	124	5	58
REPORTED GLOBAL TOTALS FROM ORIGINAL DATA SOURCES	****	****	****	****	****	****	****	****	****	****	****	****	****

Comments:

Wild fire: disaster type term comprising the two disaster subsets "forest fire" and "scrub fire".

Forest fire: fires in forest that cover extensive damage. They may start by natural causes such as volcanic eruptions or lightning, or they may be caused by arsonists or careless smokers, by those burning wood, or by clearing a forest area.

Scrub fire: fires in scrub or bush that cover extensive damage. They may start by natural causes such as volcanic eruptions or lightning, or they may be caused by arsonists or careless smokers, by those burning wood, or by clearing a forest area.

Injured: people suffering from physical injuries, trauma or an illness requiring medical treatment as a direct result of a disaster.

Wild Fires - Killed People

	1975	1980	1990	1991	1992	1993	1994	1995	1996	1997	1998	1999	2000
AFRICA	0	0	0	0	0	0	22	0	0	0	75	4	0
Central Africa	0	0	0	0	0	0	0	0	0	0	0	0	0
Eastern Africa	0	0	0	0	0	0	0	0	0	0	0	0	0
Northern Africa	0	0	0	0	0	0	22	0	0	0	47	0	0
Southern Africa	0	0	0	0	0	0	0	0	0	0	25	4	0
Western Africa	0	0	0	0	0	0	0	0	0	0	3	0	0
Western Indian Ocean	0	0	0	0	0	0	0	0	0	0	0	0	0
ASIA AND PACIFIC	0	0	0	57	56	0	4	29	25	8	2	62	2
Australia and New Zealand	0	0	0	0	0	0	4	0	0	2	0	1	0
Central Asia	0	0	0	0	0	0	0	0	0	0	0	0	0
North West Pacific and East Asia	0	0	0	0	0	0	0	29	25	0	0	23	2
South Asia	0	0	0	0	56	0	0	0	0	0	0	38	0
South East Asia	0	0	0	57	0	0	0	0	0	6	2	0	0
South Pacific	0	0	0	0	0	0	0	0	0	0	0	0	0
EUROPE	0	0	0	0	66	0	21	0	19	14	7	0	23
Central Europe	0	0	0	0	35	0	0	0	0	0	0	0	14
Eastern Europe	0	0	0	0	31	0	0	0	19	14	3	0	0
Western Europe	0	0	0	0	0	0	21	0	0	0	4	0	9
LATIN AMERICA AND CARIBBEAN	0	0	0	0	0	0	24	0	0	10	23	3	8
Caribbean	0	0	0	0	0	0	0	0	0	0	0	0	0
Meso-America	0	0	0	0	0	0	0	0	0	0	23	0	0
South America	0	0	0	0	0	0	24	0	0	10	0	3	8
NORTH AMERICA	0	0	0	28	0	3	13	0	1	0	2	1	14
North America	0	0	0	28	0	3	13	0	1	0	2	1	14
POLAR	0	0	0	0	0	0	0	0	0	0	0	0	0
Antarctic	0	0	0	0	0	0	0	0	0	0	0	0	0
Arctic	0	0	0	0	0	0	0	0	0	0	0	0	0
WEST ASIA	0	0	0	0	0	0	0	0	0	0	0	0	0
Arabian Peninsula	0	0	0	0	0	0	0	0	0	0	0	0	0
Mashriq	0	0	0	0	0	0	0	0	0	0	0	0	0
GLOBAL TOTALS	0	0	0	85	122	3	84	29	45	32	109	70	47
REPORTED GLOBAL TOTALS FROM ORIGINAL DATA SOURCES	****	****	****	****	****	****	****	****	****	****	****	****	****

Comments:

Wild fire: disaster type term comprising the two disaster subsets "forest fire" and "scrub fire".

Forest fire: fires in forest that cover extensive damage. They may start by natural causes such as volcanic eruptions or lightning, or they may be caused by arsonists or careless smokers, by those burning wood, or by clearing a forest area.

Scrub fire: fires in scrub or bush that cover extensive damage. They may start by natural causes such as volcanic eruptions or lightning, or they may be caused by arsonists or careless smokers, by those burning wood, or by clearing a forest area.

Killed: persons confirmed as dead and persons missing and presumed dead.

Wild Fires - Number

	1975	1980	1990	1991	1992	1993	1994	1995	1996	1997	1998	1999	2000
AFRICA	0	0	0	0	0	0	1	1	0	1	3	1	2
Central Africa	0	0	0	0	0	0	0	1	0	0	0	0	0
Eastern Africa	0	0	0	0	0	0	0	0	0	0	0	0	1
Northern Africa	0	0	0	0	0	0	1	0	0	0	1	0	0
Southern Africa	0	0	0	0	0	0	0	0	0	1	1	1	1
Western Africa	0	0	0	0	0	0	0	0	0	0	1	0	0
Western Indian Ocean	0	0	0	0	0	0	0	0	0	0	0	0	0
ASIA AND PACIFIC	2	5	2	1	1	0	3	2	1	5	2	7	3
Australia and New Zealand	2	1	1	0	0	0	2	0	0	2	0	1	1
Central Asia	0	0	0	0	0	0	0	0	0	1	0	0	0
North West Pacific and East Asia	0	4	1	0	0	0	0	1	1	0	0	1	1
South Asia	0	0	0	0	1	0	0	0	0	0	0	4	0
South East Asia	0	0	0	1	0	0	1	1	0	1	2	1	1
South Pacific	0	0	0	0	0	0	0	0	0	1	0	0	0
EUROPE	0	0	2	1	2	0	3	2	2	4	3	4	16
Central Europe	0	0	0	0	1	0	0	0	1	1	0	0	8
Eastern Europe	0	0	0	0	1	0	0	0	1	2	1	2	5
Western Europe	0	0	2	1	0	0	3	2	0	1	2	2	3
LATIN AMERICA AND CARIBBEAN	0	0	0	2	2	1	2	0	0	2	3	6	2
Caribbean	0	0	0	0	0	0	0	0	0	0	0	1	0
Meso-America	0	0	0	1	1	0	0	0	0	0	1	1	1
South America	0	0	0	1	1	1	2	0	0	2	2	4	1
NORTH AMERICA	0	1	0	2	3	1	4	2	2	3	5	4	7
North America	0	1	0	2	3	1	4	2	2	3	5	4	7
POLAR	0	0	0	0	0	0	0	0	0	0	0	0	0
Antarctic	0	0	0	0	0	0	0	0	0	0	0	0	0
Arctic	0	0	0	0	0	0	0	0	0	0	0	0	0
WEST ASIA	0	0	0	0	0	0	0	0	0	0	0	0	0
Arabian Peninsula	0	0	0	0	0	0	0	0	0	0	0	0	0
Mashriq	0	0	0	0	0	0	0	0	0	0	0	0	0
GLOBAL TOTALS	2	6	4	6	8	2	13	7	5	15	16	22	30
REPORTED GLOBAL TOTALS FROM ORIGINAL DATA SOURCES	****	****	****	****	****	****	****	****	****	****	****	****	****

Comments:

Wild fire: disaster type term comprising the two disaster subsets "forest fire" and "scrub fire".

Forest fire: fires in forest that cover extensive damage. They may start by natural causes such as volcanic eruptions or lightning, or they may be caused by arsonists or careless smokers, by those burning wood, or by clearing a forest area.

Scrub fire: fires in scrub or bush that cover extensive damage. They may start by natural causes such as volcanic eruptions or lightning, or they may be caused by arsonists or careless smokers, by those burning wood, or by clearing a forest area.

Wild Fires - Total Affected

	1975	1980	1990	1991	1992	1993	1994	1995	1996	1997	1998	1999	2000
AFRICA	0	0	0	0	0	0	0	85	0	0	1500	3005	1255
Central Africa	0	0	0	0	0	0	0	85	0	0	0	0	0
Eastern Africa	0	0	0	0	0	0	0	0	0	0	0	0	5
Northern Africa	0	0	0	0	0	0	0	0	0	0	0	0	0
Southern Africa	0	0	0	0	0	0	0	0	0	0	0	3005	1250
Western Africa	0	0	0	0	0	0	0	0	0	0	1500	0	0
Western Indian Ocean	0	0	0	0	0	0	0	0	0	0	0	0	0
ASIA AND PACIFIC	0	40	0	8	50000	0	3046161	3000	5061	48000	2300	6000	1055
Australia and New Zealand	0	40	0	0	0	0	46161	0	0	0	0	2000	200
Central Asia	0	0	0	0	0	0	0	0	0	8000	0	0	0
North West Pacific and East Asia	0	0	0	0	0	0	0	0	5061	0	0	0	855
South Asia	0	0	0	0	50000	0	0	0	0	0	0	4000	0
South East Asia	0	0	0	8	0	0	3000000	3000	0	32000	2300	0	0
South Pacific	0	0	0	0	0	0	0	0	0	8000	0	0	0
EUROPE	0	0	0	0	0	0	16820	2254	700	1259	101883	6	638
Central Europe	0	0	0	0	0	0	0	0	500	0	0	0	529
Eastern Europe	0	0	0	0	0	0	0	0	200	0	100683	0	0
Western Europe	0	0	0	0	0	0	16820	2254	0	1259	1200	6	109
LATIN AMERICA AND CARIBBEAN	0	0	0	0	1200	0	0	0	0	1300	12000	7300	0
Caribbean	0	0	0	0	0	0	0	0	0	0	0	0	0
Meso-America	0	0	0	0	1200	0	0	0	0	0	0	0	0
South America	0	0	0	0	0	0	0	0	0	1300	12000	7300	0
NORTH AMERICA	0	5000	0	7233	990	130	4432	6500	50	2600	49221	2519	36087
North America	0	5000	0	7233	990	130	4432	6500	50	2600	49221	2519	36087
POLAR	0	0	0	0	0	0	0	0	0	0	0	0	0
Antarctic	0	0	0	0	0	0	0	0	0	0	0	0	0
Arctic	0	0	0	0	0	0	0	0	0	0	0	0	0
WEST ASIA	0	0	0	0	0	0	0	0	0	0	0	0	0
Arabian Peninsula	0	0	0	0	0	0	0	0	0	0	0	0	0
Mashriq	0	0	0	0	0	0	0	0	0	0	0	0	0
GLOBAL TOTALS	0	5040	0	7241	52190	130	3067413	11839	5811	53159	166904	18830	39035
REPORTED GLOBAL TOTALS FROM ORIGINAL DATA SOURCES	****	****	****	****	****	****	****	****	****	****	****	****	****

Comments:

Wild fire: disaster type term comprising the two disaster subsets "forest fire" and "scrub fire".

Forest fire: fires in forest that cover extensive damage. They may start by natural causes such as volcanic eruptions or lightning, or they may be caused by arsonists or careless smokers, by those burning wood, or by clearing a forest area.

Scrub fire: fires in scrub or bush that cover extensive damage. They may start by natural causes such as volcanic eruptions or lightning, or they may be caused by arsonists or careless smokers, by those burning wood, or by clearing a forest area.

Total affected: people that have been injured affected and left homeless after a disaster are included in this category.

AGGREGATIONS

Wind Storms - Affected People

	1975	1980	1990	1991	1992	1993	1994	1995	1996	1997	1998	1999	2000
AFRICA	846199	7000	0	131000	0	1722	2500800	775	100230	520000	0	109663	110655
Central Africa	0	0	0	0	0	0	0	0	0	0	0	0	
Eastern Africa	0	0	0	0	0	0	0	775	0	0	0	30810	
Northern Africa	0	0	0	0	0	0	0	0	0	0	0	0	
Southern Africa	0	0	0	0	0	0	2000000	0	230	0	0	13000	35
Western Africa	10000	0	0	0	0	1722	0	0	0	0	0	65853	
Western Indian Ocean	836199	7000	0	131000	0	0	500800	0	100000	520000	0	0	110620
ASIA AND PACIFIC	28374	16378140	19596300	26593208	12924165	13253669	30942075	43664416	24497151	10499668	9640243	15884696	1036666
Australia and New Zealand	0	0	0	690	120	12000	2860000	431500	623500	140	0	10000	115
Central Asia	0	0	0	0	0	0	0	0	0	0	0	0	
North West Pacific and East Asia	856	35309	5698055	2595953	11175858	8631654	25445457	6254000	15207400	6637200	51000	2013800	916190
South Asia	0	1000	8013870	15000165	3700	769500	878500	2250000	7417600	2792738	4887440	13823696	47750
South East Asia	4518	16313831	5708275	8911400	1741337	3572015	1758118	34728916	1248651	1063990	4695832	37200	72609
South Pacific	23000	28000	176100	85000	3150	268500	0	0	0	5600	5971	0	
EUROPE	0	0	200000	0	3000	787700	62	1325	12060	21455	253690	3502100	26679
Central Europe	0	0	0	0	0	785000	0	0	0	0	0	1000	
Eastern Europe	0	0	0	0	1000	0	62	1325	0	21045	0	0	264740
Western Europe	0	0	200000	0	2000	2700	0	0	12060	410	253690	3501100	205
LATIN AMERICA AND CARIBBEAN	0	450000	3500	9800	0	332680	1568338	235060	1030583	757405	4273660	237357	949
Caribbean	0	450000	3500	0	0	149680	1568338	208260	302859	0	1024000	235357	
Meso-America	0	0	0	9800	0	177000	0	20800	727724	754555	3245660	2000	949
South America	0	0	0	0	0	6000	0	6000	0	2850	4000	0	
NORTH AMERICA	200	0	0	2900	30000	150000	371681	78000	518500	4670	666239	9054951	2426
North America	200	0	0	2900	30000	150000	371681	78000	518500	4670	666239	9054951	2426
POLAR	0	0	0	0	0	0	0	0	0	0	0	0	
Antarctic	0	0	0	0	0	0	0	0	0	0	0	0	
Arctic	0	0	0	0	0	0	0	0	0	0	0	0	
WEST ASIA	0	0	0	0	100000	0	0	0	0	0	0	0	0
Arabian Peninsula	0	0	0	0	0	0	0	0	0	0	0	0	
Mashriq	0	0	0	0	100000	0	0	0	0	0	0	0	
GLOBAL TOTALS	874773	16835140	19799800	26736908	13057165	14525771	35382956	43979576	26158524	11803198	14833832	28788767	1426029
REPORTED GLOBAL TOTALS FROM ORIGINAL DATA SOURCES	****	****	****	****	****	****	****	****	****	****	****	****	****

Comments:

Wind storm: disaster type comprising the following disaster subsets "cyclone", "hurricane", "storm", "tornado", "tropical storm", "typhoon", "winter storm".

Cyclone: large-scale closed circulation system in the atmosphere above the Indian Ocean and South Pacific with low barometric pressure and strong winds that rotate clockwise. Maximum wind speed of 64 knots or more.

Hurricane: large-scale closed circulation system in the atmosphere above the western Atlantic with low barometric pressure and strong winds that rotate clockwise in the southern hemisphere and counter-clockwise in the northern hemisphere. Maximum wind speed of 64 knots or more.

Storm: Wind with a speed between 48 and 55 knots.

Tornado: violently rotating storm diameter; the most violent weather phenomenon. It is produced in a very severe thunderstorm and appears as a funnel cloud extending from the base of a cumulonimbus to the ground.

Tropical storm: generic term for a non-frontal synoptic scale cyclone originating over tropical or sub-tropical waters with organised convection and definite cyclonic surface wind circulation.

Typhoon: large-scale closed circulation system in the atmosphere above the western Pacific with low barometric pressure and strong winds that rotate clockwise in the southern hemisphere and counter-clockwise in the northern hemisphere. Maximum wind speed of 64 knots or more.

Winter storm: snow (blizzard), ice or sleet storm.

Affected people: people requiring immediate assistance during a period of emergency, that is to say requiring basic survival needs such as food, water, shelter, sanitation and immediate medical assistance. Appearance of a significant number of cases of an infectious disease introduced in a region or a population that is usually free from that disease.

1 knot=1.87 km/h

AGGREGATIONS

	1975	1980	1990	1991	1992	1993	1994	1995	1996	1997	1998	1999	2000
AFRICA	0	0	0	126500	0	0	546000	2000	20	80000	0	35460	12650
Central Africa	0	0	0	0	0	0	0	0	0	0	0	0	0
Eastern Africa	0	0	0	0	0	0	0	0	0	0	0	0	10000
Northern Africa	0	0	0	0	0	0	0	0	0	0	0	0	0
Southern Africa	0	0	0	0	0	0	504500	0	20	0	0	5300	1050
Western Africa	0	0	0	0	0	0	0	2000	0	0	0	30160	1000
Western Indian Ocean	0	0	0	126500	0	0	41500	0	0	80000	0	0	600
ASIA AND PACIFIC	3992	1214479	1355217	1946855	183281	619925	2142316	1564150	14494	1640545	11097671	1009646	1176722
Australia and New Zealand	0	0	0	18	0	45	320	200	1230	0	0	702	0
Central Asia	0	0	0	0	0	0	0	0	0	0	0	1500	0
North West Pacific and East Asia	1294	2100	210195	402	135256	161363	693086	25450	10264	242000	11002376	191364	216376
South Asia	2000	0	4000	323849	1200	165000	205000	122500	0	1000000	0	809597	78000
South East Asia	698	1205379	1112020	1613586	41825	261367	1243910	1416000	3000	383045	95295	4711	882346
South Pacific	0	7000	29002	9000	5000	32150	0	0	0	15500	0	1772	0
EUROPE	0	0	0	0	0	200	30900	1200	0	300	0	940	560
Central Europe	0	0	0	0	0	200	5400	0	0	0	0	940	300
Eastern Europe	0	0	0	0	0	0	25500	1200	0	300	0	0	0
Western Europe	0	0	0	0	0	0	0	0	0	0	0	0	260
LATIN AMERICA AND CARIBBEAN	0	15500	0	200	47700	35500	131970	62037	86346	52500	123700	34077	530
Caribbean	0	15500	0	0	1700	3000	129970	16037	67596	0	122000	31077	0
Meso-America	0	0	0	200	8000	23000	0	46000	18750	52000	1700	2000	0
South America	0	0	0	0	38000	9500	2000	0	0	500	0	1000	530
NORTH AMERICA	300	400	0	0	250000	300	1051	2000	0	0	1060	9790	4179
North America	300	400	0	0	250000	300	1051	2000	0	0	1060	9790	4179
POLAR	0	0	0	0	0	0	0	0	0	0	0	0	0
Antarctic	0	0	0	0	0	0	0	0	0	0	0	0	0
Arctic	0	0	0	0	0	0	0	0	0	0	0	0	0
WEST ASIA	0	0	0	0	4000	0	0	0	0	0	0	0	0
Arabian Peninsula	0	0	0	0	0	0	0	0	0	0	0	0	0
Mashriq	0	0	0	0	4000	0	0	0	0	0	0	0	0
GLOBAL TOTALS	4292	1230379	1355217	2073555	484981	655925	2852237	1631387	100860	1773345	11222431	1089913	1194641
REPORTED GLOBAL TOTALS FROM ORIGINAL DATA SOURCES	****	****	****	****	****	****	****	****	****	****	****	****	****

Comments:

Wind storm: disaster type comprising the following disaster subsets "cyclone", "hurricane", "storm", "tornado", "tropical storm", "typhoon", "winter storm".

Cyclone: large-scale closed circulation system in the atmosphere above the Indian Ocean and South Pacific with low barometric pressure and strong winds that rotate clockwise. Maximum wind speed of 64 knots or more.

Hurricane: large-scale closed circulation system in the atmosphere above the western Atlantic with low barometric pressure and strong winds that rotate clockwise in the southern hemisphere and counter-clockwise in the northern hemisphere. Maximum wind speed of 64 knots or more.

Storm: wind with a speed between 48 and 55 knots.

Tornado: violently rotating storm diameter; the most violent weather phenomenon. It is produced in a very severe thunderstorm and appears as a funnel cloud extending from the base of a cumulonimbus to the ground.

Tropical storm: generic term for a non-frontal synoptic scale cyclone originating over tropical or sub-tropical waters with organised convection and definite cyclonic surface wind circulation.

Typhoon: large-scale closed circulation system in the atmosphere above the western Pacific with low barometric pressure and strong winds that rotate clockwise in the southern hemisphere and counter-clockwise in the northern hemisphere. Maximum wind speed of 64 knots or more.

Winter storm: snow (blizzard), ice or sleet storm.

Homeless: people needing immediate assistance in the form of shelter.

1 knot=1.87 km/h

AGGREGATIONS

Wind Storms - Injured People

	1975	1980	1990	1991	1992	1993	1994	1995	1996	1997	1998	1999	2000
AFRICA	109	0	0	300	0	0	2051	0	0	60	163	1500	20
Central Africa	0	0	0	0	0	0	0	0	0	0	0	0	0
Eastern Africa	0	0	0	0	0	0	0	0	0	0	0	0	5
Northern Africa	0	0	0	0	0	0	0	0	0	60	0	0	10
Southern Africa	0	0	0	0	0	0	2000	0	0	0	163	480	4
Western Africa	0	0	0	300	0	0	0	0	0	0	0	20	0
Western Indian Ocean	109	0	0	0	0	0	51	0	0	0	0	1000	1
ASIA AND PACIFIC	106	871	2913	151254	21081	10825	8582	8002	42431	16744	6620	9920	2343
Australia and New Zealand	0	0	0	100	0	25	94	69	76	0	0	24	0
Central Asia	0	0	0	0	0	0	0	0	0	0	0	0	0
North West Pacific and East Asia	98	109	514	7376	19926	3994	4000	2998	5556	3482	2331	4853	1144
South Asia	0	73	700	139149	400	6020	4129	1570	36195	12130	4197	5043	729
South East Asia	8	439	1696	4629	203	738	359	3365	604	912	92	0	470
South Pacific	0	250	3	0	552	48	0	0	0	220	0	0	0
EUROPE	0	0	0	3	0	20	186	10	0	145	160	36	733
Central Europe	0	0	0	3	0	0	6	0	0	60	0	0	0
Eastern Europe	0	0	0	0	0	0	80	10	0	45	160	0	710
Western Europe	0	0	0	0	0	20	100	0	0	40	0	36	23
LATIN AMERICA AND CARIBBEAN	0	64	0	0	50	505	0	221	83	353	12877	67	570
Caribbean	0	16	0	0	0	105	0	199	3	0	649	67	0
Meso-America	0	0	0	0	50	0	0	22	80	200	12228	0	570
South America	0	48	0	0	0	400	0	0	0	153	0	0	0
NORTH AMERICA	40	170	0	100	55	328	454	167	131	222	448	951	452
North America	40	170	0	100	55	328	454	167	131	222	448	951	452
POLAR	0	0	0	0	0	0	0	0	0	0	0	0	0
Antarctic	0	0	0	0	0	0	0	0	0	0	0	0	0
Arctic	0	0	0	0	0	0	0	0	0	0	0	0	0
WEST ASIA	0	0	0	0	75	0	0	0	0	0	0	0	200
Arabian Peninsula	0	0	0	0	0	0	0	0	0	0	0	0	0
Mashriq	0	0	0	0	75	0	0	0	0	0	0	0	200
GLOBAL TOTALS	255	1105	2913	151657	21261	11678	11273	8400	42645	17524	20268	12474	4318
REPORTED GLOBAL TOTALS FROM ORIGINAL DATA SOURCES	****	****	****	****	****	****	****	****	****	****	****	****	****

Comments:

Wind storm: disaster type comprising the following disaster subsets "cyclone", "hurricane", "storm", "tornado", "tropical storm", "typhoon", "winter storm".

Cyclone: large-scale closed circulation system in the atmosphere above the Indian Ocean and South Pacific with low barometric pressure and strong winds that rotate clockwise. Maximum wind speed of 64 knots or more.

Hurricane: large-scale closed circulation system in the atmosphere above the western Atlantic with low barometric pressure and strong winds that rotate clockwise in the southern hemisphere and counter-clockwise in the northern hemisphere. Maximum wind speed of 64 knots or more.

Storm: wind with a speed between 48 and 55 knots.

Tornado: violently rotating storm diameter; the most violent weather phenomenon. It is produced in a very severe thunderstorm and appears as a funnel cloud extending from the base of a cumulonimbus to the ground.

Tropical storm: generic term for a non-frontal synoptic scale cyclone originating over tropical or sub-tropical waters with organised convection and definite cyclonic surface wind circulation.

Typhoon: large-scale closed circulation system in the atmosphere above the western Pacific with low barometric pressure and strong winds that rotate clockwise in the southern hemisphere and counter-clockwise in the northern hemisphere. Maximum wind speed of 64 knots or more.

Winter storm: snow (blizzard), ice or sleet storm.

Injured: people suffering from physical injuries, trauma or an illness requiring medical treatment as a direct result of a disaster.

1 knot=1.87 km/h

AGGREGATIONS

Wind Storms - Killed People

	1975	1980	1990	1991	1992	1993	1994	1995	1996	1997	1998	1999	2000
AFRICA	28	25	0	40	0	21	490	0	26	192	18	327	157
Central Africa	0	0	0	0	0	0	0	0	0	0	0	0	0
Eastern Africa	0	0	0	0	0	0	30	0	0	0	0	0	0
Northern Africa	0	0	0	0	0	0	0	0	14	18	0	0	4
Southern Africa	11	0	0	0	0	7	246	0	0	0	18	26	15
Western Africa	1	0	0	4	0	0	0	0	0	0	0	301	4
Western Indian Ocean	16	25	0	36	0	14	214	0	12	174	0	0	134
ASIA AND PACIFIC	825	1082	4459	146818	1147	2450	2277	3275	3098	4609	4906	11128	731
Australia and New Zealand	0	0	0	1	2	0	27	8	8	3	0	7	1
Central Asia	0	0	0	0	0	0	0	112	0	0	0	0	0
North West Pacific and East Asia	71	362	761	477	878	414	1530	722	545	298	243	519	378
South Asia	515	182	2558	139140	103	1257	460	900	1601	491	3804	10541	107
South East Asia	239	520	1120	7172	160	735	260	1533	944	3741	859	17	245
South Pacific	0	18	20	28	4	44	0	0	0	76	0	44	0
EUROPE	0	0	53	48	68	32	65	148	5	62	35	193	67
Central Europe	0	0	0	1	0	9	30	0	0	12	0	40	14
Eastern Europe	0	0	0	0	0	0	15	148	0	5	8	12	10
Western Europe	0	0	53	47	68	23	20	0	5	45	27	141	43
LATIN AMERICA AND CARIBBEAN	75	337	46	13	19	194	1132	220	209	287	19395	23	49
Caribbean	26	322	8	0	4	7	1132	26	87	0	550	13	0
Meso-America	29	0	38	13	15	75	0	188	122	246	18845	9	48
South America	20	15	0	0	0	112	0	6	0	41	0	1	1
NORTH AMERICA	195	12	155	49	96	262	84	131	311	180	198	219	116
North America	195	12	155	49	96	262	84	131	311	180	198	219	116
POLAR	0	0	0	0	0	0	0	0	0	0	0	0	0
Antarctic	0	0	0	0	0	0	0	0	0	0	0	0	0
Arctic	0	0	0	0	0	0	0	0	0	0	0	0	0
WEST ASIA	0	0	0	2	25	0	0	0	0	0	0	0	9
Arabian Peninsula	0	0	0	0	0	0	0	0	0	0	0	0	0
Mashriq	0	0	0	2	25	0	0	0	0	0	0	0	9
GLOBAL TOTALS	1123	1456	4713	146970	1355	2959	4048	3774	3649	5330	24552	11890	1129
REPORTED GLOBAL TOTALS FROM ORIGINAL DATA SOURCES	****	****	****	****	****	****	****	****	****	****	****	****	****

Comments:

Wind storm: disaster type comprising the following disaster subsets "cyclone", "hurricane", "storm", "tornado", "tropical storm", "typhoon", "winter storm".

Cyclone: large-scale closed circulation system in the atmosphere above the Indian Ocean and South Pacific with low barometric pressure and strong winds that rotate clockwise. Maximum wind speed of 64 knots or more.

Hurricane: large-scale closed circulation system in the atmosphere above the western Atlantic with low barometric pressure and strong winds that rotate clockwise in the southern hemisphere and counter-clockwise in the northern hemisphere. Maximum wind speed of 64 knots or more.

Storm: wind with a speed between 48 and 55 knots.

Tornado: violently rotating storm diameter; the most violent weather phenomenon. It is produced in a very severe thunderstorm and appears as a funnel cloud extending from the base of a cumulonimbus to the ground.

Tropical storm: generic term for a non-frontal synoptic scale cyclone originating over tropical or sub-tropical waters with organised convection and definite cyclonic surface wind circulation.

Typhoon: large-scale closed circulation system in the atmosphere above the western Pacific with low barometric pressure and strong winds that rotate clockwise in the southern hemisphere and counter-clockwise in the northern hemisphere. Maximum wind speed of 64 knots or more.

Winter storm: snow (blizzard), ice or sleet storm.

Killed: persons confirmed as dead and persons missing and presumed dead.

1 knot=1.87 km/h

AGGREGATIONS

Wind Storms - Number

	1975	1980	1990	1991	1992	1993	1994	1995	1996	1997	1998	1999	2000
AFRICA	4	1	1	3	0	3	8	2	5	3	1	9	11
Central Africa	0	0	0	0	0	0	0	0	0	0	0	0	0
Eastern Africa	0	0	0	0	0	0	1	1	0	0	0	1	2
Northern Africa	0	0	0	0	0	0	0	0	1	1	0	0	1
Southern Africa	1	0	1	0	0	1	3	0	2	0	1	2	3
Western Africa	1	0	0	1	0	1	0	1	0	0	0	5	2
Western Indian Ocean	2	1	0	2	0	1	4	0	2	2	0	1	3
ASIA AND PACIFIC	16	27	47	38	37	57	36	26	32	27	34	30	50
Australia and New Zealand	1	5	2	2	4	2	4	4	7	1	0	3	4
Central Asia	0	0	0	0	0	0	0	1	0	0	0	1	0
North West Pacific and East Asia	4	7	16	14	13	26	10	7	8	8	13	12	20
South Asia	6	5	10	6	5	13	8	5	10	8	10	9	10
South East Asia	2	9	10	13	11	11	14	9	7	5	7	2	16
South Pacific	3	1	9	3	4	5	0	0	0	5	4	3	0
EUROPE	0	1	53	5	4	9	7	6	2	5	5	13	15
Central Europe	0	0	0	1	0	3	2	0	0	1	0	3	1
Eastern Europe	0	0	0	0	1	0	2	4	0	1	1	1	4
Western Europe	0	1	53	4	3	6	3	2	2	3	4	9	10
LATIN AMERICA AND CARIBBEAN	3	12	11	1	5	11	6	18	13	9	16	15	9
Caribbean	1	10	10	0	1	3	4	11	6	0	5	13	0
Meso-America	1	1	1	1	2	5	0	6	7	4	10	1	6
South America	1	1	0	0	2	3	2	1	0	5	1	1	3
NORTH AMERICA	6	4	15	14	27	20	9	7	10	23	17	18	13
North America	6	4	15	14	27	20	9	7	10	23	17	18	13
POLAR	0	0	0	0	0	0	0	0	0	0	0	0	0
Antarctic	0	0	0	0	0	0	0	0	0	0	0	0	0
Arctic	0	0	0	0	0	0	0	0	0	0	0	0	0
WEST ASIA	0	0	0	1	1	0	0	0	0	0	0	0	1
Arabian Peninsula	0	0	0	0	0	0	0	0	0	0	0	0	0
Mashriq	0	0	0	1	1	0	0	0	0	0	0	0	1
GLOBAL TOTALS	29	45	127	62	74	100	66	59	62	67	73	85	99
REPORTED GLOBAL TOTALS FROM ORIGINAL DATA SOURCES	****	****	****	****	****	****	****	****	****	****	****	****	****

Comments:

Wind storm: disaster type comprising the following disaster subsets "cyclone", "hurricane", "storm", "tornado", "tropical storm", "typhoon", "winter storm".

Cyclone: large-scale closed circulation system in the atmosphere above the Indian Ocean and South Pacific with low barometric pressure and strong winds that rotate clockwise. Maximum wind speed of 64 knots or more.

Hurricane: large-scale closed circulation system in the atmosphere above the western Atlantic with low barometric pressure and strong winds that rotate clockwise in the southern hemisphere and counter-clockwise in the northern hemisphere. Maximum wind speed of 64 knots or more.

Storm: wind with a speed between 48 and 55 knots.

Tornado: violently rotating storm diameter; the most violent weather phenomenon. It is produced in a very severe thunderstorm and appears as a funnel cloud extending from the base of a cumulonimbus to the ground.

Tropical storm: generic term for a non-frontal synoptic scale cyclone originating over tropical or sub-tropical waters with organised convection and definite cyclonic surface wind circulation.

Typhoon: large-scale closed circulation system in the atmosphere above the western Pacific with low barometric pressure and strong winds that rotate clockwise in the southern hemisphere and counter-clockwise in the northern hemisphere. Maximum wind speed of 64 knots or more.

Winter storm: snow (blizzard), ice or sleet storm.

1 knot=1.87 km/h

Wind Storms – Total Affected

	1980	1990	1991	1992	1993	1994	1995	1996	1997	1998	1999	2000
AFRICA	7000	0	257800	0	1722	3048851	2775	100250	600060	163	146623	1119228
Central Africa	0	0	0	0	0	0	0	0	0	0	0	0
Eastern Africa	0	0	0	0	0	0	775	0	0	0	30810	10005
Northern Africa	0	0	0	0	0	0	0	0	60	0	0	10
Southern Africa	0	0	0	0	0	2506500	0	250	0	163	18780	1404
Western Africa	0	0	300	0	1722	0	2000	0	0	0	96033	1000
Western Indian Ocean	7000	0	257500	0	0	542351	0	100000	600000	0	1000	1106809
ASIA AND PACIFIC	17593490	20954430	28691317	13128527	13884419	33092973	45236568	24553991	12156957	20744534	16904262	11545714
Australia and New Zealand	0	0	808	120	12070	2860414	431769	624806	140	0	10726	1150
Central Asia	0	0	0	0	0	0	0	0	0	0	1500	0
North West Pacific and East Asia	37518	5908764	2603731	11331040	8797011	26142543	6282448	15223220	6882682	11055707	2210017	9379420
South Asia	1073	8018570	15463163	5300	940520	1087629	2374070	7453710	3804868	4891637	14638336	556229
South East Asia	17519649	6821991	10529615	1783365	3834120	3002387	36148281	1252255	1447947	4791219	41911	1608915
South Pacific	35250	205105	94000	8702	300698	0	0	0	21320	5971	1772	0
EUROPE	0	200000	3	3000	787920	31148	2535	12060	21900	253850	3503076	2669203
Central Europe	0	0	3	0	785200	5406	0	0	60	0	1940	300
Eastern Europe	0	0	0	1000	0	25642	2535	0	21390	160	0	2648110
Western Europe	0	200000	0	2000	2720	100	0	12060	450	253690	3501136	20793
LATIN AMERICA AND CARIBBEAN	465564	3500	10000	47750	368685	1700308	297318	1117012	810258	4410237	271501	96010
Caribbean	465516	3500	0	1700	152785	1698308	224496	370458	0	1146649	266501	0
Meso-America	0	0	10000	8050	200000	0	66822	746554	806755	3259588	4000	95480
South America	48	0	0	38000	15900	2000	6000	0	3503	4000	1000	530
NORTH AMERICA	570	0	3000	280055	150600	373186	80167	518631	4892	667747	9065692	28899
North America	570	0	3000	280055	150600	373186	80167	518631	4892	667747	9065692	28899
POLAR	0	0	0	0	0	0	0	0	0	0	0	0
Antarctic	0	0	0	0	0	0	0	0	0	0	0	0
Arctic	0	0	0	0	0	0	0	0	0	0	0	0
WEST ASIA	0	0	0	104075	0	0	0	0	0	0	0	200
Arabian Peninsula	0	0	0	0	0	0	0	0	0	0	0	0
Mashriq	0	0	0	104075	0	0	0	0	0	0	0	200
GLOBAL TOTALS	18066624	21157930	28962120	13563407	15193346	38246466	45619363	26301944	13594067	26076531	29891154	15459254
REPORTED GLOBAL TOTALS FROM ORIGINAL DATA SOURCES	****	****	****	****	****	****	****	****	****	****	****	****

Comments:

Wind storm: disaster type comprising the following disaster subsets "cyclone", "hurricane", "storm", "tornado", "tropical storm", "typhoon", "winter storm".

Cyclone: large-scale closed circulation system in the atmosphere above the Indian Ocean and South Pacific with low barometric pressure and strong winds that rotate clockwise. Maximum wind speed of 64 knots or more.

Hurricane: large-scale closed circulation system in the atmosphere above the western Atlantic with low barometric pressure and strong winds that rotate clockwise in the southern hemisphere and counter-clockwise in the northern hemisphere. Maximum wind speed of 64 knots or more.

Storm: wind with a speed between 48 and 55 knots.

Tornado: violently rotating storm diameter; the most violent weather phenomenon. It is produced in a very severe thunderstorm and appears as a funnel cloud extending from the base of a cumulonimbus to the ground.

Tropical storm: generic term for a non-frontal synoptic scale cyclone originating over tropical or sub-tropical waters with organised convection and definite cyclonic surface wind circulation.

Typhoon: large-scale closed circulation system in the atmosphere above the western Pacific with low barometric pressure and strong winds that rotate clockwise in the southern hemisphere and counter-clockwise in the northern hemisphere. Maximum wind speed of 64 knots or more.

Winter storm: snow (blizzard), ice or sleet storm.

Total affected: people that have been injured affected and left homeless after a disaster are included in this category.

1 knot=1.87 km/h

Forest Fire Extent - Annual Average

Units: thousand hectares

Data Source: Global Forest Resources Assessment 2000

Data Provider: Food and Agriculture Organization of the United Nations (FAO)

Years: 1990-00

	1990-00
AFRICA	----
Central Africa	----
Eastern Africa	----
Northern Africa	----
Southern Africa	----
Western Africa	----
Western Indian Ocean	----
ASIA AND PACIFIC	4.46
Australia and New Zealand	0.33
Central Asia	1.80
North West Pacific and East Asia	2.33
South Asia	----
South East Asia	----
South Pacific	----
EUROPE	1050.67
Central Europe	33.45
Eastern Europe	825.71
Western Europe	191.50
LATIN AMERICA AND CARIBBEAN	556.49
Caribbean	0.42
Meso-America	66.78
South America	489.30
NORTH AMERICA	501.63
North America	501.63
POLAR	----
Antarctic	----
Arctic	----
WEST ASIA	----
Arabian Peninsula	----
Mashriq	----
GLOBAL TOTALS	----
REPORTED GLOBAL TOTALS FROM ORIGINAL DATA SOURCES	****

Comments:

Forest fire extent - annual average comprises the reported forest areas exposed to fire.

Total forest includes natural forests and forest plantations. The term is used to refer to land with a tree cover of more than 10 percent and area of more than 0.5 ha. Forests are determined both by the presence of trees and the absence of other predominant land uses. The trees should be able to reach a minimum height of 5 m. Young stands that have not yet reached, but are expected to reach, a crown density of 10 percent and tree height of 5 m are included under forest, as are temporarily unstocked areas. The term includes forests used for purposes of production, protection, multiple use or conservation (i.e. forest in national parks, nature reserves and other protected areas), as well as forest stands on agricultural lands (e.g. windbreaks and shelterbelts of trees with a width of more than 20 m) and rubberwood plantations and cork oak stands. The term specifically excludes stands of trees established primarily for agricultural production, for example fruit tree plantations. It also excludes trees planted in agroforestry systems.

2.1 Coastal and Marine

In this section, <u>only</u> those countries and territories are listed for which there is coastal and marine data. Countries for which the data value is not available (216) are not shown.

Units: square kilometres

Data Source: The Global Maritime Boundaries Database (GMBD)

Data Provider: Viridian - MRJ Technology Solutions

Year: 2000

Copyright © 2001 World Resources Institute

Exclusive Fishing Zone

Countries and Territories	2000
Albania	6210
Algeria	60515
Angola	438017
Aruba	2789
Belgium	2089
Benin	26768
Congo	41362
Cyprus	13679
Democratic Republic of the Congo	121
Ecuador	957044
El Salvador	87510
Finland	55123
France	73382
Greece	114914
Ireland	358922
Italy	155629
Jordan	100
Liberia	239100
Libyan Arab Jamahiriya	20900
Malta	7459
Netherlands	50309
Netherlands Antilles	11951
Nicaragua	94900
Peru	746525
Qatar	19788
Sierra Leone	155928
Singapore	744
Somalia	759253
South Africa	1450596
Spain	205200
Taiwan	248755
Turkey	81006
United Kingdom of Great Britain and Northern Ireland	753752
Uruguay	110543

Comments:

The exclusive fishing zone or fishery zone refers to an area beyond the outer limit of the territorial sea (12 nautical miles from the coast) in which the coastal state has the right to fish, subject to any concessions which may be granted to foreign fishermen. Some countries have made no claim beyond the territorial sea. Some states have claimed an exclusive fishing zone instead of the more encompassing 200 nautical mile exclusive economic zone (EEZ).

2.2 Atmosphere

In this section, are listed <u>only</u> countries and territories where the atmosphere variable has been registered between the year 1990 and the year 1998. Countries where the data was nil or not present are not shown.

2.2.1 Emissions of GHG, ODS, Dust, Metals

Greenhouse gases (GHGs)

The United Nations Framework Convention on Climate Change (UNFCCC) addresses such greenhouse gases as carbon dioxide (CO_2), methane (CH_4), and nitrous oxide (N_2O). The Kyoto Protocol also addresses hydrofluorocarbons (HFCs), perfluorocarbons (PFCs), and sulphur hexafluoride (SF_6).

For more detailed information about the GHG inventory database see: http://ghg.unfccc.int.

1 gigagram corresponds to 1 thousand metric tonne.

Data source

The data is based on the national greenhouse gas inventories and national communications submitted regularly to the Climate Change Secretariat by the parties to the UNFCCC.

Annex I parties

The industrialized countries listed in this annex to the Climate Change Convention tried to return their greenhouse gas emissions to 1990 levels by the year 2000. They have also accepted emission targets for the period 2008-12. They include the following members Australia, Austria, Belarus, Belgium, Bulgaria, Canada, Czech Republic, Denmark, European Economic Community, Estonia, Finland, France, Germany, Greece, Hungary, Iceland, Ireland, Italy, Japan, Latvia, Lithuania, Luxembourg, Netherlands, New Zealand, Norway, Poland, Portugal, Romania, Russian Federation, Slovakia, Slovenia, Spain, Sweden, Switzerland, Turkey[1], Ukraine, United Kingdom of Great Britain and Northern Ireland and the United States of America.

[1] As of 18/07/2002 Turkey is not party to the UNFCCC

NATIONAL DATA SETS

Emissions of CH₄ - Total Anthropogenic (UNFCCC)

Units: gigagrams of CH₄

Data Source: Greenhouse Gas (GHG) Inventory Submission 1998, 1999, and 2000

Data Provider: United Nations Framework Convention on Climate Change (UNFCCC) Secretariat

Years: 1990-1998

Copyright © 2001 United Nations Framework Convention on Climate Change (UNFCCC) Secretariat

Countries and Territories	1990	1991	1992	1993	1994	1995	1996	1997	1998
Australia	5577.2	5480.6	5475.5	5406.0	5326.4	5402.1	5444.8	5482.8	5596.1
Austria	537.6	527.1	514.4	508.3	500.0	489.4	481.5	470.0	459.4
Belarus	----	----	----	----	----	----	----	----	----
Belgium	612.1	614.0	612.8	623.7	597.5	596.9	597.4	584.3	581.3
Bulgaria	1333.8	1353.3	1247.0	1121.4	818.0	887.7	----	892.4	654.4
Canada	3541.7	3668.9	3788.8	3916.1	4018.4	4162.6	4294.2	4317.3	4260.9
Croatia	----	----	----	----	----	----	----	----	----
Czech Republic	778.5	710.8	667.9	632.7	613.4	599.5	573.0	562.3	529.4
Denmark	278.5	280.7	280.8	285.4	282.1	281.0	279.2	272.2	286.9
Estonia	105.2	102.1	91.3	79.7	79.5	67.7	63.2	103.1	101.0
Finland	291.3	271.0	247.8	227.1	221.0	222.9	215.7	207.9	197.6
France	3022.4	3024.4	2985.3	2970.1	2903.5	2844.6	2717.1	2617.9	2584.4
Germany	5571.0	5013.0	4654.0	4267.0	4022.0	3893.1	3555.1	3517.2	3484.4
Greece	451.6	455.3	459.4	464.1	473.2	479.4	493.5	500.0	508.9
Hungary	664.4	900.4	796.7	780.1	764.3	775.9	801.7	776.8	678.5
Iceland	14.0	13.9	13.7	13.7	13.8	13.6	----	----	----
Ireland	611.3	618.7	620.5	623.8	626.6	633.9	645.7	654.6	649.1
Italy	1899.5	1911.0	1830.2	1825.3	1883.1	1913.7	1922.4	1952.2	1970.4
Japan	1542.9	1523.0	1513.0	1504.5	1489.4	1476.7	1447.3	1388.6	----
Latvia	186.3	183.0	150.0	105.0	97.7	101.3	93.1	92.6	96.9
Liechtenstein	1.0	----	----	----	----	----	----	----	----
Lithuania	378.0	----	----	----	----	299.0	285.0	302.0	176.8
Luxembourg	23.9	----	----	----	21.9	22.1	----	----	----
Monaco	0.1	0.1	0.1	0.1	0.1	0.1	0.1	0.1	0.1
Netherlands	1292.3	1308.8	1256.0	1224.6	1203.0	1172.5	1164.5	1105.3	1065.4
New Zealand	1676.8	1641.8	1612.2	1614.1	1624.0	1625.9	1623.9	1593.8	1591.6
Norway	314.7	319.8	326.6	331.7	340.2	342.9	345.9	351.2	346.0
Poland	3141.0	2588.8	2474.0	2431.7	2467.0	2457.2	2252.3	2278.6	2335.5
Portugal	690.2	718.8	659.7	640.9	653.0	696.2	659.6	646.8	683.1
Romania	2357.0	1734.2	1603.3	1522.8	1460.9	----	----	----	----
Russian Federation	26500.0	----	----	----	19610.0	19064.0	18544.0	----	----
Slovakia	363.6	333.8	303.7	287.4	280.5	289.3	297.9	285.3	268.9
Slovenia	176.2	----	----	----	----	----	----	----	----
Spain	1648.9	1676.7	1720.5	1748.1	1794.7	1839.8	1936.3	1993.6	2076.9
Sweden	284.1	324.5	320.3	320.2	302.4	297.4	261.3	260.8	256.0
Switzerland	241.9	242.9	240.5	238.3	234.1	233.2	230.0	226.8	219.5
Turkey	----	----	----	----	----	----	----	----	----
Ukraine	9402.3	8653.8	8347.7	7932.2	7570.3	7294.9	7059.4	6606.1	6456.7
United Kingdom of Great Britain and Northern Ireland	3676.8	3627.2	3533.3	3178.8	2943.5	2916.9	2853.8	2762.8	2635.9
United States of America	31054.3	31020.0	31329.0	31203.2	31710.9	32147.4	31972.0	32083.7	31592.8

Comments:

Emissions of CH₄ - total anthropogenic (UNFCCC): total anthropogenic emissions of CH₄ (methane) includes emissions from the energy (fuel combustion and fugitive fuel) and agricultural (livestock and other) sectors, waste and other.

For more detailed information about the GHG inventory database see: http://ghg.unfccc.int.

Emissions of CO$_2$ - Anthropogenic Emissions and Removals from Land-Use Change and Forestry (UNFCCC)

Units: gigagrams of CO$_2$

Data Source: Greenhouse Gas (GHG) Inventory Submission 1998, 1999, and 2000

Data Provider: United Nations Framework Convention on Climate Change (UNFCCC) Secretariat

Years: 1990-1998

Copyright © 2002 United Nations Framework Convention on Climate Change (UNFCCC) Secretariat

Countries and Territories	1990	1991	1992	1993	1994	1995	1996	1997	1998
Australia	70091.9	43437.2	39942.0	37466.7	35894.6	35376.5	37945.3	36548.9	35173.0
Austria	-9214.8	-13503.9	-8656.5	-8982.4	-7861.6	-7254.0	-5385.2	-7633.4	-7633.4
Belarus	----	----	----	----	----	----	----	----	----
Belgium	-2057.0	-2057.0	-2057.0	-2057.0	-2057.0	-2057.0	-2057.0	-976.8	-976.8
Bulgaria	-4656.7	-7879.7	-7635.8	-7021.7	-6974.0	-7520.3	----	-5852.0	-6232.7
Canada	-39141.0	-57269.4	-45351.5	-34578.6	-29728.8	-21128.5	-28885.8	-23624.9	-21833.0
Croatia	----	----	----	----	----	----	----	----	----
Czech Republic	-2281.0	-5027.0	-6041.0	-5643.0	-3943.0	-5454.0	-4478.7	-4639.0	-3757.5
Denmark	-916.0	-920.0	-923.0	-927.0	-930.0	-934.0	-947.0	-959.0	-973.0
Estonia	-11317.0	----	----	----	-11125.0	-13266.0	----	7992.8	-3356.2
Finland	-23797.7	-38206.9	-31893.5	-29116.3	-17259.4	-14687.3	-21032.4	-12637.1	-9712.6
France	-59617.0	-56488.0	-61246.0	-65865.0	-67495.0	-65615.0	-67197.0	-68090.0	-69783.0
Germany	-33719.0	-33719.0	-33719.0	-33719.0	-33719.0	-33493.0	-33493.0	-33493.0	-33493.0
Greece	----	----	----	----	----	----	----	----	----
Hungary	-3097.0	-3238.8	-3822.9	-4697.9	-4820.1	-4797.9	-3931.4	-4205.0	-4410.5
Iceland	----	----	----	----	----	----	----	----	----
Ireland	-5020.1	-5175.6	-5375.5	-5586.3	-5724.0	-5913.5	-6064.6	-6271.3	-6448.4
Italy	-25614.3	-25853.5	-25083.9	-23370.5	-23494.0	-23225.5	-24286.8	-23633.3	-23633.8
Japan	-83903.1	-83865.2	-85567.8	-90084.3	-93544.8	-96705.0	----	----	----
Latvia	-10825.6	-14186.0	-14235.0	-14228.0	-14206.0	-10483.6	-14320.0	-14315.4	-10508.5
Liechtenstein	-22.0	----	----	----	----	----	----	----	----
Lithuania	-8848.0	----	----	----	----	2800.0	2800.0	2800.0	7711.9
Luxembourg	-295.0	----	----	----	-295.0	-295.0	----	----	----
Monaco	----	----	----	----	----	----	----	----	----
Netherlands	-1500.0	-1600.0	-1600.0	-1600.0	-1700.0	-1700.0	-1700.0	-1700.0	-1700.0
New Zealand	-21529.8	-20513.2	-18283.2	-16238.6	-15608.2	-16162.8	-16529.7	-18169.8	-20895.9
Norway	-9590.0	-11700.0	-13250.0	-13510.0	-15680.0	-13640.0	-17611.0	-16499.0	-17587.8
Poland	-34746.0	-42758.0	-40815.0	-40179.0	-41953.0	-42880.0	-42616.7	-40521.2	-29820.0
Portugal	-3994.0	-4119.2	-4244.4	-4369.6	-4494.8	-4620.0	-4638.0	-4656.0	-4674.0
Romania	-2925.3	-6590.1	-6590.0	-6590.0	-6590.0	----	----	----	----
Russian Federation	-392000.0	----	----	----	-568000.0	-840000.0	-840000.0	----	----
Slovakia	-2426.2	-2426.2	-2426.2	-2426.2	-3234.9	-3234.9	-4233.4	-4085.8	-1682.7
Slovenia	-2293.3	----	----	----	----	----	----	----	----
Spain	-29252.2	-29252.2	-29252.2	-29252.2	-29252.2	-29252.2	-29252.2	-29252.2	-29252.2
Sweden	-34368.0	-34368.0	-30000.0	-30000.0	-30000.0	-30000.0	-31774.4	-32296.0	-27679.7
Switzerland	-4343.0	-4404.0	-4527.0	-5718.0	-5757.0	-5729.0	-5925.0	-6169.0	-6109.0
Turkey	----	----	----	----	----	----	----	----	----
Ukraine	-52107.1	-52963.5	-52658.0	-52917.4	-52720.9	-52939.6	-66150.6	-68806.4	-68708.3
United Kingdom of Great Britain and Northern Ireland	21185.9	21175.6	20398.5	18311.3	15690.4	14526.0	14770.0	14575.0	14984.2
United States of America	-1159993.7	-1159646.4	-1159299.1	-779934.7	-778285.1	-776658.7	-774725.5	-774083.0	-773018.9

Comments:

Emissions of CO$_2$ - total anthropogenic and removals from land-use change and forestry (UNFCCC): are all the emissions of CO$_2$ associated with human activities including land-use change and forestry. These include emissions from fuel combustion, industrial processes, agriculture, waste, and other.

For more detailed information about the GHG inventory database see: http://ghg.unfccc.int.

Units: gigagrams of CO_2

Data Source: Greenhouse Gas (GHG) Inventory Submission 1998, 1999, and 2000

Data Provider: United Nations Framework Convention on Climate Change (UNFCCC) Secretariat

Years: 1990-1998

Copyright © 2002 United Nations Framework Convention on Climate Change (UNFCCC) Secretariat

Countries and Territories	1990	1991	1992	1993	1994	1995	1996	1997	1998
Australia	265289.2	267478.3	270072.8	273212.9	276771.9	288767.2	298581.8	306037.3	324203.3
Austria	46685.4	51068.2	46162.4	45827.4	46933.3	48703.7	51103.8	50909.0	51388.9
Belarus	----	----	----	----	----	----	----	----	----
Belgium	104189.8	111144.0	111196.7	109127.2	113510.9	114594.9	117995.1	113867.0	114500.8
Bulgaria	95494.7	61170.2	55064.2	57678.2	54239.3	56608.8	----	53458.3	51388.2
Canada	415689.9	405838.1	419730.1	416893.4	429922.9	440965.1	452945.3	464819.3	476426.5
Croatia	----	----	----	----	----	----	----	----	----
Czech Republic	160073.0	148807.0	135629.0	130661.0	123631.0	124647.0	129516.2	133925.2	124486.0
Denmark	51516.3	61713.0	56250.3	58039.0	61528.3	58577.9	71767.7	62173.3	58146.5
Estonia	37183.8	36342.2	27453.3	21786.0	22667.5	20637.6	21216.2	20362.0	18889.6
Finland	53888.9	53825.8	51397.0	52252.0	58337.3	55891.1	61972.8	59780.8	57403.7
France	357722.5	383061.8	375004.2	355118.0	350813.8	356884.6	370734.6	364998.9	386358.6
Germany	986832.5	951136.0	902919.0	893006.0	877158.0	876528.1	899196.0	867425.5	861181.0
Greece	77292.0	77147.4	79171.1	79383.5	81130.9	82095.6	83211.4	87763.8	91967.5
Hungary	80089.0	65256.0	58636.0	58755.0	57046.0	57567.0	58174.0	56552.0	54621.1
Iceland	1673.7	1627.9	1753.8	1810.3	1772.6	1773.6	----	----	----
Ireland	29577.5	30230.9	30856.7	30472.3	31774.8	32389.1	33626.2	35735.6	37706.9
Italy	398320.0	397160.4	395022.1	391899.4	388683.1	413302.6	409602.8	411229.8	430294.8
Japan	1052964.1	1072761.7	1085210.8	1064440.9	1133291.2	1138351.8	1153542.3	1150675.2	----
Latvia	24208.9	18836.0	16137.0	14368.0	11757.0	11899.9	10875.3	12662.9	8051.3
Liechtenstein	----	----	----	----	----	----	----	----	----
Lithuania	37332.0	----	----	----	----	14800.0	15800.0	15800.0	13982.0
Luxembourg	12133.0	----	----	----	11520.0	9109.0	----	----	----
Monaco	106.1	122.2	129.8	131.8	134.9	131.5	137.9	139.4	134.2
Netherlands	159040.0	164850.0	163440.0	165900.0	166750.0	172960.0	180030.0	178010.0	176760.0
New Zealand	22396.6	22670.1	24448.0	23735.3	23846.0	23842.2	24861.5	26822.8	25531.1
Norway	26369.7	25728.7	26413.0	27450.7	28791.4	28492.2	31214.1	31553.3	31643.9
Poland	462998.0	357661.0	360927.0	353842.0	362083.0	337942.0	363498.6	350875.9	326858.0
Portugal	39019.6	40807.6	44909.3	43613.0	44363.4	46972.3	45349.1	46590.9	49063.3
Romania	185575.4	130464.8	125498.0	122644.0	121327.0	----	----	----	----
Russian Federation	2298900.0	2123000.0	1948000.0	1805000.0	1601100.0	1550000.0	1463000.0	----	----
Slovakia	56691.0	50375.0	45667.0	43720.0	40660.0	41904.0	42494.0	41670.0	39001.5
Slovenia	13294.4	----	----	----	----	----	----	----	----
Spain	205673.5	213403.9	223784.6	211498.6	222031.1	232254.0	220255.3	239338.6	245698.3
Sweden	51328.0	51205.0	51787.0	51881.0	53946.0	53385.0	59390.6	52586.5	52717.8
Switzerland	39673.0	41854.0	41846.0	39611.0	38789.0	39764.0	40554.0	39894.0	41138.4
Turkey	----	----	----	----	----	----	----	----	----
Ukraine	672074.8	557747.4	549438.7	484956.2	390118.0	365332.9	331975.9	307090.9	298489.0
United Kingdom of Great Britain and Northern Ireland	557665.6	566206.3	552584.3	538962.3	533078.1	526632.8	544963.5	520881.5	522887.7
United States of America	4840482.8	4787926.2	4876887.1	4992123.1	5067247.8	5103837.9	5284900.8	5355899.9	5383501.6

Comments:

Emissions of CO₂ - fuel combustion (UNFCCC): emissions of CO₂ from fuel combustion include emissions from energy industries, manufacturing industries and Construction, transport and other fuel combustion sectors.

For more detailed information about the GHG inventory database see: http://ghg.unfccc.int.

NATIONAL DATA SETS

Emissions of CO$_2$ - from Fuel Combustion (UNFCCC)

Units: gigagrams of CO$_2$

Data Source: Greenhouse Gas (GHG) Inventory Submission 1998, 1999, and 2000

Data Provider: United Nations Framework Convention on Climate Change (UNFCCC) Secretariat

Years: 1990-1998

Copyright © 2002 United Nations Framework Convention on Climate Change (UNFCCC) Secretariat

Countries and Territories	1990	1991	1992	1993	1994	1995	1996	1997	1998
Australia	6654.1	6329.8	6211.4	6520.2	7277.3	7018.5	7109.9	7138.6	7823.0
Austria	12745.2	12207.4	11139.9	11298.6	11935.8	12133.3	11629.4	12727.0	11919.1
Belarus	----	----	----	----	----	----	----	----	----
Belgium	9140.1	9546.7	9665.9	9705.7	10332.0	10707.0	11261.1	7373.4	7374.8
Bulgaria	8361.0	4872.6	4118.3	4181.1	4938.9	5723.0	----	5368.8	3761.6
Canada	32724.3	33508.0	33121.4	34886.3	35785.7	36464.4	38065.8	38398.7	38065.8
Croatia	----	----	----	----	----	----	----	----	----
Czech Republic	5417.0	4335.0	4591.0	4190.0	4114.0	4170.0	2479.0	2498.0	2661.5
Denmark	1005.5	1178.1	1300.5	1311.0	1317.8	1311.0	1388.1	1539.3	1436.3
Estonia	612.9	613.8	313.1	193.0	214.8	221.5	206.4	353.5	342.3
Finland	1129.8	1000.0	1025.0	858.0	840.0	810.0	840.0	900.0	921.4
France	21253.8	19421.1	17940.0	17179.6	18228.3	18703.1	17440.9	17549.6	18008.7
Germany	27668.0	24814.0	25389.0	25262.0	26954.0	26388.0	24932.0	25000.0	25000.0
Greece	7685.8	7598.9	7510.3	7732.8	7259.7	7846.3	8074.8	8231.3	8300.4
Hungary	3587.0	1381.9	1167.9	1318.3	1396.7	1437.7	1547.6	1587.4	2151.5
Iceland	390.5	357.3	360.6	408.2	408.8	425.0	----	----	----
Ireland	1930.6	1954.2	1964.7	1877.7	2141.1	2040.7	2002.6	2264.1	2250.2
Italy	27531.6	27106.1	27502.5	23451.1	22617.5	23169.1	22877.4	23051.6	23682.3
Japan	58794.7	60381.4	60998.4	60332.7	61302.7	61236.1	61078.7	59500.6	----
Latvia	562.5	584.0	286.0	113.4	154.0	127.3	185.2	179.4	235.8
Liechtenstein	----	----	----	----	----	----	----	----	----
Lithuania	2203.0	----	----	----	----	400.0	400.0	400.0	2524.8
Luxembourg	585.0	----	----	----	447.0	406.0	----	----	----
Monaco	----	----	----	----	----	----	----	----	----
Netherlands	1900.0	1500.0	1300.0	1200.0	1400.0	1600.0	1800.0	1700.0	1500.0
New Zealand	2386.1	2510.5	2646.6	2770.2	2671.0	2736.6	2741.9	2789.0	2739.9
Norway	6548.1	6108.8	6009.3	6516.3	7061.6	7473.1	7514.0	7571.2	7777.2
Poland	13574.0	9247.0	10603.0	9248.0	9422.0	10144.0	8937.8	10664.2	10487.0
Portugal	3678.2	3728.5	3561.2	3707.7	3732.0	4073.0	3999.7	4430.8	4329.6
Romania	9243.6	5188.1	4655.0	4435.0	4263.0	----	----	----	----
Russian Federation	46300.0	43603.0	35702.0	29801.6	24000.0	23080.0	18920.0	----	----
Slovakia	5545.6	4124.6	4645.3	4261.9	4515.7	4719.6	4499.3	4734.1	4770.3
Slovenia	640.9	----	----	----	----	----	----	----	----
Spain	16667.2	16123.2	14670.7	13976.1	16139.4	17171.1	16834.6	20794.0	23057.8
Sweden	3786.3	3683.7	4123.4	4004.6	4227.3	4457.5	3711.2	3802.6	4124.2
Switzerland	3363.0	3034.0	2736.0	2548.0	2731.0	2622.0	2220.0	2207.0	2204.0
Turkey	----	----	----	----	----	----	----	----	----
Ukraine	31716.8	28730.5	28476.8	19266.1	16720.0	15595.2	14791.7	15815.6	15956.0
United Kingdom of Great Britain and Northern Ireland	13987.9	11657.4	11057.6	11191.0	14823.4	15109.3	15959.0	15321.0	15518.0
United States of America	54426.5	53197.0	53512.1	55136.7	58431.9	61735.1	63170.3	66021.4	67446.8

Comments:

Emissions of CO$_2$ - industrial processes (UNFCCC): emissions of CO$_2$ (carbon dioxide) from industrial processes of mineral products, chemical industry, metal production, other production and other.

For more detailed information about the GHG inventory database see: http://ghg.unfccc.int.

Emissions of CO_2 - Total Anthropogenic (UNFCCC) Excluding Land-Use Change and Forestry

Units: gigagrams of CO_2

Data Source: Greenhouse Gas (GHG) Inventory Submission 1998, 1999, and 2000

Data Provider: United Nations Framework Convention on Climate Change (UNFCCC) Secretariat

Years: 1990-1998

Copyright © 2002 United Nations Framework Convention on Climate Change (UNFCCC) Secretariat

Countries and Territories	1990	1991	1992	1993	1994	1995	1996	1997	1998
Australia	278669.0	280206.0	282769.4	286295.6	290366.6	302277.1	312286.5	319531.9	337972.7
Austria	62130.1	66021.6	60152.0	59899.1	61754.2	63694.3	65911.3	66788.5	66603.5
Belarus	----	----	----	----	----	----	----	----	----
Belgium	113997.1	121420.2	121591.3	119518.9	124311.9	125576.1	129654.4	121339.5	121974.8
Bulgaria	103855.7	66042.8	59182.5	61859.3	59178.2	62331.8	----	59148.3	55149.9
Canada	465755.4	456369.1	469538.6	468038.5	482059.0	493840.8	506554.6	518376.1	529430.9
Croatia	----	----	----	----	----	----	----	----	----
Czech Republic	165490.0	153142.0	140220.0	134851.0	127745.0	128817.0	132537.5	137125.3	128268.3
Denmark	52893.7	63516.2	58190.7	59928.1	63413.7	60347.9	73657.3	64346.7	60124.5
Estonia	37796.7	36956.0	27766.0	21979.0	22852.3	20859.1	21423.0	20715.5	19231.9
Finland	60771.1	60399.7	57878.8	58637.7	64813.4	62227.2	68386.5	66177.3	63945.5
France	387589.8	411325.4	401632.0	381064.7	377852.3	383816.9	396515.5	391113.1	412860.2
Germany	1014500.5	975950.0	928308.0	918268.0	904112.0	902916.1	924128.0	892425.5	886181.0
Greece	85163.9	84936.9	86868.7	87289.7	88570.1	90121.2	91466.1	96175.7	100449.1
Hungary	83676.0	66637.9	59803.9	60073.3	58442.7	59004.7	59721.6	58139.4	57600.6
Iceland	2147.2	2068.1	2197.4	2302.0	2264.7	2282.0	----	----	----
Ireland	31574.6	32256.2	32892.7	32421.1	33987.1	34501.0	35699.9	38070.8	40019.4
Italy	432565.1	430912.2	429331.4	422324.2	418104.4	443184.2	439478.7	441232.2	459460.6
Japan	1124532.1	1147844.5	1162310.4	1143843.5	1213939.8	1219418.2	1235592.6	1230831.1	----
Latvia	24771.5	19420.0	16423.0	14482.0	11910.0	12027.2	11065.1	12842.3	8287.1
Liechtenstein	208.0	----	----	----	----	----	----	----	----
Lithuania	39535.0	----	----	----	----	15200.0	16200.0	16200.0	16693.8
Luxembourg	12750.0	----	----	----	11998.0	9545.0	----	----	----
Monaco	108.1	125.4	132.9	135.4	138.4	135.0	141.4	143.1	138.0
Netherlands	161360.0	166910.0	165210.0	167450.0	168340.0	177130.0	184790.0	183230.0	181370.0
New Zealand	25397.9	25881.0	27761.9	27136.0	27197.5	27206.5	28276.3	30371.8	28941.0
Norway	35146.1	33604.8	34266.8	35917.5	37939.7	38157.3	41119.3	41426.5	41699.6
Poland	476625.0	366958.0	371591.0	363160.0	371588.0	348172.0	372530.3	361626.3	337450.0
Portugal	43132.1	44976.0	48928.2	47774.4	48587.1	51531.4	49821.5	51504.9	53890.5
Romania	194826.1	135660.2	130160.0	127086.0	125597.0	132290.2	122389.7	122305.6	117106.6
Russian Federation	2372300.0	----	----	----	1660000.0	1590420.0	1495920.0	----	----
Slovakia	62236.6	54499.6	50312.3	47981.9	45175.7	46623.6	46993.3	46404.1	43771.8
Slovenia	13935.3	----	----	----	----	----	----	----	----
Spain	226057.2	233257.4	242275.4	229514.8	242279.4	252957.6	240847.3	264117.8	273017.3
Sweden	55443.3	55218.1	56207.1	56182.3	58438.2	58107.5	63352.0	56684.1	56952.8
Switzerland	44409.0	46285.0	45990.0	43566.0	42928.0	43805.0	44212.0	43549.0	44809.4
Turkey	----	----	----	----	----	----	----	----	----
Ukraine	703791.7	586477.9	577915.5	504222.3	406838.0	380928.1	346767.6	322906.5	314445.0
United Kingdom of Great Britain and Northern Ireland	584220.4	588046.8	573655.9	559619.8	559333.4	551325.3	570278.2	544252.1	546389.9
United States of America	4914351.2	4862349.4	4951560.7	5072270.7	5150787.0	5193841.2	5376080.9	5449974.1	5478051.3

Comments:

Emissions of CO_2 - total anthropogenic (UNFCCC) excluding land-use change and forestry: are all the emissions of CO_2 associated with human activities excluding land-use change and forestry. Emissions included are emitted from fuel combustion, industrial processes, agriculture, waste, and other.

For more detailed information about the GHG inventory database see: http://ghg.unfccc.int.

Emissions of CO$_2$ - from Transport (UNFCCC)

Units: gigagrams of CO2

Data Source: Greenhouse Gas (GHG) Inventory Submission 1998, 1999, and 2000

Data Provider: United Nations Framework Convention on Climate Change (UNFCCC) Secretariat

Years: 1990-1998

Copyright © 2002 United Nations Framework Convention on Climate Change (UNFCCC) Secretariat

Countries and Territories	1990	1991	1992	1993	1994	1995	1996	1997	1998
Australia	59287.8	58576.9	60094.8	60831.9	62378.0	64870.0	66914.8	68487.6	68433.6
Austria	13569.7	15058.7	15054.2	15103.7	16163.1	15431.9	15379.4	15792.5	16752.5
Belarus	----	----	----	----	----	----	----	----	----
Belgium	20568.7	21140.9	22490.3	22514.8	22248.0	22523.6	22785.5	24334.2	24697.2
Bulgaria	12638.7	6524.6	6435.4	7443.9	6547.0	6844.6	----	5674.1	6475.2
Canada	145833.4	140611.1	144669.1	147814.2	155224.2	159440.3	163927.8	170334.7	174251.8
Croatia	----	----	----	----	----	----	----	----	----
Czech Republic	7959.0	6869.0	8143.0	8314.0	8260.0	8912.0	9895.8	11391.6	10779.0
Denmark	10741.4	11245.0	11373.4	11627.8	11339.2	11509.7	11748.9	12070.2	12421.2
Estonia	2655.6	2385.0	1432.4	1607.2	1786.0	1700.1	1534.1	1096.6	1236.4
Finland	12475.2	----	11592.0	10993.0	11414.4	11125.8	10994.1	11531.4	12299.0
France	121281.8	123991.4	126846.4	127672.0	129435.7	131313.9	131801.7	134478.4	137986.7
Germany	162280.9	165953.0	171661.0	176532.0	172899.0	176468.0	176942.0	177689.0	180874.0
Greece	15358.4	16134.9	16561.9	16760.9	16866.8	16972.1	17257.9	18040.0	19790.3
Hungary	7741.0	7383.0	7189.0	7141.0	7212.0	7001.0	6612.0	7741.0	8380.5
Iceland	720.8	726.6	728.8	738.0	742.9	748.9	----	----	----
Ireland	4960.6	5206.0	5624.5	5591.4	5828.7	6306.4	7063.5	7683.8	8768.2
Italy	95616.2	94849.3	99426.3	102031.3	103849.3	105300.3	108309.7	110188.3	110167.0
Japan	205632.6	215312.7	220473.0	222474.1	233425.3	240292.4	246873.6	251375.9	----
Latvia	5829.1	3167.0	2836.0	2577.0	2229.0	1748.8	1611.9	2178.4	2126.2
Liechtenstein	----	----	----	----	----	----	----	----	----
Lithuania	5791.0	----	----	----	----	3500.0	3800.0	4100.0	3754.4
Luxembourg	2625.0	----	----	----	3685.0	3426.0	----	----	----
Monaco	39.5	45.5	50.7	47.2	48.7	47.2	46.8	46.6	45.3
Netherlands	28560.0	28550.0	29830.0	30460.0	30800.0	32030.0	33820.0	34350.0	34720.0
New Zealand	8660.3	8661.6	9047.1	9458.2	10160.1	10868.8	10989.2	11241.6	11435.3
Norway	11645.8	11616.1	11832.9	12394.1	12180.2	12553.9	13153.9	13390.9	13752.1
Poland	28238.0	27815.0	30475.0	27675.0	29533.0	25285.0	28098.3	26662.4	28127.0
Portugal	11404.2	12132.7	13122.6	13696.6	14096.8	14646.1	15439.3	15969.0	16180.5
Romania	7893.0	7520.9	6834.0	7498.0	7744.0	----	----	----	----
Russian Federation	----	----	----	----	----	----	----	----	----
Slovakia	3179.2	3179.2	3179.2	3179.2	3179.2	3179.2	3179.2	3179.2	3179.2
Slovenia	3179.2	----	----	----	----	----	----	----	----
Spain	58003.6	60803.6	64694.7	61161.2	65755.8	66747.3	71874.0	71892.1	78390.0
Sweden	18650.0	18613.0	19099.0	18322.0	18685.0	19341.0	19572.6	18956.7	21139.7
Switzerland	14144.0	14668.0	14983.0	13933.0	14117.0	13815.0	13885.0	14462.0	14690.0
Turkey	----	----	----	----	----	----	----	----	----
Ukraine	----	----	----	----	----	----	----	----	----
United Kingdom of Great Britain and Northern Ireland	116721.5	116193.8	117646.7	118822.4	119175.2	118066.0	122679.0	123756.0	122899.2
United States of America	1413362.8	1381485.7	1420961.6	1451269.7	1494605.6	1523797.4	1570219.8	1582574.5	1607581.5

Comments:

Emissions of CO$_2$ - transport (UNFCCC): emissions of CO$_2$ from transport exclude emissions from international bunkers (aviation and marine).

For more detailed information about the GHG inventory database see: http://ghg.unfccc.int.

Emissions of CO_2, CH_4, N_2O, HFCs, PFCs and SF_6 - Aggregated (UNFCCC) Excluding CO_2 from Land-Use Change and Forestry

Units: gigagrams of CO_2

Data Source: Greenhouse Gas (GHG) Inventory Submission 1998, 1999, and 2000

Data Provider: United Nations Framework Convention on Climate Change (UNFCCC) Secretariat

Years: 1990-1998

Copyright © 2002 United Nations Framework Convention on Climate Change (UNFCCC) Secretariat

Countries and Territories	1990	1991	1992	1993	1994	1995	1996	1997	1998
Australia	423237	422863	424340	426671	428429	441702	452866	462025	484699
Austria	75452	79209	73091	72770	74515	77948	78304	78920	80315
Belarus	----	----	----	----	----	----	----	----	----
Belgium	136463	143959	143838	142234	147647	149853	154172	144510	145372
Bulgaria	157090	115679	103710	102084	92586	98083	----	84461	84317
Canada	611770	605618	620531	623834	643742	658774	676609	687282	692230
Croatia	----	----	----	----	----	----	----	----	----
Czech Republic	189837	175323	161221	154710	147291	148103	153579	157816	147777
Denmark	69567	80138	74278	76348	79624	76557	89785	80064	76144
Estonia	40719	39813	30210	24087	24925	22653	23122	23097	21756
Finland	75202	73845	69949	70328	76355	74729	80946	78711	76315
France	553778	575977	561298	535560	532545	539473	551382	545649	558726
Germany	1208807	1156666	1103602	1084324	1065610	1062021	1077143	1040351	1019745
Greece	105346	105303	106685	107713	109693	112322	114713	119895	124315
Hungary	101633	86789	78033	77895	76089	76757	78068	75783	83677
Iceland	2576	2484	2603	2704	2669	2696	----	----	----
Ireland	53497	54167	54780	54542	56439	57316	58918	61364	63718
Italy	518502	518472	514401	506719	503381	530889	527315	531043	541542
Japan	1213262	1240415	1259601	1238461	1309509	1322293	1336250	1330555	----
Latvia	35669	29401	25525	22081	19170	19196	18064	15959	11504
Liechtenstein	260	----	----	----	----	----	----	----	----
Lithuania	51548	----	----	----	----	----	----	----	23851
Luxembourg	13448	----	----	----	12674	10223	----	----	----
Monaco	111	128	136	139	142	139	145	147	142
Netherlands	217882	223803	221229	223231	225638	234404	242620	240666	236251
New Zealand	73068	72738	73998	73160	73703	73866	74927	76368	74886
Norway	52141	49915	48162	50303	52357	52387	55326	55624	56148
Poland	564286	437448	439045	429649	438895	416530	436545	426220	402477
Portugal	63858	66327	68962	67410	68317	72760	70121	71578	74870
Romania	264879	179762	172168	167187	164026	----	----	----	----
Russian Federation	3040062	----	----	----	2149969	2065711	1962441	----	----
Slovakia	76304	67006	61552	58043	55125	56940	56754	55922	52818
Slovenia	19212	----	----	----	----	----	----	----	----
Spain	305746	312445	321788	306558	324156	336705	332235	358634	369856
Sweden	69399	64888	65742	65724	67683	68403	77043	71405	73842
Switzerland	53005	54933	54607	52151	51437	52285	52624	52454	53706
Turkey	----	----	----	----	----	----	----	----	----
Ukraine	919220	784111	765090	678301	572045	538833	499634	466471	454934
United Kingdom of Great Britain and Northern Ireland	741489	742375	718732	694022	694670	685071	705714	681355	679850
United States of America	6048786	5999520	6111024	6232086	6354316	6411034	6616008	6702213	6726997

Comments:

Emissions of CO_2, CH_4, N_2O, HFCs, PFCs and SF_6 - aggregated excluding CO_2 from land-use change and forestry.

The global warming potential (GWP) is an index used to translate the level of emissions of various gases into a common measure in order to compare the relative radiative forcing of different gases without directly calculating the changes in atmospheric concentrations. GWPs are calculated as the ratio of the radiative forcing that would result from the emissions of one kilogram of a greenhouse gas to that from the emission of one kilogram of carbon dioxide over a period of time (usually 100 years). Gases involved in complex atmospheric chemical processes have not been assigned GWPs.

For more detailed information about the GHG inventory database see: http://ghg.unfccc.int.

Emissions of CO_2, CH_4, N_2O, HFCs, PFCs and SF_6 - Aggregated (UNFCCC) Including CO_2 from Land-Use Change and Forestry

Units: gigagrams of CO_2

Data Source: Greenhouse Gas (GHG) Inventory Submission 1998, 1999, and 2000

Data Provider: United Nations Framework Convention on Climate Change (UNFCCC) Secretariat

Years: 1990-1998

Copyright © 2002 United Nations Framework Convention on Climate Change (UNFCCC) Secretariat

Countries and Territories	1990	1991	1992	1993	1994	1995	1996	1997	1998
Australia	493329	466300	464282	464137	464323	477078	490811	498574	519873
Austria	66237	65705	64435	63787	66653	70695	72918	71287	72682
Belarus	----	----	----	----	----	----	----	----	----
Belgium	134406	141902	141781	140177	145590	147796	152115	143534	144396
Bulgaria	152433	107800	96074	95063	85612	90562	----	78609	78084
Canada	572628	548349	575179	589255	614013	637646	647723	663658	670396
Croatia	----	----	----	----	----	----	----	----	----
Czech Republic	187556	170296	155180	149067	143348	142649	149101	153177	144019
Denmark	68651	79218	73355	75421	78694	75622	88838	79105	75171
Estonia	29402	39813	30210	24087	13800	9387	23122	31090	18400
Finland	51404	35638	38055	41212	59096	60041	59914	66073	66602
France	494162	519488	500051	469695	465050	473858	484185	477559	488943
Germany	1175088	1122947	1069883	1050605	1031891	1028528	1043650	1006858	986252
Greece	105345	105303	106685	107713	109693	112322	114713	119895	124315
Hungary	98536	83550	74211	73197	71269	71959	74137	71578	79266
Iceland	2576	2484	2603	2704	2669	2697	----	----	----
Ireland	48477	48991	49404	48955	50715	51403	52854	55093	57269
Italy	492888	492618	489317	483349	479887	507664	503028	507409	517908
Japan	1129359	1156550	1174033	1148377	1215965	1225588	----	----	----
Latvia	24843	15215	11290	7853	4964	8712	3744	1643	995
Liechtenstein	238	----	----	----	----	----	----	----	----
Lithuania	42700	----	----	----	----	----	----	----	31563
Luxembourg	13153	----	----	----	12379	9928	----	----	----
Monaco	111	128	136	139	142	139	145	147	142
Netherlands	216382	222203	219629	221632	223938	232705	240921	238966	234551
New Zealand	51537	52225	55714	56921	58094	57703	58397	58198	53990
Norway	42551	38215	34912	36793	36677	38746	37716	39125	38561
Poland	529540	394690	398230	389470	396942	373650	393928	385699	372657
Portugal	59864	62208	64718	63040	63823	68140	65483	66922	70196
Romania	261954	173172	165578	160597	157436	----	----	----	----
Russian Federation	2648062	----	----	----	1581969	1225711	1122441	----	----
Slovakia	73878	64580	59126	55617	51890	53704	52521	51836	51136
Slovenia	16919	----	----	----	----	----	----	----	----
Spain	276493	283192	292536	277305	294904	307452	302982	329381	340604
Sweden	35031	30520	35742	35724	37683	38403	45269	39108	46162
Switzerland	48662	50529	50080	46433	45680	46556	46699	46286	47598
Turkey	----	----	----	----	----	----	----	----	----
Ukraine	867113	731147	712432	625383	519324	485893	433483	397664	386225
United Kingdom of Great Britain and Northern Ireland	762675	763551	739130	712333	710361	699597	720484	695930	694835
United States of America	4888792	4839874	4951725	5452151	5576031	5634376	5841282	5928129	5953978

Comments:

Emissions of CO_2, CH_4, N_2O, HFCs, PFCs and SF_6 - aggregated excluding CO_2 from land-use change and forestry.

The global warming potential (GWP) is an index used to translate the level of emissions of various gases into a common measure in order to compare the relative radiative forcing of different gases without directly calculating the changes in atmospheric concentrations. GWPs are calculated as the ratio of the radiative forcing that would result from the emissions of one kilogram of a greenhouse gas to that from the emission of one kilogram of carbon dioxide over a period of time (usually 100 years). Gases involved in complex atmospheric chemical processes have not been assigned GWPs. For more detailed information about the GHG inventory database see: http://ghg.unfccc.int.

NATIONAL DATA SETS

Emissions of HFCs, PFCs and SF$_6$ - Aggregated (UNFCCC)

Units: gigagrams of CO$_2$

Data Source: Greenhouse Gas (GHG) Inventory Submission 1998, 1999, and 2000

Data Provider: United Nations Framework Convention on Climate Change (UNFCCC) Secretariat

Years: 1990-1998

Copyright © 2002 United Nations Framework Convention on Climate Change (UNFCCC) Secretariat

Countries and Territories	1990	1991	1992	1993	1994	1995	1996	1997	1998
Australia	4826	4826	3581	3147	2048	1432	1292	1141	1419
Austria	----	----	----	----	----	1702	----	----	1767
Belarus	----	----	----	----	----	----	----	----	----
Belgium	----	----	----	----	----	538	624	733	733
Bulgaria	----	----	----	----	----	----	----	----	646
Canada	8845	9579	8773	9409	8949	8375	8127	8217	8423
Croatia	----	----	----	----	----	----	----	----	----
Czech Republic	----	----	----	----	----	----	----	----	----
Denmark	0	0	119	229	301	384	478	608	541
Estonia	----	----	----	----	----	----	----	----	----
Finland	55	38	19	11	20	54	123	194	277
France	7644	6207	5463	4728	4539	5050	6193	6905	7345
Germany	8930	9041	9484	11163	11440	11134	10220	10700	10885
Greece	1302	1422	1130	1736	2226	3331	3821	4024	3798
Hungary	----	----	----	----	----	----	----	----	952
Iceland	5	----	----	----	----	5	----	----	----
Ireland	----	----	----	----	----	----	----	----	----
Italy	787	816	813	829	1311	1825	1707	2140	2140
Japan	38240	43020	47800	45410	45410	52580	50190	50190	----
Latvia	----	----	----	----	----	----	----	----	----
Liechtenstein	----	----	----	----	----	----	----	----	----
Lithuania	----	----	----	----	----	----	----	----	----
Luxembourg	----	----	----	----	----	----	----	----	----
Monaco	----	----	----	----	----	----	----	----	----
Netherlands	8986	8669	8191	8737	10118	10270	11118	11502	10312
New Zealand	605	653	639	241	352	418	507	466	466
Norway	5225	4594	2714	2737	2485	2169	2083	2017	2098
Poland	----	----	----	----	----	24	----	----	----
Portugal	----	----	----	----	----	158	----	----	----
Romania	----	----	----	----	----	----	----	----	----
Russian Federation	41295	----	----	----	38603	31857	36177	----	----
Slovakia	272	267	249	156	144	148	91	114	80
Slovenia	----	----	----	----	----	----	----	----	----
Spain	3826	3468	3761	3164	4792	6539	7490	9120	9375
Sweden	----	----	----	----	----	1195	----	1601	3551
Switzerland	----	----	----	----	----	----	----	588	692
Turkey	----	----	----	----	----	----	----	----	----
Ukraine	----	----	----	----	----	----	----	----	----
United Kingdom of Great Britain and Northern Ireland	14379	14425	14138	14604	15855	17433	18466	20371	22124
United States of America	85442	80614	86288	87386	92035	106385	122923	129571	147776

Comments:

HFCs (hydrofluorocarbons) are produced commercially as a substitute for CFCs. HFCs are used largely in refrigeration and semi-conductor manufacturing. Their GWPs range from 1300 to 11700 times that of CO$_2$ (over a 100 year time horizon), depending on the HFC.

PFCs (perfluorocarbons) are a by-product of aluminum smelting and uranium enrichment. They also are the replacement for CFCs in manufacturing semiconductors. The GWP of PFCs is 6500-9200 times that of CO$_2$ (100 year time horizon).

SF$_6$ (Sulphur hexafluoride) is largely used in heavy industry to insulate high-voltage equipment and to assist in the manufacturing of cable-cooling systems. Its GWP is 23900 times that of CO$_2$ (100 year time horizon). For more detailed information about the GHG inventory database see: http://ghg.unfccc.int.

Emissions of N$_2$O - from Agricultural Soils (UNFCCC)

Units: gigagrams of N$_2$O

Data Source: Greenhouse Gas (GHG) Inventory Submission 1998, 1999, and 2000

Data Provider: United Nations Framework Convention on Climate Change (UNFCCC) Secretariat

Years: 1990-1998

Copyright © 2002 United Nations Framework Convention on Climate Change (UNFCCC) Secretariat

Countries and Territories	1990	1991	1992	1993	1994	1995	1996	1997	1998
Australia	46.9	47.3	46.9	47.7	48.0	47.1	46.7	49.7	51.8
Austria	3.3	3.3	3.3	3.3	3.3	3.3	3.3	3.3	3.3
Belarus	----	----	----	----	----	----	----	----	----
Belgium	10.9	10.9	10.9	10.7	9.4	9.4	9.4	9.3	9.3
Bulgaria	53.9	49.2	41.9	37.8	36.3	37.0	----	2.5	33.7
Canada	116.0	116.0	114.1	119.2	125.1	125.6	131.9	130.6	130.8
Croatia	----	----	----	----	----	----	----	----	----
Czech Republic	2.3	2.0	1.7	1.8	1.8	1.7	12.0	11.3	15.9
Denmark	31.6	30.8	28.8	29.0	28.1	27.9	27.0	25.9	26.1
Estonia	----	----	----	----	----	----	----	0.0	1.2
Finland	13.7	12.8	11.6	11.8	11.8	12.3	12.0	11.7	11.5
France	171.0	168.6	163.5	159.1	160.7	162.1	164.6	166.9	166.9
Germany	85.0	78.0	74.0	73.0	77.0	76.6	76.0	76.0	76.0
Greece	20.7	20.6	19.5	19.2	19.3	18.5	18.7	19.0	19.0
Hungary	4.6	1.7	1.6	1.5	1.8	1.6	1.7	1.7	32.9
Iceland	0.2	0.2	0.2	0.2	0.2	0.2	----	----	----
Ireland	20.8	20.8	20.5	21.0	21.6	22.2	22.5	21.7	23.0
Italy	65.2	68.1	68.6	69.4	68.6	67.4	66.1	69.0	69.0
Japan	3.8	3.6	3.6	3.5	3.5	3.3	2.1	1.0	----
Latvia	22.0	19.3	18.8	17.0	16.3	15.7	15.6	----	3.2
Liechtenstein	----	----	----	----	----	----	----	----	----
Lithuania	----	----	----	----	----	----	----	----	1.7
Luxembourg	0.5	----	----	----	0.5	0.5	----	----	----
Monaco	----	----	----	----	----	----	----	----	----
Netherlands	21.5	22.2	25.5	25.4	25.6	26.8	26.8	25.3	25.2
New Zealand	36.8	36.4	36.4	36.8	37.3	37.4	37.2	37.2	37.3
Norway	8.6	8.6	8.3	8.5	8.3	8.5	8.4	8.4	8.4
Poland	43.0	32.5	32.0	29.9	30.0	31.4	30.4	31.2	31.2
Portugal	8.2	7.6	7.3	7.1	6.9	7.1	7.3	7.4	7.5
Romania	25.1	6.5	----	----	----	----	----	----	----
Russian Federation	----	----	----	----	----	111.0	105.0	----	----
Slovakia	13.0	10.6	9.3	7.6	7.0	7.3	7.3	7.4	7.3
Slovenia	4.6	----	----	----	----	----	----	----	----
Spain	58.2	57.6	53.9	46.6	53.5	50.9	61.1	61.1	61.1
Sweden	14.6	0.2	0.2	0.2	----	0.2	14.6	13.8	13.8
Switzerland	7.8	7.7	7.6	7.6	7.5	7.3	7.3	7.1	7.0
Turkey	----	----	----	----	----	----	----	----	----
Ukraine	----	----	----	----	----	----	----	----	----
United Kingdom of Great Britain and Northern Ireland	95.1	94.6	89.2	87.5	89.9	90.5	90.9	93.8	90.7
United States of America	890.9	902.9	925.1	913.7	987.6	951.5	974.9	996.3	991.9

Comments:

Emissions of N$_2$O (nitrous oxide) from agricultural soils include direct and indirect soil emissions, animal production emissions, and other emissions.

For more detailed information about the GHG inventory database see: http://ghg.unfccc.int.

Emissions of N$_2$O - from Agriculture (UNFCCC)

Units: gigagrams of N$_2$O

Data Source: Greenhouse Gas (GHG) Inventory Submission 1998, 1999, and 2000

Data Provider: United Nations Framework Convention on Climate Change (UNFCCC) Secretariat

Years: 1990-1998

Copyright © 2002 United Nations Framework Convention on Climate Change (UNFCCC) Secretariat

Countries and Territories	1990	1991	1992	1993	1994	1995	1996	1997	1998
Australia	60.7	61.2	60.6	62.1	62.9	63.1	63.3	66.6	70.7
Austria	3.3	3.3	3.3	3.3	3.3	3.3	3.3	3.3	3.3
Belarus	----	----	----	----	----	----	----	----	----
Belgium	10.9	10.9	10.9	10.7	9.8	9.8	10.8	10.8	10.8
Bulgaria	57.4	52.2	44.4	39.8	38.0	38.6	----	4.5	35.2
Canada	129.7	130.0	128.2	133.8	140.5	141.6	148.1	147.1	147.0
Croatia	----	----	----	----	----	----	----	----	----
Czech Republic	2.3	2.0	1.7	1.8	1.8	1.7	20.6	19.3	17.4
Denmark	33.1	32.4	30.3	30.6	29.7	29.5	28.5	27.4	27.6
Estonia	0.9	0.9	0.7	0.5	0.5	0.4	0.4	0.0	1.2
Finland	15.4	14.3	13.1	13.3	13.3	13.7	13.4	13.2	12.9
France	181.7	179.1	173.8	169.4	171.0	172.5	175.0	177.2	177.2
Germany	96.0	87.0	83.0	82.0	86.0	85.2	85.0	84.0	84.0
Greece	21.2	21.2	20.0	19.7	19.9	19.0	19.3	19.6	19.7
Hungary	4.6	1.7	1.6	1.5	1.8	1.6	1.7	1.7	34.6
Iceland	0.2	0.2	0.2	0.2	0.2	0.2	----	----	----
Ireland	22.9	22.9	22.6	23.1	23.7	24.4	24.8	24.1	25.3
Italy	77.6	80.5	80.5	81.3	80.4	79.6	78.3	81.3	81.3
Japan	9.3	9.1	9.0	9.0	8.8	8.5	7.6	6.5	----
Latvia	22.0	19.3	18.8	17.0	16.3	15.7	15.6	3.1	3.2
Liechtenstein	----	----	----	----	----	----	----	----	----
Lithuania	10.8	----	----	----	----	----	----	----	1.7
Luxembourg	0.5	----	----	----	0.5	0.5	----	----	----
Monaco	----	----	----	----	----	----	----	----	----
Netherlands	22.2	22.9	26.2	26.2	26.4	27.6	27.5	25.9	25.9
New Zealand	37.1	36.7	36.7	37.2	37.7	37.8	37.6	37.6	37.6
Norway	8.6	8.6	8.3	8.5	8.3	8.5	8.4	8.4	8.4
Poland	43.0	32.6	32.0	30.0	30.0	31.4	30.5	31.2	31.2
Portugal	8.4	7.9	7.5	7.4	7.1	7.4	7.5	7.6	7.7
Romania	25.3	6.7	6.8	6.8	6.8	----	----	----	----
Russian Federation	200.0	----	----	----	110.0	111.0	105.0	----	----
Slovakia	16.6	13.8	12.1	10.0	9.3	9.6	9.5	9.4	9.1
Slovenia	4.6	----	----	----	----	----	----	----	----
Spain	108.1	106.1	103.6	97.9	102.4	98.3	113.5	113.5	113.5
Sweden	16.8	0.2	0.2	0.2	0.2	0.2	16.6	15.7	15.7
Switzerland	9.2	9.2	9.1	9.0	8.9	8.7	8.6	8.5	8.3
Turkey	----	----	----	----	----	----	----	----	----
Ukraine	28.0	27.4	21.3	15.7	9.6	9.4	5.9	6.5	6.4
United Kingdom of Great Britain and Northern Ireland	100.4	99.9	94.3	92.5	95.0	95.5	96.0	98.9	95.7
United States of America	931.8	946.2	968.0	958.0	1033.4	996.8	1021.5	1044.0	1040.6

Comments:

Emissions of N$_2$O (nitrous oxide) from agriculture include emissions from manure management, agricultural soils, prescribed burning of savannas, field burning of agricultural residues and other.

For more detailed information about the GHG inventory database see: http://ghg.unfccc.int.

Emissions of N₂O - from Fuel Combustion (UNFCCC)

Units: gigagrams of N₂O

Data Source: Greenhouse Gas (GHG) Inventory Submission 1998, 1999, and 2000

Data Provider: United Nations Framework Convention on Climate Change (UNFCCC) Secretariat

Years: 1990-1998

Copyright © 2002 United Nations Framework Convention on Climate Change (UNFCCC) Secretariat

Countries and Territories	1990	1991	1992	1993	1994	1995	1996	1997	1998
Australia	265289.2	267478.3	270072.8	273212.9	276771.9	288767.2	298581.8	306037.3	324203.3
Austria	46685.4	51068.2	46162.4	45827.4	46933.3	48703.7	51103.8	50909.0	51388.9
Belarus	----	----	----	----	----	----	----	----	----
Belgium	104189.8	111144.0	111196.7	109127.2	113510.9	114594.9	117995.1	113867.0	114500.8
Bulgaria	95494.7	61170.2	55064.2	57678.2	54239.3	56608.8	----	53458.3	51388.2
Canada	415689.9	405838.1	419730.1	416893.4	429922.9	440965.1	452945.3	464819.3	476426.5
Croatia	----	----	----	----	----	----	----	----	----
Czech Republic	160073.0	148807.0	135629.0	130661.0	123631.0	124647.0	129516.2	133925.2	124486.0
Denmark	51516.3	61713.0	56250.3	58039.0	61528.3	58577.9	71767.7	62173.3	58146.5
Estonia	37183.8	36342.2	27453.3	21786.0	22667.5	20637.6	21216.2	20362.0	18889.6
Finland	53888.9	53825.8	51397.0	52252.0	58337.3	55891.1	61972.8	59780.8	57403.7
France	357722.5	383061.8	375004.2	355118.0	350813.8	356884.6	370734.6	364998.9	386358.6
Germany	986832.5	951136.0	902919.0	893006.0	877158.0	876528.1	899196.0	867425.5	861181.0
Greece	77292.0	77147.4	79171.1	79383.5	81130.9	82095.6	83211.4	87763.8	91967.5
Hungary	80089.0	65256.0	58636.0	58755.0	57046.0	57567.0	58174.0	56552.0	54621.1
Iceland	1673.7	1627.9	1753.8	1810.3	1772.6	1773.6	----	----	----
Ireland	29577.5	30230.9	30856.7	30472.3	31774.8	32389.1	33626.2	35735.6	37706.9
Italy	398320.0	397160.4	395022.1	391899.4	388683.1	413302.6	409602.8	411229.8	430294.8
Japan	1052964.1	1072761.7	1085210.8	1064440.9	1133291.2	1138351.8	1153542.3	1150675.2	----
Latvia	24208.9	18836.0	16137.0	14368.0	11757.0	11899.9	10875.3	12662.9	8051.3
Liechtenstein	----	----	----	----	----	----	----	----	----
Lithuania	37332.0	----	----	----	----	14800.0	15800.0	15800.0	13982.0
Luxembourg	12133.0	----	----	----	11520.0	9109.0	----	----	----
Monaco	106.1	122.2	129.8	131.8	134.9	131.5	137.9	139.4	134.2
Netherlands	159040.0	164850.0	163440.0	165900.0	166750.0	172960.0	180030.0	178010.0	176760.0
New Zealand	22396.6	22670.1	24448.0	23735.3	23846.0	23842.2	24861.5	26822.8	25531.1
Norway	26369.7	25728.7	26413.0	27450.7	28791.4	28492.2	31214.1	31553.3	31643.9
Poland	462998.0	357661.0	360927.0	353842.0	362083.0	337942.0	363498.6	350875.9	326858.0
Portugal	39019.6	40807.6	44909.3	43613.0	44363.4	46972.3	45349.1	46590.9	49063.3
Romania	185575.4	130464.8	125498.0	122644.0	121327.0	----	----	----	----
Russian Federation	2298900.0	2123000.0	1948000.0	1805000.0	1601100.0	1550000.0	1463000.0	----	----
Slovakia	56691.0	50375.0	45667.0	43720.0	40660.0	41904.0	42494.0	41670.0	39001.5
Slovenia	13294.4	----	----	----	----	----	----	----	----
Spain	205673.5	213403.9	223784.6	211498.6	222031.1	232254.0	220255.3	239338.6	245698.3
Sweden	51328.0	51205.0	51787.0	51881.0	53946.0	53385.0	59390.6	52586.5	52717.8
Switzerland	39673.0	41854.0	41846.0	39611.0	38789.0	39764.0	40554.0	39894.0	41138.4
Turkey	----	----	----	----	----	----	----	----	----
Ukraine	672074.8	557747.4	549438.7	484956.2	390118.0	365332.9	331975.9	307090.9	298489.0
United Kingdom of Great Britain and Northern Ireland	557665.6	566206.3	552584.3	538962.3	533078.1	526632.8	544963.5	520881.5	522887.7
United States of America	4840482.8	4787926.2	4876887.1	4992123.1	5067247.8	5103837.9	5284900.8	5355899.9	5383501.6

Comments:

Emissions of CO₂ - fuel combustion (UNFCCC): emissions of CO₂ from fuel combustion include emissions from energy industries, manufacturing industries and construction, transport and other fuel combustion sectors.

For more detailed information about the GHG inventory database see: http://ghg.unfccc.int.

NATIONAL DATA SETS

Emissions of N$_2$O - from Industrial Processes (UNFCCC)

Units: gigagrams of N$_2$O

Data Source: Greenhouse Gas (GHG) Inventory Submission 1998, 1999, and 2000

Data Provider: United Nations Framework Convention on Climate Change (UNFCCC) Secretariat

Years: 1990-1998

Copyright © 2002 United Nations Framework Convention on Climate Change (UNFCCC) Secretariat

Countries and Territories	1990	1991	1992	1993	1994	1995	1996	1997	1998
Australia	1.6	1.5	1.8	1.6	1.4	1.4	1.6	1.6	1.7
Austria	0.6	0.6	0.5	0.6	0.6	0.5	0.5	0.5	0.5
Belarus	----	----	----	----	----	----	----	----	----
Belgium	11.5	11.2	10.1	10.9	12.6	13.8	15.1	14.5	15.1
Bulgaria	7.8	5.2	4.3	3.7	4.3	6.2	----	6.9	3.1
Canada	37.1	34.7	34.6	31.8	37.9	37.1	39.6	34.4	18.8
Croatia	----	----	----	----	----	----	----	----	----
Czech Republic	3.3	2.8	3.5	2.7	3.0	3.4	3.3	3.6	3.9
Denmark	----	----	----	----	----	----	----	----	----
Estonia	----	----	----	----	----	----	----	----	----
Finland	5.1	4.5	4.0	4.2	4.4	4.5	4.5	4.5	4.3
France	89.6	89.9	83.8	74.9	78.6	81.6	79.9	80.6	51.9
Germany	82.0	83.0	93.0	84.0	81.0	82.3	87.4	73.0	29.0
Greece	2.3	1.9	2.0	1.9	1.8	1.8	2.1	1.8	1.8
Hungary	----	0.0	0.1	0.0	0.0	0.0	0.0	0.0	0.0
Iceland	0.2	0.2	0.1	0.1	0.1	0.1	----	----	----
Ireland	3.3	2.6	2.6	2.6	2.6	2.6	2.6	2.6	2.6
Italy	23.5	24.6	22.9	21.8	20.6	23.5	22.7	22.8	20.0
Japan	23.9	21.8	21.6	21.2	24.0	23.8	26.6	28.1	----
Latvia	----	----	----	----	----	----	----	----	----
Liechtenstein	----	----	----	----	----	----	----	----	----
Lithuania	1.4	----	----	----	----	----	----	----	9.3
Luxembourg	----	----	----	----	----	----	----	----	----
Monaco	----	----	----	----	----	----	----	----	----
Netherlands	31.5	32.3	30.4	30.0	31.6	31.6	31.7	35.0	33.8
New Zealand	----	----	----	----	----	----	----	----	----
Norway	6.7	6.1	4.2	5.0	5.4	5.3	5.2	4.8	5.4
Poland	20.0	13.1	13.0	13.1	14.0	15.8	16.2	15.7	12.9
Portugal	1.9	1.9	2.0	1.6	1.2	2.0	2.0	2.0	2.0
Romania	24.4	6.4	6.8	6.5	5.6	----	----	----	----
Russian Federation	3.0	----	----	----	1.2	1.0	1.0	----	----
Slovakia	1.9	1.7	1.6	1.3	2.1	2.3	0.2	0.3	0.2
Slovenia	----	----	----	----	----	----	----	----	----
Spain	9.3	8.4	7.1	5.9	7.0	7.4	7.9	7.5	7.0
Sweden	2.7	2.7	2.3	2.3	2.3	2.3	2.8	2.2	2.6
Switzerland	0.3	0.3	0.3	0.3	0.3	0.3	0.3	0.3	0.3
Turkey	----	----	----	----	----	----	----	----	----
Ukraine	22.9	23.1	15.3	7.5	9.1	4.7	7.4	8.5	8.6
United Kingdom of Great Britain and Northern Ireland	93.9	87.8	70.8	60.1	71.3	61.1	65.3	66.8	59.4
United States of America	116.6	119.5	115.9	121.4	128.7	129.7	133.9	123.7	91.4

Comments:

Emissions of N$_2$O (nitrous oxide) from industrial processes include emissions from mineral products, chemical industry, metal production, other production and other.

For more detailed information about the GHG inventory database see: http://ghg.unfccc.int.

Emissions of N$_2$O - Total Anthropogenic (UNFCCC)

Units: gigagrams of N$_2$O

Data Source: Greenhouse Gas (GHG) Inventory Submission 1998, 1999, and 2000

Data Provider: United Nations Framework Convention on Climate Change (UNFCCC) Secretariat

Years: 1990-1998

Copyright © 2002 United Nations Framework Convention on Climate Change (UNFCCC) Secretariat

Countries and Territories	1990	1991	1992	1993	1994	1995	1996	1997	1998
Australia	73.0	73.4	74.2	76.5	77.9	79.2	80.5	84.6	89.6
Austria	6.6	6.8	6.9	7.1	7.3	7.3	7.4	7.3	7.4
Belarus	----	----	----	----	----	----	----	----	----
Belgium	31.0	31.1	30.2	31.0	34.8	36.1	36.6	32.8	33.7
Bulgaria	81.4	68.4	59.2	53.8	52.4	55.2	----	21.2	47.7
Canada	202.6	202.0	202.1	206.9	220.5	223.0	231.4	225.9	209.3
Croatia	----	----	----	----	----	----	----	----	----
Czech Republic	25.8	23.4	22.5	21.2	21.5	21.6	29.1	28.7	27.1
Denmark	34.9	34.6	32.5	32.9	32.2	32.0	31.6	30.3	30.5
Estonia	2.3	2.3	1.7	1.4	1.3	1.2	1.2	0.7	1.3
Finland	26.6	24.9	22.1	22.3	22.2	25.1	25.5	25.7	25.6
France	306.7	306.2	295.2	281.9	287.7	293.1	295.5	298.9	271.8
Germany	220.6	214.2	219.6	210.6	211.6	213.6	219.8	204.4	159.7
Greece	30.3	30.3	29.2	28.8	28.9	28.4	29.2	29.7	30.3
Hungary	12.9	4.0	4.8	4.6	5.2	4.7	4.9	4.3	35.1
Iceland	0.4	0.4	0.4	0.4	0.4	0.4	----	----	----
Ireland	29.3	28.8	28.6	29.1	30.0	30.7	31.2	30.8	32.5
Italy	146.0	150.4	147.8	145.9	143.3	147.4	147.6	150.6	124.4
Japan	58.4	56.7	57.2	56.8	60.9	62.2	64.8	65.7	----
Latvia	22.5	19.8	19.2	17.4	16.8	16.3	16.3	3.8	3.8
Liechtenstein	0.1	----	----	----	----	----	----	----	----
Lithuania	13.2	----	----	----	----	----	----	----	11.1
Luxembourg	0.6	----	----	----	0.7	0.7	----	----	----
Monaco	0.0	0.0	0.0	0.0	0.0	0.0	0.0	0.0	0.0
Netherlands	65.8	66.9	69.2	68.8	70.7	72.2	71.8	73.3	71.6
New Zealand	38.2	37.8	37.9	38.3	38.9	39.0	38.8	38.9	38.9
Norway	16.6	16.1	13.9	15.1	15.4	15.7	15.7	15.5	16.4
Poland	70.0	52.0	50.0	49.8	50.0	54.0	53.9	54.0	51.6
Portugal	20.1	20.2	19.9	19.9	19.4	20.8	20.8	20.9	21.4
Romania	66.3	24.8	26.9	26.2	25.0	----	----	----	----
Russian Federation	225.7	----	----	----	127.6	139.0	132.0	----	----
Slovakia	19.9	16.9	14.9	12.5	12.6	13.2	11.0	11.0	10.7
Slovenia	5.1	----	----	----	----	----	----	----	----
Spain	133.0	130.7	127.8	119.9	127.1	124.4	139.5	140.4	141.4
Sweden	25.8	9.2	9.1	9.1	9.3	9.2	26.5	24.7	25.7
Switzerland	11.3	11.4	11.5	11.6	11.6	11.6	11.6	11.5	11.6
Turkey	----	----	----	----	----	----	----	----	----
Ukraine	58.0	51.3	38.3	24.2	20.1	15.2	14.9	15.6	15.8
United Kingdom of Great Britain and Northern Ireland	211.9	205.6	183.0	171.1	186.0	177.6	184.0	189.4	180.6
United States of America	1280.2	1306.9	1339.6	1345.7	1437.3	1405.5	1437.4	1448.1	1412.0

Comments:

Emissions of N$_2$O (nitrous oxide) - total anthropogenic are all the emissions of N$_2$O associated with human activities. These include emissions from fuel combustion, industrial processes, agriculture, waste, and other.

For more detailed information about the GHG inventory database see: http://ghg.unfccc.int.

NATIONAL DATA SETS

Units: ozone depleting potential (ODP) tonnes

Data Source: Production and Consumption of Ozone Depleting Substances 1986-2000

Data Provider: Secretariat for the Vienna Convention and the Montreal Protocol (the Ozone Secretariat)

Years: 1989-2000

Copyright © 2001 United Nations Environment Programme (UNEP)

Countries and Territories	1989	1990	1991	1992	1993	1994	1995	1996	1997	1998	1999	2000
Argentina	5291.00	1595.00	0.00	0.00	0.00	0.00	0.00	-3300.00	0.00	0.00	0.00	----
Australia	0.00	----	----	-11485.10	-11464.12	-5639.81	-3377.80	0.00	0.00	0.00	0.00	----
Belarus	0.00	0.00	0.00	0.00	0.00	0.00	-1.72	0.00	0.00	0.00	0.00	----
Belgium	26592.50	----	----	23257.30	0.00	0.00	0.00	0.00	0.00	0.00	----	----
Brazil	46871.00	29993.70	----	48400.00	48367.00	18133.50	11462.00	15646.40	-1245.20	16525.30	11350.90	----
Canada	29309.50	11731.50	21648.00	6847.50	6978.40	3643.20	2553.10	-700.70	0.00	0.00	0.00	----
China	29810.00	----	-9900.00	-11220.00	-3096.50	-36791.70	-40429.40	-35479.40	0.00	35090.00	0.00	----
Croatia	0.00	0.00	0.00	0.00	0.00	0.00	0.00	-49.61	0.00	0.00	----	----
Cuba	0.00	0.00	----	----	----	----	0.00	0.00	0.00	0.00	0.00	----
Cyprus	0.00	0.00	----	0.00	0.00	0.00	0.00	0.00	0.00	0.00	----	----
Czech Republic	5285.50	----	----	----	-57.09	0.44	0.55	0.00	6.05	44.44	12.10	----
Finland	0.00	0.00	0.00	0.00	0.00	0.00	0.00	-14.30	0.00	0.00	----	----
France	5119.40	----	----	767.80	1551.00	787.60	983.40	506.00	692.12	737.00	-18.70	----
Germany	8067.40	----	----	514.80	180.40	0.00	0.00	0.00	-369.60	-576.40	0.00	----
Hungary	0.00	0.00	0.00	0.00	-16.50	0.00	0.00	0.00	0.00	0.00	0.00	----
India	4757.50	----	----	1958.00	-1036.20	8432.60	-21787.70	-19786.80	7876.01	6614.39	15896.90	----
Italy	8769.20	1445.40	320.10	4165.70	456.50	0.00	0.00	0.00	341.00	0.00	828.30	----
Jamaica	0.00	----	----	----	0.00	0.00	0.00	0.00	0.00	0.00	0.00	----
Japan	19602.00	----	8173.00	5170.00	4081.00	2943.60	2462.90	539.00	40.70	34.10	38.50	----
Lithuania	0.00	----	----	----	----	0.00	-0.04	0.00	0.00	0.00	0.00	----
Mexico	4668.40	-5470.30	-6512.00	-6546.10	-10472.00	-9212.50	-9887.90	-8890.20	0.00	0.00	0.00	----
Netherlands	0.00	0.00	2790.70	0.00	-13377.10	0.00	-14648.70	-21873.50	158.40	887.70	150.70	----
Poland	3960.00	----	----	----	4963.20	4915.90	----	9.02	0.00	0.00	0.00	----
Romania	11878.52	----	----	----	7646.10	6283.20	4665.10	2878.70	-198.58	1949.18	-684.80	----
Russian Federation	103290.00	1980.00	-26950.00	2200.00	2200.00	3372.60	2735.15	743.60	0.00	33.00	346.61	----
South Africa	12697.30	10264.10	----	7139.00	4774.00	6340.40	4931.30	0.00	0.00	0.00	0.00	----
Spain	40634.00	23007.60	19017.90	4347.20	4400.00	6600.00	0.00	0.00	0.00	0.00	0.00	----
Sweden	0.00	0.00	-178.20	-178.20	0.00	0.00	0.00	0.00	0.00	0.00	----	----
Switzerland	0.00	----	-33.00	0.00	-38.50	-58.30	-62.70	-52.80	0.00	0.00	0.00	----
Ukraine	0.00	0.00	0.00	0.00	0.00	0.00	16.50	3.30	-2178.55	2820.73	0.00	----
United States of America	56036.20	----	----	12126.40	16225.00	15225.10	8932.00	11.00	14.30	2.20	18.70	----
Venezuela	0.00	-4682.04	-3426.94	-9096.78	-7256.48	-6639.60	-6816.18	-6363.50	0.00	0.00	0.00	----

Comments:

The data is reported to the secretariat by the Parties to the Montreal Protocol on Substances that Deplete the Ozone Layer. Annex B, group II: carbon tetrachloride comprises CCl_4.

Production means the amount of controlled substances produced, minus the amount destroyed by technologies approved by the parties (to the Montreal Protocol on Substances that Deplete the Ozone Layer) and minus the amount entirely used as feedstock in the manufacture of other chemicals. The amount recycled and reused is not to be considered as production. The data forms prescribe reporting of feedstock use and of quantities destroyed separately, and reporting of total production without deduction. The Secretariat would make the necessary deduction. Some of the figures may be negative since the figures are for each calendar year, it is quite possible that in some years the feedstock figure may exceed the production figure of that year, if the feedstock use is from a carry-over stock. The production could be negative in such cases. For the same reason, the consumption could also be negative.

Units: ozone depleting potential (ODP) tonnes

Data Source: Production and Consumption of Ozone Depleting Substances 1986-2000

Data Provider: Secretariat for the Vienna Convention and the Montreal Protocol (the Ozone Secretariat)

Years: 1986, 1989-2000

Copyright © 2001 United Nations Environment Programme (UNEP)

Countries and Territories	1986	1989	1990	1991	1992	1993	1994	1995	1996	1997	1998	1999	2000
Argentina	5574.00	2960.00	3201.00	3257.00	1650.00	1536.00	1260.00	3308.40	2632.00	2804.00	2954.00	3101.00	----
Australia	15385.40	17613.00	8263.00	7485.40	6752.55	6644.00	4452.42	3849.93	0.00	0.00	0.00	0.00	----
Belgium	0.00	0.00	0.00	0.00	0.00	0.00	0.00	0.00	0.00	-17.56	-37.58	----	----
Brazil	10218.00	9109.68	8538.80	9551.40	9345.00	13012.00	11860.25	11750.52	9434.00	9362.00	7986.00	11286.00	----
Canada	19104.20	17895.40	11959.20	8330.00	13694.00	1135.00	0.00	0.00	0.00	0.00	0.00	0.00	----
China	11540.00	20700.00	20687.60	26017.80	24941.00	31658.00	50809.00	46671.60	44016.20	50323.80	55401.80	44739.40	----
Czech Republic	1977.60	2122.20	----	----	----	897.30	231.26	319.98	6.96	12.08	6.08	11.20	----
Denmark	0.00	0.00	0.00	0.00	0.00	0.00	0.00	0.00	0.00	0.00	-39.00	-21.00	----
Finland	0.00	0.00	0.00	0.00	0.00	0.00	0.00	0.00	-8.00	-3.00	-6.00	----	----
France	71018.40	55205.60	38988.80	22896.00	3756.80	3061.20	3687.80	244.40	-35.20	0.00	-131.00	0.00	----
Germany	123652.80	104095.80	78470.00	63400.60	57698.00	51258.40	15997.20	0.00	0.00	0.00	0.00	520.80	----
Greece	14045.00	12372.00	8559.00	11397.00	12635.00	11667.00	3505.00	2453.00	1450.00	1530.00	765.00	----	----
India	2202.00	4317.00	----	----	6096.80	11438.80	16646.00	21779.60	22459.60	23658.03	20012.79	22498.60	----
Italy	56656.40	48840.20	36394.80	35087.40	40996.80	36035.60	9842.20	6192.60	8474.80	7011.40	7578.20	6422.80	----
Japan	119997.80	146744.20	109311.40	99361.60	65670.00	51213.80	21593.20	29757.40	704.80	164.60	-59.00	0.00	----
Liechtenstein	0.00	0.00	0.00	0.00	0.00	0.00	0.00	0.00	0.00	0.00	-0.20	-0.20	----
Lithuania	0.00	0.00	0.00	0.00	0.00	----	0.00	-6.00	0.00	0.00	0.00	0.00	----
Mexico	8608.95	9346.00	10576.00	9784.00	9964.00	12525.00	15417.00	15737.00	8959.00	8431.00	5252.00	5530.00	----
Netherlands	42330.80	41293.60	33288.00	22432.00	30777.20	34712.60	21013.00	12245.00	13293.00	14844.00	15049.00	15721.00	----
Norway	0.00	0.00	0.00	0.00	0.00	0.00	0.00	0.00	0.00	0.00	-19.00	-62.50	----
Romania	0.00	----	----	----	----	508.05	191.00	22.00	0.00	0.00	0.00	0.00	----
Russian Federation	105296.00	105046.00	103696.00	84289.00	62127.40	40580.40	42526.00	39322.40	16770.00	14731.70	13807.59	18416.72	----
South Africa	10800.00	9500.00	6639.00	4748.00	3437.00	3722.00	1947.00	1627.00	0.00	0.00	0.00	0.00	----
Spain	33728.00	30833.40	23596.40	25292.20	30752.00	27172.60	18729.00	5435.00	5424.00	6405.00	5570.00	5839.00	----
Sweden	0.00	0.00	0.00	0.00	-5.00	0.00	0.00	0.00	0.00	0.00	0.00	----	----
Switzerland	0.00	0.00	0.00	0.00	0.00	0.00	0.00	0.00	-49.96	-54.00	-58.00	-58.30	----
United States of America	311021.20	320436.20	199696.60	######	152730.00	######	78208.40	34727.60	675.60	739.00	243.20	436.20	----
Venezuela	4789.80	4211.20	4337.80	4456.80	5284.50	5303.10	4619.90	4284.90	4412.83	5662.95	3652.00	2859.09	----

Comments:

The data is reported to the secretariat by the parties to the Montreal Protocol on Substances that Deplete the Ozone Layer. The Annex A, group I: chlorofluorocarbons (CFCs) comprise $CFCl_3$, CF_2Cl_2, $C_2F_3Cl_3$, $C_2F_4Cl_2$, C_2F_5Cl.

Production of Ozone-Depleting Substances (ODS) / Annex C, Group I: Hydrochlorofluorocarbons (HCFCs)

Units: ozone depleting potential (ODP) tonnes

Data Source: Production and Consumption of Ozone Depleting Substances 1986-2000

Data Provider: Secretariat for the Vienna Convention and the Montreal Protocol (the Ozone Secretariat)

Years: 1989-2000

Copyright © 2001 United Nations Environment Programme (UNEP)

Countries and Territories	1989	1990	1991	1992	1993	1994	1995	1996	1997	1998	1999	2000
Argentina	26.73	----	----	----	8.97	10.45	0.00	0.00	0.00	0.00	0.00	----
Australia	132.36	----	----	123.86	170.01	95.39	84.48	0.00	0.00	0.00	0.00	----
Brazil	184.03	167.37	----	220.00	235.00	232.03	210.43	122.43	131.07	133.60	17.93	----
Canada	245.85	90.15	127.82	96.55	42.54	76.83	59.01	64.78	66.70	57.86	238.81	----
China	621.50	----	313.50	305.80	698.56	1308.04	687.50	896.50	1526.32	971.60	4044.00	----
Finland	0.00	0.00	0.00	0.00	0.00	0.00	0.00	-0.44	0.00	0.00	----	----
France	791.69	----	----	2046.13	2889.81	4723.80	5797.72	5672.83	5334.70	6314.02	6299.39	----
Germany	511.15	----	----	594.72	479.66	610.53	641.47	793.71	804.93	681.95	528.50	----
Greece	75.35	109.89	130.08	115.01	188.82	251.96	376.81	440.28	465.41	512.38	----	----
India	118.93	----	----	205.76	265.04	265.98	314.19	280.45	0.00	0.00	0.00	----
Italy	336.60	426.86	433.80	451.96	498.91	551.23	462.82	799.20	652.21	700.51	775.75	----
Japan	1639.01	----	2383.77	2590.51	4555.56	3408.49	6296.16	4428.36	4224.42	3994.89	4658.37	----
Lithuania	0.00	0.00	0.00	0.00	----	0.00	-0.12	0.00	0.00	0.00	0.00	----
Mexico	208.45	138.27	160.49	102.96	158.13	126.39	117.76	296.67	303.32	238.54	328.96	----
Netherlands	514.02	604.13	0.00	700.67	822.25	883.10	844.13	974.63	926.70	1144.61	984.45	----
Norway	0.00	0.00	0.00	0.00	0.00	0.00	0.00	0.00	0.00	0.00	-0.30	----
Poland	0.00	----	----	----	0.00	0.00	----	-46.26	0.00	0.00	0.00	----
Russian Federation	1194.00	436.50	425.50	267.16	172.00	198.12	184.20	74.41	72.26	67.10	146.27	----
South Africa	0.00	0.00	0.00	56.54	84.48	87.01	56.76	0.00	0.00	0.00	0.00	----
Spain	482.24	422.46	316.97	399.85	412.50	97.74	741.24	249.37	948.59	915.37	916.36	----
Switzerland	0.00	----	0.00	0.00	0.00	0.00	0.00	-6.01	0.00	0.00	0.00	----
United States of America	6564.82	----	----	5560.82	7185.27	#####	14892.74	12542.76	#####	14985.64	14489.80	----
Venezuela	80.31	85.81	102.52	112.83	113.00	84.17	87.18	85.86	88.97	66.40	25.07	----

Comments:

The data is reported to the secretariat by the parties to the Montreal Protocol on Substances that Deplete the Ozone Layer. The The Annex C, group I: Hydrochlorofluorocarbons (HCFCs) comprise $CHFCl_2$, CHF_2Cl, CH_2FCl, C_2HFCl_4, $C_2HF_2Cl_3$, $CHCl_2CF_3$, C_2HF_4Cl, $CHFClCF_3$, $C_2H_2FCl_3$, $C_2H_2F_2Cl_2$, $C_2H_2F_3Cl$, $C_2H_3FCl_2$, CH_3CFCl_2, $C_2H_3F_2Cl$, CH_3CF_2Cl, C_2H_4FCl, C_3HFCl_6, $C_3HF_2Cl_5$, $C_3HF_3Cl_4$, $C_3HF_4Cl_3$, $C_3HF_5Cl_2$, $CF_3CF_2CHCl_2$, CF_2ClCF_2CHClF, C_3HF_6Cl, $C_3H_2FCl_5$, $C_3H_2F_2Cl_4$, $C_3H_2F_3Cl_3$, $C_3H_2F_4Cl_2$, $C_3H_2F_5Cl$, $C_3H_3FCl_4$, $C_3H_3F_2Cl_3$, $C_3H_3F_3Cl_2$, $C_3H_3F_4Cl$, $C_3H_4FCl_3$, $C_3H_4F_2Cl_2$, $C_3H_4F_3Cl$, $C_3H_5FCl_2$, $C_3H_5F_2Cl$, C_3H_6FCl.

Production of Ozone-Depleting Substances (ODS) / Annex B, Group III: Methyl Chloroform

Units: ozone depleting potential (ODP) tonnes

Data Source: Production and Consumption of Ozone Depleting Substances 1986-2000

Data Provider: Secretariat for the Vienna Convention and the Montreal Protocol (the Ozone Secretariat)

Years: 1989-2000

Copyright © 2001 United Nations Environment Programme (UNEP)

Countries and Territories	1989	1990	1991	1992	1993	1994	1995	1996	1997	1998	1999
Belarus	0.00	0.00	0.00	0.00	0.00	0.00	-5.99	0.00	0.00	0.00	0.00
Brazil	1130.00	660.00	----	750.00	764.00	39.20	0.00	0.00	0.00	0.00	97.20
Canada	1132.10	823.00	508.00	0.00	0.00	0.00	0.00	0.00	0.00	0.00	0.00
China	0.00	44.00	50.60	20.00	19.45	-74.80	-98.90	99.40	104.40	134.90	122.50
Finland	0.00	0.00	0.00	0.00	0.00	0.00	0.00	-0.04	0.00	0.00	----
France	6169.50	----	----	5427.20	2422.40	1704.50	1439.00	71.00	223.00	184.30	259.50
Germany	6895.20	----	----	5534.20	3005.90	1135.60	0.00	0.00	0.00	0.00	0.00
India	46.70	----	----	48.30	55.90	0.00	0.00	0.00	0.00	0.00	0.00
Japan	15636.40	----	17068.50	15705.40	7146.00	463.70	5248.00	867.60	1078.70	898.80	1047.80
Romania	27.20	----	----	----	9.90	7.80	6.05	12.90	2.91	0.00	0.00
Russian Federation	330.00	310.00	310.00	400.00	100.00	196.57	202.89	0.00	0.00	0.00	0.00
Switzerland	0.00	----	-15.00	0.00	-3.50	0.00	0.00	-4.00	0.00	0.00	-0.10
United States of America	31517.00	----	----	25722.60	20637.10	5794.60	4598.60	447.50	437.30	262.30	245.80

Comments:

The data is reported to the Secretariat by the parties to the Montreal Protocol on Substances that Deplete the Ozone Layer. The Annex B, group III: methyl chloroform comprises $C_2H_3Cl_3$ (this formula does not refer to 1,1,2-trichloroethane).

Production of Ozone-Depleting Substances (ODS) / Annex B, Group I: Other fully Halogenated CFCs

Units: ozone depleting potential (ODP) tonnes

Data Source: Production and Consumption of Ozone Depleting Substances 1986-2000

Data Provider: Secretariat for the Vienna Convention and the Montreal Protocol (the Ozone Secretariat)

Years: 1989-2000

Copyright © 2001 United Nations Environment Programme (UNEP)

Countries and Territories	1989	1990	1991	1992	1993	1994	1995	1996	1997	1998	1999	2000
Argentina	0.00	0.00	0.00	10.80	0.00	0.00	0.00	0.00	0.00	0.00	0.00	----
Belarus	0.00	0.00	0.00	0.00	0.00	0.00	-0.22	0.00	0.00	0.00	0.00	----
China	0.00	----	0.00	0.00	0.00	0.00	35.00	17.00	27.00	26.00	27.00	----
Czech Republic	0.00	----	----	----	0.00	0.00	-0.25	0.00	0.00	0.00	0.00	----
Germany	61.00	----	----	20.00	32.00	13.00	0.00	0.00	0.00	0.00	0.00	----
Italy	0.00	0.00	0.00	0.00	0.00	-184.00	-282.00	-146.00	0.00	0.00	0.00	----
Japan	2342.00	----	1585.00	1600.00	808.00	136.00	0.00	0.00	0.00	0.00	0.00	----
Lithuania	0.00	----	----	----	----	0.00	-0.30	0.00	0.00	0.00	0.00	----
Netherlands	23.00	32.00	0.00	0.00	19.00	0.50	0.00	0.00	0.00	0.00	0.00	----
Russian Federation	300.00	300.00	250.00	17.00	0.60	25.00	25.00	20.00	75.00	13.05	16.50	----
United States of America	577.00	----	----	75.00	106.00	101.00	38.00	0.00	0.00	-1.00	0.00	----

Comments:

The data is reported to the Secretariat by the parties to the Montreal Protocol on Substances that Deplete the Ozone Layer. Annex B, group I: other fully halogenated CFCs comprise CF_3Cl, C_2FCl_5, $C_2F_2Cl_4$, C_3FCl_7, $C_3F_2Cl_6$, $C_3F_3Cl_5$, $C_3F_4Cl_4$, $C_3F_5Cl_3$, $C_3F_6Cl_2$, C_3F_7Cl.

2.3 Disasters

In this section, only those countries are listed for which there is data available. Countries for which the data value is zero or not available are not shown.

2.3.1 Environmental Hazards

If not specified otherwise, for all variables:

Units: number of people

Data Source: EM-DAT: The OFDA/CRED International Disaster Database (data as of December 2001)

Data Provider: The OFDA/CRED International Disaster Database –
http://www.cred.be/emdat - Université Catholique de Louvain - Brussels - Belgium

Years: 1975-2000

Copyright © 2001 Université Catholique de Louvain - Brussels - Belgium

Extreme Temperatures - Total Affected

Countries and Territories	1975	1980	1990	1991	1992	1993	1994	1995	1996	1997	1998	1999	2000
Afghanistan	0	0	0	200	0	0	0	0	0	0	0	0	0
Argentina	0	0	0	0	0	0	0	25000	0	0	0	0	300
Australia	0	0	0	0	0	3000500	1100184	500100	0	0	0	0	0
Bangladesh	0	0	0	0	0	0	0	0	0	0	34000	0	0
Bolivia	0	0	0	0	0	0	0	0	0	0	0	0	25277
Brazil	600	0	0	0	0	0	0	0	0	0	0	0	0
Bulgaria	0	0	0	0	0	0	0	0	0	0	323	0	0
Chile	0	0	0	0	0	0	0	10000	0	0	0	0	0
China	0	0	0	0	0	0	0	0	0	3180	0	0	0
Croatia	0	0	0	0	0	0	0	0	0	0	0	0	200
Cyprus	0	0	0	0	0	0	0	0	0	0	100	0	400
Egypt	0	0	0	0	0	0	0	0	0	0	0	0	105
France	0	0	0	0	0	0	0	0	0	10000	0	0	0
Greece	0	0	0	0	0	0	0	0	0	0	0	0	12
Israel	0	0	0	0	240	0	0	0	0	0	0	0	0
Kazakhstan	0	0	0	0	0	0	0	0	0	600000	0	0	0
Liberia	0	0	1000000	0	0	0	0	0	0	0	0	0	0
Mexico	0	0	0	0	16000	0	0	0	0	1400	0	0	0
Pakistan	0	0	0	250	0	0	0	0	0	0	0	0	0
Romania	0	0	0	0	0	0	0	0	200	0	1700	0	100
Russian Federation	0	0	0	0	0	260	0	108	0	0	263	725246	346
Spain	0	0	0	0	0	0	0	70	0	0	0	0	0
Turkey	0	0	0	0	0	0	8000	0	0	0	0	0	300
Uruguay	0	0	0	0	0	0	0	0	0	0	0	0	400
Yugoslavia	0	0	0	0	0	0	0	0	0	0	0	0	70

Comments:

Extreme temperature: disaster type term comprising the two disaster subsets "heat wave" and "cold wave" (long lasting period with extremely high or low surface temperature).

Total affected: people that have been injured affected and left homeless after a disaster are included in this category.

Countries and Territories	1975	1980	1990	1991	1992	1993	1994	1995	1996	1997	1998	1999	2000
Afghanistan	0	0	0	95000	0	0	100330	0	3250	0	149753	124867	0
Algeria	0	930407	0	0	2250	0	22789	0	0	0	0	32675	0
Australia	0	0	0	0	0	0	5025	0	0	0	0	0	0
Azerbaijan	0	0	0	0	0	0	0	0	0	0	700000	15265	3294
Bangladesh	0	0	0	0	0	0	0	0	0	200	0	15200	1000
Bolivia	0	0	0	0	0	0	0	0	0	0	18150	0	0
Chile	0	0	0	0	0	0	0	1888	0	40098	0	0	0
China	0	0	64335	283358	16000	30066	600	735456	1857029	3106	938527	71607	2159765
Colombia	0	0	0	867	3560	3071	37416	15641	0	0	0	1205933	0
Costa Rica	0	0	14609	14349	0	240	0	0	0	0	0	0	0
Croatia	0	0	0	0	0	0	0	0	2000	0	0	0	0
Cuba	0	0	0	0	7050	0	0	0	0	0	0	0	0
Cyprus	0	0	0	0	0	0	0	3115	0	0	0	0	0
Ecuador	0	40	10	0	0	0	0	890	30180	0	2040	0	0
Egypt	0	0	0	0	34929	0	0	69	0	0	0	0	0
Georgia	0	0	0	266700	0	0	0	0	0	0	0	0	0
Germany	0	0	0	0	1525	0	0	0	0	0	0	0	0
Greece	0	17	61	0	0	1516	0	28920	1500	0	0	172000	600
Guam	0	0	0	0	0	71	0	0	0	0	0	0	0
Guatemala	0	0	0	23890	0	0	0	0	0	0	0	280	0
Iceland	0	0	0	0	0	0	0	0	0	0	0	0	33
India	0	44	0	54383	0	195566	0	0	0	126500	0	477394	0
Indonesia	0	20000	9208	5581	99603	0	260596	185924	16761	3105	5088	17930	205713
Iran (Islamic Republic of)	0	100	626100	1501	4050	4049	16641	0	0	196607	8765	7057	3015
Iraq	0	0	0	500	0	0	0	0	0	0	0	0	0
Italy	0	407700	3216	0	0	0	0	0	0	168100	0	0	0
Jamaica	0	0	0	0	0	0	0	0	0	0	0	0	0
Japan	0	67	0	0	32	14755	1738	1838447	0	0	0	0	7232
Kazakhstan	0	0	21516	0	0	0	0	0	0	0	0	0	0
Kyrgyzstan	0	0	0	0	196900	0	0	0	0	0	0	0	0
Mexico	0	6360	0	0	0	0	0	67310	0	17000	0	40778	0
Myanmar	0	0	0	0	0	0	0	100136	0	0	0	0	0
Nepal	0	275600	0	0	0	0	0	0	0	0	0	0	0
Netherlands	0	0	0	0	620	0	0	0	0	0	0	0	0
Nicaragua	0	0	0	0	40989	0	0	0	0	0	0	0	7477
Pakistan	0	0	246	205244	6600	0	0	0	0	10100	0	0	0
Panama	0	0	0	18060	0	0	0	0	0	0	0	0	0
Papua New Guinea	0	0	0	0	0	20200	0	0	0	0	0	0	5000
Peru	0	3000	100800	47285	0	145	0	0	85598	0	0	8130	0
Philippines	0	0	1766863	0	0	0	200195	0	0	0	0	190	0
Portugal	0	21900	0	0	0	0	0	0	0	0	1100	0	0
Romania	0	0	700	4005	0	0	0	0	0	0	0	0	0
Russian Federation	0	0	0	0	0	0	2342	750	0	0	0	11260	19108
Slovenia	0	0	0	0	0	0	0	0	0	0	700	0	0
Spain	0	0	0	0	0	0	0	0	0	0	0	20	0
Sudan	0	0	10000	0	0	15	0	0	0	0	0	0	0
Taiwan	0	0	0	0	0	0	0	0	0	0	0	108987	0
Tajikistan	0	0	0	3006	0	0	0	0	0	0	0	0	17000
Turkey	53372	0	0	0	322000	0	0	50348	0	0	53433	1569792	24481
Uganda	0	0	0	0	0	0	50000	0	0	0	0	0	0
United Republic of Tanzania	0	0	0	0	0	0	0	0	0	0	0	0	791
United States of America	0	0	30	104	647	0	33788	0	0	0	0	0	41
Uzbekistan	0	0	0	0	50000	0	0	0	0	0	0	0	0
Vanuatu	0	0	2	0	0	0	0	0	0	0	0	14100	0
Venezuela	0	0	0	0	0	0	0	0	0	28683	0	0	0
Wallis and Futuna	0	0	0	0	0	20	0	0	0	0	0	0	0
Yemen	0	0	0	367040	0	0	0	0	0	0	0	0	0
Yugoslavia	0	5100	120	0	0	0	0	0	0	0	0	0	0

Comments:

Earthquake: sudden break within the upper layers of the earth, sometimes breaking the surface, resulting in the vibration of the ground, which where strong enough will cause the collapse of buildings and destruction of life and property.

Total affected: people that have been injured affected and left homeless after a disaster are included in this category.

Countries and Territories	1975	1980	1990	1991	1992	1993	1994	1995	1996	1997	1998	1999	2000
Afghanistan	0	0	0	0	0	0	0	174	0	0	0	0	0
Algeria	0	0	0	0	0	696	0	0	0	0	0	0	0
Australia	0	0	0	0	0	0	0	0	101	0	0	0	0
Austria	0	0	0	0	0	0	0	0	0	0	0	10000	0
Bolivia	0	0	0	0	600	0	165000	0	0	0	23	0	0
Bosnia and Herzegovina	0	0	0	0	0	0	0	0	0	0	0	0	403
Brazil	0	0	0	600	60	0	0	2000	7105	0	0	9	143000
Chile	0	0	0	82811	0	0	0	0	0	0	0	0	0
China	0	0	5115	500	0	65	0	0	631	1851	15	45	2027
Colombia	0	0	0	221	0	3492	0	0	110	0	0	3400	0
Costa Rica	0	0	0	0	0	0	0	0	0	0	0	0	200
Ecuador	0	0	0	0	0	75066	0	0	0	0	0	0	385
Egypt	0	0	0	0	0	300	0	0	0	0	0	0	0
Ethiopia	0	0	0	0	0	0	29	0	0	0	0	0	165
France	0	0	0	0	0	0	0	0	0	0	21	260	0
French Polynesia	0	0	0	0	0	0	0	0	0	0	511	0	0
Iceland	0	0	0	0	0	0	0	83	0	0	0	0	0
India	0	0	0	0	0	25	0	1119535	0	2030	200156	0	100
Indonesia	0	3010	0	0	37000	0	0	0	4	0	0	2	56749
Iran (Islamic Republic of)	0	0	0	0	0	0	0	0	0	0	5040	0	0
Italy	0	0	0	0	0	0	0	0	100	0	0	0	0
Japan	0	0	0	0	0	70	0	0	3	15	130	0	0
Kyrgyzstan	0	0	0	0	0	0	58500	0	0	0	0	0	0
Malaysia	0	0	0	0	0	0	0	23	262	0	0	0	0
Mexico	0	0	0	0	0	0	0	0	0	0	120	0	0
Mozambique	0	0	0	0	0	0	0	0	0	0	2500	0	0
Nepal	75000	0	0	0	0	200	0	534	0	0	0	0	0
Nigeria	0	0	0	0	0	0	0	0	0	0	0	0	300
Pakistan	0	0	0	0	0	314	0	0	0	0	3000	0	0
Papua New Guinea	0	0	0	5000	0	0	0	0	8	0	0	0	0
Peru	0	0	0	0	0	0	0	0	0	30000	2607	1200	0
Philippines	0	0	0	0	0	0	73843	0	8	0	0	45	2800
Portugal	0	0	0	0	0	0	0	0	0	55	0	0	0
Romania	0	0	0	0	0	0	0	0	0	0	0	330	0
Russian Federation	0	0	0	0	1750	0	0	0	300	0	0	0	508
Saint Lucia	0	0	0	0	0	0	0	0	175	0	0	0	0
Spain	0	0	0	0	0	0	0	0	129	0	0	0	0
Sri Lanka	0	0	0	0	0	130	0	0	0	0	0	0	0
Switzerland	0	0	0	0	0	0	0	0	0	0	0	0	1500
Turkey	0	0	0	0	1084	0	0	0	0	0	0	0	0
Uzbekistan	0	0	0	0	400	0	0	0	0	0	0	0	0
Viet Nam	0	0	0	0	38000	0	1034	0	0	0	0	0	39

Comments:

Slide: disaster type comprising the two disaster subsets "avalanche" (rapid and sudden sliding and flowage of masses of usually unsorted mixtures of snow/ice/rock material) and "landslide" (in general, all varieties of slope movement, under the influence of gravity. More strictly refers to down-slope movement of rock and/or earth masses along one or several slide surfaces).

Total affected: people that have been injured affected and left homeless after a disaster are included in this category.

Volcanic Eruptions - Total Affected

Countries and Territories	1975	1980	1990	1991	1992	1993	1994	1995	1996	1997	1998	1999	2000
Argentina	0	0	0	63200	0	0	0	0	0	0	0	0	0
Cameroon	0	0	0	0	0	0	0	0	0	0	0	3010	0
Cape Verde	0	0	0	0	0	0	0	6306	0	0	0	0	0
Chile	0	0	0	63200	0	350	0	0	0	0	0	0	0
Comoros	0	0	0	200	0	0	0	0	0	0	0	0	0
Costa Rica	0	0	0	0	0	0	0	0	0	0	450	0	0
Ecuador	0	0	0	0	0	0	0	0	0	0	0	24200	0
Guatemala	0	0	0	0	5000	0	0	0	743	0	600	0	800
Indonesia	0	52235	42851	7679	0	3464	8048	0	0	3000	6000	0	0
Italy	0	0	0	7000	0	0	0	0	0	0	0	0	0
Japan	0	0	0	10020	40000	4930	0	0	0	0	0	0	16400
Mexico	0	0	0	0	0	0	75700	0	0	0	758	650	41000
Montserrat	0	0	0	0	0	0	0	5000	4000	4000	0	0	0
Nicaragua	0	0	0	0	310075	0	0	12000	0	0	0	6195	0
Papua New Guinea	0	0	0	0	0	0	152002	0	1829	0	0	0	0
Peru	0	0	4000	0	0	0	0	0	0	0	0	0	0
Philippines	0	0	0	1041865	1578	165009	0	0	0	0	0	0	60796
Trinidad and Tobago	0	0	0	0	0	0	0	0	0	200	0	0	0
United States of America	0	2500	0	0	0	0	0	0	0	0	0	0	0

Comments:

Volcanic eruption: discharge (aerially explosive) of fragmentary ejecta, lava and gases from a volcanic vent.

Total affected: people that have been injured affected and left homeless after a disaster are included in this category.

Wild Fires - Total Affected

Countries and Territories	1975	1980	1990	1991	1992	1993	1994	1995	1996	1997	1998	1999	2000
Australia	0	40	0	0	0	0	46161	0	0	0	0	2000	200
Bolivia	0	0	0	0	0	0	0	0	0	0	0	6300	0
Brazil	0	0	0	0	0	0	0	0	0	0	12000	0	0
Bulgaria	0	0	0	0	0	0	0	0	0	0	0	0	167
Canada	0	5000	0	0	0	0	3000	6500	0	1600	8000	1500	0
Central African Republic	0	0	0	0	0	0	0	85	0	0	0	0	0
Chile	0	0	0	0	0	0	0	0	0	300	0	1000	0
Costa Rica	0	0	0	0	1200	0	0	0	0	0	0	0	0
France	0	0	0	0	0	0	0	0	0	1259	0	6	0
Greece	0	0	0	0	0	0	0	2014	0	0	600	0	109
Guinea-Bissau	0	0	0	0	0	0	0	0	0	0	1500	0	0
Indonesia	0	0	0	8	0	0	3000000	0	0	32000	2000	0	0
Israel	0	0	0	0	0	0	0	240	0	0	0	0	0
Italy	0	0	0	0	0	0	300	0	0	0	0	0	0
Kazakhstan	0	0	0	0	0	0	0	0	0	8000	0	0	0
Malaysia	0	0	0	0	0	0	0	3000	0	0	0	0	0
Mongolia	0	0	0	0	0	0	0	0	5061	0	0	0	0
Nepal	0	0	0	0	50000	0	0	0	0	0	0	4000	0
Papua New Guinea	0	0	0	0	0	0	0	0	0	8000	0	0	0
Peru	0	0	0	0	0	0	0	0	0	1000	0	0	0
Philippines	0	0	0	0	0	0	0	0	0	0	300	0	0
Republic of Korea	0	0	0	0	0	0	0	0	0	0	0	0	855
Russian Federation	0	0	0	0	0	0	0	0	200	0	100683	0	0
South Africa	0	0	0	0	0	0	0	0	0	0	0	3005	1250
Spain	0	0	0	0	0	0	16520	0	0	0	600	0	0
Turkey	0	0	0	0	0	0	0	0	500	0	0	0	350
United States of America	0	0	0	7233	990	130	1432	0	50	1000	41221	1019	36087

Comments:

Wild fire: disaster type term comprising the two disaster subsets "forest fire" and "scrub fire".

Forest fire: fires in forest that cover extensive damage. They may start by natural causes such as volcanic eruptions or lightning, or they may be caused by arsonists or careless smokers, by those burning wood, or by clearing a forest area.

Scrub fire: fires in scrub or bush that cover extensive damage. They may start by natural causes such as volcanic eruptions or lightning, or they may be caused by arsonists or careless smokers, by those burning wood, or by clearing a forest area.

Total affected: people that have been injured affected and left homeless after a disaster are included in this category.

Forest Fire Extent - Annual Average

Units: thousand hectares

Data Source: Global Forest Resources Assessment 2000

Data Provider: Food and Agriculture Organization of the United Nations (FAO)

Years: 1990-00

Copyright © 2001 Food and Agriculture Organization of the United Nations (FAO)

Countries and Territories	1990-00
Albania	0.49
Argentina	465.23
Armenia	0.02
Austria	0.12
Azerbaijan	0.04
Belarus	4.12
Belgium	0.11
Bosnia and Herzegovina	0.76
Bulgaria	3.57
Canada	501.63
Chile	24.07
Croatia	3.89
Czech Republic	0.88
Denmark	0.02
Estonia	0.21
Finland	0.78
France	31.50
Germany	1.44
Greece	19.74
Ireland	0.19
Italy	20.51
Japan	2.33
Kazakhstan	1.80
Latvia	0.84
Lithuania	0.28
Mexico	66.78
Moldova, Republic of	0.02
Netherlands	0.03
New Zealand	0.33
Norway	0.46
Poland	6.99
Portugal	45.78
Romania	0.34
Russian Federation	799.90
Slovakia	0.11
Slovenia	0.36
Spain	68.28
Sweden	1.63
Switzerland	0.48
The former Yugoslav Republic of Macedonia	3.36
Trinidad and Tobago	0.42
Turkey	8.45
Ukraine	21.60
United Kingdom of Great Britain and Northern Ireland	0.43
Yugoslavia	2.93

Comments:

Forest fire extent - annual average comprises the reported forest areas exposed to fire.

Total forest includes natural forests and forest plantations. The term is used to refer to land with a tree cover of more than 10 percent and area of more than 0.5 ha. Forests are determined both by the presence of trees and the absence of other predominant land uses. The trees should be able to reach a minimum height of 5 m. Young stands that have not yet reached, but are expected to reach, a crown density of 10 percent and tree height of 5 m are included under forest, as are temporarily unstocked areas. The term includes forests used for purposes of production, protection, multiple use or conservation (i.e. forest in national parks, nature reserves and other protected areas), as well as forest stands on agricultural lands (e.g. windbreaks and shelterbelts of trees with a width of more than 20 m) and rubberwood plantations and cork oak stands. The term specifically excludes stands of trees established primarily for agricultural production, for example fruit tree plantations. It also excludes trees planted in agroforestry systems.

BKB	Patent Fuel and Brown Coal/Peat Briquettes
CDIAC	Carbon Dioxide Information Analysis Centre (http://cdiac.esd.ornl.gov/home.html)
CFC	Chlorofluorocarbon
CIESIN	Centre for International Earth Science Information Network (http://www.ciesin.org)
CIFOR	Centre for International Forestry Research (http://www.cifor.cgiar.org)
cm	Centimetre
CRED	Centre for Research on the Epidemiology of Disasters (http://www.cred.be)
EDGAR	Emission Database for Global Atmospheric Research
EEZ	Exclusive Economic Zone
EM-DAT	The OFDA/CRED International Disaster Database
FAO	Food and Agriculture Organization of the United Nations (http://www.fao.org)
FAOSTAT	Food and Agriculture Organization of the United Nations Statistical Database (http://apps.fao.org)
FDI	Foreign direct investment
FSC	Forest Stewardship Council (http://www.fscoax.org)
g	Gram
GDP	Gross Domestic Product
GEO	Global Environment Outlook
GHG	Green House Gases
GIS	Geographic Information System
GMBD	Global Maritime Boundaries Database (www.maritimeboundaries.com)
GRID	Global Resource Information Database
GWh	Gigawatt-hour
GWP	Global Warming Potential
ha	Hectare
HFC	Hydrofluorocarbon
IEA	International Energy Agency (http://www.iea.org)
IMF	International Monetary Fund (http://www.imf.org)
IPCC	Intergovernmental Panel on Climate Change (http://www.ipcc.ch)
IRF	International Road Federation (http://www.irfnet.org)
ISIC	International Standard Industrial Classification
ISP	Internet Service Provider
ISSCAAP	FAO International Standard Statistical Classification of Aquatic Animals and Plants
ITU	International Telecommunication Union's (http://www.itu.org)
IUCN	The International Union for the Conservation of Nature / The World Conservation Union (www.iucn.org)
JMP	United Nations Children's Fund Joint Monitoring Programme for Water Supply and Sanitation
UNICEF	United Nations Children's Fund (http://www.unicef.org)
kg	Kilogram
km	Kilometre
km^2	Square kilometre
ktoe	Thousand Tonnes of Oil Equivalent
kWh	Kilowatt per hour
LPG	Liquefied Petroleum Gas
Mtoe	Million Tonnes of Oil Equivalent
MWe	Megawatt-electricity
NATO	North Atlantic Treaty Organization (http://www.nato.int)
NER	Net enrolment ratio
NGL	Natural Gas Liquids
NMVOC	Non-Methane Volatile Organic Compounds
ODA	Official Development Assistance
ODP	Ozone Depleting Potential
OECD	Organization for Economic Co-operation and Development (http://www.oecd.org)
OFDA	Office of U.S Foreign Disaster Assistance (http://www.usaid.gov/ofda)
PFC	Perfluorocarbon
PIN	Production Index Number
RIVM	National Institute of Public Health and the Environment (Netherlands) (http://www.rivm.nl)
SSC	Species Survival Commission (http://www.iucn.org/themes/ssc)
TFC	Total final energy consumption
TJ	Terajoules
TNO	Netherlands Organisation for Applied Scientific Research (http://www.tno.nl)
toe	Tonnes of Oil Equivalents
TPES	Total primary energy supply
UIS	Institute for Statistics of the United Nations Organization for Education, Science and Culture
UN	United Nations (http://www.un.org)
UNCLOS	United Nations Convention on the Law of the Sea (http://www.unclos.com)
UNCTAD	United Nations Conference on Trade and Development (http://www.unctad.org)
UNEP-WCMC	United Nations Environment Programme World Conservation Monitoring Centre (www.unep-wcmc.org)
UNESCO	United Nations Organization for Education, Science and Culture (http://www.unesco.org)
UNFCCC	United Nations Framework Convention on Climate Change (http://www.unesco.org)
US$	US dollar
WHO	World Health Organization (http://www.who.int)